Biological Concepts and Techniques in Toxicology

Biological Concepts and Techniques in Toxicology

An Integrated Approach

edited by

Jim E. Riviere
North Carolina State University
Raleigh, North Carolina, U.S.A.

Taylor & Francis
Taylor & Francis Group
New York   London

Published in 2006 by
Taylor & Francis Group
270 Madison Avenue
New York, NY 10016

© 2006 by Taylor & Francis Group, LLC

No claim to original U.S. Government works
Printed in the United States of America on acid-free paper
10 9 8 7 6 5 4 3 2 1

International Standard Book Number-10: 0-8247-2979-X (Hardcover)
International Standard Book Number-13: 978-0-8247-2979-0 (Hardcover)
Library of Congress Card Number 2005056861

Library of Congress Cataloging-in-Publication Data

Biological concepts and techniques in toxicology : an integrated approach / edited by Jim E. Riviere.
    p. ; cm.
  Includes bibliographical references and index.
  ISBN-13: 978-0-8247-2979-0 (hardcover : alk. paper)
  ISBN-10: 0-8247-2979-X (hardcover: alk. paper)
  1. Toxicology. 2. Molecular toxicology. 3. Drugs--Toxicology. I. Riviere, J. Edmond (Jim Edmond)
  [DNLM: 1. Toxicology--methods. 2. Drug Toxicity. QV 600 B6146 2006]

RA1191.B52 2006
615.9--dc22   2005056861

Taylor & Francis Group
is the Academic Division of Informa plc.

**Visit the Taylor & Francis Web site at
http://www.taylorandfrancis.com**

Preface

Toxicology is a rapidly developing discipline whose advances are often based on utilizing concepts and techniques developed from basic biomedical research. Earlier revolutions in molecular biology and mathematical modeling have led to routine use of in vitro models and toxicokinetics developed from these disciplines. In order for toxicology to remain current and continue making significant advances, it is important that the latest developments be presented in a context focused on how they can be applied to different fields of investigation. How do concepts and new paradigms migrate into routine practice? Several recent toxicology issues have engendered attention and are deserving of a review that illustrates how new techniques have been applied to actual problems of toxicological interest. How did these migrate from the basic science laboratory to routine toxicology test protocols? How will today's revolution in the "-omics" affect the practice of toxicology in the next decade? This book is meant to provide a timely introduction to these techniques focused on how the latest science can be applied to existing problems in toxicology. It is also meant to overview some recent areas where progress has been made and a story can begin to be told.

The initial chapters review new approaches or concepts that will have a major impact on the practice of toxicology in the coming decade. These include the "*omics*" which together comprise systems biology as well as mathematical approaches designed to link chemical structure to activity. This section ends with a cogent presentation of hormesis, a concept that has begun to alter the way toxicologists and others interpret dose–response relationships, the cornerstone to the practice of classical toxicology. The four chapters comprising the next section of the book deal with how biological data and test systems are integrated into the actual practice of toxicology. What is the validation process that allows a novel idea to become a standard test? The final section of chapters reviews applications of toxicology to specific and more focused topics that represent current issues facing society today, including discussions of genetically modified food, nanomaterials, and pharmaceutics. Specific disciplines, including inhalational and forensic toxicology are discussed, as are current military issues that continue to draw attention.

The goal of this book is not to reproduce comprehensive toxicology texts covering basic mechanisms of toxicity or target organ toxicology; nor is it a methods text overviewing the nuts and bolts of new techniques. The goal is to highlight new methods and concepts that may have a major impact on toxicology and present how such concepts and techniques migrate into the mainstream of toxicology.

Jim E. Riviere
Raleigh, North Carolina

Contents

PART III. EXAMPLES APPLIED TO PROBLEMS

Contributors

Beat Aebi   Institute of Legal Medicine, University of Berne, Berne, Switzerland

Cynthia A. Afshari   National Institute of Environmental Health Sciences, Research Triangle Park, North Carolina, U.S.A.

Rupesh P. Amin   National Institute of Environmental Health Sciences, Research Triangle Park, North Carolina, U.S.A.

J. Todd Auman   National Institute of Environmental Health Sciences, Research Triangle Park, North Carolina, U.S.A.

Subhash C. Basak   Natural Resources Research Institute, University of Minnesota Duluth, Duluth, Minnesota, U.S.A.

Ronald E. Baynes   Center for Chemical Toxicology Research and Pharmacokinetics, College of Veterinary Medicine, North Carolina State University, Raleigh, North Carolina, U.S.A.

Lee Bennett   National Institute of Environmental Health Sciences, Research Triangle Park, North Carolina, U.S.A.

Werner Bernhard   Institute of Legal Medicine, University of Berne, Berne, Switzerland

Pierre R. Bushel   National Institute of Environmental Health Sciences, Research Triangle Park, North Carolina, U.S.A.

Jennifer L. Buur   Center for Chemical Toxicology Research and Pharmacokinetics, College of Veterinary Medicine, North Carolina State University, Raleigh, North Carolina, U.S.A.

Edward J. Calabrese   Environmental Health Sciences, University of Massachusetts, Amherst, Massachusetts, U.S.A.

J. Christopher Corton   U.S. Environmental Protection Agency, Research Triangle Park, North Carolina, U.S.A.

J. Dennison   Colorado State University, Fort Collins, Colorado, U.S.A.

C. T. De Rosa   Division of Toxicology and Environmental Medicine, Agency for Toxic Substances and Disease Registry, Atlanta, Georgia, U.S.A.

Rakesh Dixit   Toxicology Department, Johnson and Johnson Pharmaceutical Research and Development, L.L.C., San Diego, California, U.S.A.

John Farmer   Section of Cardiology, Baylor College of Medicine, Houston, Texas, U.S.A.

B. A. Fowler   Division of Toxicology and Environmental Medicine, Agency for Toxic Substances and Disease Registry, Atlanta, Georgia, U.S.A.

Kevin J. Greenlees   USFDA Center for Veterinary Medicine, Rockville, Maryland, U.S.A.

Julian L. Griffin   Department of Biochemistry, University of Cambridge, Cambridge, U.K.

Brian D. Gute   Natural Resources Research Institute, University of Minnesota Duluth, Duluth, Minnesota, U.S.A.

Hisham K. Hamadeh   National Institute of Environmental Health Sciences, Research Triangle Park, North Carolina, U.S.A.

Joseph P. Hanig   Food and Drug Administration, Center for Drug Evaluation and Research, Silver Spring, Maryland, U.S.A.

John C. Matheson III   USFDA Center for Veterinary Medicine, Rockville, Maryland, U.S.A.

Roger O. McClellan   Inhalation Toxicology and Human Health Risk Analysis, Albuquerque, New Mexico, U.S.A.

Michele A. Medinsky   Toxicon, Durham, North Carolina, U.S.A.

Denise Mills   Natural Resources Research Institute, University of Minnesota Duluth, Duluth, Minnesota, U.S.A.

Nancy A. Monteiro-Riviere   Center for Chemical Toxicology Research and Pharmacokinetics, College of Veterinary Medicine, North Carolina State University, Raleigh, North Carolina, U.S.A.

M. M. Mumtaz   Division of Toxicology and Environmental Medicine, Agency for Toxic Substances and Disease Registry, Atlanta, Georgia, U.S.A.

Robert E. Osterberg   Food and Drug Administration, Center for Drug Evaluation and Research, Silver Spring, Maryland, U.S.A.

Richard S. Paules   National Institute of Environmental Health Sciences, Research Triangle Park, North Carolina, U.S.A.

Jim E. Riviere   Center for Chemical Toxicology Research and Pharmacokinetics, Biomathematics Program, North Carolina State University, Raleigh, North Carolina, U.S.A.

Larisa Rudenko   USFDA Center for Veterinary Medicine, Rockville, Maryland, U.S.A.

P. Ruiz   Division of Toxicology and Environmental Medicine, Agency for Toxic Substances and Disease Registry, Atlanta, Georgia, and Oak Ridge Institute for Science and Education, Oak Ridge, Tennessee, U.S.A.

Jessica P. Ryman-Rasmussen   Center for Chemical Toxicology Research and Pharmacokinetics, College of Veterinary Medicine, North Carolina State University, Raleigh, North Carolina, U.S.A.

M. Burton Snipes   Inhalation Toxicology Research Institute, Tijeras, New Mexico, U.S.A.

William S. Stokes   National Toxicology Program Interagency Center for the Evaluation of Alternative Toxicological Methods, National Institute of Environmental Health Sciences, National Institutes of Health, Department of Health and Human Services, Research Triangle Park, North Carolina, U.S.A.

M. Whittaker   ToxServices, Washington, D.C., U.S.A.

Mark L. Witten   Department of Pediatrics, University of Arizona Health Sciences Center, Tucson, Arizona, U.S.A.

Frank A. Witzmann   Department of Cellular and Integrative Physiology, Biotechnology Research and Training Center, Indiana University School of Medicine, Indianapolis, Indiana, U.S.A.

Simon S. Wong   Department of Pediatrics, University of Arizona Health Sciences Center, Tucson, Arizona, U.S.A.

1
Introduction and Overview

Jim E. Riviere
Center for Chemical Toxicology Research and Pharmacokinetics,
Biomathematics Program, North Carolina State University, Raleigh,
North Carolina, U.S.A.

Toxicology is the science of poisons and dates back to the earliest annals of recorded history. This history is steeped in both using and developing animal- and plant-derived poisons, as well as treating the victims who succumbed to these practices. Early practitioners were also integrated into the developing arts of medicine and pharmacy where chemicals, mostly isolated from plant sources, began to be used to treat human and animal diseases. In the early 15th century, the basic tenet of the dose–response relationship was articulated by the alchemist–physician Paracelsus when he stated "All substances are poisons; there is none which is not a poison. The proper dose separates a poison from a remedy." To the present, this rubric has guided the practice of toxicology in most of its varied applications.

With the explosion of chemistry in the 19th and 20th centuries, toxicology likewise blossomed. A similar phase of growth is now upon us. It started mid-century with the advent of molecular biology, which was facilitated by the revolution in computer technology, and is now propelled by the growth of genomics that we are presently experiencing. Toxicology has come of age. The public views toxicology not from these expanding scientific roots, but rather from its application to practical problems. In recent decades toxicology has become synonymous with the science focused on assessing the safety of new drugs and determining the risk to environmental or occupational chemical exposure. The working definition of toxicology is generally defined as the study of adverse effects of chemicals on biological systems.

Toxicology has been viewed by many as an orphan discipline, because it relies so heavily on fundamental advances in chemistry, biology, and mathematics. However, this view is oversimplified because modern toxicology has evolved into a number of niches where elucidating the basic mechanisms of chemical interactions with living systems is only conducted by scientists who can best be classified as toxicologists. These include studies of the mechanism of cancer, chemical biotransformation, inhalational and dermal exposure analysis, as well as defining the biological response to specific chemical classes including metals and pesticides. Subdisciplines focused on natural product toxicology, drug safety, forensics, as well as occupational and environmental risk assessment have emerged. The use of physiologically based pharmacokinetic models, once applied to pharmaceutics, has now been largely developed and put to practice in the field of toxicology.

Toxicology is thus a vibrant discipline in which novel techniques of other sciences (e.g., analytical chemistry, molecular biology, computer-aided modeling, and genomics/proteomics/metabonomics) are explored and then become mainstream tools in its practice. The nature of this migration is the focus of this present text. How does a basic technique or insight into mechanism of action move from a science-discipline laboratory to the laboratory of a toxicologist exploring chemical effects to an approved tool in the risk assessor's arsenal? When do these migrations work and when do they not? How are products of one discipline (e.g., genomics or analytical chemistry) interfaced to another (e.g., drug-induced alteration of genetic function, coupling of low-level chemical detection to a biological effect, etc.)?

Three different approaches to exploring this phenomenon are taken in this text. The first is to closely examine new disciplines from within to see where potential applications to toxicology may occur. These include detailed presentations in genomics, proteomics, and metabonomics. Exciting developments are occurring in all of these disciplines. What is now required is an understanding of how to interpret them, define their true relevance, and incorporate mechanistic insights gleaned from these studies into the theoretical foundation of the underlying science of toxicology. Two other areas have had a major impact on this discipline. The first is the general area of quantitative structure–activity relationships. This field bridges chemistry to toxicological effects. The second, hormesis, a concept that many consider a challenge to both Paracelsus and the dose–response paradigm itself, suggests that at very low doses, some chemicals demonstrate a beneficial rather than the expected lack of response. Incorporation of hormesis into the risk assessment paradigm that has evolved over toxicology's history is truly a challenging problem.

The next section examines these same issues from a slightly different perspective. In these cases, the fields of toxicology that actually accomplish this incorporation of science into practice are discussed. Insights into how new approaches to assessing biological effects are used in chemical and drug risk assessments are explored. In addition, how specific toxicological testing systems, developed from basic toxicology research studies, are validated for regulatory use provide a different perspective to the hurdles that face widespread integration of science into the practice of toxicology.

The third approach examines this issue by looking at specific technologies, drugs, chemicals, issues, and disciplines to see how new scientific techniques have been incorporated into these areas. The first involves two exciting areas of technology that have potential for widespread exposure to humans. These are genetically modified foods and nanomaterials. Although both come from very different advances in science and technology, they are now being examined by toxicologists using the tools that have evolved to assess chemical toxicity. Because they have been developed using many of the modern techniques of biotechnology and material engineering, they are also being examined by some of the newer toxicology techniques that were not available to earlier workers.

The remainder of the text deals with more specific applications. Current issues confronting the military, including jet fuel toxicology are reviewed and provide a good overview of how the various tools of toxicology and exposure assessment have been employed. Two chapters deal with the application of toxicology to drug safety. Additional chapters are subdiscipline-oriented, where recent advances in inhalational and forensic toxicology are presented.

Reading this text gives one a true feeling of the intellectual diversity of modern toxicology and how various seasoned practitioners have integrated basic chemistry, biology, and mathematical models into specific applications. These developing fields and issues define the scope of toxicology in the 21st century as it evolves to assure that novel products of technology benefit and do no harm to either our species or environment.

2

Toxicogenomics: Gene Expression Analysis and Computational Tools

Rupesh P. Amin, Hisham K. Hamadeh, J. Todd Auman, Lee Bennett, Cynthia A. Afshari, Pierre R. Bushel, and Richard S. Paules
National Institute of Environmental Health Sciences, Research Triangle Park, North Carolina, U.S.A.

J. Christopher Corton
U.S. Environmental Protection Agency, Research Triangle Park, North Carolina, U.S.A.

INTRODUCTION

Technologies that probe genomic responses to chemical exposure are expected to improve our understanding of molecular mechanisms of toxicity. Accelerated by advances in high-throughput DNA sequencing, a number of animals that are important as models of human disease (e.g., yeast, worm, fruit fly) have been completely sequenced. A draft of the human genome released in February 2001 and a draft of the mouse genome released in December 2002 have provided a wealth of genetic information much sooner than initially anticipated (1–3). A high-quality draft covering more than 90% of the Brown Norway rat genome has also been recently reported (4). The availability of genomic sequences has provided revolutionary opportunities to monitor gene expression changes across an entire genome. This flood of genomic data has also led to the identification and quantitation of sequence differences between individual humans or between animal strains in the form of deletions, insertions, and single nucleotide polymorphisms. Dual sources of information about global gene expression and genetic differences will facilitate an understanding of the molecular basis of disease in general and, in particular, how chemicals interact with genes, proteins, and other biological components to cause adverse or beneficial effects.

Changes in the abundance of messenger ribonucleic acid (mRNA) are some of the earliest events that occur after chemical exposure and usually precede observable perturbations in cellular and tissue homeostasis. Elucidating the molecular mechanism(s) of chemical exposure using the old paradigm of hypothesis-driven research in which a gene or group of related genes was examined for association with toxicity has been difficult. Availability of complementary sequences derived from expressed genes has allowed construction of tools that can be used to simultaneously interrogate the expression changes in hundreds or thousands of genes upon chemical exposure

5

(5–7). Transcript profiling has provided an avenue for simultaneously testing multiple hypotheses of chemical toxicity, thereby accelerating the identification of pathways mechanistically linked to toxicity. The use of these tools is facilitating novel insights into genome-wide effects of environmentally relevant chemicals of concern to society as well as new chemical entities of interest to the pharmaceutical industry. Recent applications of expression profiling have provided insights into compound-specific and mode of action-related gene expression changes relevant to xenobiotic-induced organ system toxicity, such as hepatotoxicity (8–14) and nephrotoxicity (15–17). Identification of specific gene networks perturbed by classes of xenobiotics holds tremendous potential for understanding responses to toxicants and will be greatly facilitated by the use of bioinformatics tools to allow for improved biological integration and interpretation of vast amounts of expression data (18–22).

This chapter provides an overview of microarray data analysis of gene expression applied to toxicology (toxicogenomics). We discuss the basics of performing a microarray experiment, including statistical analysis of expression data and computational image analysis. We use as an example a study comparing gene expression profiles between different classes of chemicals to illustrate the types of analyses that can be useful in toxicogenomics studies.

MICROARRAY-BASED GENE EXPRESSION PROFILING

The National Center for Toxicogenomics (NCT) was created by National Institute of Environmental Health Sciences (NIEHS) to foster the development and applications of toxicogenomics. One of the goals of the NCT is to develop a public database that will be populated with toxicity data associated with exposures to a variety of xenobiotics representing various classes of industrial chemicals, environmental toxicants, and pharmaceutical compounds, as well as the corresponding gene expression profiles that result in multiple species and tissues (23). A database containing gene expression information, along with classical toxicity parameters (e.g., clinical chemistry, urinalysis, histopathology, etc.), will create an opportunity to mine the data for identifying novel mechanisms of toxicities or for associating traditional toxicity endpoints with hallmark gene profiles. This database would be generally useful for academic, industrial, and government toxicologists interested in understanding mode of action and relevance of human exposure more quickly. The NCT has initiated a large program to generate xenobiotic-induced gene expression profiles in appropriate tissues, to house the profiles in a publicly available database, and to continue to develop algorithms for the comparison of "unknowns" to the database. More information about this effort can be obtained from the cited website (24). The usefulness of the database will depend in part on the analysis tools used to facilitate the classification of unknown compounds and to predict their toxicity based on similarities to profiles of well-studied compounds. Information derived from database queries (e.g., biomarkers of exposure) will provide opportunities to conduct exposure assessment studies in workplace or environmental exposure scenarios. Identification of subgroups with genetic susceptibility to environmentally induced disease will help to monitor and prevent disease in those individuals.

The NIEHS Microarray Group of the NCT developed a gene expression analysis platform based on the cDNA microarray approach of Brown and Botstein (6). DNA sequences for the microarrays are generated by polymerase chain reaction amplification from purified plasmid DNA encoding cloned cDNA inserts from expressed sequence tag (EST) libraries. ESTs are sequences of DNA derived from a gene's mRNA and are generated from an mRNA population isolated from a tissue or cell population. Many ESTs are uncharacterized genes without an official name or known function. Even "known" genes often have only inferred or, at best, partly characterized functions. A computer-driven high-speed robotic arrayer is used to create high-density spotted cDNA arrays printed on glass slides (5–7), thus allowing for critical quality control aspects of manipulating thousands of bacterial clones and DNA fragments to be controlled more easily. The arrayer selects predetermined cDNAs from multiwell plates and spots the DNA onto poly-L-lysine–coated glass slides at designated locations. The location of each sequence arrayed onto a chip is computer tracked by a "gene in plate order" file. Presynthesized oligonucleotides that are commercially available can also be printed on glass slides using a similar technique with a few modifications. Other commercial methods for producing microarrays for gene expression studies include ink-jet fabrication (25) and photolithography (26).

cDNA arrays representing several distinct genomes, i.e., human, mouse, rat, and yeast (*Saccharomyces cerevisiae*) have been printed and used at the NCT for a wide variety of toxicogenomics studies (27). Each chip contains between 1700 and 20,000 spots depending on the genome being represented. Having chips containing an entire set of genes in a genome will help identify the most informative genes for a particular biological response or process, possibly leading to the development of smaller, focused microarrays specialized for monitoring an optimal subset of genes involved in a biological pathway or mechanistic response.

The NIEHS microarray efforts began with the development and application of the NIEHS Human ToxChip (28), a collection of clones representing approximately 2000 genes proposed by a variety of experts to be particularly relevant to cellular responses linked to toxicity. The genes on the NIEHS Human ToxChip fall into a number of categories, including genes implicated in apoptosis, DNA replication and repair, cell cycle control, and oxidative stress, as well as oncogenes, tumor suppressor genes, peroxisome proliferator, aryl hydrocarbon receptor-, and estrogen-responsive genes, genes for transcription factors, receptors, kinases, phosphatases, and heat shock proteins, and cytochrome P450 genes. The NIEHS rat and mouse chips contain genes representative of these categories from their respective genomes. In addition, the laboratory utilizes mouse and human oligo chips, each containing approximately 17,000 oligonucleotides 70 nucleotides long, as well as commercial oligonucleotide chips, with the goal to monitor as broad a representation of the expressed genome (i.e., transcriptome) as possible.

An overview flowchart for a typical microarray study used by NCT is depicted in Figure 1. The approach involves isolating total or polyA$^+$ RNA from experimental samples (e.g., control or treated, normal or diseased) that will be compared for changes in gene expression. Isolation of intact mRNA is critical to the success of a microarray experiment, and optimized protocols for isolation of high-quality RNA for gene expression experiments are available (29). An indication of the quality of a RNA sample can be obtained from gel electrophoresis analysis, or from a Bioanalyzer (Agilent Technologies, Palo Alto, California, U.S.A.). Results from the Bioanalyzer are displayed in an electropherogram, which is a computer-generated gel-like image of the sample that provides the 28S/18S ribosomal RNA ratios and the RNA concentration in the sample. These methods help to determine whether there is RNA degradation in a sample. Poor quality RNA can affect subsequent labeling reactions and lead to erroneous results in expression profiling experiments.

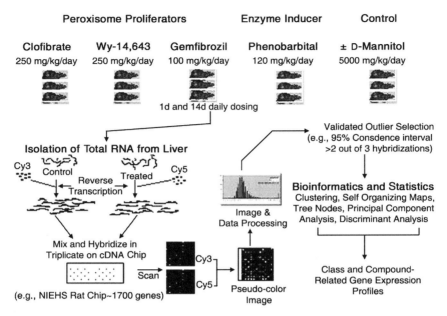

Figure 1 Experimental design and flowchart for microarray analysis. This illustration provides an overview of a toxicogenomics study, including experimental design. Rats ($n = 3$ per group) were treated with either a peroxisome proliferator (clofibrate, Wy-14,643, or gemfibrozil) or an enzyme inducer (phenobarbital) for 1 and 14 days with daily dosing. D-mannitol served as a negative control in this study. Steps involved in a cDNA microarray experiment and the data analysis approach are illustrated.

The identification and exclusion of poor quality RNA from microarray analysis will increase the chances that a database is populated by high-quality data.

Isolated mRNA is converted to cDNA using reverse transcriptase (RT) in the presence of one of two fluorescently tagged nucleotides, commonly Cy3-dUTP or Cy5-dUTP. The two populations of labeled cDNAs are mixed in equal amounts and hybridized for 18 to 24 hours onto glass slides containing the spotted cDNAs. After washing off labeled cDNAs not specifically hybridized to the slide, the amount of the two dyes on each spot is quantitated using a specialized laser scanner that captures the fluorescence intensity of hybridized dye-labeled cDNA at each spot on an array. The Cy3 and Cy5 dyes exhibit different wavelengths for excitation (532 and 635 nm) and emission (575 and 675 nm), respectively. The images derived from the Cy3 or Cy5 dyes are used for subsequent image analysis and data acquisition, making it possible to detect the relative abundance of mRNA.

Analysis of microarray images is complex and can be subdivided into array target segmentation, background intensity extraction, target detection, target intensity extraction, normalization and ratio analysis, and measurement quality assurance (29). The NIEHS Microarray Group uses a variety of software tools, including the ArraySuite image-processing software originally written by Chen et al. (30) at the National Human Genome Research Institute (NHGRI). This software is available as MicroArray Suite (Scanalytics Inc., Fairfax, Virginia, U.S.A.) and carries out a number of functions, including detection of targets, i.e., spots, and associates these spots with identified genes on an x–y coordinate map of the chip. Another program in MicroArray Suite processes mean pixel intensity for each spot as well as the intensity of the background region, the latter being used for local background subtraction of the images (30). Once the background-subtracted intensities are determined, a ratio of the intensities at 635 and 532 nm (Cy5/Cy3) is calculated for each spot. However, first the ratio intensities need to be normalized to correct for scanning imbalances. To do this, the software fits all of the ratio intensity data from a chip to a probability distribution and calculates an estimated normalization constant, which is then used to normalize the chip-to-chip ratio intensities. Another program in Micro-Array Suite facilitates superimposing the images generated by scanning the chip at two different wavelengths, creating one composite overlaid image, called a pseudo-image. The pseudoimage is a visual representation of the normalized ratio intensities, with different colored spots indicating different levels of expression, ranging from red (Cy5/Cy3 ratio >1) to green (Cy5/Cy3 ratio < 1). Thus, a red spot indicates greater abundance of transcripts from the experimental sample (provided the experimental RNA was tagged with Cy5), whereas a green spot indicates greater abundance of transcripts from the control sample. A yellow spot indicates roughly equivalent transcript levels in control and experimental samples (Cy5/Cy3 $\cong$ 1), whereas no spot indicates that neither sample contains detectable amounts of that transcript.

SELECTION OF DIFFERENTIALLY EXPRESSED GENES

A microarray experiment can monitor the expression changes of thousands of genes simultaneously, yet only a fraction of the genes in any given experiment will exhibit a statistically significant level of differential expression. The genes that remain, presumably, unchanged can thus be used to normalize the intensity ratios across the entire chip to approximately one and then are frequently excluded from further analysis. As research into the analysis of microarray expression data has progressed, many different methods have been developed for identifying differentially expressed genes, ranging from simple fold-change cutoffs to sophisticated stochastic models.

Chen et al. (30) developed one of the first statistical treatments of microarray expression-ratio data from two-color experiments. This method allows the researcher to specify a confidence level for the analysis, and a confidence interval is calculated for the ratio intensity values on the chip; genes having ratio values outside of this interval are considered differentially expressed with a high degree of confidence. The first version of this method used the distribution of ratio values from housekeeping genes to produce fixed-width confidence intervals, but an extension of the method allows for gene-specific adaptive confidence intervals that account for differences in the signal-to-noise ratio between spots (31). The ratio distribution models developed by Chen et al. (30) are available in the MicroArray Suite package and are one of the types of gene selection methods used by members of the NCT.

If the ratio model fits the data well and independent replicate hybridizations are performed, a binomial probability distribution can be used to determine the probability of a gene being detected as differentially expressed multiple times strictly by chance. To diminish experimental error in microarray analysis, hybridizations can be performed using technical replicates from individual samples; a binomial distribution can be used to model the results of the analyses at given confidence levels (32). For example, scoring genes as differentially expressed at the 95% confidence level ($p = 0.05$) four or more times ($k \geq 4$) out of nine replicate experiments ($n = 9$) has a binomial probability (p) of 0.00064 of being detected by chance. In addition, it is advisable to reverse the fluorescent molecules tagged to the control and treated samples by tagging RNA from experimental samples with Cy3 instead of Cy5 and control samples with Cy5 instead of Cy3, which in microarray parlance is called fluor-flipping or dye-reversal. By utilizing fluor-flips in the experimental design, false positives in gene selection due to dye biases can be avoided. The use of fluor-flipping in microarray experiments including minimizing the number of replicates needed has been recently addressed (33). An examination of all data will yield a list of genes that are consistently identified as differentially expressed according to criteria specified by the researcher generating a statistically validated list of differentially expressed genes. Genes with highly variable expression changes across hybridizations are flagged owing to a large coefficient of variation or a modified Z-score computation to detect outliers (32,34). Thus, a simple calculation can be used to convey some measure of confidence for the results of an entire experiment.

In addition to using fold-change models, a variety of statistical techniques have been developed or adapted for the analysis of microarray data to detect differential gene expression, including the use of linear models and analysis of variance (ANOVA) (35,36). These models incorporate all of the replicate hybridization data into one analysis, and provide a measure of statistical significance, a p-value, for each individual gene. Additionally, models of this type allow for estimates of the sources of experimental variation, such as effects due to the dyes, differences between chips, and even spatial variability within chips. The mixed linear model method developed by Wolfinger et al. (35) uses two separate ANOVA models to identify genes that are significantly changed by a treatment. First, a model is fit to normalize the data across arrays; systematic effects due to the dyes are identified and the variation due to chip-to-chip differences is quantified. The difference between the observed data and the fitted values given by this model are called the residual values, and these serve as the input for the second set of ANOVA models, which are fit on a gene-by-gene basis. These models are designed to assess effects due to the treatment while accounting for gene-specific dye biases as well as spot variation between arrays. In both these models, the arrays are considered to be a "random effect"; that is, it is assumed that the chips used in the experiment were drawn from a large population of available chips. Doing this allows us to make inferences about the treatments that are derived from the analysis, to be generalized to a larger population of chips rather than restricted to only the arrays in the experiment.

These ANOVA models rely on typical statistical assumptions for linear models. First, it is assumed that the "error" terms in the individual gene models, which measure the variation that the model cannot account for, are normally distributed with a mean of zero and constant variance. Second, the random effects due to arrays are assumed to have a normal distribution with a mean of zero. The limited number of observations for each gene in a microarray experiment can make it difficult to assess the validity of these assumptions, but some diagnostic tools are available to verify that the models fit the data adequately. The inference results are sometimes displayed in the form of a "volcano plot" (Fig. 2), which shows graphically the relationship between fold-change estimates and statistical significance values, reported as the negative $\log_{10}(p\text{-value})$. The horizontal line on the plot indicates the cutoff value for statistical significance while the vertical lines represent a twofold change in gene expression, both of which are predetermined by the researcher. Spots above the horizontal line in the center area of this plot indicate genes that have small expression changes but are nonetheless determined to be statistically significant by mixed linear modeling. Owing to the multiple testing problem created by the large number of genes on an array, multiple comparison correction procedures such as the Bonferonni

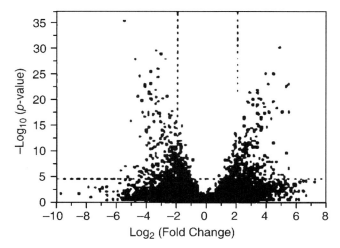

Figure 2 Example of a volcano plot from mixed model analysis. Fold-change values are compared to statistical significance values for each gene on a microarray. The dashed vertical lines represent fourfold induction and repression, as measured on the x-axis. The horizontal line at $-\log(p) = 5$ represents a p-value of 0.00001. Genes falling above this line exhibit differential expression that is statistically significant based on this p-value.

method are necessary to reduce false positives but are often considered to be overly conservative (37).

Another type of method to select differentially expressed genes from microarray data is the error model-based approach. Hughes et al. (38) utilized a model for the uncertainties in individual array experiments that took into account both additive and multiplicative error components in the channels of a two-color experiment. In addition, a p-value and the minimum-variance–weighted average used to compute the mean $\log_{10}$ intensity ratio of each reported gene are computed to allow the analyst to select differentially expressed genes with more reliability. In the case where the distribution of the gene expression data is not known or desired to be assumed for analysis, the nonparametric gene selection of Callow et al. (39), which assumes no specific parametric form for the distribution of the expression levels and employs a permutation procedure to estimate the joint null distribution of the t-test statistic for each gene, is a reasonable option. Significance analysis of microarrays (40) uses permutations of repeated measures of microarray data as well to estimate the false discovery rate—the percent of genes identified by chance.

FINDING EXPRESSION PATTERNS

Many researchers who use microarrays to monitor gene expression are interested in identifying groups of genes that exhibit similar expression profiles. These genes may provide insight into the biological mechanisms at work as cells or tissues respond to chemical or physical perturbations. Cluster analysis, an exploratory data analysis technique familiar to biologists because of its use in phylogenetic analysis, has become a popular tool for finding patterns in gene expression data. Since its first application to microarray expression data by Eisen et al. (41), clustering results have appeared regularly in the gene expression literature. The use of clustering by scientists at the NCT Microarray Group is illustrated later in this chapter using gene expression data generated from a rat in vivo study.

A popular type of cluster analysis is hierarchical cluster analysis, which produces output in the form of a "tree" or dendrogram (41). This type of analysis is a form of unsupervised clustering, meaning that an informed classifier is not used; also labels (categories) or prior information about the samples are not used for classification. A similarity measure, such as Pearson correlation or Euclidean distance, is used to determine the level of dissimilarity or similarity between two clusters. A method for assigning the distance between the clusters, termed "a linkage method," must be selected by the researcher. Single, complete, and average linkage methods define the distance between clusters as the minimum, maximum, and average pairwise distance between members of the clusters, respectively. The agglomerative process of comparing and merging clusters continues until only one cluster remains, and the resulting dendrogram illustrates the relationships between clusters at each step.

Another method is k-means clustering, which requires a priori knowledge of the number of clusters in the data. k-means clustering iteratively assigns objects to a cluster according to minimization of an objective function that measures the distance of the object to the mediod (center of a cluster). The process continues for a large number of iterations, until there are no more movements (reassignments) of objects to clusters or there is no more change in the objective functions after some predetermined number of iterations. Self-organizing maps (SOMs), a technique developed by Kohonen (42) in the early 1990s, is based on neural network theory and was first used by Tamayo et al. (43) to reveal unsupervised nodes (clusters) of genes that distinguish acute myeloid leukemia from acute lymphocytic leukemia biological samples.

Other clustering algorithms are known as supervised, where the labels of the samples and/or prior information about the samples (learning set) are provided to the classifier. For example, k-nearest neighbors (KNN) assign samples to the class of the nearest neighbor to it by majority vote, and support vector machines (SVMs) use the labels of the training samples near or at the decision boundaries separating two classes to classify the unknown/unlabeled sample. Steiner et al. (44) used multiple SVMs to separate classes and subclasses of toxicants, separate nonresponders from hepatotoxicants, identify sets of genes that discriminate hepatotoxicants from nonhepatotoxicants, and reveal how a predictive model built from one strain of rats can be used to classify treatment of another strain of rats.

One application of toxicogenomics analysis is the comparison of different treatment groups for classification of chemicals based on gene expression profiles. Biological samples derived from toxicant or control-treated animals can be represented as gene expression patterns consisting of fingerprints or profiles. These profiles are analyzed with automated pattern recognition analyses aimed at determining similarity between datasets rather than probing the genes for mechanistic information. For example, one goal of an in vivo study by Hamadeh et al. (11) was to use expression profiling to classify blinded samples derived from livers of xenobiotic-treated rats. Linear discriminant analysis (LDA) is one classical approach used to solve a task of this nature, but this method often relies on the assumption that the data follows a multivariate normal distribution. For gene expression data, Li et al. (45,46) have utilized an innovative nonparametric approach that combines a genetic algorithm (GA) for gene selection with the KNN method for sample classification (47). Subsets of candidate genes, "chromosomes," are generated and tested for their ability to separate known classes of samples using the KNN criterion. Through probabilistic "mutations," the best subsets of genes evolve as the process goes through several iterations. When a large number of these subsets have been selected, the frequency with which each gene appears in the subsets is tallied. Intuitively, the most frequently occurring genes should be the most informative in terms of discriminating between the queried classes. It is important to note that, Ooi and Tan (48), who developed the GA/maximum likelihood (GA/MLHD) method for multiclass prediction of gene expression data, indicate that the GA/KNN method may not be optimal for multiclass classification purposes because of limitations with using KNN with a high dimensional dataset; the computational complexity of the approach and the gene selection component of the algorithm is best suited for binary classification purposes. In any case, the flexibility of combined algorithms for gene selection and classification of microarray gene expression data is appealing because different components of the algorithm can be replaced with more improved approaches as they become available.

Principal component analysis (PCA) has also been used with microarray expression data (49). The goal of this type of analysis is to reduce the complexity of a dataset through the creation of new variables, called the principal components. The principal components retain a large percentage of the information that is present in the original variables, but are uncorrelated with each other. Depending on the goal of the analysis, either the genes or the arrays can be considered the original variables in a microarray experiment. If genes are used, the resulting principal components may provide insight into the particular features of the genes that explain the experimental response (50). The visualization of high dimensional data in two- or three-dimensional principal components space may reveal groups in the dataset. This information can then be used as input for classification methods, such as clustering, SOM or KNN. If there is significant separation between the clusters, PCA may also be used to assist with classification problems. Unknown samples can be "scored" using the results of a PCA on known samples, and these scores may be used to place the unknown into one of the previously identified clusters.

DATA MANAGEMENT

One of the primary challenges that has arisen with the development of microarray technology is that of efficiently managing the large volumes of information that these experiments produce. The NIEHS Microarray Group of the NCT has integrated two web-based applications that allow users to track and analyze their experimental information from start to finish—MicroArray Project System (MAPS), a laboratory information management system, and ArrayDB, an analysis information management system that allows users to store data and analyze individual array experiments.

The MAPS application, developed by Bushel et al. (32) at NIEHS, is a tool that allows researchers to store information about projects, samples, hybridizations, and quality control parameters. Originally designed as a Microsoft Access database with a ColdFusion web interface, MAPS has since been moved to an Oracle database for faster and more robust performance. A key feature of MAPS is the ability to store information about genes that were identified as differentially expressed using the confidence interval model described above, thus enabling the experimenter to analyze multiple hybridizations easily. For any combination of arrays, a list of differentially expressed, statistically valid "signature" genes that appear in some specified number of those arrays can be produced. Additionally, MAPS provides some statistical details about the genes that were chosen to give a measure of the quality of the data. For example, genes that show large variation in expression level across hybridizations can be flagged for further investigation or exclusion. In addition, MAPS also tracks information on RNA quality, scanner details, and other relevant parameters that vary and affect the quality of microarray data.

ArrayDB, originally developed at the NHGRI, provides an interactive system for the analysis of microarray experiments, and has been modified for use with an Oracle database at NIEHS. Once the image analysis is performed on the scanned array slide, the images and data are stored in the database for access via the web. An experimenter can view the pseudocolor array image, check intensity and expression values for particular spots on the array or for particular genes, and perform a differential expression analysis using the same method as the ArraySuite software described above. Because ArrayDB stores data for every gene on an array instead of just those that are differentially expressed, it provides a platform from which other programs can query data for analysis. The Microarray Group is currently in the process of interfacing several other analysis tools with ArrayDB to facilitate new analysis of existing datasets without reformatting.

BIOLOGICAL DATABASES USEFUL FOR TOXICOGENOMICS

Making biological sense of gene expression data generated by microarray analysis is a major rate-limiting step, yet probably the most critical and exciting component of a toxicogenomics study. Bioinformatics tools help visualize and dissect toxicant-induced gene expression data; however, interpreting and making biological sense of the coordinated regulation of the hundreds or thousands of genes requires tremendous scientific input. Xenobiotics can affect multiple molecular pathways and it is not uncommon to expect tissue-specific regulation of genes involved in various cellular processes to be simultaneously modulated by a chemical. Understanding and elucidating mechanism(s) of toxicity requires identification of coordinately regulated genes in common pathways, developing hypotheses as to how changes occur and the toxicological significance of the changes, and testing those hypotheses by conducting additional mechanistic studies. Table 1 provides some genomics-related web resources available for investigating the function of genes and gene products. These resources include those that allow characterization of ESTs, identification of biological and biochemical function of gene products as well as their cellular location, and identification of orthologous genes in other species and methods to search promoter sequences for common elements that may coordinately regulate gene expression. New resources are rapidly evolving and coming online.

EXAMPLES FROM AN IN VIVO TOXICOGENOMICS STUDY

The utility of using microarrays to perform a toxicogenomics study can be illustrated by describing the results of pioneering studies performed at the NCT (10,11). An in vivo rat model system was used to investigate the hypothesis that treatment with different xenobiotics results in chemical-specific patterns of altered gene expression. This hypothesis was based on the premise that genes that altered following exposure to different classes of chemicals can differentiate one class from another. Alterations

Table 1   Resources for Toxicogenomics

Category	Database	Web location[a]
Public EST sequences	GenBank	www.ncbi.nlm.nih.gov/Genbank/ index.html
	DbEST	www.ncbi.nlm.nih.gov/dbEST/index.html
cDNA databases	UniGene	www.ncbi.nlm.nih.gov/entrez/ query.fcgi?db=unigene
	TGIR gene indices	www.tigr.org
	DoTS	www.allgenes.org
	Eugene	eugenes.org/
Gene annotation and pathway(s)	DAVID	http://david.niaid.nih.gov/david/
	Genecard	www.genecards.org
	DRAGON	pevsnerlab.kennedykrieger.org/dragon.htm
	S.O.U.R.C.E.	http://source.stanford.edu
	KEGG	www.genome.jp/kegg/
	OMIM	www.ncbi.nlm.nih.gov/entrez/ query.fcgi?db=OMIM
	Ensembl-Human Genome Server	www.ensembl.org/
	SPAD	www.grt.kyushu-u.ac.jp/spad/index.html
	Signal transduction maps	stke.sciencemag.org/search/ searchpage.dtl?search_my=all
Other	ExPASy	us.expasy.org
	GO	www.geneontology.org/
	Pubmed	www.ncbi.nlm.nih.gov/entrez
	Mouse, rat, and human comparative maps	www3.ncbi.nlm.nih.gov/Homology/ index.html
	Locuslink	http://www.ncbi.nih.gov/entrez/ query.fcgi?db=gene
	Transcription element search system	www.cbil.upenn.edu/tess/
	Links to mammalian genomes	www.genelynx.org/
	Genome bioinformatics	genome.ucsc.edu/

[a]Website addresses as of December, 2005.
Abbreviations: DAVID, database for annotation, visualization, and integrated discovery; DoTS, database of transcribed sequences; DRAGON, database referencing of array genes online; EST, expressed sequence tag; ExPASy, expert protein analysis system; GO, gene ontology; KEGG, kyoto encyclopedia of genes and genomes; OMIM, online mendelian inheritance in man; SPAD, signalling pathways database.

in hepatic gene expression of Sprague-Dawley rats were studied 1 and 14 days after treatment with nongenotoxic rodent hepatocarcinogens (Fig. 1) (10). The compounds studied represent two classes of chemicals that alter gene expression through nuclear receptors. The enzyme inducer, phenobarbital, activates the constitutive androstane (or activated) receptor (51) and three known peroxisome proliferators (clofibrate, Wy-14,643, and gemfibrozil) activate the peroxisome proliferator-activated receptor α (PPARα) (52). D-Mannitol, which caused no detectable pathological effects in the liver, was used as a negative control. Microarray analysis was performed using the NIEHS rat chip containing approximately 1700 genes, as described above.

Transcript changes observed in rat livers after treatment were analyzed using a number of computational approaches. Clustering analysis was useful for determining similarities and differences in expression ratios between the genes and samples being analyzed, as visualized by the dendrogram in Figure 3. The three peroxisome proliferators clustered together, indicating a similar pattern of gene expression that was different from the gene expression profile for phenobarbital. Many of the changes in gene expression were in agreement with known responses to these agents. Genes reported to be involved in peroxisomal or mitochondrial fatty acid β-oxidation were upregulated by peroxisome proliferators, but suppressed by phenobarbital. Genes upregulated by phenobarbital treatment, but not by peroxisome proliferator treatment, included genes known to be involved in the detoxification and metabolism of xenobiotics. Hamadeh et al. (10) also observed ESTs with unknown functions that were coordinately regulated with named genes with known functions [Fig. 3 (node II)]. Clustering of microarray data could be used in toxicological studies to associate poorly characterized genes with expression profiles similar to genes involved in well-defined pathways or in the mechanism(s) of toxicity of chemicals. This has been termed "guilt by association" and will likely increase our understanding of gene function in general (38,53).

Detailed analysis of gene expression patterns of each compound revealed time-dependent and time-independent changes. Following phenobarbital treatment, the expression of 57 and 81 genes were significantly altered at 1 and 14 days, respectively (Fig. 4), with 38 genes altered at both time points, 19 genes specifically expressed at one day and 43 genes specific at 14 days. Cdc2 was upregulated only one day after phenobarbital treatment, while lipoprotein lipase was upregulated only at 14 days, indicating that phenobarbital treatment resulted in specific early effects related to hepatocyte hyperplasia and late time-dependent changes related to lipid metabolism.

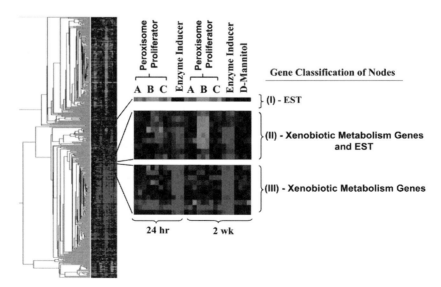

Figure 3 Hierarchical clustering of validated genes. Hierarchical tree depicts the grouping of correlated genes in specific nodes. I and II are nodes depicting classes of correlated genes.

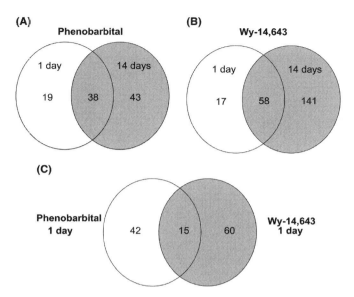

Figure 4 Gene alterations by phenobarbital and Wy-14,643. Validated gene outliers for each compound at 1 and 14 days, obtained using a binomial distribution analysis at 95% confidence interval, were compared and the results are presented using a Venn diagram. (**A**) Time-dependent and time-independent gene alterations by phenobarbital at 1 and 14 days, (**B**) time-dependent and time-independent gene alterations by Wy-14,643 at 1 and 14 days, and (**C**) partial overlap in genes regulated by phenobarbital or Wy-14,643 at one day are illustrated.

Phenobarbital-induced increases in lipoprotein lipase and hepatic triglyceride lipase activities have been reported with concomitant increases in hepatic synthesis of triglyceride, but lower serum concentration of triglyceride (54,55). Expression of CYP2B2 and glutathione-S-transferase were elevated at both time points. Liu et al. (56) demonstrated that a 163-bp fragment of the rat CYP2B2 gene contains sequences that mediate phenobarbital responsiveness, and mutations within this region reduced responsiveness to phenobarbital.

A similar analysis of gene expression altered by Wy-14,643 also revealed time-dependent and time-independent patterns of gene expression, with a total of 75 and 199 genes significantly modulated in rat liver 1 and 14 days after treatment, respectively (Fig. 4B). Of these, 58 genes were common to both time points, while 17 genes were differentially expressed only at one day and 141 genes were only differentially expressed at 14 days after treatment. Acyl-CoA oxidase and acyl-CoA dehydrogenase, genes involved in the first and second steps of fatty acid β-oxidation, were 2 of the 17 genes that were observed to be uniquely modulated by Wy-14,643 at one day. One of the 141 genes modulated by Wy-14,643 at 14 days was lipid-binding protein. Thiolase and stearyl-CoA desaturase exhibited increased expression at both time points. All of these genes are known to be transcriptionally regulated by PPARα (57,58). An understanding of early and late changes in gene expression may not only be indicative of altered biological processes that arise from xenobiotic exposure, but may also be useful for chemical classification.

Alterations in gene expression induced by compounds in these two different chemical classes were compared one day after treatment. The data from this comparison showed that Wy-14,643 and phenobarbital significantly modulated the expression of 75 and 57 genes, respectively, of which only 15 genes were common to both compounds (Fig. 4C). Therefore, Wy-14,643 and phenobarbital treatment uniquely modulated 60 and 42 genes, respectively, suggesting that these genes may help define chemical class-specific "gene signatures." For example, the expression of CYP2B2 was increased by phenobarbital but not by Wy-14,643, while the expression of thiolase was increased by Wy-14,643 but not by phenobarbital. In contrast, both Wy-14,643 and phenobarbital increased transcript levels of uridine diphosphate (UDP)–glucose dehydrogenase, suggesting that increasing the expression of this enzyme is a common response to xenobiotic exposure because this enzyme furnishes UDP-glucuronic acid for Phase II xenobiotic metabolism (59).

Comparison of the altered gene expression profiles after treatment with the three peroxisome proliferators (clofibrate, Wy-14,643, and gemfibrozil) revealed 12 genes that were regulated in the same manner by all three compounds one day after treatment and 13 genes commonly regulated 14 days after treatment (Fig. 5A and B). One of the genes overexpressed by all three peroxisome proliferators at both time points was rat liver stearyl-CoA desaturase. In addition to changes in gene expression shared by the three compounds, the peroxisome proliferators caused changes in gene expression that were unique for each compound (Fig. 5A and B). Further analyses of these observations will hopefully provide evidence that gene expression studies can be used to identify subtle similarities and differences in mechanism(s) of action of compounds within a chemical class, which may, in part, be attributed to differences in chemical structure, receptor–ligand interactions, drug metabolism, and/or gene targets.

As the first study was able to identify genes that distinguish two classes of compounds, the next study by Hamadeh et al. (11) sought to determine if it was possible

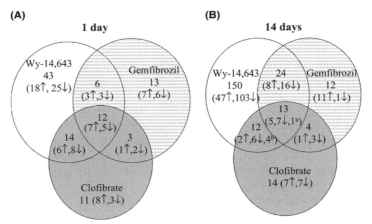

[a]A gene was induced by both clofibrate and gemfibrozil and repressed by Wy-14,643 but occurred as a validated outlier for all three compounds.
[b]Three genes were repressed by clofibrate but induced by Wy-14,643 and an additional gene was induced by clofibrate but repressed by Wy-14,643; all four of these genes are validated outliers.

Figure 5 Gene expression changes by peroxisome proliferators. Validated gene outliers for each peroxisome proliferator compound (i.e., clofibrate, Wy-14,643, and gemfibrozil), at the (**A**) one-day time point and (**B**) 14-day time points, obtained using a binomial distribution analysis and 95% confidence interval are compared and the results are presented using a Venn diagram.

to assign an unknown compound to a specific chemical class based on changes in gene expression elicited by treatment. Multiple approaches were used to find highly discriminatory and informative genes whose expression pattern could distinguish RNA samples derived from livers exposed to different chemicals. Two methods, LDA and GA/KNN discussed above were useful in revealing genes that could separate known samples based on the class of chemical involved in the exposure in a time-independent manner. Using these procedures, 22 highly informative genes that clearly exhibited different patterns of expression between enzyme inducers and peroxisome proliferators were identified. A blinded study was conducted using rat liver RNA from animals treated with various compounds, comparing the gene expression alterations in the blinded samples, using computational tools, to the gene expression changes elicited by compounds from the two chemical classes, as well as the negative control studied. The results were extremely encouraging because this approach made it possible to successfully classify 22 out of 23 unknown compounds based on the expression signatures of known compounds (11).

A number of other groups have published studies that categorize chemicals based on gene expression patterns (Table 2). Most of these studies examined pattern changes after chemical exposure in the rat liver because the liver is the primary site for toxicity and much is known about mechanisms of chemical toxicity in this organ, allowing for preliminary chemical classification based on mode of action. In some studies, unsupervised methods were initially used for clustering the compounds; success was limited because of the lack of reproducible gene responses likely due to biological variability and limitations imposed by lack of replicates (60) or due to incomplete understanding of the types of liver toxicity induced by the queried compounds (63). Most studies used supervised methods that required prior knowledge of the type of toxicity induced by chemical exposure. Although these studies focused on high doses that led to clear toxicity as assessed by conventional end-points, a major challenge will be to include in the model building process expression profiles after treatment with doses of chemicals that do not induce the conventional toxicology endpoints. For example, a recent study analyzing gene expression changes in the livers of rats treated with acetaminophen showed that gene expression changes predicted toxic effects that were not observed by conventional endpoints at higher doses (14). Earlier time points should also be considered to determine if gene expression changes are early predictors of developing toxicities before standard tests could detect them. Together these findings indicate that it will be possible to obtain compound-related gene expression signatures that are useful in chemical class prediction.

FUTURE PROSPECTS FOR TOXICOGENOMICS

Advances in toxicogenomics are expected to facilitate pharmaceutical and industrial lead compound development by identifying, much earlier than is presently possible, which compounds have the propensity to cause human toxicity and, perhaps, to predict the target population for either the pharmacological or toxicological effect. An era of genetic medicine in which therapeutic strategies will be tailored to the needs of individuals with known genome sequence variations is likely to emerge in the future. The potential for monitoring subtle gene expression changes will facilitate the development of biomarkers of exposure and effect, providing future opportunities to screen ongoing molecular changes in accessible human tissues, i.e., blood, urine,

Table 2 Chemical Classification Studies

References	Experimental parameters (#cmpds; system; treatment time; dose)	Discriminant method	Supporting data	Results
Burczynski et al., 2000 (60)	100; HepG2 cells; 24 hr; various doses	Unsupervised; supervised computational algorithm	None	Discriminated between DNA damaging and anti-inflammatory drugs
Waring et al., 2001 (8)	15; rats; 3 days; known hepatotoxic doses	Unsupervised methods: hierarchical, divisive hierarchical, k-means, and SOMs	Histopathology and clinical chemistry	Profiles correlated to clinical chemistry and pathology
Waring et al., 2001 (61)	15; rat primary hepatocytes; 24hr; 20 μM	Unsupervised unweighted pair-group method with arithmetic mean	None	Compounds with similar toxic mechanisms formed clusters
Hamadeh et al., 2002 (10,11)	4; rats; 1 and 14 days; various doses	Supervised; LDA GA/KNN	Histopathology, RT-PCR	Correctly identified 22 out of 23 blinded samples
De Longueville et al., 2003 (62)	11; rat primary hepatocytes; 24 hr; below toxic dose	Unsupervised classical agglomerative hierarchical clustering	RT-PCR	Steatosis inducers clustered together
McMillian et al., 2004 (63)	100; rats; 24 hr; MTD	Unsupervised; supervised discriminant analysis and cross-validation	RT-PCR of cytokines	Discriminated between macrophage activators and peroxisome proliferators
Steiner et al., 2004 (44)	31; rats; 6 hr to 14 days; various doses	Supervised SVMs with recursive feature elimination	Serum chemistry; histopathology	Discriminated between hepatotoxic and nonhepatotox; predicted class in most cases

Abbreviations: cmpds, compounds; GA, genetic algorithm; KNN, *k*-nearest neighbors; LDA, linear discriminant analysis; MTD, maximum tolerated dose; RT-PCR, reverse transcriptase-polymerase chain reaction; SOMs, self-organizing maps; SVMs, support vector machines.

buccal scrapings. Not only will environmental and occupational health physicians and scientists be able to identify toxic compounds and their mechanism(s), but it is also expected that genomic and technological advances will propel medical advances and provide opportunities for physicians to intervene during disease development and progression.

REFERENCES

1. Venter JC, Adams MD, Myers EW, et al. The sequence of the human genome. Science 2001; 291:1304–1351.
2. Lander ES, Linton LM, Birren B, et al. Initial sequencing and analysis of the human genome. Nature 2001; 409:860–921.
3. Waterston RH, Lindblad-Toh K, Birney E, et al. (Mouse Genome Sequencing Consortium) Initial sequencing and comparative analysis of the mouse genome. Nature 2002; 420:520–562.
4. Gibbs RA, Weinstock GM, Metzker ML, et al. (Rat Genome Sequening Project evolution) Genome sequence of the Brown Norway rat yields insights into mammalian evolution. Nature 2004; 428:493–521.
5. Duggan DJ, Bittner M, Chen Y, Meltzer P, Trent JM. Expression profiling using cDNA microarrays. Nat Genet 1999; 21:10–14.
6. Brown PO, Botstein D. Exploring the new world of the genome with DNA microarrays. Nat Genet 1999; 21:33–37.
7. Bowtell DD. Options available—from start to finish—for obtaining expression data by microarray. Nat Genet 1999; 21:25–32.
8. Waring JF, Jolly RA, Ciurlionis R, et al. Clustering of hepatotoxins based on mechanism of toxicity using gene expression profiles. Toxicol Appl Pharmacol 2001; 175:28–42.
9. Hamadeh HK, Knight BL, Haugen AC, et al. Methapyrilene toxicity: anchorage of pathologic observations to gene expression alterations. Toxicol Pathol 2002; 30:470–482.
10. Hamadeh HK, Bushel PR, Jayadev S, et al. Gene expression analysis reveals chemical-specific profiles. Toxicol Sci 2002; 67:219–231.
11. Hamadeh HK, Bushel PR, Jayadev S, et al. Prediction of compound signature using high density gene expression profiling. Toxicol Sci 2002; 67:232–240.
12. Ruepp SU, Tonge RP, Shaw J, Wallis N, Pognan F. Genomics and proteomics analysis of acetaminophen toxicity in mouse liver. Toxicol Sci 2002; 65:135–150.
13. Ueda A, Hamadeh HK, Webb HK, et al. Diverse roles of the nuclear orphan receptor CAR in regulating hepatic genes in response to phenobarbital. Mol Pharmacol 2002; 61:1–6.
14. Heinloth AN, Irwin RD, Boorman GA, et al. Gene expression profiling of rat livers reveals indicators of potential adverse effects. Toxicol Sci 2004; 80:193–202.
15. Huang Q, Dunn RT II, Jayadev S, et al. Assessment of cisplatin-induced nephrotoxicity by microarray technology. Toxicol Sci 2001; 63:196–207.
16. Kramer JA, Pettit SD, Amin RP, et al. Overview on the application of transcription profiling using selected nephrotoxicants for toxicology assessment. Environ Health Perspect 2004; 112:460–464.
17. Amin RP, Vickers AE, Sistare F, et al. Identification of putative gene based markers of renal toxicity. Environ Health Perspect 2004; 112:465–479.
18. Henry CJ, Phillips R, Carpanini F, et al. Use of genomics in toxicology and epidemiology: findings and recommendations of a workshop. Environ Health Perspect 2002; 110:1047–1050.
19. Amin RP, Hamadeh HK, Bushel PR, Bennett L, Afshari CA, Paules RS. Genomic interrogation of mechanism(s) underlying cellular responses to toxicants. Toxicology 2002; 181–182:555–563.

20. Gant TW. Application of toxicogenomics in drug development. Drug News Perspect 2003; 16:217–221.
21. Suter L, Babiss LE, Wheeldon EB. Toxicogenomics in predictive toxicology in drug development. Chem Biol 2004; 11:161–171.
22. Irwin RD, Boorman GA, Cunningham ML, Heinloth AN, Malarkey DE, Paules RS. Application of toxicogenomics to toxicology: basic concepts in the analysis of microarray data. Toxicol Pathol 2004; 32(suppl 1):72–83.
23. Waters M, Boorman G, Bushel P, et al. Systems toxicology and the chemic effects in biological systems (CEBS) knowledge base. EHP Toxicogenomics 2003; 111:15–28.
24. www.niehs.nih.gov/nct/.
25. Hughes TR, Mao M, Jones AR, et al. Expression profiling using microarrays fabricated by an ink-jet oligonucleotide synthesizer. Nat Biotechnol 2001; 19:342–347.
26. Lipshutz RJ, Fodor SP, Gingeras TR, Lockhart DJ. High density synthetic oligonucleotide arrays. Nat Genet 1999; 21:20–24.
27. www.dir.niehs.nih.gov/microarray/.
28. Nuwaysir EF, Bittner M, Trent J, Barrett JC, Afshari CA. Microarrays and toxicology: the advent of toxicogenomics. Mol Carcin 1999; 24:153–159.
29. www.nhgri.nih.gov/DIR/Microarray.
30. Chen Y, Dougherty ER, Bittner ML. Ratio-based decisions and the quantitative analysis of cDNA microarray images. J Biomed Opt 1997; 2:364–374.
31. Chen Y, Kamat V, Dougherty ER, Bittner ML, Meltzer PS, Trent JM. Ratio statistics of gene expression levels and applications to microarray data analysis. Bioinformatics 2002; 18:1207–1215.
32. Bushel PR, Hamadeh H, Bennett L, et al. MAPS: a microarray project system for gene expression experiment information and analysis. Bioinformatics 2001; 17:564–565.
33. Rosenzweig BA, Pine PS, Domon OE, Morris SM, Chen JJ, Sistare FD. Dye bias correction in dual-labeled cDNA microarray gene expression measurements. Environ Health Perspect 2004; 112:480–487.
34. Bushel PR, Hamadeh HK, Bennett L, et al. Computational selection of distinct class- and subclass-specific gene expression signatures. J Biomed Inform 2003; 35:160–170.
35. Wolfinger RD, Gibson G, Wolfinger ED, et al. Assessing gene significance from cDNA microarray expression data via mixed models. J Comp Biol 2001; 8:625–637.
36. Kerr MK, Martin M, Churchill GA. Analysis of variance for gene expression microarray data. J Comp Biol 2000; 7:819–837.
37. Lin DY. An efficient Monte Carlo approach to assessing statistical significance in genomic studies. Bioinformatics 2004; 21(6):781–787.
38. Hughes TR, Marton MJ, Jones AR, et al. Functional discovery via a compendium of expression profiles. Cell 2000; 102:109–126.
39. Callow MJ, Dudoit S, Gong EL, Speed TP, Rubin EM. Microarray expression profiling identifies genes with altered expression in HDL-deficient mice. Genome Res 2000; 10:2022–2029.
40. Tusher VG, Tibshirani R, Chu G. Significance analysis of microarrays applied to the ionizing radiation response. Proc Natl Acad Sci USA 2001; 98:5116–5121.
41. Eisen MB, Spellman PT, Brown PO, Botstein D. Cluster analysis and display of genome-wide expression patterns. Proc Natl Acad Sci USA 1998; 95:14,863–14,868.
42. Kohonen T. Self-organizing maps. In: Kohonen T, ed. Springer Series in Information Sciences. Vol. 30. Berlin: Springer, 2001:501.
43. Tamayo P, Slonim D, Mesirov J, et al. Interpreting patterns of gene expression with self-organizing maps: methods and application to hematopoietic differentiation. Proc Natl Acad Sci USA 1999; 96:2907–2912.
44. Steiner G, Suter L, Boess F, et al. Discriminating different classes of toxicants by transcript profiling. Environ Health Perspect 2004; 112:1236–1248.

45. Li L, Darden TA, Weinberg CR, Levine AJ, Pedersen LG. Gene assessment and sample classification for gene expression data using a genetic algorithm/k-nearest neighbor method. Comb Chem High Throughput Screen 2001; 4:727–739.
46. Li L, Weinberg CR, Darden TA, Pedersen LG. Gene selection for sample classification based on gene expression data: study of sensitivity to choice of parameters of the GA/KNN method. Bioinformatics 2001; 17:1131–1142.
47. Pei M, Goodman ED, Punch WF. Feature extraction using genetic algorithms. Proceeding of International Symposium on Intelligent Data Engineering and Learning' 98 IDEAL October 1998:371–384.
48. Ooi CH, Tan P. Genetic algorithms applied to multi-class prediction for the analysis of gene expression data. Bioinformatics 2003; 19:37–44.
49. Hilsenbeck SG, Friedrichs WE, Schiff R, et al. Statistical analysis of array expression data as applied to the problem of tamoxifen resistance. J Natl Cancer Inst 1999; 91:453–459.
50. Raychaudhuri S, Stuart JM, Altman RB. Principal components analysis to summarize microarray experiments: applications to sporulation time series. Pac Symp Biocomput 2000:455–466.
51. Kakizaki S, Yamamoto Y, Ueda A, Moore R, Sueyoshi T, Negishi M. Phenobarbital induction of drug/steroid-metabolizing enzymes and nuclear receptor CAR. Biochim Biophys Acta 2003; 1619:239–242.
52. Corton JC, Anderson SP, Stauber A. Central role of peroxisome proliferator-activated receptors in the actions of peroxisome proliferators. Annu Rev Pharmacol Toxicol 2000; 40:491–518.
53. Stuart JM, Segal E, Koller D, Kim SK. A gene-coexpression network for global discovery of conserved genetic modules. Science 2003; 302:249–255.
54. Goldberg DM, Parkes JG. Lipolytic enzymes as markers of induction and differentiation. Clin Biochem 1987; 20:405–413.
55. Goldberg DM, Roomi MW, Yu A. Modulation by phenobarbital of lipolytic activity in postheparin plasma and tissues of the rat. Can J Biochem 1982; 60:1077–1083.
56. Liu S, Rivera-Rivera I, Bredemeyer AJ, Kemper B. Functional analysis of the phenobarbital-responsive unit in rat CYP2B2. Biochem Pharmacol 2001; 62:21–28.
57. Anderson SP, Dunn C, Laughter A, et al. Overlapping transcriptional programs regulated by the nuclear receptors peroxisome proliferator-activated receptor alpha, retinoid X receptor and liver X receptor in mouse liver. Mol Pharmacol 2004; 66(6):1440–1452.
58. Anderson SP, Howroyd P, Liu J, et al. The transcriptional response to a peroxisome proliferator-activated receptor alpha (PPAR alpha) agonist includes increased expression of proteome maintenance genes. J Biol Chem 2004; 279:52,390–52,398.
59. Sheweita SA. Drug-metabolizing enzymes: mechanisms and functions. Curr Drug Metab 2000; 1:107–132.
60. Burczynski ME, McMillian M, Ciervo J, et al. Toxicogenomics-based discrimination of toxic mechanism in HepG2 human hepatoma cells. Toxicol Sci 2000; 58:399–415.
61. Waring JF, Ciurlionis R, Jolly RA, Heindel M, Ulrich RG. Microarray analysis of hepatotoxins in vitro reveals a correlation between gene expression profiles and mechanisms of toxicity. Toxicol Lett 2001; 120:359–368.
62. de Longueville F, Atienzar FA, Marcq L, et al. Use of a low-density microarray for studying gene expression patterns induced by hepatotoxicants on primary cultures of rat hepatocytes. Toxicol Sci 2003; 75:378–392.
63. McMillian M, Nie AY, Parker JB, et al. Inverse gene expression patterns for macrophage activating hepatotoxicants and peroxisome proliferators in rat liver. Biochem Pharmacol 2004; 67:2141–2165.

3
Proteomics

Frank A. Witzmann
Department of Cellular and Integrative Physiology, Biotechnology Research and Training Center, Indiana University School of Medicine, Indianapolis, Indiana, U.S.A.

INTRODUCTION

It is well known that xenobiotics exert their biological effects via the alteration of protein expression, through up- and down-regulation, alteration of protein synthesis and degradation rates, or chemically modifying existing proteins. Either way, chemical toxicants can have very complicated cellular effects—those that can be studied by proteomics. Proteomics measures the quantitative and qualitative changes in cellular or tissue protein expression and explores protein–protein and protein–ligand interactions. Toxicoproteomics is an approach for the identification and characterization of proteins whose expression is altered by chemical exposure, and it is complementary to toxicogenomics.

There seems to be general agreement among toxicologists that differential protein expression information is a critically important component of a comprehensive biomolecular (panomics) approach in characterizing, understanding, and even predicting chemical toxicity (1–3). Exactly how one should obtain such proteomic information remains a matter of diverse opinion, given the unique design of toxicological experiments and the often complicated responses exhibited by cells and tissues to chemical exposure. Nonetheless, the assumption that changes in protein expression may provide evidence of specific mechanism of toxic effect, by increased expression, decreased expression, or even subtle post-translational modifications, is based on years of published observation. The prospect of using protein expression alterations as diagnostic markers of exposure or effect, often in such hugely complex samples as serum, plasma, cerebrospinal fluid, or urine is an additional driving force empowering the application of proteomic methodologies in toxicity studies.

Notwithstanding the logical combination of proteomics and toxicology, this technological union seems to be underrepresented in the scientific literature. As Figure 1 illustrates, despite nearly 30 years of quasi global protein expression analysis (initially by two-dimensional electrophoresis or 2-DE) and well over 20,000 published citations, relatively few papers combining 2-DE and tox* have been published (approximately 2.5%). The same is true for the more recent appellation "proteomics," e.g., the global analysis of the proteins encoded (and many not encoded) by the genome.

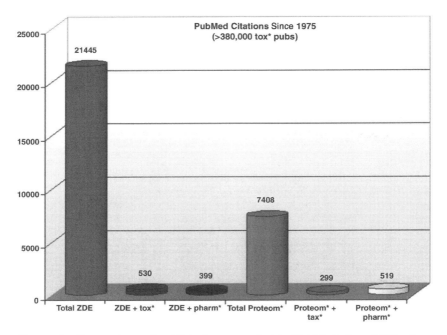

Figure 1 Comparison of published papers cited in PubMed since 1975 using search terms tox*, two-dimensional electrophor*, 2-DE, 2-DE, 2-D gel, pharm*, proteom* in various combinations, and searching all fields. The results suggest that 2-DE, and other proteomic techniques have not been exploited fully in toxicology-related studies. *Abbreviation*: 2-DE, two-dimensional electrophoresis.

Since the proteomics explosion began during the last decade (4), this combination of various protein analytical techniques has been cited in PubMed nearly 7500 times, while its combination with tox* appears in only approximately 4% of those papers. Whatever the underlying reasons for this discontinuity, the comparatively small number of tox-related citations may be due to the difficulty toxicologists encounter in applying the various proteomic technologies in their experiments.

Unique Challenges for Toxicoproteomics

The bulk of toxicity testing is conducted using animal models with subsequent extrapolation to humans. Therefore, to best exploit the power of toxicoproteomics, it is necessary to establish and fully characterize the unique target tissue proteomes of various species (5), including normal and sensitive human populations. This also applies to their response to intoxication. Furthermore, proteomic toxicity tests frequently take place in the context of other "omics" (notably functional genomics and metabolomics), the dose–response, and the time–course. The latter two place a unique burden on the proteomic technology selected to study differential protein expression, and its implementation in toxicology has proven to be challenging (6).

Perhaps the most significant difficulty associated with toxicoproteomics relates to the nature of the typical toxicologic experiment. As alluded to above, the design often includes the dose–response relationship, a paradigm made even more complicated by the single versus repeated–dose response, the time–course, interaction assessment, complex synergisms or antagonisms, potentiation, metabolic biotrans-formation reactions, or various combinations of these. Moreover, a toxicologic investigation frequently includes various classes of tests—acute, subacute, subchronic, chronic, and even mutagenic. Consequently, the typical experimental design is complicated and includes large numbers of samples per experiment. This poses tremendous challenges to proteomic application, regardless of which strategy or platform one employs.

The Goal of Toxicoproteomics

Ultimately, the question that must be asked is "What is it that we really want to know about protein expression in response to chemical exposure?" Typically, the answer lies in globally assessing quantitative and qualitative changes in the proteome. This necessitates examining increases or decreases in protein expression with exposure, requiring relative quantitation at least and absolute quantitation at best. Secondly, abnormal protein posttranslational modification (PTM) must be examined, requiring the assessment of the extent, chemical nature, and specific location of the chemical modification. Any alterations in protein expression noted by the above must be related to relevant toxic end points, rendering these as "toxicologically relevant" proteins. Finally, one must ascertain that the observed protein changes occur in both the animal model and the human, requiring the investigation of homologous systems. The greatest challenge to toxicologists lies in selecting a proteomic platform that best addresses these issues and provides proteomic information with interspecies relevance.

In a toxicologic target cell or tissue, each expressed protein is unique and the final, fully functional protein product rarely resembles the gene(s) from which it was coded. Therefore, a comprehensive approach that includes a variety of proteomic techniques is necessary to detect, quantify, and identify individual proteins, to elucidate protein activities, and to determine protein–protein interactions. Unlike mRNA analysis, protein analysis is severely limited by the tremendous dynamic range of protein expression and the large number of relevant proteins whose cellular expression or abundance in body fluids is minuscule.

This chapter will address several of the core technologies that comprise contemporary differential expression proteomics most relevant to toxicology, focusing on those whose protein analyses are quantitative and where they have been applied successfully in toxicologic investigations.

DIFFERENTIAL EXPRESSION PROTEOMIC TECHNOLOGIES

2-DE

It is still considered by many to be the most powerful and readily applicable of the proteomic techniques, 2-DE involves the polyacrylamide-based separation of complex protein mixtures first by protein charge (i.e., isoelectric point or pI) via isoelectric focusing (IEF) in a pH gradient, followed by mass separation in the presence of sodium dodecyl sulfate (7). In 2-DE, resolved proteins are visualized by any of a variety of stains or dyes, and generally, comparative quantitation of protein expression is conducted by image and statistical analyses. Distinctively, this technique is simultaneously analytical and preparative. Separated protein spots can be cut from the gel, digested proteolytically, and the resulting peptides analyzed further by mass spectrometry (MS) (8). Online peptide-mass or -sequence database comparisons then identify and characterize the proteins [see section on MS and Tandem MS (MS/MS)].

PTMs are frequently accompanied by changes in protein charge and are easily observed by their altered pI. In 2-DE, protein phosphorylation, glycosylation, and chemical adduct formation—three of the most relevant PTM in toxicologic studies—frequently generate multiple spots of differing pI so that individual proteins may be represented numerous times on a 2-D gel. Consequently, a gel depicting 2000 detected protein spots should not be misinterpreted as having resolved 2000 unique proteins. Instead, a significant number of spots are charge isoforms representing far fewer actual gene products. Nonetheless, 2-DE is one of few proteomic approaches capable of readily detecting PTM, quantitatively comparing the extent of PTM, and enabling the determination of the PTMs' chemical nature.

The combination of 2-DE and MS constitutes a powerful platform for protein expression analysis. Unfortunately, 2-DE has several shortcomings that have spawned a number of chromatographic and mass spectrometric approaches designed to overcome these deficiencies and render protein expression analysis truly global. The major weaknesses of the 2-DE approach include an inability to resolve very hydrophobic proteins and those with extremes of pI (particularly basic proteins). The hydrophobicity issue is difficult to overcome, as it is in all proteomic approaches, while the pI issues continue to be addressed by technical developments. Another major problem that has limited 2-DE's applicability is its poor dynamic range. This is due to a combination of factors, including limited physical space for protein separation (gel-format) and protein detection (stain or dye sensitivity). If the range of the putatively expressed $> 100,000$ different cellular protein forms resembles that postulated for plasma, greater than nine orders of magnitude (9), researchers using this approach will remain incapable of analyzing physiologically relevant, low abundance proteins altered by chemical exposures. Finally, toxicologists who choose 2-DE to make relative quantitative comparisons between numerous individual groups of protein samples quickly realize that gel-to-gel variation is an issue. While the underlying reasons for this are numerous, this and other difficulties can be overcome.

Addressing 2-DE's Limitations

2-D DIGE and Multiplexing.   To enable the separation and differential expression analysis of two or more samples on a single 2-D gel, fluorescent two-dimensional difference gel electrophoresis (2-D DIGE) was developed (10). Despite having its own limitations (11), this approach overcomes most gel variability issues and can be useful in small experiments with limited numbers of samples for comparison. In DIGE (Fig. 2), complex protein samples are labeled with fluorescent cyanine dyes Cy2, Cy3, and Cy5 and the samples mixed so that they can be examined in a single 2-D gel with each dye giving an independent channel of measurement. For each spot detected in the gel pattern, the intensities in the respective dye channels are compared, and ratios are calculated from these intensities to indicate the extent of differential protein expression. With DIGE, the uncertainty of coordinate gel matching across two or more gels is overcome, for the most part. With the advent of saturation labeling (12,13), 2-D DIGE has demonstrated unprecedented sensitivity. For instance, Stühler et al. (14) recently used this approach to detect approximately 2400 protein spots on 2-D gels loaded with protein from approximately 5000 cells (2 to 3 μg) microdissected from precursor lesions of pancreatic adenocarcinoma.

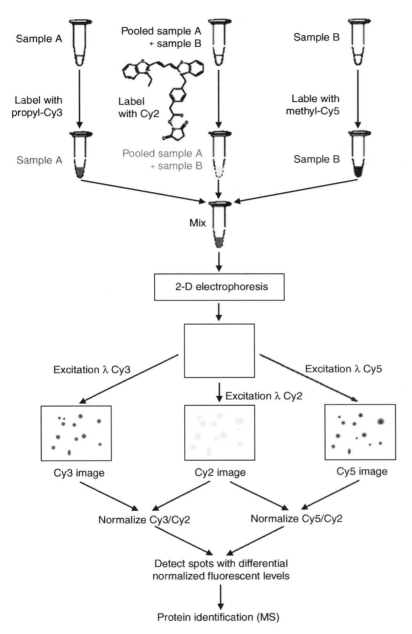

Figure 2 Schematic of a 2-D DIGE experiment with an internal pooled standard and three fluorescent dyes. Samples A and B are labeled with either Cy3 or Cy5, and a pooled internal standard is also constructed from equal amounts of all the samples in the experiment and labeled with Cy2. After mixing these protein samples and performing 2-DE, the protein spot patterns can be visualized by illuminating the gel with the specific excitation wavelengths. Samples A and B protein spot intensities are then normalized by dividing each by the corresponding spot intensity of the pooled internal standard. Analyzing the normalized spot intensities enables the detection of subtle differences in protein expression levels with a higher statistical confidence. *Abbreviations*: 2-D DIGE, 2-D gel, fluorescent two-dimensional difference gel electrophoresis; 2-DE, two-dimensional electrophoresis. *Source*: From Ref. 11.

DIGE has been used effectively in several toxicology studies. The mechanism of acetaminophen hepatotoxicity has been characterized and the DIGE approach optimized (15,16), and preclinical studies resulted in the discovery of predictive biomarkers of compounds with a propensity to induce liver steatosis (17). Taking advantage of running liver samples from hydrazine-treated and control rats on a single 2-D gel, Kleno et al. (18) applied multivariate analysis to discover new biomarkers of hydrazine hepatotoxicity. Finally, combining the power of DIGE and rodent airway epithelial cell isolation, Wheelock et al. (19) improved the applicability of 2-DE-based proteomics to respiratory toxicology by demonstrating the enrichment (by 36%) of epithelial cell-specific proteins and resolving 2365 detectable spots.

A related approach avoids Cy dyes but incorporates the multiplexing strategy by means of multiple, repetitive staining of individual gels to detect all proteins in the 2-D gel pattern, along with phosphoproteins (20) and glycoproteins (21). This methodology enables parallel determination of altered glycosylation and phosphorylation patterns and protein expression level changes without running three gels.

Large-Scale, Highly Parallel 2-DE. An alternative to multiplexing samples in 2-DE by DIGE is found in highly parallel 2-DE separations. While current maximal capacities of commercially available 2-DE apparatus are limited to 12 gels per run, the ISO-DALT System (22,23), in which 20 large-format 2-D gels can be cast and run simultaneously, is uniquely well suited to overcome gel-to-gel variability. The separation of up to 20 samples per run (or 10 replicates per run) directly addresses the demands placed on 2-D gel–based proteomics by complicated toxicologic experimental design. For example, recently, we have analyzed the effect of various concentrations of hydrazine and cadmium exposure in vitro in rat hepatocyte primary culture (Witzmann, unpublished data) using multiple runs of the 20-gel ISO-DALT System. Using PDQuest 2-D Gel Analysis Software (BioRad), 144 individual sample gels representing 144 individual wells from 24 six-well culture plates were analyzed. Figure 3 illustrates the composition of this matchset and the remarkable pattern reproducibility achieved. An average of 1100 proteins were matched in this gel set, 415 of them in every single pattern. Whether the toxicological experiment is larger or smaller, using a highly parallel platform, technical limitations now seem to rest in detection sensitivity and image analysis capacity, not in the reproducibility of electrophoretic separation.

Increasing Analytical "Depth of Field"—Sample Prefractionation. With the ability to run literally hundreds of gels with acceptable consistency, 2-DE analysis in toxicoproteomics is poised to take advantage of another trend in proteomics—the reduction of sample complexity. It is clear that global protein analysis across the range of protein expression is currently impossible, no matter which proteomics platform one applies. Consequently, the strategy of examining subsets of the proteome has gained significant momentum, and methods to reduce sample complexity in all manner of proteomic studies are becoming routine (25).

In toxicoproteomics, reducing sample complexity generally involves decreasing sample heterogeneity and improving the analytical "depth of field" by digging deeper into the proteome. For example, rather than analyzing protein expression in whole liver, kidney, or brain samples where numerous cell types reside, one uses carefully microdissected regions obtained by laser capture (26) or cells isolated by traditional collagenase treatment and centrifugation. Increased depth of field is achieved through the analysis of cell organelles, specific membrane fractions (27), or multiple subproteomes (28). One can also study specifically enriched proteomic subsets generated by depletion of highly abundant proteins, as in serum or plasma (29,30) and urine (31), by preparative solution phase sIEF (32–34) or by chromatography (35).

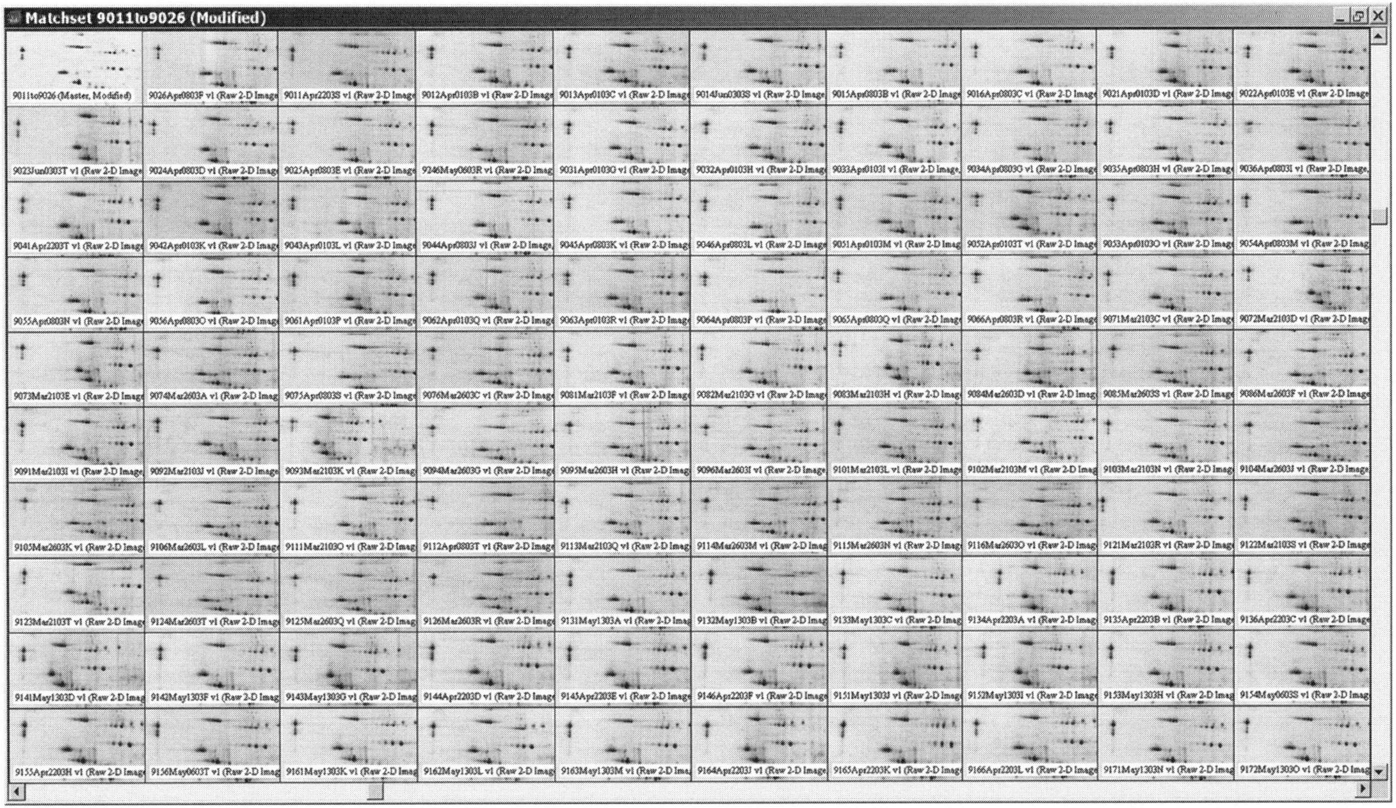

Figure 3 Screen-capture of PDQuest montage illustrating 99 of 144 rat hepatocyte 2-D gel images (plus 1 reference pattern, *first frame*) analyzed in a single matchset (100 frames/display window is maximum). Hepatocytes were exposed to hydrazine or cadmium in a range of exposures in six-well plates, solubilized in situ after removal of media, separated by 2-DE using the ISO-DALT System (20 gels/run), and stained with colloidal Coomassie blue, as referred to in the text. Each frame illustrates the same region of the 2-DE gel pattern, containing calreticulin precursor (*upper left in each frame*) and hsp60 (*middle right*), among others. Approximately 1100 protein spots were matched in each pattern, with 415 spots matched in every pattern. *Abbreviation:* 2-DE, two-dimensional electrophoresis. *Source:* From Ref. 24.

In photography, the phrase "depth of field" can be defined as "the distance range between the nearest and farthest objects that appear in acceptably sharp focus." Its use (or misuse) here is intended as a point of emphasis with respect to the vast dynamic range of protein expression and our desire to "sharply focus" on and accurately analyze expression of the least abundantly expressed protein alongside the most abundant protein, and everything in between. To improve analytical depth for 2-DE, sIEF provides perhaps the best and most attractive remedy. This approach subdivides a complex protein mixture into well-resolved fractions based on the pI of each protein in the mixture. This is accomplished in a conveniently low-volume platform (32,36) using series of chambers connected in tandem and separated by thin membranes that contain covalently attached buffers of defined pH (immobilines). The protein sample is loaded into the chambers separated by these disks and spacers and subjected to sIEF. The result is a significant reduction in original sample complexity in the form of a set of five highly resolved protein fractions suitable for further analysis by 2-DE.

In our laboratory's 2-D gel–based toxicoproteomics efforts, we have begun to address this issue in various proof-of-concept studies by prefractionating samples using a combination of subcellular fractionation (by differential centrifugation) combined with further fractionation using sIEF. We hypothesize that by combining the two fractionation techniques, a significantly greater portion of a cell's or tissue's protein complement can be analyzed. Figure 4 illustrates the components of this approach and emphasizes the need for a 2-DE platform capable of handling the large number of samples. The successful application of this approach is heavily dependent on reproducibility in the fractionation technique as well as in 2-DE. Using a large capacity, highly parallel gel system makes this a viable approach. It is unlikely that a more limited-capacity platform or multiplexing strategy would be sufficient.

In a preliminary study comparing baseline protein expression in rat hippocampus and nucleus accumbens in various strains of rats determined by 2-DE (37), it was readily apparent that substantive differences between these proteomes occurring at low levels of protein expression were not detectable, mainly because of a lack of depth of field. To rectify this limitation, we have proposed the approach illustrated in Figure 5. Recent results in cerebral synaptosomes (38) and sIEF fractions of nucleus accumbens (Bai et al., unpublished results) support the final protein spot numbers speculated in Figure 4, where 20,000 to 30,000 protein spots are putatively resolved from a single brain region sample (albeit on 20 individual gels per sample).

For example, in an initial experiment, 200 µg synaptosomal and cerebral cytosol protein loading resulted in the detection of roughly 1000 and 1500 protein spots on broad range 2-DE, respectively. Heavier gel loading (approximately 1 mg) would have increased that total significantly. In a separate analysis using the commercial rendition of Zuo and Speicher's (33) original concept of sIEF fractionation, the Zoom® IEF Fractionator (Invitrogen), a rat brain cerebral protein fraction pH 4.6 to 5.4 resolved by narrow-range 2-DE (24 cm, pH 4.5–5.5), yielded 1250 protein spots. By combining the two fractionations and loading significant protein amounts, an optimistic estimate of 20,000 to 30,000 resolvable protein spots per brain region seems plausible, corresponding to a significant improvement in depth of field. Studies are currently underway to substantiate these predictions.

In another preliminary assessment of the prefractionation strategy, in the prokaryote *Escherichia coli*, we subjected cell lysates to sIEF, and separated the resulting fractions by large-format, narrow-range 2-DE (Fig. 6) (24). Consistent with previous observations in *E. coli* 2-DE in the literature, broad range IEF (pH 3–10) resulted in

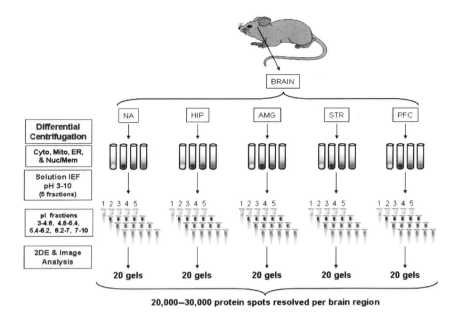

Figure 4 Proposed application of a multi-step sample prefractionation approach for complexity reduction and increased proteomic depth of field in assessing the toxic effect of alcohol exposure in the brain. The approach includes (1) brain region dissection NA, HIP, AMG, STR, and PFC, (2) subcellular fractionation by differential centrifugation, and (3) solution phase isoelectric focusing. Consequently, the proteome of each brain region is represented on 20 individual 2-D gel patterns. Based on preliminary results, we anticipate the resolution and analysis of 20,000 to 30,000 protein spots—a significant improvement over single sample analysis. *Abbreviations*: 2-D, two-dimensional; NA, nucleus accumbens; HIP, hippocampus; AMG, amygdala; STR, striatum; PFC, prefrontal cortex. *Source*: From Ref. 24.

the detection of 1577 protein spots. In contrast, the sum of five sIEF fraction gels totaled 5525 proteins, another improvement in depth of field. These kinds of studies suggest that 2-DE, when applied to eukaryotic systems in the manner just described, continues to have powerful utility in toxicoproteomics.

MS and Tandem MS

Gel-separated protein spots can be identified following proteolytic digestion and analysis by any of several MS methods. For instance, peptides resulting from a tryptic digest can be analyzed by matrix-assisted laser desorption ionization time-of-flight (MALDI-TOF) MS, a process referred to as peptide mass fingerprinting (PMF) (39). The measured and optimized monisotopic mass data are then compared with theoretically derived peptide mass databases generated by applying specific enzymatic cleavage rules to predicted or known protein sequences. MALDI-TOF–based PMF enables high-throughput, accurate, and sensitive mass detection. For unambiguous identification of 2-D gel–separated proteins, peptide fragment spectra generated by collision-induced dissociation (CID) and detected as MS/MS spectra can be compared to spectra predicted from sequence databases using search engines, such as Mascot, (40) and algorithms, such as SEQUEST (41). With the advent of the TOF/TOF (42,43) and FT-ICR (44) MS, protein quantitation, identification, and

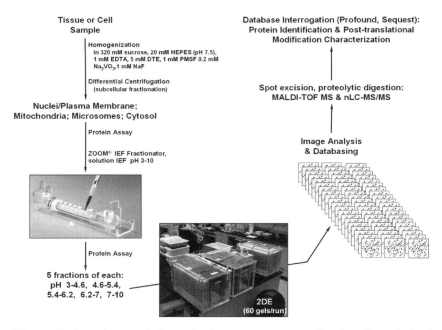

Figure 5 Sample complexity reduction strategy. By prefractionating whole tissue or cell lysates by a sequential combination subcellular fractionation and solution phase isoelectric focusing significantly more proteins per sample can be analyzed. This approach requires a large capacity, highly parallel 2-DE platform to minimize gel–gel variation and enable robust application to toxicology experiments with large numbers of samples. *Abbreviation*: 2-DE, two-dimensional electrophoresis.

characterization are entering a new era in which MS-based proteomic methods will become faster and more accurate.

Quantitative Proteomics Using MS

Isotope-Coded Affinity Tagging

A mass spectrometric technique known as isotope-coded affinity tagging (ICAT) has been developed to improve and expand relative protein quantitation in differential expression proteomics studies. Using light or heavy isotope–labeled peptide "tagging" reagents that differ only in molecular mass, proteins derived from normal or diseased or untreated or treated samples can be quantified, compared, and identified using LC–MS/MS (45). The tagging reagents contain a cysteine-reactive alkylating group at one end and a biotin tag at the other. After mixing the differently labeled proteins together, they are digested by the addition of trypsin, and the biotin-tagged, cysteine-containing peptides, are purified over an avidin column. These peptides are then separated on a C18 reversed phase column that is directly coupled to an MS/MS instrument. As Figure 7 illustrates, the relative amounts of the various peptides in the original sample are determined from the ratio of the isotope-labeled ion pairs, and proteins are identified from the fragmentation pattern.

A number of variations to the standard ICAT method have been developed. To reduce both the complexity of the sample and the computing resources necessary for data analysis, intact ICAT-labeled proteins have been fractionated initially on a

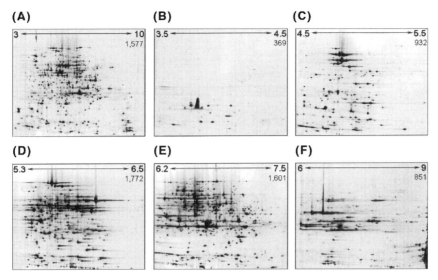

Figure 6 Large-format 2-D gels ($20 \times 25 \times 0.15$ cm; stained with CBB) of (**A**) whole *E. coli* lysate and (**B–F**) five solution sIEF (by ZoomIEF™, Invitrogen) fractions. A 24 cm IPG strips (Amersham) with (**A**) broad-range pH 3 to 10 NL and (as labeled above in **B–F**) five narrow-range pH gradients were used for first-dimension separations. ZoomIEF fractions roughly correspond to the bracketed narrow-range strips: **B** = pH 3 to 4.6, **C** = pH 4.6 to 5.4, **D** = pH pH 5.4 to 6.2, **E** = pH 6.2 to 7, and **F** = pH 7 to 10. Sample prefractionation by sIEF resulted in the resolution and detection of a total of 5525 proteins in the five sIEF 2-D gels, compared to only 1577 in the whole lysate, broad-range pattern. *Abbreviation*: sIEF, isoelectric focusing. *Source*: From Ref. 38.

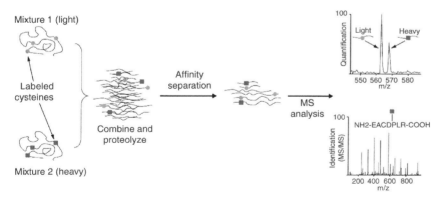

Figure 7 The ICAT reagent strategy for quantifying differential proteins. Two protein mixtures representing two different cell states are treated with the isotopically light or heavy ICAT reagents. The labeled protein mixtures are then combined and proteolyzed; tagged peptides are selectively isolated and analyzed by MS. The relative abundance is determined by the ratio of signal intensities of the tagged peptide pairs. Every other scan is devoted to fragmenting a peptide. The CID spectra are recorded and searched against large protein sequence databases to identify the protein. Therefore, in a single operation, the relative abundance and sequence of a peptide are determined. *Abbreviations*: ICAT, isotope-coded affinity tagging; MS, mass spectrometry; CID, collision-induced dissociation. *Source*: From Ref. 46.

2-D gel and then digested, quantitated, and identified by MS (47,48). Because deuterated moieties are known to alter the peptide retention time in reversed-phase high performance liquid chromatography (RP-HPLC), ^{13}C-containing ICAT-labels have been incorporated to avoid the differential chromatographic behavior of the "light" and the "heavy" peptides, thereby improving relative quantitation (49). This, along with the development of second-generation acid cleavable ICAT reagent (50), has helped to overcome some of the technical limitations of the original reagents and expand their utility. For instance, new generation ICATs have been used for quantitative proteomics in hepatic laser-capture samples (51) and in evaluating an antitumor drug in the brain (49). In a rare toxicologic application, acetaminophen hepatotoxicity was studied (52) using improved chromatography achieved by an RP-microLC-ESI column-emitter along with optimized microLC conditions. Continuous development and improvement of this technology along with parallel improvements in MS (53) will continue to enhance its applicability in toxicity testing.

Isobaric Tag Reagents

The recent development and commercialization of amine-reactive labeling reagents for multiplexed relative and absolute protein quantification called Isobaric Tag Reagents (iTRAQ™), along with ProQUANT™ software (Applied Biosystems), promises to be a realistic alternative to the limitations of 2-DE and cysteine-restricted labeling reagents like ICAT.

The utility of the iTRAC approach has been demonstrated in a study where this quantitative methodology was to identify global protein expression changes in a set of isogenic yeast strains (54). The investigators used of a set of four isobaric peptide derivatization reagents, which provide MS/MS spectra for both peptide identification and quantitation. As a general description of the approach used in that study, proteins from the different samples are extracted, tryptically digested, and peptides in each sample labeled with a different reagent (Fig. 8). The combined samples are then fractionated by cation exchange followed by conventional capillary reversed-phase LC–MS/MS coupled to MALDI or electrospray-based MS. Although the derivatized peptides are indistinguishable by their MS spectra or MS/MS ion series, the variable mass reporter ions that are freed by CID provide readily detectable, low-mass MS/MS profiles that permit quantitation of members of the multiplex set and thus accurate quantitation of the original peptide source (Fig. 5). If synthetic isobaric peptide standards are incorporated into the analysis, absolute levels of a target protein can be determined accurately.

Whereas the use of the iTRAQ approach in the example above enabled the simultaneous comparison of protein expression in multiple yeast strains, more importantly, ratio measurements for all the identified peptides was 100% for all strains. Expression ratios were highly consistent, and intra-protein peptide mean and standard deviations were highly reproducible (15–17%). Because the tagging chemistry is global, and any peptide with a free amine can be labeled and measured, the iTRAQ approach should find tremendous utility in toxicoproteomics.

Other Quantitative MS Approaches

Additional methods that obviate the use of gels exploit unique stable isotope-labeling approaches. Differential protein labeling of paired samples (e.g., control vs. exposed) can be carried out during cell culture using various other stable isotopes, such as $^{18}O, ^{13}C, ^{2}H$ or ^{15}N versus a control, nonisotopic medium (55). Separately cultured

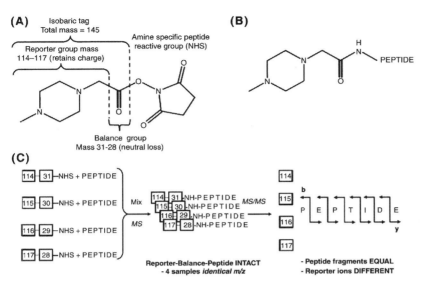

Figure 8 (A) Components of the multiplexed isobaric tagging chemistry. The complete molecule consists of a reporter group (based on N-methylpiperazine), a mass balance group (carbonyl) and a peptide reactive group (NHS ester). The overall mass of reporter and balance components of the molecule are kept constant using differential isotopic enrichment with ^{13}C and ^{18}O atoms, thus avoiding problems with chromatographic separation seen with enrichment involving deuterium substitution. The reporter group ranges in mass from m/z 114.1 to 117.1, while the balance group ranges in mass from 28 to 31 Da, such that the combined mass remains constant (145.1 Da) for each of the four reagents. (B) When reacted with a peptide, the tag forms an amide linkage to any peptide amine (N-terminal or epsilon amino group of lysine). These amide linkages fragment in a similar fashion to backbone peptide bonds when subjected to CID. Following fragmentation of the tag amide bond, however, the balance (carbonyl) moiety is lost (neutral loss) while charge is retained by the reporter group fragment. (C) Isotopic tagging used to arrive at four isobaric combinations with four different reporter group masses. A mixture of four identical peptides each labeled with one member of the multiplex set appears as a single, unresolved precursor ion in MS (identical m/z). Following CID, the four reporter group ions appear as distinct masses (114–117 Da). All other sequence-informative fragment ions (b-, y-, etc.) remain isobaric. The relative concentration of the peptides is determined from the relative intensities of the corresponding reporter-ions. *Abbreviations*: CID, collision-induced dissociation; MS, mass spectrometry. *Source*: From Ref. 60.

cells are combined after treatment and processed for relative mass spectrometric quantitation. In this strategy, all detected peptides derived from the stable isotope–labeled sample will have correspondingly higher masses than "normal" samples (56), proportional to the number of O, C, H, or N atoms within identical peptides.

For instance, in a recent study, stable isotope-labeling of amino acids in culture was used to analyze the difficult-to-isolate lipid raft proteome, representing the first functional proteomic analysis of these physiologically significant membrane components (57). In essence, these investigators used quantitative MS by encoding all of the proteins in one of two HeLa cell populations by metabolically labeling with deuterium-substituted leucine. One of the populations was treated with nystatin or methyl-β-cyclodextrin to disrupt the lipid rafts in those cells, and then the treated and untreated cells were combined and biochemically fractionated. Because of raft disruption, labeled lipid-raft proteins from that population were absent (or severely reduced) in the analyzed fraction, as reflected in the quantitative isotopic ratios between treated and untreated peptides. Peptides (and corresponding identified proteins) with large ratios thus comprised the lipid raft proteome, while other, nonraft proteins were present in roughly equal proportions.

Because this approach has been shown effective in multicellular organisms (58), it has significant potential in in vitro toxicology. A similar, in vivo version of this method where whole animals are ^{15}N-labeled (59) and protein expression compared to unlabeled counterparts also has tremendous potential as a quantitative differential expression proteomics tool in toxicology.

While metabolic labeling requires cell growth and isotope incorporation into synthesized proteins, a unique feature of ^{18}O incorporation relies on the universal incorporation of the isotope into the carboxyl termini of all peptides produced by tryptic cleavage (60) of the proteins in one of two paired samples. The second sample is digested similarly, with the exception that ^{16}O atoms are used. After pooling the two peptide mixtures, the sample is then subjected to liquid chromatography (LC) and analyzed by MS and MS/MS for 4 kDa mass differences in corresponding peptide pairs, after which relative quantitation is possible.

Finally, another isotope labeling strategy for relative protein quantification, global internal standard technology (GIST), has been developed (61). The GIST protocol includes tryptic digestion of proteins from paired samples followed by differential isotopic labeling of the resulting tryptic peptides, mixing the differentially labeled control and experimental digests, fractionation of the peptide mixture by reversed-phase chromatography, and isotope ratio analysis by MS. In this manner, each peptide in the sample can be labeled.

For example, N-acetoxysuccinimide (light label) and N-acetoxy-[^{2}H$_3$] succinimide (heavy label) are used to differentially derivatize primary amine groups (N-termini and epsilon nitrogens of lysine side chains) in peptides from experimental and control samples, respectively. Once derived with the appropriate label, the paired peptides have a mass difference of 3 Da and elute at the same point in reversed-phase chromatography. The ratio of light-to-heavy pairs is determined by integrating the area under the respective peaks from mass spectra. The peptide masses and sequences obtained by MS/MS are then used to identify proteins. Like iTRAQ, this approach may find great utility in toxicoproteomics.

Surface-Enhanced Laser-Desorption Ionization Protein Profiling

Another useful MS-based method for proteomic analysis makes use of the ProteinChip® System from Ciphergen Biosystems. The effectiveness of this system resides in the surface-enhanced laser desorption ionization (SELDI) ProteinChip surface technology that performs separation, detection, and analysis of proteins at femtomole levels, directly from complex (unfractionated) biological samples. In essence, specifically modified slides bearing various surface chemistries (cationic, anionic, hydrophobic, hydrophilic, etc.) or biochemistries (antibody, receptor, DNA, etc.), which bind and selectively purify proteins from a complex biological sample. For a given ProteinChip, various buffer and elution conditions can be used to further fractionate the sample. The slide then can be analyzed using a SELDI mass spectrometer (essentially a MALDI-TOF instrument) as portrayed in Figure 6. In a classic paper illustrating the potential of SELDI in the study of cancer, Petricoin et al. (62) identified a proteomic pattern in serum that is diagnostic of ovarian cancer. Based on this study and the database it generated, a new and improved version of this approach was recently published (63). The ease and speed of screening samples makes this a potentially useful method for biomarker detection in toxicologic samples (64–68). Despite its clever utility, SELDI suffers from several limitations. SELDI–MS instrumentation usually is capable of accurately detecting proteins with molecular weights less than 45,000; the detected proteins cannot be identified using this technique alone, and reproducibility in complicated experiments is suspect (69). Nevertheless, a recent study by Petricoin et al. (70), where serum samples from rat models of drug-induced cardiotoxicity were investigated, reinforces the potential utility of diagnostic proteomic patterns where the low molecular weight peptides and protein fragments detected by SELDI may have higher accuracy than traditional biomarkers of cardiotoxicity. Continued improvement in next-generation instruments using ProteinChip MS/MS techniques for direct protein identification (71), improved surface chemistries (72), and improved experimental design (73) should all greatly enhance SELDI's effectiveness as a powerful toxicoproteomic tool.

A 2-D GEL-BASED TOXICOPROTEOMIC ANALYSIS OF JET FUEL EXPOSURE

A typical toxicoproteomic application can be found in a series of studies investigating the toxicity of JP-8 jet fuel. The summary that follows demonstrates that, while 2-DE is well suited for toxicologic studies, the limitations of the approach clearly confine the scope of the toxicoproteomic analysis to a rather minor segment of the various target-tissue proteomes studied. While this is not to minimize the importance of the results, it emphasizes the relevance of the recent developments in proteomic technology presented earlier in this chapter, new strategies in protein separation that are likely to improve the applicability of 2-DE, and the expansion of proteomic technology available to the toxicologist.

JP-8 (Jet-A) jet fuel is the dominant military and civilian aviation fuel. This kerosene-like fuel is a complex mixture containing more than 228 hydrocarbon constituents including performance additives and solvents. Because of its high flashpoint and low vapor pressure, JP-8 has characteristics that reduce risk of fire and explosion but increase its availability for exposure. Based on evidence from human and rodent studies, jet fuel exposure is associated with consequential effects and in some cases pathology (74,75).

Lung Proteomics

Given the tendency for inhalation of JP-8 fuel vapor and aerosol, the logical target is the lung. A study using an animal model simulating occupational jet fuel exposure was conducted (76) to investigate the effect of JP-8 toxicity in the lung, specifically relating to lung epithelial cell apoptosis and edema. To contrast the effect of high- and low-dose exposure (mice exposed to 1 hr/day aerosolized JP-8 jet fuel at concentrations of 250 and 2500 mg/m^3 for seven days), the following observations were made.

Significant quantitative differences in lung protein expression were found as a result of JP-8 exposure. At 2500 mg/m^3, 30 proteins were elevated while 135 were decreased, relative to the sham-exposed controls. Some of these included cytoplasmic proteins hsc70, Hop, transitional endoplasmic reticulum ATPase, and Nedds (septin). The implications for impaired protein synthesis and protein misfolding are supported by the nearly identical percent decline in Hop and hsc70 expression. Hop (hsc70/hsp90-organizing protein) is required by hsc70 (and hsp70) for protein-folding activity (77). The potential increase in the number of misfolded proteins that might result from JP-8 intoxication leads quite naturally to the more classic response to toxic stress found in the observed upregulation of hsp84 (murine hsp90-β).

Additional impairment in protein processing was suggested by a decline in total transitional endoplasmic reticular ATPase and Nedds, both involved in vesicular trafficking. With the altered expression discussed above, these protein alterations strongly imply a JP-8–mediated disturbance of posttranslational protein and lipid processing in the lung and may underlie the appearance of vacuolated type II alveolar cells (78), as evidence of JP-8–mediated disruption of surfactant processing.

Additional observations suggest ultrastructural damage (decreased Ulip2, Nedds, and laminin receptor protein), toxic or metabolic stress and upregulation of detoxification systems [increased glutathione S-transferase (GSTP1)], hsp84 (murine hsp90), and thioether S-methyltransferase), and altered functional responses to CO_2 handling, acid–base homeostasis, and fluid secretion (increased carbonic anhydrase II).

In direct contrast to the high level JP-8 exposures described earlier, not only were far fewer proteins affected by the lower dose, but also 42 proteins were induced, and only five were down-regulated. Of the altered proteins, several identified proteins were found to be significant markers of JP-8–induced stress on lung epithelial cells. A decrease in 1-antitrypsin, and with it, decreased antiproteinase activity, has significant implications in potential damage to lung ultrastructure (79). In what may be a related response, lung proteins involved in cell remodeling activities, including actin, keratin type 1 cytoskeleton 9, desmin fragments, and cytokeratin endo A, were all up-regulated. Furthermore, a generalized stress response to the intoxication was indicated by the upregulation of the mitochondrial stress protein hsp60, and the apoptotic mediator—Fas-associated factor 1.

The overall response to the 250 mg/m³ exposure in the lung, lower in magnitude and opposite in the direction of expression changes observed with 2500 mg/m³, underscores the utility of quantitative expression analysis by 2-DE. The results also reflect what may be an "adaptive or compensatory response" to the JP-8 perturbation, e.g., protein upregulation and mobilization, in contrast to the high-dose exposures where significant injury processes seem to dominate, and where widespread protein downregulation manifests the "injury response" as protein expression changes match histological evidence of JP-8's damaging effects.

Testis Proteomics

Given the considerable interest in the potential reproductive toxicity of jet fuel, a study was conducted in a rodent model, examining JP-8 vapor exposures (at 250, 500, or 1000 mg/m³, for 6 hr/day for 91 consecutive days) and their effects on differential protein expression determined by 2-DE (80). Similar to observations made in the lung, JP-8 vapor exposure at 250 mg/m³ resulted in fivefold more up-regulated proteins than did exposure to the highest dose (1000 mg/m³). Exposure to the highest dose level resulted in twice as many downregulated proteins as the lowest dose exposure, and individual protein expression, plotted as a function of JP-8 vapor dose, seldom reflected a strict linear dose relationship.

Identified proteins whose expression increased as a function of dose included Hsp86, nicotinic acetylcholine receptor alpha subunit, serum albumin, and

T-complex protein 1. Other proteins altered in various dose groups but not necessarily in linear fashion included mitochondrial aldehyde dehydrogenase, heat shock factor 2, hsp70t, Grp78, Tat binding protein-1, interleukin-18, lamin B, protein kinase C (PKC) binding protein Zetai, A-kinase anchor protein, and β7 integrin—all of which increased. The only down-regulated protein identified was the 150-kDa oxygen-regulated protein, and the significance of its decline remains unclear.

So the observation that repeated low-dose JP-8 vapor exposure resulted in far more testis proteins with increased expression than did exposure to the highest dose, and that exposure to the highest dose level resulted in significantly reduced expression of twice as many proteins as the low-dose exposure. While this phenomenon is difficult to explain, it is not unusual, having been observed consistently with JP-8 vapor and aerosol exposures in other organs (e.g., lung). Whether the alterations are related to injury versus adaptive or repair mechanisms as mentioned earlier or are perhaps due to a more complex hormetic (81) response remains to be determined.

Renal Proteomics

Although the most-studied exposure route for JP-8 has been via the lung, where one might expect to see the greatest effects, the testis study mentioned above suggests that generalized systemic effects can be expected. Surprisingly, the kidney responds dramatically to pulmonary jet fuel exposure, to an extent greater than the liver, at least from a proteomic standpoint (82). 2-DE of cytoplasmic proteins prepared from whole kidney homogenates from mice exposed to $1000 \, mg/m^3$ JP-8 aerosol or vapor for one hour per day for five days, revealed that JP-8 exposure had significantly altered ($P < 0.05$) the expression of 56 proteins spots (21 up-regulated and 35 down-regulated).

Some of the altered proteins identified by PMF were categorized functionally as ultrastructural abnormalities (tropomyosin 4 and high mobility group 1 protein decreased while ezrin increased); altered protein processing (aminopeptidase decreased and Rab GDP-dissociation inhibitor beta increased); metabolic effects (alpha enolase and phosphoglycerate kinase decreased while lactate dehydrogenase and pyruvate carboxylase increased); and detoxification system response [amine N-sulfotransferase (ST3A1) increased while thioether S-methyltransferase and superoxide dismutase were down-regulated].

CONCLUSION

The various studies of JP-8 jet fuel toxicity summarized above have led to several conclusions. First, in rodent models where occupational exposures to JP-8 aerosol and vapor have been simulated and protein expression analyzed by 2-D gel–based proteomics, in each case, a moderate yet significant change in protein expression resulted. Depending on the tissue and exposure conditions, the protein alterations were either quantitative (up- or downregulation) or qualitative (posttranslational), and based on protein identification, the functional consequences of the altered proteomes could be postulated. Second, the utility of 2-D gel–based approaches in analyzing differential protein expression, across many samples or dose groups was demonstrated in these studies. Using a large-scale, highly parallel approach to 2-D gel electrophoresis of numerous samples, the resulting gel–gel uniformity enabled accurate comparisons of protein expression.

In contrast, these results also clearly demonstrate that only a fraction of the various target tissue proteomes were analyzed. One can only wonder what important protein alterations associated with JP-8 exposure were missed as a result of the major foibles of 2-DE, e.g., analytical comprehensiveness and dynamic range. Before the true analytical power of 2-D gel–based proteomics can be realized and appreciated by the toxicological community, the traditional approach must be modified and improved. Those non–gel-based approaches briefly addressed in this chapter (e.g., track and GIST) also must be optimized and incorporated into a comprehensive toxicoproteomic platform, one that provides truly global assessment of exposure-induced quantitative and qualitative changes in the proteome.

REFERENCES

1. Xirasagar S, Gustafson S, Merrick BA, et al. CEBS object model for systems biology data, SysBio-OM. Bioinformatics 2004; 20(13):2004–2015.
2. Merrick BA, Tomer KB. Toxicoproteomics: a parallel approach to identifying biomarkers. Environ Health Perspect 2003; 111(11):A578–A579.
3. Waters M, Boorman G, Bushel P, et al. Systems toxicology and the chemical effects in biological systems (CEBS) knowledge base. EHP Toxicogenomics 2003; 11(1T):15–28.
4. Wasinger VC, Cordwell SJ, Cerpa-PoljakA, et al. Progress with gene-product mapping of the Mollicutes: Mycoplasma genitalium. Electrophoresis 1995; 16(7):1090–1094.
5. Thome-Kromer B, Bonk I, Klatt M, et al. Toward the identification of liver toxicity markers: a proteome study in human cell culture and rats. Proteomics 2003; 3(10):1835–1862.
6. Moller A, Soldan M, Volker U, Maser E. Two-dimensional gel electrophoresis: a powerful method to elucidate cellular responses to toxic compounds. Toxicology 2001; 160(1–3): 129–138.
7. Görg A, Weiss W, Dunn MJ. Current two-dimensional electrophoresis technology for proteomics. Proteomics 2004; 4(12):3665–3685.
8. Lim H, Eng J, Yates JR III, et al. Identification of 2D-gel proteins: a comparison of MALDI/TOF peptide mass mapping to mu LC-ESI tandem mass spectrometry. J Am Soc Mass Spectrom 2003; 14(9):957–970.
9. Anderson NL, Anderson NG. The human plasma proteome: history, character, and diagnostic prospects. Mol Cell Proteom 2002; 1(11):845–867.
10. Unlu M, Morgan ME, Minden JS. Difference gel electrophoresis: a single gel method for detecting changes in protein extracts. Electrophoresis 1997; 18(11):2071–2077.
11. Van den Bergh G, Arckens L. Fluorescent two-dimensional difference gel electrophoresis unveils the potential of gel-based proteomics. Curr Opin Biotechnol 2004; 15(1):38–43.
12. Shaw J, Rowlinson R, Nickson J, et al. Evaluation of saturation labelling two-dimensional difference gel electrophoresis fluorescent dyes. Proteomics 2003; 3(7):1181–1195.
13. Evans CA, Tonge R, Blinco D, et al. Comparative proteomics of primitive hematopoietic cell populations reveals differences in expression of proteins regulating motility. Blood 2004; 103(10):3751–3759.
14. Stühler K, Sitek B, Pfeiffer K, Luttges J, Hahn SA, Meyer HE. Differential display of pancreatic adenocarcinoma precursor lesions using DIGE saturation labelling. Mol Cell Proteom 2004; 3:S48.
15. Tonge R, Shaw J, Middleton B, et al. Validation and development of fluorescence two-dimensional differential gel electrophoresis proteomics technology. Proteomics 2001; 1(3):377–396.
16. Ruepp SU, Tonge RP, Shaw J, Wallis N, Pognan F. Genomics and proteomics analysis of acetaminophen toxicity in mouse liver. Toxicol Sci 2002; 65(1):135–150.
17. Meneses-Lorente G, Guest PC, Lawrence J, et al. Proteomic investigation of drug-induced steatosis in rat liver. Chem Res Toxicol 2004; 17(5):605–612.

18. Kleno TG, Leonardsen LR, Kjeldal HO, Laursen SM, Jensen ON, Baunsgaard D. Mechanisms of hydrazine toxicity in rat liver investigated by proteomics and multivariate data analysis. Proteomics 2004; 4(3):868–880.
19. Wheelock AM, Zhang L, Tran MU, et al. Isolation of rodent airway epithelial cell proteins facilitates in vivo proteomics studies of lung toxicity. Am J Physiol Lung Cell Mol Physiol 2004; 286(2):L399–L410.
20. Steinberg TH, Agnew BJ, Gee KR, et al. Global quantitative phosphoprotein analysis using multiplexed proteomics technology. Proteomics 2003; 3(7):1128–1144.
21. Schulenberg B, Beechem JM, Patton WF. Mapping glycosylation changes related to cancer using the multiplexed proteomics technology: a protein differential display approach. J Chromatogr B Analyt Technol Biomed Life Sci 2003; 793(1):127–139.
22. Anderson NL, Anderson NG. Analytical techniques for cell fractions. XXII. Two-dimensional analysis of serum and tissue proteins: multiple gradient-slab gel electrophoresis. Anal Biochem 1978; 85(2):341–354.
23. Anderson NL, Hickman BJ. Analytical techniques for cell fractions. XXIV. Isoelectric point standards for two-dimensional electrophoresis. Anal Biochem 1979; 93(2):312–320.
24. Witzmann FA, Bai F, Hong SM, et al. Gels and more gels: probing toxicity. Curr Opin Mol Ther 2004; 6(6):608–615.
25. Stasyk T, Huber LA. Zooming in: fractionation strategies in proteomics. Proteomics 2004; 4(12):3704–3716.
26. Liao L, Cheng D, Wang J, et al. Proteomic characterization of postmortem amyloid plaques isolated by laser capture microdissection. J Biol Chem 2004; 279(35):37061–37068.
27. Babu GJ, Wheeler D, Alzate O, Periasamy M. Solubilization of membrane proteins for two-dimensional gel electrophoresis: identification of sarcoplasmic reticulum membrane proteins. Anal Biochem 2004; 325(1):121–125.
28. Abdolzade-Bavil A, Hayes S, Goretzki L, Kroger M, Anders J, Hendriks R. Convenient and versatile subcellular extraction procedure, that facilitates classical protein expression profiling and functional protein analysis. Proteomics 2004; 4(5):1397–1405.
29. Greenough C, Jenkins RE, Kitteringham NR, Pirmohamed M, Park BK, Pennington SR. A method for the rapid depletion of albumin and immunoglobulin from human plasma. Proteomics 2004; 4(10):3107–3111.
30. Pieper R, Su Q, Gatlin CL, Huang ST, Anderson NL, Steiner S. Multi-component immunoaffinity subtraction chromatography: an innovative step towards a comprehensive survey of the human plasma proteome. Proteomics 2003; 3(4):422–432.
31. Pieper R, Gatlin CL, McGrath AM, et al. Characterization of the human urinary proteome: a method for high-resolution display of urinary proteins on two-dimensional electrophoresis gels with a yield of nearly 1400 distinct protein spots. Proteomics 2004; 4(4):1159–1174.
32. Zuo X, Speicher DW. Microscale solution isoelectrofocusing: a sample prefractionation method for comprehensive proteome analysis. Meth Mol Biol 2004; 244:361–375.
33. Zuo X, Speicher DW. A method for global analysis of complex proteomes using sample prefractionation by solution isoelectrofocusing prior to two-dimensional electrophoresis. Anal Biochem 2000; 284(2):266–278.
34. Fountoulakis M, Juranville JF, Tsangaris G, Suter L. Fractionation of liver proteins by preparative electrophoresis. Amino Acids 2004; 26(1):27–36.
35. Lescuyer P, Hochstrasser DF, Sanchez JC. Comprehensive proteome analysis by chromatographic protein prefractionation. Electrophoresis 2004; 25(7–8):1125–1135.
36. Zuo X, Echan L, Hembach P, et al. Towards global analysis of mammalian proteomes using sample prefractionation prior to narrow pH range two-dimensional gels and using one-dimensional gels for insoluble and large proteins. Electrophoresis 2001; 22(9):1603–1615.
37. Witzmann FA, Li J, StrotherWN, et al. Innate differences in protein expression in the nucleus accumbens and hippocampus of inbred alcohol-preferring and nonpreferring rats. Proteomics 2003; 3(7):1335–1344.

38. Witzmann FA, Arnold RJ, Bai F, et al. A proteomic survey of rat cerebral cortical synaptosomes. Proteomics 2005; 5(8):2177–2201.
39. Cottrell JS, Sutton CW. The identification of electrophoretically separated proteins by peptide mass fingerprinting. Meth Mol Biol 1996; 61:67–82.
40. Perkins DN, Pappin DJ, Creasy DM, Cottrell JS. Probability-based protein identification by searching sequence databases using mass spectrometry data. Electrophoresis 1999; 20(18):3551–3567.
41. Eng JK, McCormack AL, Yates JR III. An approach to correlate tandem mass spectral data of peptides with amino acid sequences in protein database. J Am Soc Mass Spectrom 1994; 5:976–989.
42. Medzihradszky KF, Campbell JM, Baldwin MA, et al. The characteristics of peptide collision-induced dissociation using a high-performance MALDI-TOF/TOF tandem mass spectrometer. Anal Chem 2000; 72(3):552–558.
43. RejtarT, Chen HS, Andreev V, Moskovets E, Karger BL. Increased identification of peptides by enhanced data processing of high-resolution MALDI TOF/TOF mass spectra prior to database searching. Anal Chem 2004; 76(20):6017–6028.
44. Page JS, Masselon CD, Smith RD. FTICR mass spectrometry for qualitative and quantitative bioanalyses. Curr Opin Biotechnol 2004; 15(1):3–11.
45. Gygi SP, Rist B, Gerber SA, Turecek F, Gelb MH, Aebersold R. Quantitative analysis of complex protein mixtures using isotope-coded affinity tags. Nat Biotechnol 1999; 17(10):994–999.
46. Tao WA, Aebersold R. Advances in quantitative proteomics via stable isotope tagging and mass spectrometry. Curr Opin Biotechnol 2003; 14(1):110–118.
47. Smolka MB, Zhou H, Purkayastha S, Aebersold R. Optimization of the isotope-coded affinity tag-labeling procedure for quantitative proteome analysis. Anal Biochem 2001; 297(1):25–31.
48. Smolka M, Zhou H, Aebersold R. Quantitative protein profiling using two-dimensional gel electrophoresis, isotope-coded affinity tag labeling, and mass spectrometry. Mol Cell Proteom 2002; 1(1):19–29.
49. Yu LR, Conrads TP, Uo T, Issaq HJ, Morrison RS, Veenstra TD. Evaluation of the acid-cleavable isotope-coded affinity tag reagents: application to camptothecin-treated cortical neurons. J Proteom Res 2004; 3(3):469–477.
50. Zhou H, Ranish JA, Watts JD, Aebersold R. Quantitative proteome analysis by solid-phase isotope tagging and mass spectrometry. Nat Biotechnol 2002; 20(5):512–515.
51. Li C, Hong Y, Tan YX, et al. Accurate qualitative and quantitative proteomic analysis of clinical hepatocellular carcinoma using laser capture microdissection coupled with isotope-coded affinity tag and two-dimensional liquid chromatography mass spectrometry. Mol Cell Proteom 2004; 3(4):399–409.
52. Lee H, Yi EC, Wen B, et al. Optimization of reversed-phase microcapillary liquid chromatography for quantitative proteomics. J Chromatogr B Analyt Technol Biomed Life Sci 2004; 803(1):101–110.
53. Bogdanov B, Smith RD. Proteomics by FTICR mass spectrometry: top down and bottom up. Mass Spectrom Rev 2004.
54. Ross PL, Huang YN, Marchese J, et al. Multiplexed protein quantitation in saccharomyces cerevisiae using amine-reactive isobaric tagging reagents. Mol Cell Proteom 2004:M400129–MCP20u.
55. Wang YK, Quinn DF, Ma Z, Fu EW. Inverse labeling-mass spectrometry for the rapid identification of differentially expressed protein markers/targets. J Chromatogr B Analyt Technol Biomed Life Sci 2002; 782(1–2):291–306.
56. Conrads TP, Alving K, Veenstra TD, et al. Quantitative analysis of bacterial and mammalian proteomes using a combination of cysteine affinity tags and 15N-metabolic labeling. Anal Chem 2001; 73(9):2132–2139.

57. Foster LJ, De Hoog CL, Mann M. Unbiased quantitative proteomics of lipid rafts reveals high specificity for signaling factors. Proc Natl Acad Sci, USA 2003; 100(10):5813–5818.

58. Krijgsveld J, Ketting RF, et al. Metabolic labeling of C. elegans and D. melanogaster for quantitative proteomics. Nat Biotechnol 2003; 21(8):927–931.

59. Wu CC, MacCoss MJ, Howell KE, Matthews DE, Yates JR III. Metabolic labeling of mammalian organisms with stable isotopes for quantitative proteomic analysis. Anal Chem 2004; 76(17):4951–4959.

60. Schnblzer M, Jedrzejewski P, Lehmann WD. Protease-catalyzed incorporation of 180 into peptide fragments and its application for protein sequencing by electrospray and matrix-assisted laser desorption/ionization mass spectrometry. Electrophoresis 1996; 17(5):945–953.

61. Chakraborty A, Regnier FE. Global internal standard technology for comparative proteomics. J Chromatogr A 2002; 949(1–2):173–184.

62. Petricoin EF, Ardekani AM, Hitt BA, et al. Use of proteomic patterns in serum to identify ovarian cancer. Lancet 2002; 359(9306):572–577.

63. Alexe G, Alexe S, Liotta LA, Petricoin E, Reiss M, Hammer PL. Ovarian cancer detection by logical analysis of proteomic data. Proteomics 2004; 4(3):766–783.

64. Dare TO, Davies HA, Turton JA, Lomas L, Williams TC, York MJ. Application of surface-enhanced laser desorption/ionization technology to the detection and identification of urinary parvalbumin-alpha: a biomarker of compound-induced skeletal muscle toxicity in the rat. Electrophoresis 2002; 23(18):3241–3251.

65. Hogstrand C, Balesaria S, Glover CN. Application of genomics and proteomics for study of the integrated response to zinc exposure in a non-model fish species, the rainbow trout. Comp Biochem Physiol B Biochem Mol Biol 2002; 133(4):523–535.

66. He QY, Yip TT, Li M, Chiu JF. Proteomic analyses of arsenic-induced cell transformation with SELDI-TOF ProteinChip technology. J Cell Biochem 2003; 88(1):1–8.

67. Seibert V, Wiesner A, Buschmann T, Meuer J. Surface-enhanced laser desorption ionization time-of-flight mass spectrometry (SELDI TOF-MS) and ProteinChip technology in proteomics research. Pathol Res Pract 2004; 200(2):83–94.

68. Yuan M, Carmichael WW. Detection and analysis of the cyanobacterial peptide hepatotoxins microcystin and nodularin using SELDI-TOF mass spectrometry. Toxicon 2004; 44(5):561–570.

69. Baggerly KA, Morris JS, Coombes KR. Reproducibility of SELDI-TOF protein patterns in serum: comparing datasets from different experiments. Bioinformatics 2004; 20(5):777–785.

70. Petricoin EF, Rajapaske V, Herman EH, et al. Toxicoproteomics: serum proteomic pattern diagnostics for early detection of drug induced cardiac toxicities and cardioprotection. Toxicol Pathol 2004; 32(suppl 1):122–130.

71. Weinberger SR, Viner Rl, Ho P. Tagless extraction-retentate chromatography: a new global protein digestion strategy for monitoring differential protein expression. Electrophoresis 2002; 23(18):3182–3192.

72. Thuiasiraman V, Wang Z, Katrekar A, Lomas L, Yip TT. Simultaneous monitoring of multiple kinase activities by SELDI-TOF mass spectrometry. Meth Mol Biol 2004; 264:205–214.

73. Cordingley HC, Roberts SL, Tooke P, et al. Multifactorial screening design and analysis of SELDI-TOF ProteinChip array optimization experiments. Biotechniques 2003; 34(2):364–365, 8–73.

74. Ritchie G, Still K, Rossi J III, Bekkedal M, Bobb A, Arfsten D. Biological and health effects of exposure to kerosene-based jet fuels and performance additives. J Toxicol Environ Health B Crit Rev 2003; 6(4):357–451.

75. Ritchie GD, Still KR, Alexander WK, et al. A review of the neurotoxicity risk of selected hydrocarbon fuels. J Toxicol Environ Health B Crit Rev 2001; 4(3):223–312.

76. Drake MG, Witzmann FA, Hyde J, Witten ML. JP-8 jet fuel exposure alters protein expression in the lung. Toxicology 2003; 191(2–3):199–210.

77. Johnson BD, Schumacher RJ, Ross ED, Toft DO. Hop modulates Hsp70/Hsp90 interactions in protein folding. J Biol Chem 1998; 273(6):3679–3686.

78. Witzmann FA, Bauer MD, Fieno AM, et al. Proteomic analysis of simulated occupational jet fuel exposure in the lung. Electrophoresis 1999; 20(18):3659–3669.

79. Coakley RJ, Taggart C, O'Neill S, McElvaney NG. Alpha-antitrypsin deficiency: biological answers to clinical questions. Am J Med Sci 2001; 321(1):33–41.

80. Witzmann FA, Bobb A, Briggs GB, et al. Analysis of rat testicular protein expression following 91-day exposure to JP-8 jet fuel vapor. Proteomics 2003; 3(6):1016–1017.

81. Calabrese EJ. Hormesis: a revolution in toxicology, risk assessment and medicine. EMBO Rep 2004; 5(suppl 1):S37–S40.

82. Witzmann FA, Bauer MD, Fieno AM, et al. Proteomic analysis of the renal effects of simulated occupational jet fuel exposure. Electrophoresis 2000; 21(5):976–984.

4

The Potential of Metabonomics in Toxicology

Julian L. Griffin

Department of Biochemistry, University of Cambridge, Cambridge, U.K.

INTRODUCTION

With recent increased interests in systems biology, a number of analytical approaches have been developed to globally profile a tier of organization in a cell, tissue, or organism. Genomics, the first of the "-omic" technologies to embrace global analysis, describes the genes present in an organism. This approach is notably compared with other -omic approaches in that it is not context dependent, with a genome not being directly influenced by the environment. Large-scale genome projects are now being complemented by other -omic strategies, including transcriptomics, proteomics, and metabolomics or metabonomics to profile all the mRNA, proteins, or small molecule metabolites in a tissue, cell, or organism. One of the major challenges of these "-omic" technologies is that the transcriptome, proteome, and metabolome are all context dependent and will vary with pathology, development stage, and environmental factors. Thus, the possibility of globally profiling the transcriptome, proteome, or metabolome of an organism is a real analytical challenge, because by definition these efforts must also take into consideration all factors that influence metabolism. However, one major advantage that metabonomics has over the other "-omic" approaches is that the analytical approaches are relatively cheap on a per sample basis, suggesting that databases which embrace both environmental and genomic influences on the metabolism of a given cell, tissue, organ, or even organism may be possible.

The terms metabolomics as well as metabonomics have been widely used to describe "the quantitative measurement of metabolic responses to pathophysiological stimuli or genetic modification" (1–3). Some researchers have distinguished these two terms suggesting that metabolomics deals with metabolism at the cellular level while metabonomics addresses the complete system (3). Furthermore, other researchers have distinguished terms according to the technology used to generate a "metabolic profile" [largely used for mass spectrometry (MS)–based approaches] or "metabolic fingerprint" [addressing nuclear magnetic resonance (NMR) spectroscopy and other approaches that only detect the high concentration metabolites in a tissue or biofluid] (4). However, these distinctions seem rather artificial and do not reflect the inclusiveness of definitions for transcriptomics and proteomics. In this article the term metabonomics will be used because this is currently the most widely used term in toxicology, although the reader should note that other researchers may favor the use of *metabolomics, metabolic profiling*, or *metabolic fingerprinting*.

The concept of the approach is to measure all the small molecule concentrations through a global analytical approach, and then to apply pattern recognition techniques to define metabolism in multidimensional space (3,5). The pattern recognition process is an integral part of the analytical approach, because metabolism is influenced by normal metabolic variation across a population as well as metabolic processes linked to a disease or toxicological insults. Clusterings of data associated with disease status or drug dose are identified and separated from "intersubject variation" by pattern recognition, and the resultant metabolic biomarkers used to define a metabolic phenotype or "metabotype" for that disease or drug intervention (6). This information can then be used either to define an end point, such as the presence of drug toxicity or detection of disease, or to data mine another "-omic" technology (Fig. 1).

The most extensively used analytical approach for metabonomics in toxicology is ^{1}H NMR spectroscopy, largely as a result of work in this area by Prof. Jeremy Nicholson of Imperial College, London (2,3,5). This approach has analyzed biofluids and tissue extracts using solution state ^{1}H NMR (7,8), intact tissues using high resolution magic angle spinning (HRMAS) ^{1}H NMR spectroscopy (9,10), and even tissues within living organisms using in vivo magnetic resonance spectroscopy (11). Solution state NMR is also amenable to high throughput, capable of running

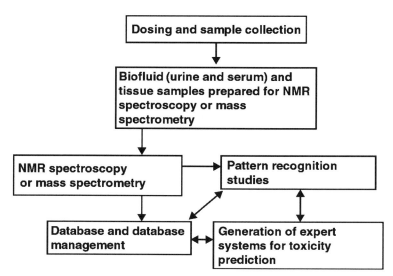

Figure 1 A systematic method for defining a metabolic phenotype. The initial phase usually involves the acquisition of a large data set in terms of both the variables (metabolites) and observations (subjects), commonly using either ^{1}H NMR spectroscopy or MS. However, the use of pattern recognition techniques is an integral part of the approach if the innate metabolic variation associated with different individuals is to be separated from that induced by the insult or physiological stimulus. Following the generation of a database from metabolic responses of known drugs or toxins, this can then be used to build a predictive pattern recognition model that is capable of predicting toxicity from drugs or toxins with unknown action. *Abbreviations*: ^{1}H NMR, ^{1}H nuclear magnetic resonance; MS, mass spectrometry.

hundreds of samples in a day, and relatively robust, making it ideal as a large-scale profiling tool for metabolites. However, NMR spectroscopy has relatively low sensitivity and can only detect the high concentration metabolites. This currently limits the approach to detecting approximately 100 to 200 metabolites in urine and approximately 20 to 30 metabolites in most mammalian tissue extracts. The major competition to NMR spectroscopy for analytical approaches involves a range of MS-based approaches, in particular gas chromatography (GC)–MS and liquid chromatography (LC)–MS. These approaches are more sensitive, and hence potentially truly global in terms of metabolic profiling. This has led to a number of recent manuscripts exploring the potential of metabonomics in toxicology (12,13). However, both GC–MS and LC–MS depend critically on the reproducibility of the chromatography, and in the case of LC–MS, results may be impaired by ion suppression. In addition to these approaches, others have employed Fourier transform–infrared spectroscopy, thin layer chromatography, metabolite arrays (analogous to lab-on-a-chip arrangements), and even automated biochemical assays to provide a global description of metabolism (14).

As previously mentioned, a major part of the metabonomic approach is the application of pattern recognition techniques to identify the metabolites most correlated with a pathology or toxic insult. Unsupervised techniques require no prior information about class membership and use the innate variation in a data set to map samples measured into multidimensional space. Examples of unsupervised techniques include principal components analysis (PCA) and hierarchical cluster analysis. Supervised techniques correlate variation in the data set with an external variable such as disease status, age, or drug response. Examples of supervised techniques include partial least squares (PLS), orthogonal signal correction (OSC), neural networks, and genetic algorithms. PCA is one of the most widely used approaches, and this approach is illustrated in Figure 2. For a review of these techniques the reader is referred to two comprehensive recent reviews on this subject (15,16).

In the subsequent sections of this chapter metabonomics will be considered with respect to several other applications and technologies, to place recent advances in metabonomics in context. In each section, attempts are made to illustrate the versatility of metabonomics within these subject areas.

METABONOMICS AND GENOMICS

Metabolic profiling techniques have been used to phenotype a wide range of animals, plants, and microbes (for this area of application the term "metabolomics" has been most widely used). One of the first applications of the approach was to genotype *Arabidopsis thaliana* leaf extracts (17). Plant metabolomics is a huge analytical challenge because, despite typical plant genomes containing 20,000 to 50,000 genes, it is estimated that there are currently 50,000 identified metabolites with this number set to rise to approximately 200,000. The current preferred technique for metabolic profiling of plants uses GC–MS and in the seminal manuscript in this area, Fiehn et al. (17) quantified 326 distinct compounds in *A. thaliana* leaf extracts, further elucidating the chemical structure of half of these compounds. Applying PCA to this data set, the approach was able to separate out four different genotypes. Since this paper, the authors have used GC–time of flight–MS to detect and characterize approximately 1000 metabolites, as well as used the detected metabolites to generate metabolic cliques associated with given genotypes (18).

(A)   (B)

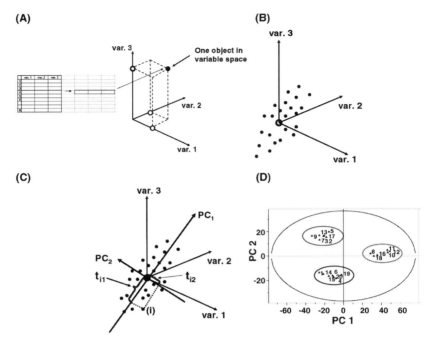

(C)   (D)

Figure 2 Pattern recognition tools are a vital part of the process of metabonomics and are increasingly being used to fully analyze the large multivariate data sets that are produced by other "-omic" technologies. This figure demonstrates the basis of PCA, one of the most widely used data reduction tools. PCA investigates the variation across the variables to produce loading scores and variation across samples to produce a scores plot. The process involves the mapping of a given sample according to the values of the variables measured. (**A** and **B**) The mapping of a sample representing three variables. (**C**) If this is repeated for all the variables, correlates can be investigated. The most amount of correlated variation is found along PC1 and the second most amount of variation is found along PC2. This is repeated until the variation in the data set is described by new latent variables represented by PCs. (**D**) An example of PCA plot that results from such a process. *Abbreviations*: PCA, principal components analysis; PCs, principal components.

In addition to *Arabidopsis*, yeast, the other workhorse of functional genomics, has also been examined by metabolomics (19). Yeast was the first eukaryote to be sequenced, and mutant strains for the approximately 6000 genes in yeast can be examined from cell banks such as EUROFAN (1). This suggests that researchers could potentially phenotype all the genes in yeast. This could have a significant impact on human disease through comparison of gene sequence similarities between yeast and man.

The standard method to phenotype yeast strains is to see how rapidly a strain grows on a given substrate mixture. If the mutation does not alter the rate of growth, it is said to be a silent mutation, and thus no function can be deduced from this gene deletion. However, Raamsdonk et al. (19) have used ^{1}H NMR–based metabolomics to distinguish these silent phenotypes. Applying a combination of PCA and Discriminate Function Analysis, they were able to cocluster strains with deletions of similar genes together. This included one cluster consisting of mutants related to oxidative phosphorylation and another cluster involving 6-phosphofructo-2-kinase. Since this original paper, a range of analytical techniques have been used to further characterize yeast mutants, including LC–MS of the cell extracts and MS analysis of the yeast media, with the latter approach being referred to as "metabolic footprinting" and providing a method of phenotyping yeast effectively through their waste products (20).

METABONOMICS AND DISEASE

A key advantage of NMR spectroscopy–based metabonomics is that the approach gives high throughput, allowing the rapid acquisition of large data sets. This makes it ideal as a screening tool, particularly for human populations where there may be significant environmental and dietary influences on tissue and biofluid "metabolomes" necessitating the collection of large data sets to fully define the innate variation within a population as well as the metabolic effects of a disease. A range of diseases have been investigated including Duchenne muscular dystrophy, multiple sclerosis, and cancer (Fig. 3) (21,23,24). As metabonomics makes no prior assumption as to the metabolic events that accompany a disease, it is particularly appropriate for diseases where conventional approaches to date have drawn a blank.

NMR-based metabonomics has also been used for screening human populations for both the presence and severity of coronary artery disease (CAD) using blood plasma (25). To reduce the variation in the data set not correlated with disease presence or severity, OSC was used as a data filter to subtract variation orthogonal to that associated with disease severity. The resultant postprocessed data was then analyzed using PLS–discriminate analysis to produce a pattern-recognition model that was greater than 90% accurate in predicting disease severity. Such an intelligent pattern-recognition model could produce significant financial savings by reducing the need for angiography, currently the gold standard for CAD diagnosis. This approach has also been used to correlate blood pressure with the ^{1}H NMR–derived metabolic profile of blood plasma (26). The researchers involved in these studies have since been extending the results to include transcriptional analysis using DNA microarrays, in the hope that a joint transcriptional and metabonomic study of blood plasma could compete with the approximately 99% correct prediction of angiography.

METABONOMICS AND TRANSCRIPTOMICS

While it is currently not possible to analyze all the metabolites in a biofluid or tissue extract using one analytical approach (indeed even a combination of technologies will not produce complete coverage of a metabolome), it is now possible to analyze the mRNA content present within a tissue against all the mRNA products from the entire genome of an organism using DNA microarrays. Hence, there is currently a great deal of interest in toxicology studies in combining the high throughput screening approach of metabonomics, with the potentially more truly global profiling capability of DNA microarrays. This has led to a number of studies in which the two approaches have been used to build up a combined mRNA and metabolite description of drug-induced pathology simultaneously.

One such study has examined orotic acid–induced fatty liver in the rat (27). Supplementation of orotic acid to normal food intake is known to induce fatty liver in the rat, producing symptoms very similar to alcohol-induced fatty liver disease. Although it has been known since the 1950s that disruption of the various apo proteins occurs during orotic acid exposure, it is still not known how the disruption of nucleotide metabolism, the primary effect of orotic acid, results in the impaired production of the various apo proteins important for the transport of lipids around the body. To investigate this, Griffin et al. (27) applied a transcriptomic and metabonomic analysis to the Kyoto and Wistar strains of rat. The Wistar rat is an outbred strain of rat and has been classically used to follow orotic acid–induced fatty liver disease. However, the Kyoto rat, an in-bred strain, is particularly susceptible to fatty liver disease. These two strains provided a pharmacogenomic model of the drug insult and illustrate a common problem with the development of many drugs—that the induced

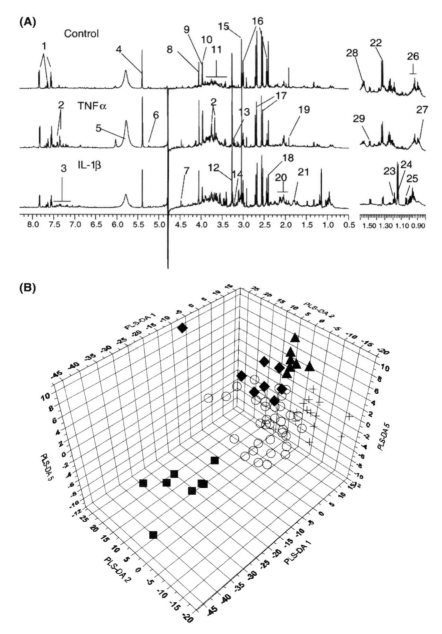

Figure 3 (*Caption on facing page*)

response may vary depending on the population it is administered to. To fully characterize the systemic changes as well as to analyze the metabolite and mRNA composition of the liver, blood and urine were also analyzed using NMR spectroscopy. Analysis of blood plasma demonstrated the expected decrease in circulating low-density lipoprotein and very low–density lipoprotein lipids, demonstrating the disruption of ApoB and ApoC production, but it also identified an increase in β-hydroxybutyrate, suggesting diabetes-like response to the lesion. The use of NMR-based metabonomics to follow systemic metabolism was possible because of the relative inexpensiveness of the approach on a per sample basis. This ensured that some information was obtained outside the liver, despite transcriptional analysis being confined to the liver because of cost or in the case of urine, a lack of mRNA to analyze.

Using PLS to cross-correlate transcriptional and metabonomic data in the liver, both data sets identified pathways concerning uridine production, choline turnover, and stress responses as being perturbed by the drug (Fig. 4). By careful analysis of these pathways it was possible to trace the metabolic perturbations from the initial exposure to orotic acid to disruption of fatty acid metabolism in the liver. Furthermore, the approach modeled the pharmacogenomics of the drug, showing that the metabolome of the Kyoto rat was more profoundly influenced by orotic acid. This approach also identified a number of key transcriptional changes that could be targeted for drug intervention including stearyl-CoA desaturase, indicating that metabonomics has a distinct role in the drug discovery process.

Transcriptional analysis, using real time (RT)–PCR, has also been used to assist in unraveling the metabolic changes detected by a metabonomic study of peroxisome proliferation in the rat using ^{1}H NMR spectroscopy of urine (28). In this study Ringeissen et al. found that peroxisome proliferator–activated receptor (PPAR) ligands induced large increases in urinary N-methylnicotinamide (NMN) and N-methyl-4-pyridone-3-carboxamide (4PY) concentrations. These two metabolites are intermediates in the tryptophan–NAD$^+$ pathway, suggesting a global upregulation of this pathway. Furthermore, these biomarkers were shown to correlate with peroxisome proliferation as measured by electron microscopy, suggesting that NMN and 4PY could be used as biomarkers for peroxisome proliferation in cases

---

Figure 3 (*Figure on facing page*) A range of diseases have been investigated including the two examples shown here: (**A**) multiple sclerosis and (**B**) cardiac disease. (**A**) 600.13 MHz spectra of rat urine from either a control animal or animals exposed to focal lesions from Ad5IL-1β or Ad5TNF-α$_m$. Spectra were acquired using the NOESYPR1D pulse sequence for solvent suppression. *Key*: 1, hippurate; 2, phenyacetylglycine (with traces of phenylalanine in aromatic region); 3, tyrosine; 4, allantoin; 5, urea; 6, α-glucose; 7, N-methylnicotinamide; 8, creatinine; 9, hippurate; 10, creatine; 11, glucose and amino acid CH protons; 12, TMAO; 13, phosphocholine; 14, choline; 15, creatine and creatinine; 16, 2-oxoglutarate; 17, citrate; 18, succinate; 19, acetate; 20, N-acetyl glycoproteins; 21, tentatively assigned to bile acids; 22, lactate; 23, ethanol (contaminant); 24, isobutyrate; 25, valine; 26, leucine, isoleucine, and valine; 27, n-butyrate; 28, n-butyrate; 29, alanine. (**B**) An orthogonal signal corrected PLS–DA analysis of four mouse models of cardiac disease using metabolic profiles from ^{1}H NMR analysis of cardiac tissue extracts. *Key*: (○) control animals (three different strains), (+) mouse model of Duchenne muscular dystrophy, (◇) mouse model of cardiac hypertrophy (muscle LIM protein knock out mouse), and two models of cardiac arrhythmia: a cardiac sodium channel knock out mouse (Scn$^{-/+}$), (■) and a model where the closure of the cardiac sodium channel is impaired (Scn$^{Δ/+}$) (▲). *Abbreviations*: NMR, nuclear magnetic resonance; PLS–DA, partial least squares–discriminate analysis; TMAO, trimethylamine N-oxide. *Source*: From Refs. 21, 22.

(A)

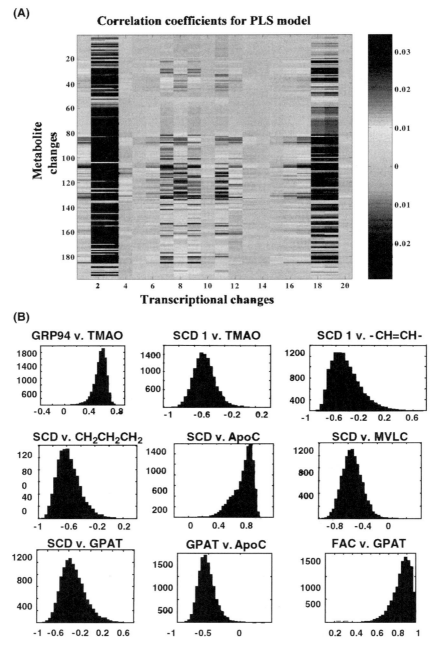

Figure 4   Two methods for cross-correlating data from different -omic approaches. (**A**) A heat map plot of correlation coefficients from a PLS model of 20 transcripts (along the *x*-axis) and 200 metabolites (along the *y*-axis). Darker regions highlight correlated metabolites and transcripts. This correlation can be either positive or negative. (**B**) Histograms of bootstraps for correlation coefficients between key metabolite regions and transcripts. The *x*-axis represents the correlation coefficients while the *y*-axis represents the number of times this correlation was returned during 10,000 iterations. *Key*: SCD, stearyl-CoA desaturase 1; ApoC, apolipoprotein C III; MVLC, mitochondrial very long chain acyl CoA thioesterase; GPAT, glycerol 3 phosphate acyltransferase; FAC, fatty acid CoA ligase 4; CH=CH, unsaturated lipid resonance; $CH_2CH_2CH_2$, saturated lipid resonance. *Abbreviation*: PLS, partial least squares. *Source*: From Ref. 26.

where the biopsy of liver tissue was not possible, such as clinical trials. This has great relevance to the drug safety assessment of these PPAR ligands because currently a number of these ligands are being investigated as drugs for treating dyslipidemia and type 2 diabetes; but there is no current clinical method for assessing peroxisome proliferation in humans despite this being a potential side effect of some drugs of this class in rodents. RT-PCR of key enzymes also identified transcriptional changes involved in tryptophan–NAD$^+$ pathway, indicating that the results of both transcriptional and metabonomic analyses of the tissue agreed in terms of the major metabolic pathway targeted. The two biomarkers were then measured by a HPLC assay as part of a high throughput specific screen for peroxisome proliferation in the rat. This study demonstrates how metabonomics can be used to identify biomarkers, which can be validated through transcriptomics, and these biomarkers can ultimately be used to assess drug toxicity in situations relevant to drug safety assessment.

In a similar manner, paracetamol toxicity has also been investigated using a combination of affymetrix gene arrays and NMR-based metabonomics of urine, blood plasma, liver tissue, and liver extracts following acetaminophen hepatotoxicity (29). This demonstrates one of the major benefits of using metabonomics in conjunction with DNA microarrays. While it can be too expensive to carry out global transcriptional analysis of several tissues, the analytical techniques used in metabonomics are relatively cheap on a per sample basis allowing the simultaneous examination of several tissues and biofluids. Metabolic profiling detected a decrease in glucose and glycogen and an increase in lipids in intact liver tissue, in conjunction with glucose, pyruvate, alanine, and lactate increases in blood plasma. These changes were indicative of an increased rate of glycolysis, with changes in this pathway also confirmed by transcriptomics. This again demonstrates that the two technologies can be used in parallel for drug toxicity studies, either as a strategy to generate more specific biomarker combinations for monitoring a given toxic insult or to provide a better understanding of the underlying perturbations that accompany toxicity.

MS AND METABONOMICS

To date, there have been relatively few examples of the use of MS to assess drug action and toxicity, and these have largely been confined to application papers that discuss the potential of this technology (12,13). However, given the exceptional sensitivity of the approach and the widespread use of MS already in the pharmaceutical industry, for example, during drug metabolism and pharmacokinetic investigations, this is set to change. Furthermore, the potential of MS has already been demonstrated by those using the technology to monitor plant metabolomics. In plant metabolomics (4,18), the challenge of profiling all the metabolites in a given tissue is even greater than that in the mammalian systems. Despite plant genomes typically containing 20,000 to 50,000 genes, currently 50,000 metabolites have been identified in the plant kingdom with the number predicted to rise to about 200,000 (30) compared with the estimated 300 to 600 metabolites in a mammalian cell.

The current detection limits for MS-based approaches are of the order of 100 nM, allowing the detection of approximately 1000 metabolites with typical acquisition times of approximately 30 minutes. The commonest approaches currently being used in metabonomics are GC–MS and LC–MS. GC–MS requires the prior derivatization of metabolites to make them volatile at a relatively low temperature. While this is quite straight forward, it does limit analysis to those metabolites that can be made volatile. However, the approach is both robust, in terms of chromatography as well as sample preparation, and relatively cheap to set up. While LC–MS can be used to analyze biofluids directly, much more care is required to maintain good and reproducible chromatography and MS may be significantly affected by ion suppression. However, LC–MS is the favored method for global profiling of complex lipids, such as triglycerides and phosphatidylcholines, suggesting that this technique may be ideal for monitoring changes in lipophilic environments within the cell, for example, in cell membranes. Furthermore, because of the innate sensitivity of the approaches, these techniques are likely to become more and more important in toxicology-based metabonomics.

METABONOMICS AND SYSTEMS BIOLOGY

To fully understand such complex neurological disorders as schizophrenia, a complete systems biology approach may be required to unravel all the inter-related pathways that are perturbed by such diseases where complex environmental risk factors interact with as yet unknown genetic risk factors. Furthermore, in these diseases where the current hypothesis-driven research appears to be making slow progress, a series of hypothesis-generating approaches may prove to be more fruitful. Thus, a number of researchers are using functional genomics tools for studying a range of neurological disorders. Using human brain tissue from a brain bank of tissue from sufferers of schizophrenia and bipolar disorder, Prabakaran et al. (31) have examined the disease through all three tiers of biological organization. Transcriptomics, proteomics, and metabonomics all indicated that schizophrenia was associated with a series of metabolic deficits (Fig. 5). Intriguingly, these changes were much more apparent in the white matter compared with the grey matter. To analyze this data, the researchers used a range of bioinformatics approaches, including multivariate statistics. This was particularly important because the brain tissue had been influenced by a number of postmortem effects, such as brain pH and time taken to freeze sections after death, and variation among the patients, for example, due to drug treatment history, sex, and age. Hence, it was important to rule out that sample degradation and patient demographics had an important influence on the brain tissue being analyzed. By applying multivariate statistics to analyze the data, it was possible to model the influence these confounding factors had on the data acquired, validating the biomarkers produced by the combined transcriptomics, proteomics, and metabonomics analysis. As these technologies become more widespread it is likely that more researchers will use this all-encompassing approach.

FUTURE DEVELOPMENTS

The primary drive in metabonomics is to improve analytical techniques to provide an ever-increasing coverage of the complete metabolome of an organism. To date, NMR-based techniques have focused on using ^{1}H NMR spectroscopy, but this approach suffers from a small chemical shift range, producing significant overlap of the resonances of a number of different metabolites. While ^{13}C NMR spectroscopy has a much larger chemical shift range allowing the resolution of a wider range of metabolites, the approach is intrinsically less sensitive compared with ^{1}H NMR spectroscopy, as a result of the lower gyromagnetic ratio of the ^{13}C nucleus compared with the ^{1}H. However, in CryoProbes, an NMR probe in which the receiver and transmitter coil is cooled using liquid helium to approximately 4 K, a significant improvement in sensitivity can be achieved by removing the impact of noise on the spectra. This allows the rapid

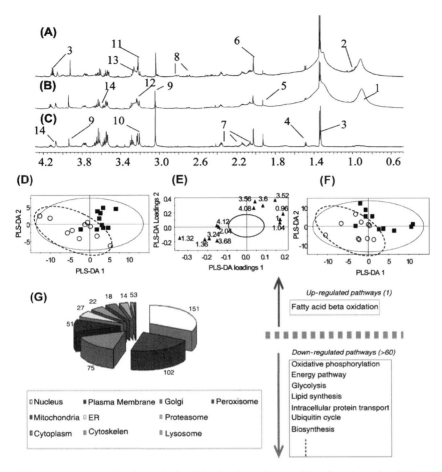

Figure 5 Metabolomic analysis of brain tissue from schizophrenics using HRMAS ^{1}H NMR spectroscopy. Tissue was profiled using both a solvent-suppressed pulse sequence (**A**: grey matter, **B**: white matter) and a T_2 relaxation–weighted pulse sequence (A Carr Purcell Meiboom and Gill sequence) to selectively attenuate lipid resonances relative to aqueous metabolites (**C**: white matter). Spectra were acquired at 700.13 MHz at 3°C and at a rotor spinning speed of 7000 Hz. This data formed was examined using multivariate data analysis including PLS–DA, a supervised regression extension of PCA. (**D**) The observation plot of the PLS–DA model demonstrated that spectra of white matter from schizophrenic patients (○) formed a cluster apart from tissue from control patients (■) (cluster highlighted with broken line). The metabolic differences causing this was identified in (**E**) the loadings plot of the PLS–DA model, and was largely caused by increases in concentration of lactate (δ 1.32, 4.12), –$CH_2CH_2CH_2$ lipid group (δ 1.36–1.32), phosphocholine (δ 3.24, 3.68) and NAA (δ 2.04), and decreases in CH_3– terminal lipid groups (δ 0.96–1.04) and myo-inositol (δ 3.52–3.60, 4.08), where δ signifies the center of the 0.04 ppm chemical shift region used as a variable in the multivariate analysis. (**F**) A similar PLS–DA model could be built for solvent suppressed spectra from grey matter. *Key*: 1, –CH_2CH_3 lipid group; 2, leucine, isoleucine, and valine; 3, lactate (sharp doublet) superimposed on –$CH_2CH_2CH_2$– lipid resonance (broad resonance); 4, alanine; 5, acetate; 6, NAA; 7, glutamate and glutamine; 8, citrate; 9, creatine; 10, choline; 11, phosphocholine; 12, phosphatidylcholine and glycerophosphocholine; 13, taurine; 14, myo-inositol (series of resonances from 3.52–3.60). (**G**) A diagrammatic summary of the transcriptional changes identified alongside the metabolomic analysis. Numbers signify the number of transcripts identified as increased or decreased in expression in each group. *Abbreviations*: HRMAS ^{1}H NMR, high resolution magic angle spinning ^{1}H nuclear magnetic resonance; PCA, principal components analysis; PLS–DA, partial least squares–discriminate analysis; NAA, *N*-acetyl aspartate. *Source*: From Ref. 31.

acquisition of ^{13}C NMR spectra using such CryoProbes, even from samples where there is no enrichment of the ^{13}C above the natural level of 1%. Keun et al. (32) have already applied this approach to studying hydrazine toxicity through biofluid ^{13}C NMR spectroscopy of unenriched samples. While in this particular study the biomarkers of hydrazine toxicity were already largely known, the use of ^{13}C spectroscopy did allow the identification of these metabolites from one-dimensional spectroscopy, without the need to identify the metabolites responsible for key resonances from a series of two-dimensional approaches. Furthermore, this approach may be particularly good for quantifying concentrations of metabolites that only produce singlets in ^{1}H NMR spectra and hence would not have cross-peaks in two-dimensional spectra such as correlation spectroscopy, total correlation spectroscopy, and J-coupling resolved spectroscopy. While ^{1}H NMR may not conclusively identify the metabolite, the extra chemical shift range for ^{13}C NMR is usually enough to allow unambiguous assignment of a given singlet.

Another approach to improve the sensitivity of the NMR experiment in terms of the metabolites detectable is to hyphenate the NMR spectroscopy with LC. This improves sensitivity by two mechanisms. First, high- and low-concentration metabolites are separated by the LC, reducing the likelihood of coresonant peaks and also improving the dynamic range of the NMR experiment for the low concentration metabolites. Second, metabolites are concentrated by chromatography, further aiding the detection of low concentration metabolites. Bailey et al. (33,34) have used this approach of LC–NMR spectroscopy to metabolically profile a number of plants. This may be a particularly useful approach if hyphenated further with cryoprobe technology to allow Cryo–LC–NMR.

LC–MS and GC–MS are increasingly being used as profiling tools for diseases and toxicology studies. This is set to increase as the chromatography becomes more reliable, and the software for matching mass fragmentation patterns is improved. Indeed a number of manufacturers are developing systems that are designed to work in tandem with NMR to provide LC–NMR–MS analysis of biofluids, thus, reaping the benefits of these two technologies while avoiding many of the pitfalls, and providing a more truly global description of metabolism in a tissue or biofluid.

With all the techniques discussed above there is a need to form tissue extracts if the toxicologist is to examine the metabolomic changes in a tissue directly. However, if high resolution MAS ^{1}H NMR spectroscopy could be automated, this would provide a viable alternative to laborious tissue extraction procedures. Finally, there is an urgent need for improved pattern-recognition processes for integrating the information produced by a variety of analytical approaches to provide a fuller coverage of the metabolome as well as a means to disseminate this information. Currently there is no consensus as to what information should be reported alongside a metabonomic study so that the data can be interpreted by other researchers. Such "metadata," data about the data, will be vital if researchers are to generate databases akin to those being produced by the microarray and proteomic communities using such conventions as the minimum information about a microarray experiment (MIAME) (35).

CONCLUSIONS

Metabonomics is being performed using a variety of analytical approaches and pattern recognition techniques to answer the questions to a diverse range of problems relevant to the pathologist and toxicologist. At its most immediate level it provides a rapid screen for large-scale drug toxicity testing and screening of large human populations for such common diseases as CAD. These metabolic profiles are also ideal for following changes in transcriptional and proteomic profiles, and there are already a number of mathematical tools for this sort of data fusion. This rapid increase in interest in metabonomics, as well as a need to disseminate information about drug safety or toxicity issues necessitates the development of metabonomic databases and ultimately a standardization of approaches to provide an equivalent of the transcriptomic MIAME protocol for metabonomics.

REFERENCES

1. Oliver SG, Winson MK, Kell DB, Baganz F. Systematic functional analysis of the yeast genome. Trends Biotechnol 1998; 16:373–378.
2. Nicholson JK, et al. 1998.
3. Nicholson JK, Connelly J, Lindon JC, Holmes E. Metabonomics: a platform for studying drug toxicity and gene function. Nat Rev Drug Discov 2002; 1:153–161.
4. Fiehn O. Metabolomics—the link between genotypes and phenotypes. Plant Mol Biol 2002; 48:155–171.
5. Nicholson JK, Wilson I. Understanding 'global' systems biology: metabonomics and the continuum of metabolism. Nat Rev Drug Discov 2003; 2:668–676.
6. Gavaghan CL, Holmes E, Lenz E, Wilson ID, Nicholson JK. An NMR-based metabonomic approach to investigate the biochemical consequences of genetic strain differences: application to the C57BL10J and Alpk:ApfCD mouse. FEBS Lett 2000; 484 (3):169–174.
7. Beckwith-Hall BM, Nicholson JK, Nicholls AW, et al. Nuclear magnetic resonance spectroscopic and principal components analysis investigations into biochemical effects of three model hepatotoxins. Chem Res Toxicol 1998; 11(4):260–272.
8. Holmes E, Nicholls AW, Lindon JC, et al. NMR Biomed 1998; 11(4–5):235–244.
9. Garrod S, Humpfer E, Spraul M, et al. Magn Reson Med 1999; 41(6):1108–1118.
10. Griffin JL, Walker LA, Troke J, Osborn D, Shore RF, Nicholson JK. The initial pathogenesis of cadmium induced renal toxicity. FEBS Lett 2000; 478(1–2):147–150.
11. Griffin JL, Lehtimaki KK, Valonen PK, et al. Assignment of 1H nuclear magnetic resonance visible polyunsaturated fatty acids in BT4C gliomas undergoing ganciclovir-thymidine kinase gene therapy–induced programmed cell death. Cancer Res 2003; 63(12):3195–3201.
12. Plumb RS, Stumpf CL, Gorenstein MV, et al. Metabonomics: the use of electrospray mass spectrometry coupled to reversed-phase liquid chromatography shows potential for the screening of rat urine in drug development. Rapid Commun Mass Spectrum 2002; 16(20):1991–1996.
13. Plumb R, Granger J, Stumpf C, Wilson ID, Evans JA, Lenz EM. Metabonomic analysis of mouse urine by liquid chromatography–time of flight mass spectrometry (LC–TOFMS): detection of strain, diurnal and gender differences. Analyst 2003; 128:819–823.
14. Harrigan GC, Goodacre R, eds. Metabolic Profiling—Its Role in Biomarker Discovery and Gene Function Analysis. Kluwer Academic Publishing, 2003:83–94.
15. Lindon JC, Holmes E, Nicholson JK. Pattern recognition methods and applications in biomedical magnetic resonance. Prog Nucl Magn Reson Spectrosc 2001; 39:1–40.
16. Valafar F. Pattern recognition techniques in microarray data analysis. Ann NY Acad Sci 2002; 980:41–64.
17. Fiehn O, Kopka J, Dormann P, Altmann T, Trethewey RN, Willmitzer L. Nat Biotechnol 2000; 18:1157–1161.
18. Fiehn O. Combining genomics, metabolome analysis and biochemical modeling to understand metabolic networks. Comp Funct Genomics 2001; 2:155–168.

19. Raamsdonk LM, Teusink B, Broadhurst D, et al. A functional genomics strategy that uses metabolome data to reveal the phenotype of silent mutations. Nat Biotechnol 2001; 19(1):45–50.
20. Allen J, Davey HM, Broadhurst D, et al. High-throughput classification of yeast mutants for functional genomics using metabolic footprinting. Nat Biotechnol 2003; 21:692–696.
21. Griffin JL, Anthony DC, Campbell SJ, et al. Study of cytokine induced neuropathology by high resolution proton NMR spectroscopy of rat urine. FEBS Lett 2004; 568(1–3):49–54.
22. Jones GL, Sang E, Goddard C, et al. A functional analysis of mouse models of cardiac disease through metabolic profiling. J Biol Chem 2004 (Epub ahead of print).
23. Griffin JL, Sang E, Evens T, Davies K, Clarke K. FEBS Lett 2002; 530(1–3):109–116.
24. Griffin JL, Shockcor JP. Metabolic profiles of cancer cells. Nat Rev Cancer 2004; 4(7):551–561.
25. Brindle JT, Antii H, Holmes E, et al. Rapid and noninvasive diagnosis of the presence and severity of coronary heart disease using ^{1}H NMR-based metabonomics. Nat Med 2002; 8(12):1439–1444.
26. Brindle JT, Nicholson JK, Schofield PM, Grainger DJ, Holmes E. Analyst 2003; 128(1):32–36.
27. Griffin JL, Bonney SA, Mann C, et al. An integrated reverse functional genomic and metabolic approach to understanding orotic acid-induced fatty liver. Physiol Genomics 2004; 17(2):140–149.
28. Ringeissen S, Connor SC, Brown HR, et al. Biomarkers 2004; 8:240–271.
29. Coen M, Ruepp SU, Lindon JC, et al. Integrated application of transcriptomics and metabonomics yields new insight into the toxicity due to paracetamol in the mouse. J Pharm Biomed Anal 2004; 35(1):93–105.
30. De Luca & St. Pierre, 2000.
31. Prabakaran S, Swatton JE, Ryan MM, et al. Mitochondrial dysfunction in schizophrenia: evidence for compromised brain metabolism and oxidative stress. Mol Psychiat 2004; 9(7):643.
32. Keun HC, Beckonert O, Griffin JL, et al. Crogenic probe 13C NMR spectroscopy of urine for metabonomic studies. Anal Chem 2002; 74:4588–4593.
33. Bailey NJ, Oven M, Holmes E, Nicholson JK, Zenk MH. Metabolomic analysis of the consequences of cadmium exposure in *Silene cucubalus* cell cultures via ^{1}H NMR spectroscopy and chemometrics. Phytochemistry 2003; 62:851–858.
34. Bailey NJ, Sampson J, Hylands PJ, Nicholson JK, Holmes E. Multi-component metabolic classification of commercial fever few preparations via high-field ^{1}H-NMR spectroscopy and chemometrics. Planta Med 2002; 68:734–738.
35. Brazma A, Hingamp P, Quackenbush J, et al. Minimum information about a microarray experiment (MIAME)-toward standards for microarray data. Nat Genet 2001; 29(4):365–371.

5

Quantitative Structure–Toxicity Relationships Using Chemodescriptors and Biodescriptors

Subhash C. Basak, Denise Mills, and Brian D. Gute
Natural Resources Research Institute, University of Minnesota Duluth, Duluth, Minnesota, U.S.A.

INTRODUCTION

A contemporary interest in computational toxicology is the prediction of potential toxicity of drugs, industrial chemicals, and xenobiotics using properties of molecules that can be calculated directly from their structure alone without the input of any other experimental data (1–14). This field of research is popularly known as quantitative structure–activity relationship (QSAR), where "activity" is sometimes used generally to represent toxicity [quantitative structure–toxicity relationships (QSTR)], physicochemical property [quantitative structure–property relationships (QSPR)], or biological activity [quantitative structure–activity relationships (QSAR)]. QSAR is based on the paradigm expressed by Eq. (1):

$$P = f(S) \tag{1}$$

where P represents any toxicological, physicochemical, or biological property of interest produced by chemical–biological interactions at any level of biological organization, e.g., molecular (receptor, enzyme), cellular, tissue or organ, whole organism, etc., and S symbolizes quantifier(s) of salient features of molecular structure related to the property of interest P.

It may be mentioned that biological effects of chemicals, such as toxicity, are necessarily produced by chemical–biological interactions. In that sense, Eq. (1) can be looked upon as a special case of a more comprehensive relationship given by Eq. (2):

$$P = f(S, Biol) \tag{2}$$

where $Biol$ symbolizes the critical biological target; S and P have the same meaning as in Eq. (1). In standard bioassays of chemicals for toxicity, the biological system is the same. In that case, Eq. (2) becomes Eq. (1) because only the variation in the structure of the toxicant determines differences in hazard between one molecule and another.

The growing interest in QSAR studies arises out of the fact that the majority of them are based on calculated molecular descriptors, although prediction methods

Table 1 Toxicologically Relevant Properties

Physicochemical	Biological
Molar volume	Receptor binding (K_D)
Boiling point	Michaelis constant (K_m)
Melting point	Inhibitor constant (K_i)
Vapor pressure	Biodegradation
Aqueous solubility	Bioconcentration
Dissociation constant (pK_a)	Alkylation profile
Partition coefficient	Metabolic profile
Octanol–water (log p)	Chronic toxicity
Air–water	Carcinogenicity
Sediment–water	Mutagenicity
	Acute toxicity
	LD_{50}
Reactivity (Electrophile)	LC_{50}
	EC_{50}

based on linear free energy relationships (LFER), linear solvation energy relationships (LSER), and experimental properties, such as octanol–water partition coefficient or binding of a chemical to plasma proteins (15), can be data intensive. Traditionally, hazards posed by chemicals have been estimated by toxicologists from a suite of experimental physicochemical and biological or toxicological data at different levels of biological organization. Table 1 provides a partial list of such properties that have been used by practitioners of predictive toxicology for decades.

The principal problem with the use of properties listed in Table 1 is that they are not available for most of the candidate chemicals. The Toxic Substances Control Act Inventory of the United States Environmental Protection Agency has approximately 75,000 chemical substances (16), more than 50% of which have no data at all. Recently, the American Chemistry Council attempted to collate/determine toxicologically relevant data for nearly 3000 high production volume chemicals, the cost of which was estimated to be at 500 to 700 million dollars.

It is clear that estimation of toxicity posed by chemicals will have to be carried out in a data-poor situation in the foreseeable future. It may be mentioned that many chemicals are metabolized or environmentally degraded to other products that might contribute to toxicity. Very little data are available for such chemicals. There is also an interest in estimating potential toxicity of chemicals not yet synthesized. Toxicity estimation methods based on laboratory data are of no help in that case. Also, there is a growing aversion worldwide to the use of animals in assessing the safety of chemicals used for pharmaceutical, industrial, and cosmetic purposes. In silico QSAR studies based on computed molecular descriptors alone can help in the prioritization of chemicals for more exhaustive testing that are demanding in terms of time, animals, chemicals, and testing facilities.

THE MOLECULAR STRUCTURE CONUNDRUM

The simplistic Eq. (1) states that one should be able to predict toxicity of a chemical from its structure. But the moot point is "Do we know definitely which aspects of molecular structure are relevant to the specific toxicity endpoint in which we are interested?" The answer to this question is not definite; in fact, it is often equivocal. The main reason for this quagmire is that the term "molecular structure" means different things to different people. The structure of a chemical can be represented by molecular graphs, various three-dimensional (3-D) models based on molecular mechanics, and by various quantum chemical (QC) formalisms practiced in semiempirical as well as ab initio quantum chemistry. Each one of them is a valid representation in its own domain and has been found to be useful in QSAR studies. Subsequently, for each one of the above representations, one can have multiple mathematical models for the quantification of molecular structure (17,18). The large number of calculable descriptors, together with the various shades of molecular structure representation, provides a large and nontrivial list of indices to choose from for QSAR studies. To help the average practitioner of QSARs, our research team has developed a novel approach called hierarchical QSAR (HiQSAR) modeling in which calculated descriptors of increasing computational complexity and demand for resources are used in a graduated manner.

THE HIQSAR APPROACH

Molecular descriptors can be partitioned into hierarchical classes based on level of complexity and demand for computational resources. The approach in HiQSAR is to include the more complex and resource-intensive descriptors only if they result in significant improvement in the predictive quality of the model. We begin by building a model using only the simplest class of descriptors, followed by the creation of additional models based on the successive inclusion of increasingly complex descriptor classes. By comparing the resulting models, the contribution of each descriptor class is elucidated. In addition, the hierarchical approach enables us to determine whether or not the higher-level descriptors are necessary for the data under consideration. In situations where they are not useful, we can avoid spending the time required for their calculation. For comparative purposes, we typically develop models based on single classes of molecular descriptors in addition to the hierarchical models. While chemodescriptors are based on molecular structure alone, biodescriptors are derived from DNA sequences, genomics, and proteomics. Figure 1 illustrates the hierarchical nature of the various classes of chemodescriptors and biodescriptors.

Chemodescriptors

The topostructural (TS) descriptors are at the low end of the hierarchy, based solely on the connectedness of atoms within a molecule and devoid of any chemical information. The topochemical (TC) descriptors are more complex, encoding such chemical information as atom type and bond type, in addition to information on the connectedness of the atoms. The TS and TC descriptors are collectively referred to as topological descriptors and are based on a two-dimensional (2-D) representation of the molecule. More complex yet are the geometrical (3-D) descriptors that encode information on the 3-D aspects of molecular structure. The most complex CDs are the QC descriptors that encode electronic aspects of molecular structure (Fig. 1). The QC class can be subdivided into semiempirical and ab initio classes, with the latter often being prohibitive in terms of computation time. The time required to calculate hundreds of topological descriptors for a set of 100 compounds,

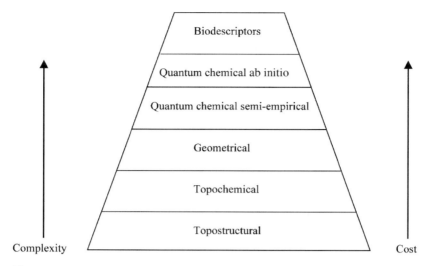

Figure 1 Hierarchical descriptor classes.

for example, may be on the order of seconds, while computing a single ab initio descriptor for one large molecule may require days!

The structure-based chemodescriptors used in our QSAR modeling studies were calculated using various software programs including POLLY 2.3 (19), Triplet (20), Molconn-Z 3.5 (21), MOPAC 6.00 (22), and Gaussian 98W (23). From POLLY 2.3 and associated software, a set of 102 topological descriptors is available, including a large group of connectivity indices (24,25), path-length descriptors (26), J indices (27), and information theoretic (28,29) and neighborhood complexity indices (30). An additional 100 topological descriptors are available from the Triplet program. An extended set of connectivity indices, along with descriptors of polarity and hydrogen bonding and a large set of electrotopological state indices (31), are calculated using Molconn-Z 3.5. Semiempirical QC descriptors, such as the Austin Model 1 (AM1) descriptors, are obtained using MOPAC 6.00, whereas ab initio QC descriptors are calculated using Gaussian 98W. We have used ab initio calculations based on the STO-3G, 6–31G(d), 6–311G, 6–311G(d), and cc-pVTZ basis sets. A list of theoretical descriptors typically calculated for our QSAR studies, along with brief descriptions and hierarchical classification, is provided in (Table 2).

Biodescriptors: A New Class of Descriptors in the Postgenomic Era

As described in Eq. (2), the biological action of a chemical is the result of its interactions with the relevant biotarget. In the postgenomic era, technologies of genomics, proteomics, and metabolomics are providing sophisticated and high-dimensional information relevant to toxicity of chemicals. Our research team has been involved in the development of biodescriptors from DNA sequence information and proteomics patterns of cells exposed to pollutants (33–43). The expression of the genomic information into functionalities of the cell can be modulated both qualitatively and quantitatively by cellular dysfunctions, injury, diseases, or exposure to chemicals including drugs and xenobiotics. The approach, called "functional genomics, at the

Table 2 Symbols, Definitions, and Classification of Calculated Molecular Descriptors

Topostructural	
I_D^W	Information index for the magnitudes of distances between all possible pairs of vertices of a graph
$\overline{I_D^W}$	Mean information index for the magnitude of distance
W	Wiener index = half-sum of the off-diagonal elements of the distance matrix of a graph
I^D	Degree complexity
H^V	Graph vertex complexity
H^D	Graph distance complexity
IC	Information content of the distance matrix partitioned by frequency of occurrences of distance h
M_1	A Zagreb group parameter = sum of square of degree over all vertices
M_2	A Zagreb group parameter = sum of cross-product of degrees over all neighboring (connected) vertices
$^h\chi$	Path connectivity index of order $h = 0$–10
$^h\chi_C$	Cluster connectivity index of order $h = 3$–6
$^h\chi_{PC}$	Path–cluster connectivity index of order $h = 4$–6
$^h\chi_{Ch}$	Chain connectivity index of order $h = 3$–10
P_h	Number of paths of length $h = 0$–10
J	Balaban's J index based on topological distance
n_{rings}	Number of rings in a graph
n_{circ}	Number of circuits in a graph
DN^2S_y	Triplet index from distance matrix, square of graph order (number of non-H atoms), and distance sum; operation $y = 1$–5
DN^21_y	Triplet index from distance matrix, square of graph order, and number 1; operation $y = 1$–5
$AS1_y$	Triplet index from adjacency matrix, distance sum, and number 1; operation $y = 1$–5
$DS1_y$	Triplet index from distance matrix, distance, sum, and number 1; operation $y = 1$–5
ASN_y	Triplet index from adjacency matrix, distance sum, and graph order; operation $y = 1$–5
DSN_y	Triplet index from distance matrix, distance sum, and graph order; operation $y = 1$–5
DN^2N_y	Triplet index from distance matrix, square of graph order, and graph order, operation $y = 1$–5
ANS_y	Triplet index from adjacency matrix, graph order, and distance sum; operation $y = 1$–5
$AN1_y$	Triplet index from adjacency matrix, graph order, and number 1; operation $y = 1$–5
ANN_y	Triplet index from adjacency matrix, graph order, and graph order again; operation $y = 1$–5
ASV_y	Triplet index from adjacency matrix, distance sum, and vertex degree; operation $y = 1$–5
DSV_y	Triplet index from distance matrix, distance sum, and vertex degree; operation $y = 1$–5
ANV_y	Triplet index fromfrom adjacency matrix, graph order, and vertex degree; operation $y = 1$–5
Topochemical	
O	Order of neighborhood when IC_r reaches its maximum value for the hydrogen-filled graph

(Continued)

Table 2 Symbols, Definitions, and Classification of Calculated Molecular Descriptors (*Continued*)

O_{orb}	Order of neighborhood when IC_r reaches its maximum value for the hydrogen-suppressed graph neighborhood of vertices
I_{orb}	Information content or complexity of the hydrogen-suppressed graph at its maximum neighborhood of vertices
IC_r	Mean information content or complexity of a graph based on the rth ($r = 0$–6) order neighborhood of vertices in a hydrogen-filled graph
SIC_r	Structural information content for rth ($r = 0$–6) order neighborhood of vertices in a hydrogen-filled graph
CIC_r	Complementary information content for rth ($r = 1$–6) order neighborhood of vertices in a hydrogen-filled graph
$^h\chi^b$	Bond path connectivity index of order $h = 0$–6
$^h\chi_C^b$	Bond cluster connectivity index of order $h = 3$–6
$^h\chi_{Ch}^b$	Bond chain connectivity index of order $h = 3$–6
$^h\chi_{PC}^b$	Bond path–cluster connectivity index of order $h = 4$–6
$^h\chi^v$	Valence path connectivity index of order $h = 0$–10
$^h\chi_C^v$	Valence cluster connectivity index of order $h = 3$–6
$^h\chi_{Ch}^v$	Valence chain connectivity index of order $h = 3$–10
$^h\chi_{PC}^v$	Valence path-cluster connectivity index of order $h = 4$–6
J^B	Balaban's J index based on bond types
J^x	Balaban's J index based on relative electronegativities
J^Y	Balaban's J index based on relative covalent radii
AZV_y	Triplet index from adjacency matrix, atomic number, and vertex degree; operation $y = 1$–5
AZS_y	Triplet index from adjacency matrix, atomic number, and distance sum; operation $y = 1$–5
ASZ_y	Triplet index from adjacency matrix, distance sum, and atomic number; operation y = 1–5
AZN_y	Triplet index from adjacency matrix, atomic number, and graph order; operation $y = 1$–5
ANZ_y	Triplet index from adjacency matrix, graph order, and atomic number; operation $y = 1$–5
DSZ_y	Triplet index from distance matrix, distance sum, and atomic number; operation $y = 1$–5
DN^2Z_y	Triplet index from distance matrix, square of graph order, and atomic number; operation $y = 1$–5
nvx	Number of nonhydrogen atoms in a molecule
nelem	Number of elements in a molecule
Fw	Molecular weight
Si	Shannon information index
totop	Total topological index, t
sumI	Sum of the intrinsic state values, I
sumdelI	Sum of delta-I values
tets2	Total topological state index based on electrotopological state indices
phia	Flexibility index (kpl $*$ kp2/nvx)
IdCbar	Bonchev–Trinajstic information index
IdC	Bonchev–Trinajstic information index
Wp	Wienerp
Pf	Plattf
Wt	Total Wiener number

(*Continued*)

Table 2   Symbols, Definitions, and Classification of Calculated Molecular Descriptors (*Continued*)

knotp	Difference of chi-cluster-3 and path/cluster-4
knotpv	Valence difference of chi-cluster-3 and path/cluster-4
n_{class}	Number of classes of topologically (symmetry) equivalent graph vertices
numHBd	Number of hydrogen bond donors
numwHBd	Number of weak hydrogen bond donors
numHBa	Number of hydrogen bond acceptors
SHCsats	E-State of C sp^3 bonded to other saturated C atoms
SHCsatu	E-State of C sp^3 bonded to unsaturated C atoms
SHvin	E-State of C atoms in the vinyl group, =CH–
SHtvin	E-State of C atoms in the terminal vinyl group, =CH$_2$
SHavin	E-State of C atoms in the vinyl group, =CH–, bonded to an aromatic C
SHarom	E-State of C sp^2 that are part of an aromatic system
SHHBd	Hydrogen bond donor index, sum of hydrogen E-State values for –OH, =NH –NH2, –NH–, –SH, and #CH
SHwHBd	Weak hydrogen bond donor index, sum of C–H hydrogen E-State values for hydrogen atoms an a C to which an F and/or Cl are also bonded
SHHBa	Hydrogen bond acceptor index, sum of the E-State values for–OH, = NH, –NH2, –NH–, N–, –O–, –S–, along with –F and –Cl
Qv	General polarity descriptor
NHBint$_y$	Count of potential internal hydrogen bonders ($y = 2$–10)
SHBint$_y$	E-State descriptors of potential internal hydrogen bond strength ($y = 2$–10) Electrotopological state index values for atoms types: ShsOH, SHdNH, SHsNH2, SHssNH, SHtCH, Shother, SHCHnX, Hmax, Gmax, Hmin, Gmin, Hmaxpos, Hminneg, SsLi, SssBe, SssssBem, SssBH, SsssBm, SsCH3, SdCH2, SssCH2, StCH, SdsCH, SaaCH, SsssCH, SddC, StsC, SaasC, SaaaC, SssssC, SsNH3p, SsNH2, SssNH2p, SdNH, SssNH, SaaNH, StN, SsssNHp, SdsN, SaaN, SsssN, SddsN, SaasN, SssssN$_P$, SsoH, Sdo, SssO, SaaO, SsF, SsSiH3, SssSiH2, SsssSiH, SssssSi, SsPH2, SssPH, SsssP, SdsssP, SssssssP, SsSH, SdS, SssS, SaaS, SdssS, SddssS, SssssssS, SsCl, SsGeH3, SssGeH2, SsssGeH, SssssGe, SsAsH2, SssAsH, SsssAs, SdsssAs, SssssssAs, SsSeH, SdSe, SssSe, SaaSe, SdssSe, SddssSe, SsBr, SsSnH3, SssSnH2, SsssSnH, SssssSn, SsI, SsPbH3, SssPbH2, SsssPbH, SssssPb

Geometrical/Shape (3-D)

kp0	Kappa zero
kpl–kp3	Kappa simple indices
kal–ka3	Kappa alpha indices
V_w	Van der Waals volume
$^{3\text{-}D}W_H$	3-D Wiener number based on the hydrogen-filled grometric distance matrix
$^{3\text{-}D}W$	3-D Wiener number based on the hydrogen-suppressed geometric distance matrix

Quantum Chemical

E_{HOMO}	Energy of the highest occupied molecular orbital
E_{HOMO-1}	Energy of the second highest occupied molecular orbital
E_{LUMO}	Energy of the lowest unoccupied molecular orbital
E_{LUMO+1}	Energy of the second lowest unoccupied molecular orbital
ΔH_f	Heat of formation
μ	Dipole moment

Abbreviations: IC, Information Content; 3-D, three-dimensional.
Source: From Ref. 32.

level of mRNA gene transcripts (the transcriptome), is poorly correlated with the corresponding protein patterns (44,45). Consequently, it has been suggested that protein profiling of expressed genes in cells and tissues will lead to a better understanding of cellular phenomena and provide molecular mechanisms of action of drugs, xenobiotics, and various disease processes (46). The number of functionally unique proteins in the human proteome expressed from the different types of cells, each with nearly 30,000 genes, could be 100,000. The number could easily approach about a million if we take into account the multiple forms of each protein produced via mRNA splicing and co- and post-translational modification. The branch of proteomics encompasses 2-D gel electrophoresis (2DE), mass spectrometry, and bioinformatics tools for the separation and purification of proteins and their function.

The 2DE method of proteomics is now capable of detecting and characterizing a few thousand proteins from a cell, tissue, or animal. One can then study the effects of well-designed structural or mechanistic classes of chemicals on animals or specialized cells and use these proteomics data to classify the molecules or predict their biological action. But there is a problem: With 1000 to 2000 protein spots per gel, how do we make sense of the complex pattern of proteins? Our research group has attacked this problem through the formulation of novel mathematical biodescriptors, applying the techniques of discrete mathematics to proteomics maps. Described below are four classes of techniques developed by our research team at the Natural Resources Research Institute and collaborators for the quantitative calculation of biodescriptors of proteomics maps; unfortunately, not much data are available to test the utility of such approaches exhaustively.

The Spectrum-Like Approach

In each 2-D gel, the proteins are separated by charge and mass. Also associated with each protein spot is a value representing abundance, which quantifies the amount of that particular protein or another closely related class of proteins gathered in one spot. Mathematically, the data generated by 2DE may be looked upon as points in a 3-D space, with the axes described by charge, mass, and spot abundance. One can then have projections of the data to three planes, i.e., XY, YZ, and XZ. The spectrum-like data so derived can be converted into vectors, and similarity of proteomics maps can be computed from these map descriptors (38). We have calculated the similarity or dissimilarity of toxicants based on the set of spectrum-like biodescriptors (38).

The Graph Invariant Approach

In this approach, different types of graphs are associated with the proteomics maps, and descriptors calculated on the structural characteristics of such graphs are used as "biodescriptors" to compare maps derived from normal and treated cells or animals (33–36,39,41–43).

The Spot Variable Approach

In this method, the individual spots are looked upon as independent variables, and statistical methods are applied to find which of the spots are related to the particular situation, e.g., cellular stress, disease, or effects of drugs or xenobiotics on biological systems. A preliminary statistical analysis of 1401 spots derived from exposing primary hepatocytes to 14 halocarbon toxicants led to the arrangement of the spots in decreasing order of ability to discriminate among the toxicants (Hawkins DM, Basak SC, Witzmann F, et al. Unpublished Work, 2004). The most important spots so derived can be used as biodescriptors in predictive toxicology or to generate toxicologically relevant hypotheses regarding the modes of action of toxicants.

The Mass Distribution Approach

A proteomics map may be looked upon as a pattern of protein mass distributed over a 2-D space. The distribution may vary depending on the functional state of the cell under various developmental and pathological conditions as well as under the influence of exogenous toxicants, drugs, or xenobiotics. One indicator of the distribution is the center of mass. Preliminary work has been initiated in this area (Basak SC, Guo X, Zhang F, et al. Unpublished Work, 2004). Further research is needed to test the utility of this approach in the characterization of proteomics maps.

All four types of methods discussed above provide quantitative biodescriptors based on mutually different viewpoints. Currently, we are carrying out research to find the relative utility of these different classes of biodescriptors in predictive toxicology.

STATISTICAL METHODS

In the studies reported in this chapter, three linear regression methodologies were used comparatively to relate structural descriptors to the property, activity, or toxicity of interest. Ridge regression (RR) (47), principal components regression (PCR) (48), and partial least squares (PLS) (49) are appropriate modeling methods that can be used in situations where the number of independent variables (descriptors) exceeds the number of observations (chemicals), i.e., rank deficient data sets, and when the independent variables are highly intercorrelated. Each of these regression methods makes use of the entire pool of available descriptors in modeling. In contrast, subsetting is commonly seen in QSAR studies, where the number of descriptors is reduced via some variable selection technique prior to modeling. However, subsetting has been found to be inferior to retaining all of the available descriptors and dealing with the problem of rank deficiency in some other way, such as utilizing alternative regression techniques including RR, PCR, or PLS (49). In some of our studies, we have compared our structure-based models with property-based models. With respect to the latter, only one or two experimentally determined properties were used as descriptors, in which case rank deficiency does not exist, and ordinary least squares (OLS) regression was used.

RR, like PCR, transforms the descriptors to their principal components (PCs) and uses the PCs as descriptors. However, unlike PCR, RR retains all of the PCs, and "shrinks" them differentially according to their eigenvalues (47). As with PCR and RR, PLS also involves new axes in predictor space, however, they are based on both the independent and dependent variables (50,51).

For the sake of brevity, we have not reported the highly parameterized QSAR models within this chapter. Rather, we have reported summary statistics for the models, including the cross-validated R^2 and the prediction sum of squares (PRESS). The cross-validated R^2 is calculated using the leave-one-out approach wherein each compound is removed, in turn, from the data set and the regression is fitted based on the remaining $n-1$ compounds. The cross-validated R^2 mimics the

results of applying the final regression to a future compound; large values can be interpreted unequivocally and without regard to the number of compounds or descriptors,
indicating that the model will as accurately predict the activity of a compound of the
same chemical type as those used to calibrate the regression. Although some QSAR
practitioners routinely recommend partitioning the available data into training and
test sets, where the model is developed based on the training set compounds and the
activity of the test compounds is then predicted by the model; this is unnecessary and
wasteful when one is working with small data sets, and the leave-one-out cross-
validation approach should be used (52). The cross-validated R^2 is defined by

$$R_{cv}^2 = 1 - \frac{PRESS}{SSTotal} \tag{3}$$

where *SSTotal* is the total sum of squares. Unlike the conventional R^2, the cross-
validated R^2 may be negative if the model is very poor. It should be stressed that the
conventional R^2 is unreliable in assessing modeling predictability when rank deficiency
exists. In fact, the R^2 value will increase upon the addition of any descriptor, even those
that are irrelevant. In contrast, the cross-validated R^2 will decrease upon the addition
of irrelevant descriptors, thereby providing a reliable measure of model quality.

Prior to model development, any descriptor with a constant value for all compounds within a given data set was omitted. In addition, only one descriptor of any
perfectly correlated pair (i.e., $r = 1.0$), as identified by the correlation procedure of
the statistical analysis system statistical package (53), was retained. Owing to the fact
that the variable scales differ from one another by many orders of magnitude, they
were transformed by the natural logarithm prior to modeling.

APPLICATIONS IN TOXICOKINETICS AND TOXICODYNAMICS

It is well known that the ultimate toxic chemical effect is determined not only by toxicodynamics, i.e., the susceptibility of the biological target macromolecules to the
toxicant in the critical biophase, but also by the toxicokinetic profile that is determined principally by the absorption, distribution, metabolism, and excretion
(ADME) properties of a molecule. We have used our HiQSAR modeling methods
in both of these aspects of chemical toxicology.

Tissue–Air Partitioning

Biological partition coefficients (PCs) are used as inputs in physiologically based
pharmacokinetic (PBPK) models, which are used to estimate the effective dose of
toxicants in the various toxicologically relevant biological compartments. Experimental data have been used to estimate blood–air PCs with some success. But generating such data for the large numbers of chemicals to be evaluated is impractical.
Therefore, we were interested to see how far we could use our HiQSAR approach
based solely on computed molecular descriptors in the prediction of various
tissue–air partitioning behavior of diverse sets of chemicals.

Rat tissue–air PCs namely, fat–air, liver–air, and muscle–air PCs, along with
olive oil–air and saline–air PCs, for a set of diverse low-molecular-weight chemicals
were obtained experimentally by Gargas et al. (54). We developed structure-based
models for a set of 41 diverse chemicals and a subset of 26 halocarbons. The objective of this study was threefold: (a) to determine whether structure-based models

that are comparable or superior in quality to the more expensive property-based models can be developed, (b) to determine whether high-quality structure-based models can be developed for a congeneric set as well as a more diverse set of chemicals, and (c) to determine whether we can gain some mechanistic insight by comparing the descriptors that are important for partitioning between air and the various tissues.

The results indicate that the structure-based models are comparable to models in which experimental properties are used as descriptors (Table 3). The TC descriptors, alone, are capable of providing high-quality RR models, as evidenced by R^2_{cv} values of 0.939, 0.942, and 0.863 for fat–air, liver–air, and muscle–air partitioning, respectively, based on the set of 46 diverse compounds, and values of 0.972, 0.964, and 0.906 based on the set of 26 haloalkanes. Note that the addition of neither the 3-D descriptors nor the calculated log P values results in model improvement. As we have generally observed that RR outperforms PLS and PCR, we have reported only the RR results in Table 3 for the sake of brevity. For comparative purposes, model descriptors for the various tissue–air endpoints were ranked with respect to $|t|$ value, where t represents the descriptor coefficient divided by its standard error; thus descriptors with high $|t|$ values are important in predicting the property under consideration. In doing so, it was noted that descriptors of hydrogen bonding and polarity are more important in the prediction of liver–air and muscle–air partitioning than fat–air partitioning. This is not surprising in that liver and muscle tissue contain higher amounts of such polar lipids as phospholipids, gangliosides, sulfolipids, etc., as compared to the fatty tissues.

Aryl Hydrocarbon Receptor Binding Affinity

The aryl hydrocarbon (*Ah*) receptor is well documented in toxicology, with the toxicity of certain classes of persistent pollutants being guided primarily through their toxicodynamics. Dibenzofurans fall into this category. We developed HiQSAR models based on a set of 32 dibenzofurans with *Ah* receptor binding potency values obtained from the literature (Table 4) (55). Statistical results are provided in (Table 5). The TS + TC descriptors provide a high-quality RR model as evidenced by an R^2_{cv} value of 0.852. Note that the addition of the 3-D and the STO-3G ab initio QC descriptors does not result in significant model improvement. The binding affinity potency values as predicted by the TS + TC RR model are also provided in Table 4, along with the differences between the experimental and predicted values.

CELL LEVEL TOXICITY ESTIMATION

The ultimate goal of predictive toxicology is the estimation of potential hazard of myriads of chemicals to human and ecological health, and experimental model systems at various levels of biological organization have been used for this purpose. Unicellular organisms and cultured prokaryotic as well as eukaryotic cells have been extensively used in toxicity estimation. Even though they are less expensive and less time consuming than whole animal bioassays, they cannot always be tested in the lab. Rapid, real-time toxicity estimation methods for determination of such toxicity end points would be desirable. Therefore, we applied our HiQSAR approach in estimating cellular toxicity of diverse sets of molecules.

Table 3 Summary Statistics of Predictive Models for Rat Fat–Air, Liver–Air, and Muscle–Air Partition Coefficient Based on Experimental Properties and Theoretical Structural Descriptors

| | R^2_{cv} | | | | | |
| | Fat–air | | Liver–air | | Muscle–air | |
Independent variables	RR	OLS	RR	OLS	RR	OLS
46 Diverse Chemicals						
Structural descriptors						
TS	0.325		0.226		0.203	
TS+TC	0.947		0.917		0.849	
TS+TC+3-D	0.943		0.916		0.855	
TS+TC+3-D+log P^a	0.941		0.920		0.855	
TS	0.325		0.226		0.203	
TC	0.939		0.942		0.863	
3-D	0.253		0.165		0.089	
Properties						
log $P_\text{olive oil–air}$		0.927				
log $P_\text{olive oil–air}$+log $P_\text{saline–air}$		0.932		0.894		0.891
26 Haloalkanes						
Structural descriptors						
TS	0.143		0.029		−0.014	
TS+TC	0.971		0.954		0.902	
TS+TC+3-D	0.969		0.950		0.900	
TS+TC+3-D+log P^a	0.969		0.957		0.899	
TS	0.143		0.029		−0.014	
TC	0.972		0.964		0.906	
3-D	0.926		0.826		0.820	
Properties						
log $P_\text{olive oil–air}$		0.960				
log $P_\text{olive oil–air}$ + log $P_\text{saline–air}$		0.958		0.851		0.876

[a] Calculated log $P_{n\text{-octanol:water}}$ (www.logP.com).

Abbreviations: RR, ridge regression; OLS, ordinary least squares; TS, topostructural; TC, topochemical; 3-D, three-dimensional.

Table 4 Experimental and Cross-Validated Predicted *Ah* Receptor Binding Potency, Based on the TS + TC RR Model

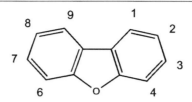

No.	Chemical	Experimental pEC$_{50}$	Predicted pEC$_{50}$	Exp—Pred
1	2-Cl	3.553	3.169	0.384
2	3-Cl	4.377	4.199	0.178
3	4-Cl	3.000	3.692	−0.692
4	2,3-diCl	5.326	4.964	0.362
5	2,6-diCl	3.609	4.279	−0.670
6	2,8-diCl	3.590	4.251	−0.661
7	1,2,7,-trCl	6.347	5.646	0.701
8	1,3,6-trCl	5.357	4.705	0.652
9	1,3,8-trCl	4.071	5.330	−1.259
10	2,3,8-trCl	6.000	6.394	−0.394
11	1,2,3,6-teCl	6.456	6.480	−0.024
12	1,2,3,7-teCl	6.959	7.066	−0.107
13	1,2,4,8-teCl	5.000	4.715	0.285
14	2,3,4,8-teCl	6.456	7.321	−0.865
15	2,3,4,6-teCl	7.602	7.496	0.106
16	2,3,4,8-teCl	6.699	6.976	−0.277
17	2,3,6,8-teCl	6.658	6.008	0.650
18	2,3,7,8-teCl	7.387	7.139	0.248
19	1,2,3,7,8-peCl	7.128	7.213	−0.085
20	1,2,3,7,8-peCl	7.128	7.213	−0.085
21	1,2,3,7,9-peCl	6.398	5.724	0.674
22	1,2,4,6,7-peCl	7.169	6.135	1.035
23	1,2,4,7,8-peCl	5.886	6.607	−0.720
24	1,2,4,7,9-peCl	4.699	4.937	−0.238
25	1,3,4,7,8-peCl	6.999	4.937	0.186
26	2,3,4,7,8-peCl	7.824	7.479	0.345
27	2,3,4,7,9-peCl	6.699	6.509	0.190
28	1,2,3,4,7,8-heCl	6.638	6.802	−0.164
29	1,2,3,6,7,8-heCl	6.569	7.124	−0.555
30	1,2,3,6,7,8-heCl	5.081	5.672	−0.591
31	2,3,4,6,7,8-heCl	7.328	7.019	0.309
32	Dibenzofuran	3.000	2.765	0.235

Abbreviations: *Ah*, aryl hydrocarbon; TS, topostructural; TC, topochemical; RR, ridge regression.
Source: From Ref. 5.

Mutagenicity

Mutagenicity is an important endpoint for the assessment of potential genotoxicity of drugs and xenobiotics. The Ames test using *Salmonella typhimurium* is an important test in the battery of genotoxicity assay. Therefore, we have developed HiQSAR models based on purely calculated molecular descriptors for different congeneric and structurally diverse sets of mutagens (9–11,56–58). In one study, a diverse set of 508 chemicals classified as mutagen or nonmutagen was taken from the CRC Handbook of Identified Carcinogens and Noncarcinogens (58). Only those compounds having a

Table 5 Summary Statistics for Predictive *Ah* Receptor Binding Affinity Models

Model type	RR		PCR		PLS	
	R^2_{cv}	*PRESS*	R^2_{cv}	*PRESS*	R^2_{cv}	*PRESS*
TS	0.731	16.9	0.690	19.4	0.701	18.7
TS + TC	0.852	9.27	0.683	19.9	0.836	10.3
TS + TC + 3-D + STO-3G	0.862	8.62	0.595	25.4	0.862	8.67
TS	0.731	16.9	0.690	19.4	0.701	18.7
TC	0.820	11.3	0.694	19.1	0.749	15.7
3-D	0.508	30.8	0.523	29.9	0.419	36.4
STO-3G	0.544	28.6	0.458	33.9	0.501	31.3

Abbreviations: A*h*, aryl hydrocarbon; RR, rigid regression; PCR, principal components regression; PLS, partial least squares; R^2_{cv}, cross–validated R^2; *PRESS*, prediction sum of squares; TS, topostructural; TC, topochemical, 3-D, three dimensional.
Source: From Ref. 5.

positive or negative response to the Ames mutagenicity test were selected. Of the 508 compounds, 256 were identified as mutagens and 252 were identified as nonmutagens. The diversity of the chemicals in this database is apparent from (Table 6). Dependent variable values of 1 and 0 were assigned to mutagens and nonmutagens, respectively. Ridge linear discriminant analysis (59) was used to develop classification models for (a) the entire diverse set of 508 chemicals and (b) three relatively homogenous subsets,

Table 6 Major Chemical Class (Not Mutually Exclusive) Within the Mutagen/Nonmutagen Database

Chemical class	Number of compounds
Aliphatic alkanes, alkenes, alkynes	124
Monocyclic compounds	260
Monocyclic carbocycles	186
Monocyclic heterocycles	74
Polycyclic compounds	192
Polycyclic carbocycles	119
Polycyclic heterocycles	73
Nitro compounds	47
Nitroso compounds	30
Alkyl halides	55
Alcohols, thiols	93
Ethers, sulfides	38
Ketones, ketenes, imines, quinines	39
Carboxylic acids, peroxy acids	34
Esters, lactones	34
Amides, imides, lactams	36
Carbamates, ureas, thioureas, guanidines	41
Amines, hydroxylamines	143
Hydrazines, hydrazides, hydrazones, traizines	55
Oxygenated sulfur and phosphorus	53
Epoxides, peroxides, aziridines	25

Source: From Ref. 10.

Table 7 Summary of Mutagen/Nonmutagen Classification Results Based on Ridge Linear Discriminant Analysis

Model type	Cross-validated correct classification rate (%)			
	Diverse set ($n = 508$)	Monocycles ($n = 260$)	Polycycles ($n = 192$)	Aliphatics ($n = 124$)
TS	66.5	65.8	67.7	65.3
TS + TC	74.6	74.2	72.4	74.2
TS + TC + 3-D	74.4	74.6	74.4	75.0
TS + TC + 3-D + AM1	76.0	74.2	70.8	74.2
TS	66.5	65.8	67.7	65.3
TC	73.4	73.5	72.4	71.8
3-D	59.1	58.5	69.8	62.9
AM1	63.2	65.0	66.2	56.5

Abbreviations: TS, topostructural; TC, topochemical; 3-D, three-dimensional; AM1, Austin model 1.
Source: From Ref. 10.

utilizing the hierarchical descriptor classes (10). Note that the cross-validated correct classification rates for the diverse set of 508 compounds are comparable to those obtained for the relatively homogeneous subsets, indicating that our diverse collection of theoretical descriptors is capable of representing diverse structural features (Table 7). In addition, we again find that adding the more complex descriptors to the topological descriptors does not result in significant model improvement.

Cellular Toxicity of Halocarbons

Halocarbons are important industrial chemicals being used worldwide as solvents and synthetic intermediates. Crebelli et al. (60,61) and Benigni et al. (62) have developed data for chromosomal malsegregation, lethality, and mitotic growth arrest in *Aspergillus nidulans* diploid strain P1. They have also conducted experimental analysis of these chemicals, and developed predictive QSAR models using a combination of physicochemical properties and QC indices calculated using the STO-3G basis set.

We have carried out HiQSAR model development using TS, TC, and 3-D descriptors, in addition to Austin model 1 (AM1) semiempirical QC descriptors obtained using MOPAC 6.00 (22) and ab initio QC descriptors calculated with Gaussian 98W (23) using the STO-3G, 6–311G, and aug-cc-pVTZ basis sets, and the results are provided in Table 8.

The results show that, for the set of 55 halocarbons, a very high level of ab initio calculation was required before there was any significant improvement in model quality over and above the models derived from easily calculable TS and TC descriptors. We have also formulated HiQSARs for toxicity of hepatocytes tested in vitro for a subset of 20 of these chemicals (63).

MOLECULAR SIMILARITY AND PREDICTIVE TOXICOLOGY

Molecular similarity can be useful in predicting toxicity when QSAR fails. Such situations may arise when the molecules under consideration are very complex, or

Table 8 HiQSAR Model Results for Toxicity of the 55 Halocarbons to *Aspergillus nidulans*

Model	RR		PCR		PLS	
	R^2_{cv}	PRESS	R^2_{cv}	PRESS	R^2_{cv}	PRESS
TS	0.290	90.00	0.240	96.38	0.285	90.64
TS + TC	0.770	29.13	0.426	72.84	0.644	45.13
TS + TC + 3-D	0.780	27.87	0.438	71.23	0.645	44.98
TS + TC + 3-D + AM1	0.775	28.49	0.492	64.37	0.753	21.29
TS + TC + 3-D + STO-3G	0.772	28.95	0.489	64.78	0.613	49.02
TS + TC + 3-D + 6-311G	0.777	28.26	0.510	62.14	0.631	46.75
TS + TC + 3-D + cc-pVTZ	0.838	20.59	0.507	62.49	0.821	22.67

Abbreviations: RR, ridge regression; PCR, principal components regression; PLS, partial least squares; R^2_{cv}, cross-validated R^2; *PRESS*, prediction sum of squares; TS, topostructural; TC, topochemical; 3-D, three-dimensional; AM1, Austin model 1.

the data set contains diverse, noncongeneric chemicals. Analogs of a chemical of interest can be used in predicting its property or toxicity. Most molecular similarity methods are user defined, i.e., parameters used for computing intermolecular similarity are not objective. Quantitative molecular similarity analysis (QMSA) methods based on computed molecular descriptors as well as experimental properties have been used by our research group in analog selection and property estimation, with reasonable results (64–68). More recently, we have developed the concept of "tailored similarity," where the structure space is chosen based on the specific property to be predicted. In contrast to the arbitrary similarity method, the tailored similarity space will vary from property to property, even for the same set of chemicals. Our limited studies with tailored QMSA methods show that they outperform the arbitrary QMSA techniques (69,70). We have also used similarity spaces to cluster large sets of structures such as the components of jet petroleum 8 (JP8) jet fuel into smaller subsets for toxicological evaluation. (Basak SC, Gute BD, Mills D, Unpublished Work, 2002).

DISCUSSION

The major objective of this chapter was to review our research in the formulation of QSTR models for predicting toxicity of chemicals. In view of the fact that the majority of candidate chemicals to be evaluated do not have adequate laboratory test data for their hazard assessment, it is desirable that we use computed structural descriptors for toxicity estimation. Results of our QSTR studies with toxicokinetics show that a combination of calculated TS and TC descriptors give good quality models comparable with those formulated using physicochemical properties (Table 3). One interesting aspect of our work on QSTR for tissue–air partitioning is that the most important parameters picked up by the RR method fit well with our understanding of the biochemical composition of the various tissues; hydrogen bonding parameters are less important for the prediction of fat–air partitioning as compared with muscle–air or liver–air partitioning. It is well known that fatty adipose tissue contains more hydrophobic lipids than does muscle or liver. A perusal of the toxicodynamics models at the level of enzyme and receptor (Table 5), and toxicity models for both prokaryotic and eukaryotic cells (Tables 7 and 8), show that a combination of TS and TC descriptors is capable of producing predictive models of reasonable quality. HiQSAR for the cellular toxicity of halocarbons revealed that the addition of 3-D or QC (semiempirical or ab initio) descriptors resulted in either little or marginal improvement in the predictive power of the models. This is good news for computational toxicology because TS and TC descriptors can be calculated very quickly and inexpensively. An inspection of the descriptors included in the TS and TC classes shows that each group contains a broad spectrum of parameters that quantify slightly different aspects of molecular structure. It is tempting to speculate that this diversity is the reason why the TS + TC combination has been very successful in quantitative structure–activity, –property, and –toxicity relationship studies (13,71), with structural and mechanistic diversity requiring a diversity of descriptors. In the post-genomic era, relevant genomic sequence and proteomics information can be condensed into descriptors and used in QSTR model development. We have done some preliminary studies in the development of biodescriptors and their use, along with chemodescriptors, in QSTRs. Analog selection of toxicants by way of computational molecular similarity methods is routinely used in the hazard assessment of pollutants (72). We have developed user-defined (arbitrary) and property-specific (tailored) molecular similarity methods for toxicological evaluation of chemicals.

As shown in (Fig. 2), there are many alternative pathways to the toxicological evaluation of chemicals. Starting with a chemical set C, one can (a) conduct laboratory toxicity testing (α), (b) carry out toxicity estimation from toxicologically relevant properties P measured in the lab ($\theta_1\theta_2$), (c) estimate toxicity from chemodescriptors that are calculated directly from structure ($\beta_1\beta_2$), and (d) finally, in the postgenomic era, estimate toxicity from biodescriptors that are extracted from DNA sequence,

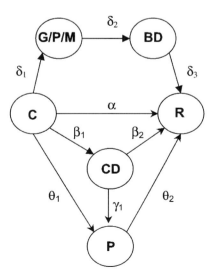

Figure 2 Structure–activity relationship (SAR) and property–activity relationship (PAR) pathways. *Abbreviations*: BDs, biodescriptors; C, chemicals; CDs, chemical descriptors; G/P/M, genomics, proteomics, or metabolomics data; P, properties; R, real numbers.

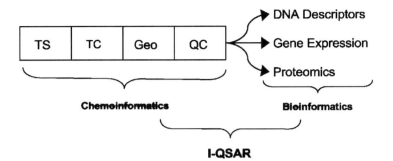

Figure 3 Integrated QSAR, combining chemodescriptors and biodescriptors. *Abbreviations*: 3-D, three-dimensional; TC, topochemical; TS, topostructural; QC, quantum chemical; QSAR, quantitative structure–activity relationship; I-QSAR, integrated approach.

genomics, or proteomics data ($\delta_1\delta_2\delta_3$). Our experience with biodescriptors, and 2-D gel proteomics–based biodescriptors in particular, shows that such descriptors are capable of discriminating among molecules that are closely related either biochemically or structurally. We also found that a combination of chemodescriptors and biodescriptors gives better predictive models as compared with either class of descriptors alone. A perusal of the history of QSAR would show that the field started with models that were based primarily on experimental data, e.g., the LFER approach. Later on, the emphasis shifted to the use of calculated descriptors, and this change served us well because LFER parameters are not available for all structures within structurally or mechanistically diverse and large data sets. Now, genomics, proteomics, and metabolomics technologies are providing us information, which might not always be revealed via the chemical structure–driven approach of chemoinformatics. Therefore, we feel that the field of computational toxicology needs a shift in its paradigm (73). Instead of developing models based on chemodescriptors derived from molecular structure (chemoinformatics) or biodescriptors derived from genomics or proteomics (bioinformatics) alone, we need to develop an integrated approach (I-QSAR) with a combination of both classes of descriptors as shown in (Fig. 3). More detailed studies are needed to test the validity of this conjecture.

ACKNOWLEDGMENTS

This chapter is publication number 374 from the Center for Water and the Environment of the Natural Resources Research Institute. The research reported herein was supported by the Air Force Office of Scientific Research, Air Force Material Command, USAF, under grant number F49620-02-1-0138, in addition to Cooperative Agreement 572112 from the Agency for Toxic Substances and Disease Registry. The U.S. Government is authorized to reproduce and distribute reprints for Governmental purposes notwithstanding any copyright notation thereon. The views and conclusions contained herein are those of the authors and should not be interpreted as necessarily representing the official policies or endorsements, either expressed or implied, of the Air Force Office of Scientific Research or the U.S. Government.

REFERENCES

1. Cronin MTD, Livingstone DJ. Predicting Chemical Toxicity and Fate. Boca Raton, FL: CRC Press, 2004.
2. Basak SC, Mills D, El-Masri HA, et al. Predicting blood:air partition coefficients using theoretical molecular descriptors. Environ Toxicol Pharmacol 2004; 16:45–55.
3. Basak S, Mills D, Hawkins DM, El-Masri HA. Prediction of human blood:air partition coefficient: a comparison of structure-based and property-based methods. Risk Anal 2003; 23(6):1173–1184.
4. Basak SC, Mills D, Hawkins DM, El-Masri HA. Prediction of tissue:air partition coefficients: a comparison of structure-based and property-based methods. SAR QSAR Environ Res 2002; 3:649–665.
5. Basak SC, Mills D, Mumtaz MM, et al. Use of topological indices in predicting aryl hydrocarbon receptor binding potency of dibenzofurans: a hierarchical QSAR approach. Indian J Chem 2003; 42A:1385–1391.
6. Basak SC, Gute BD, Drewes LR. Predicting blood–brain transport of drugs: a computational approach. Pharm Res 1996; 13:775–778.
7. Basak SC, Gute BD. Use of graph theoretic parameters in predicting inhibition of microsomal hydroxylation of anilines by alcohols: a molecular similarity approach. In: Johnson BL, Xintaras C, Andrews JS, eds. Proceedings of the International Congress on Hazardous Waste: Impact on Human and Ecological Health. Princeton: Scientific Publishing Co Inc., 1997:492–504.
8. Gute BD, Grunwald GD, Basak SC. Prediction of the dermal penetration of polycyclic aromatic hydrocarbons (PAHs): a hierarchical QSAR approach. SAR QSAR Environ Res 1999; 10:1–15.
9. Basak SC, Gute BD, Grunwald GD. Assessment of the mutagenicity of aromatic amines from theoretical structural parameters: a hierarchical approach. SAR QSAR Environ Res 1999; 10:117–129.
10. Basak SC, Mills D, Gute BD, et al. Predicting mutagenicity of congeneric and diverse sets of chemicals using computed molecular descriptors: a hierarchical approach. In: Benigni R, ed. Quantitative Structure-Activity Relationship (QSAR) Models of Mutagens and Carcinogens. Boca Raton, FL: CRC Press, 2003:207–234.
11. Basak SC, Mills DR, Balaban AT, Gute BD. Prediction of mutagenicity of aromatic and heteroaromatic amines from structure: a hierarchical QSAR approach. J Chem Inf Comput Sci 2001; 41:671–678.
12. Basak SC, Bertelsen S, Grunwald GD. Use of graph theoretic parameters in risk assessment of chemicals. Toxicol Lett 1995; 79:239–250.
13. Basak SC, Mills D, Gute BD, et al. Applications of topological indices in property/bioactivity/toxicity prediction of chemicals. In: Rouvray DH, King RB, eds. Topology in Chemistry: Discrete Mathematics of Molecules. Chichester, England: Horwood Publishing Limited, 2002:113–184.
14. Basak SC, Grunwald GD, Niemi GJ. Use of graph-theoretic and geometrical molecular descriptors in structure-activity relationships. In: Balaban AT, ed. From Chemical Topology to Three-dimensional Geometry. New York: Plenum Press, 1997:73–116.
15. Poulin P, Krishnan K. A mechanistic algorithm for predicting blood:air partition coefficients of organic chemicals with the consideration of reversible binding in hemoglobin. Toxicol Appl Pharmacol 1996; 136:131–137.
16. United States Environmental Protection Agency. What is the TSCA Chemical Substance Inventory? http://www.epa.gov/opptintr/newchems/invntory.htm, accessed 9/30/2004.
17. Basak SC, Niemi GJ, Veith GD. Optimal characterization of structure for prediction of properties. J Math Chem 1990; 4:185–205.
18. Basak SC, Niemi GJ, Veith GD. Predicting properties of molecules using graph invariants. J Math Chem 1991; 7:243–272.

19. Basak SC, Harriss DK, Magnuson VR. POLLY, Version 2.3. Copyright of the University of Minnesota, 1988.

20. Filip PA, Balaban TS, Balaban AT. A new approach for devising local graph invariants: derived topological indices with low degeneracy and good correlational ability. J Math Chem 1987; 1:61–83.

21. Hall Associates Consulting. Molconn-Z Version 3.5, Quincy, MA, 2000.

22. Stewart JJP. MOPAC Version 6.00, QCPE #455. Frank J Seiler Research Laboratory, US Air Force Academy, Colorado, 1990.

23. Gaussian 98W (Revision A.11.2). Pittsburgh, PA: Gaussian, Inc., 1998.

24. Kier LB, Murray WJ, Randic M, Hall LH. Molecular connectivity V. Connectivity series concept applied to diversity. J Pharm Sci 1976; 65:1226–1230.

25. Randic M. On characterization of molecular branching. J Am Chem Soc 1975; 97: 6609–6615.

26. Kier LB, Hall LH. Molecular Connectivity in Structure-Activity Analysis. Letchworth, Hertfordshire, U.K.: Research Studies Press, 1986.

27. Balaban AT. Highly discriminating distance-based topological indices. Chem Phys Lett 1982; 89:399–404.

28. Raychaudhury C, Ray SK, Ghosh JJ, et al. Discrimination of isomeric structures using information theoretic topological indices. J Comput Chem 1984; 5:581–588.

29. Bonchev D. Information theoretic indices for characterization of chemical structures. Letchworth, Hertfordshire, U.K.: Research Studies Press, 1983.

30. Roy AB, Basak SC, Harriss DK, et al. Neighborhood complexities and symmetry of chemical graphs and their biological applications. In: Avula XJR, Kalman RE, Liapis AI, Rodin EY, eds. Mathematical Modelling in Science and Technology. Pergamon Press, 1983:745–750.

31. Kier LB, Hall LH, Frazer JW. An index of electrotopological state for atoms in molecules. J Math Chem 1991; 7:229–241.

32. Basak SC, Mills D. Prediction of mutagenicity utilizing a hierarchical QSAR approach. SAR QSAR Environ Res 2001; 12(6):481–496.

33. Randic M, Zupan J, Novic M, Gute BD, Basak SC. Novel matrix invariants for characterization of changes of proteomics maps. SAR QSAR Environ Res 2002; 13:689–703.

34. Bajzer Z, Randic M, Plavsic D, Basak SC. Novel map descriptors for characterization of toxic effects in proteomics maps. J Mol Graph Model 2003; 22:1–9.

35. Randic M, Witzmann F, Vracko M, et al. On characterization of proteomics maps and chemically induced changes in proteomes using matrix invariants: application to peroxisome proliferators. Med Chem Res 2001; 10:456–479.

36. Randic M, Lers N, Plavsic D, et al. On invariants of a 2-D proteome map derived from neighborhood graphs. J Proteom Res 2004; 3:778–785.

37. Vracko M, Basak SC, Geiss K, et al. Proteomics maps-toxicity relationship of halocarbons studied with similarity index and genetic algorithm. J Chem Inf Model, in press.

38. Vracko M, Basak SC. Similarity study of proteomic maps. Chemometr Intell Lab Syst 2004; 70(1):33–38.

39. Randic M, Basak SC. A comparative study of proteomics maps using graph theoretical biodescriptors. J Chem Inf Comput Sci 2002; 42:983–992.

40. Nandy A, Basak SC. A simple numerical descriptor for quantifying effect of toxic substances on DNA sequences. J Chem Inf Comput Sci 2000; 40:915–919.

41. Randic M, Basak SC. Characterization of DNA primary sequences based on the average distance between bases. J Chem Inf Comput Sci 2001; 41(3):561–568.

42. Randic M, Guo X, Basak SC. On the characterization of DNA primary sequences by triplet of nucleic acid bases. J Chem Inf Comput Sci 2001; 41(3):619–626.

43. Randic M, Vracko M, Nandy A, et al. On 3-D graphical representation of DNA primary sequences and their numerical characterization. J Chem Inf Comput Sci 2000; 40: 1235–1244.

44. Witzmann FA, Li J. Cutting-edge technology II. Proteomics: core technologies and applications in physiology. Am J Physiol Gastrointest Liver Physiol 2002; 282:G735–G741.

45. Witzmann FA. Proteomic applications in toxicology. In: Vanden Heuvel JP, Perdew GH, Mattes WB, Greenlee WF, eds. Comprehensive Toxicology. Elsevier Science BV, 2002:539–558.

46. VanBogelen RA, Schiller EE, Thomas JD, et al. Diagnosis of cellular states of microbial organisms using proteomics. Electrophoresis 1999; 20:2149–2159.

47. Hoerl AE, Kennard RW. Ridge regression: biased estimation for nonorthogonal problems. Technometrics 1970; 8:27–51.

48. Massy WF. Principal components regression in exploratory statistical research. J Am Stat Assoc 1965; 60:234–246.

49. Frank IE, Friedman JH. A statistical view of some chemometrics regression tools. Technometrics 1993; 35(2):109–135.

50. Hoskuldsson A. A combined theory for PCA and PLS. J Chemometrics 1995; 9:91–123.

51. Hoskuldsson A. PLS regression methods. J Chemometrics 1988; 2:211–228.

52. Hawkins DM, Basak SC, Mills D. Assessing model fit by cross-validation. J Chem Inf Comput Sci 2003; 43:579–586.

53. SAS Institute Inc. SAS/STAT User Guide Release 6.03 Edition. Cary, NC, 1988.

54. Gargas ML, Burgess RJ, Voisard DE, et al. Partition coefficients of low molecular weight volatile chemicals in various tissues and liquids. Toxicol Appl Pharmacol 1989; 98:87–99.

55. So SS, Karplus M. Three-dimensional quantitative structure-activity relationships from molecular similarity matrices and genetic neural networks. 2. Applications. J Med Chem 1997; 40:4360–4371.

56. Basak SC, Frane CM, Rosen ME, et al. Molecular topology and mutagenicity: a QSAR study of nitrosamines. IRCS Med Sci 1986; 14:848–849.

57. Vracko M, Mills D, Basak SC. Structure-mutagenicity modeling using counter propagation neural networks. Environ Toxicol Pharmacol 2004; 16:25–36.

58. Soderman JV. CRC Handbook of Identified Carcinogens and Noncarcinogens: Carcinogenicity-Mutagenicity Database. Vol. I. Boca Raton, FL: CRC Press, 1982.

59. Campbell NA. Shrunken estimators in discriminant and canonical variate analysis. Appl Stat 1980; 29:5–14.

60. Crebelli R, Andreoli C, Carere A, et al. The induction of mitotic chromosome malsegregation in Aspergillus nidulans. Quantitative structure activity relationship (QSAR) analysis with chlorinated aliphatic hydrocarbons. Mutat Res 1992; 266:117–134.

61. Crebelli R, Andreoli C, Carere A, et al. Toxicology of halogenated aliphatic hydrocarbons: structural and molecular determinants for the disturbance of chromosome segregation and the induction of lipid peroxidation. Chem Biol Interact 1995; 98:113–129.

62. Benigni R, Andreoli C, Conti L, et al. Quantitative structure–activity relationship models correctly predict the toxic and aneuploidizing properties of halogenated methanes in *Aspergillus nidualans*. Mutagenesis 1993; 8:301–305.

63. Gute BD, Balasubramanian K, Geiss K, et al. Prediction of halocarbon toxicity from structure: a hierarchical QSAR approach. Environ Toxicol Pharmacol 2004; 16:121–129.

64. Basak SC, Gute BD, Grunwald GD. Use of graph invariants in QMSA and predictive toxicology, DIMACS Series 51. In: Hansen P, Fowler P, Zheng M, eds. Discrete Mathematical Chemistry. Providence, Rhode Island: American Mathematical Society, 2000:9–24.

65. Basak SC, Gute BD, Grunwald GD. Characterization of the molecular similarity of chemicals using topological invariants. In: Carbo-Dorca R, Mezey PG, eds. Advances in Molecular Similarity Vol. 2. Stanford, Connecticut: JAI Press, 1998:171–185.

66. Basak SC, Magnuson VR, Niemi GJ, et al. Determining structural similarity of chemicals using graph-theoretic indices. Discrete Appl Math 1988; 19:17–44.

67. Basak SC, Grunwald GD. Molecular similarity and risk assessment: analog selection and property estimation using graph invariants. SAR QSAR Environ Res 1994; 2:289–307.

68. Basak SC, Grunwald GD. Molecular similarity and estimation of molecular properties. J Chem Inf Comput Sci 1995; 35:366–372.

69. Basak SC, Gute BD, Mills D. Quantitative molecular similarity analysis (QMSA) methods for property estimation: a comparison of property-based arbitrary and tailored similarity spaces. SAR QSAR Environ Res 2002; 13:727–742.
70. Basak SC, Gute BD, Mills D, et al. Quantitative molecular similarity methods in the property/toxicity estimation of chemicals: a comparison of arbitrary versus tailored similarity spaces. J Mol Struct (Theochem) 2003; 622(1–2):127–145.
71. Basak SC, Mills D, Gute BD. Prediciting bioactivity and toxicity of chemicals from mathematical descriptors: a chemical-cum-biochemical approach. In: Kelin DJ, Brandas E, eds. Advances in Quantum Chemistry, Elsevier, in press.
72. Auer CM, Nabholz JV, Baetcke KP. Mode of action and the assessment of chemical hazards in the presence of limited data: use of structure-activity relationships (SAR) under TSCA section 5. Environ Health Perspect 1990; 87:183–197.
73. Kuhn TS. The Structure of Scientific Revolutions. Chicago: The University of Chicago Press, 1962.

6

Hormesis: A Key Concept in Toxicology

Edward J. Calabrese
Environmental Health Sciences, University of Massachusetts, Amherst, Massachusetts, U.S.A.

Throughout the 20th century, the field of toxicology settled into a discipline that would be generally characterized as a high dose–testing methodology that also employed very few doses. This "high dose–few doses" approach was designed to provide information on what may be a safe exposure level and what types of toxicities the agent could cause once the safe exposure level (i.e., threshold dose) was sufficiently exceeded.

This toxicological approach was built upon the assumption, indeed, strong belief, in the existence of the threshold dose–response model. This model grew out of substantial practical experience in which apparent dose–response thresholds were universally observed. Further, the threshold model was consistent with the sigmoidal nature of the dose–response. In this case, the extremes of the dose–response are approached asymptotically while there is rapid change in response in the in-between exposure levels (Fig. 1).

The belief in the dominating presence of the threshold dose–response model had widespread significance for the fields of toxicology and risk assessment. It affected hazard assessment goals that included the derivation of the no observed adverse effect level (NOAEL) (i.e., highest dose not statistically significantly different from control) and the lowest observed adverse effect level (LOAEL) (i.e., lowest dose statistically significantly different from control), the study design including the number and spacing of doses, the selection of dose–response modeling methods, and the parameters of the models and biological characteristics of the animal models especially as it related to background disease incidence.

The concept of a threshold dose–response model also affected the focus of toxicological research on the occurrence of toxic end points, as well as their characterization and quantification. For example, such biomarkers of toxicity as increase in serum enzymes, such as alanine aminotransferase and aspartate aminotransferase, became widely used to estimate liver and other organ toxicities. As central as this is to the field of toxicology and the hazard assessment process, it led to a notable under emphasis of adaptive and/or repair responses to the toxicant effects and how this may affect the final outcome of the entire dose–response continuum (1).

Despite its broad-based acceptance within toxicology, the threshold model, however, fell out of favor due to concerns about radiation-induced chronic health

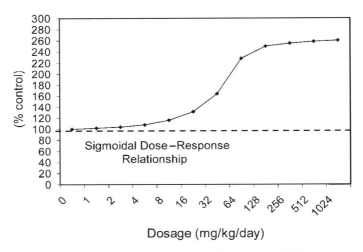

Figure 1 Schematic representation of the sigmoidal dose–response.

effects during the 1950s as the fear of radiation-induced mutation and cancer began to gain a strong societal foothold, especially in the aftermath of World War II. Such fears were reinforced by the earlier findings of Muller, who in 1928, published results showing that X-rays could induce mutations in fruit flies in a linear manner (2). These observations and others suggested that the long-held belief in the existence of thresholds for radiation effects may not be true for biological effects, such as cancer, that could theoretically be initiated by a few mutations in a single cell. Thus, by the late 1950s to early 1960s, a decided shift in the risk assessment of radiation occurred. No longer was the threshold model accepted, it was replaced by stochastic modeling that assumed a linear nonthreshold (LNT) response. This change in modeling led to a rejection that "safe" exposure levels were possible. It was replaced with the concept of "acceptable" levels of risk [i.e., one tumorous person per number of people (e.g., 10^6) over number of years (e.g., 70)]. The view that carcinogens (i.e., radioactivity) were different from noncarcinogenic agents subsequently became transferred to the assessment of chemical carcinogens. By 1977, the Safe Drinking Water Committee of the U.S. National Academy of Science (NAS) made the recommendation that chemical carcinogens should be similarly handled as radiation as far as cancer risk assessment is concerned (3). Soon after this recommendation, the Environment Protection Agency (EPA) then applied the LNT modeling concept to chemical carcinogen risk assessment. From the late 1970s to the present, the field of risk assessment has viewed carcinogens as having no threshold (4,5), with every exposure posing at least a theoretical risk even down to the immeasurable single molecule.

Risks of one cancer per million people per 70 years lifetime became a standard feature in the risk assessment jargon since the early 1980s. Doses that could cause such theoretical risks were at the forefront of what was an acceptable level of risk and were often the target zone for environmental remediation. Such levels of exposure were often deemed by affected parties as extremely low levels of exposures based on what was argued as being very conservative extrapolation models, procedures, and assumptions and very expensive to remediate and/or implement. After years of dealing with the financial challenges of low-level pollutant cleanups and the near impossibility of proving the LNT model wrong, industry, especially lead by the nuclear industry, began to explore a long-abandoned view of the dose–response by

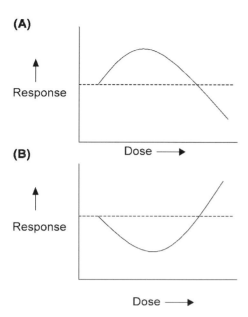

(A)

Response

Dose ⟶

(B)

Response

Dose ⟶

Figure 2 (**A**) The most common form of the hormetic dose–response curve depiciting low-dose stimulatory and high-dose inhibitory response, the β- or inverted U-shaped curve. End-points displaying this curve include growth, fecundity, and longevity. (**B**) The hormetic dose–response curve depicting low-dose reduction and high-dose enhancement of adverse effects. Endpoints displaying this curve include carcinogenesis, mutagenesis, and disease incidence.

the name of hormesis. The hormetic dose–response model asserted that the most fundamental nature of the dose–response was neither threshold nor linear, but is either an inverted U- or J-shape depending on the endpoint measured (Fig. 2). In the case of the inverted U-shaped dose–response, this is generally seen for end points such as growth, longevity, and cognitive functions. In the case of the J-shaped dose–response, this generally applies to disease incidence such as cancer, birth defects, cardiovascular disease, and others. Regardless of whether the shape of the dose–response is either an inverted U-shape or a J-shape, they are both considered hormetic dose responses having similar quantitative features with the shape simply being a function of the end point measured.

The hormetic dose–response model was potentially important to the regulated community. However, it was not because of the hormetic "bump" or "dip" (or potential health benefit) in the dose–response. Rather, it was because this dose–response model provided the best opportunity to conclude that carcinogens most likely act via thresholds under normal circumstances because it is essentially impossible to confidently distinguish between a threshold and the LNT models in animal bioassay studies with only two to four doses, which typify what is usually utilized. In such instances, where the data are consistent with both models, regulatory agencies such as EPA would almost always default to that model estimating greater risk (i.e., the LNT model) as a matter of public health policy. Hormesis offered a potential way past this revolving regulatory risk assessment door.

During this period, Luckey (6) published a book on ionizing radiation and hormesis, documenting many apparent examples of the hormesis phenomenon. However, this early work of Luckey oddly did not include any consideration of cancer—the principal risk assessment challenge for industry. Nonetheless, the work of Luckey caught the attention of Japanese scientists who then collaborated with the Electric Power Research Institute (Palo Alto, California, U.S.A.) to create the first conference on Radiation Hormesis in 1985 (Oakland, California, U.S.A.) (Conference proceedings, Health Physics, 1987). It was the "fall out" from this initial meeting that has resulted in the substantial reassessment of the nature of the dose–response in the low-dose zone and has led to the striking, and indeed, revolutionary conclusion that the toxicology community made an error of significant proportions during the middle decades of the 20th century concerning the nature of the dose–response. During this crucial period of concept consolidation (or toxicological intellectual imprinting), the threshold dose–response model became accepted while the hormesis model (as known at the time by the terms Arndt–Schulz Law or Hueppe's Rule) became fully marginalized. That is, the hormetic dose–response model that had developed a reasonably large database, was not to be included in the major textbooks on toxicology and pharmacology, was omitted from academic courses, was never the focus of a session at national conferences of major professional societies, and was excluded from consideration by regulatory agencies. This marginalization was quite successful, continuing to the present time and represents a clear form of intellectual censorship.

The credit for discovering hormesis goes to Schulz [see a historical biographical remembrance by Schulz (1923)] and his assistant Hofman, at the University Greiswald in Northern Germany, who reported that low doses of various chemical disinfectants stimulated the metabolism of yeasts at low doses while being inhibitory at higher doses (7–9). In this reflective statement of his career, Schulz recounts the excitement of his original observation of the low-dose stimulatory response, his initial thoughts that the stimulatory response was the result of a spurious background variability, and his repeated replication of the findings via additional experimentation which led him to confidently conclude that the biphasic dose–response in his experimental system was reproducible. Many similar observations were subsequently recorded by others in the late 19th and early 20th centuries (10–14). Typically, these findings were reported with toxic metals and other agents such as phenol and formaldehyde and their effects on bacteria, yeasts, and many plant species. Following the discovery of radiation in 1896, similar hormesis-like biphasic dose–response observations were reported with X-rays and gamma rays. So reliable and extensive were these biphasic dose–response findings that they became incorporated into major botanical (15–17) and microbiological (18–20) texts. In fact, the demonstration of the hormetic dose–response in microbiology became a routine laboratory exercise (20). Notable amongst early investigators publishing research on hormetic dose responses were Townsend (21) of Columbia who developed the concept of hormesis being a modest overcompensation response to a disruption in homeostasis, Branham (22) of Columbia University who replicated the findings of Schulz in a detailed and convincing dose–time response manner, Jensen (23) of Stanford University with research on plant responses to toxic metals, and C.E. Winslow of Yale University who directed a number of bacteriological Ph.D. dissertations that greatly extended the initial concepts of Schulz. Of particular interest was the work of Hotchkiss (24,25), one of Winslow's students, who demonstrated the hormetic phenomenon in bacteria for a wide range of agents (Fig. 3).

The problem with research in this era (i.e., prior to the 1930s) was that it was scattered and never effectively organized and assessed. Likewise, the magnitude of the stimulatory response was invariably modest, generally only 30% to 60% greater

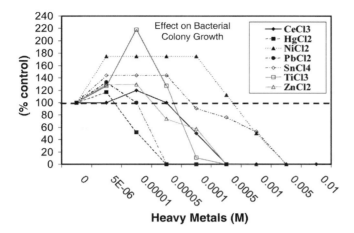

Figure 3 Hormetic-like biphasic dose–response relationships induced by heavy metals in bacteria. *Source*: From Ref. 24.

than controls at maximum, making replication of findings challenging and commercial applications tenuous at best. At the same time, there was a great interest in determining safe levels of exposure for industrial workers—chemical exposures that could kill insects and other pests—changing the focus from the low-dose stimulatory response to that of the more practical high-dose zone which demanded the derivation of LOAELs and NOAELs that would be used to derive exposure standards for workers in industry.

While such practical considerations drew the toxicological community away from the hormesis concept during the 1920s to 1930s, the hormesis concept had a focus within the biomedical community because Schulz had closely linked it to the medical practice of homeopathy. As can be discerned from Schulz's detailed biographical statement (9), this association with homeopathy occurred immediately upon his research discovery and its publication, which was claimed by Schulz to provide the underlying explanatory principle of homeopathy. Schulz made this association because homeopathy focuses on the body's adaptive response (as reflected in clinical symptoms) to the stressor agent (microbe, chemical). Because his data suggested that low doses of toxic agents enhance adaptive responses, Schulz felt that inducing adaptive responses could be used to prevent disease and/or enhance the curative process. This close association with homeopathy had a crippling effect on the scientific acceptance of the hormesis concept given the long-standing antipathies between traditional medicine and homeopathy, preventing a normal assimilation into the texture of scientific literature and concept influence.

Later significant publications by the leading pharmacologist, Clark (26), of the University of Edinburgh, continued to challenge the concept of hormesis. Such criticisms were never effectively countered by adherents of the hormesis concept even though they lacked substance and should have been far from convincing (27). Likewise, the threshold dose–response model was being consolidated within the toxicological community. Quite significant was the fact that efforts in biostatistical modeling of dose–response data by highly influential associates of Clark (i.e., J.H. Gaddum, C.I. Bliss, and R.A. Fisher) excluded the possibility of negative (i.e., below control) data to be estimated with the probit model, a model used extensively by subsequent generations of toxicologists (28). Such a decision, in effect, denied the biological possibility of hormesis because below control data was believed to be background variation. Thus, at this time the hormesis concept was being marginalized and structural steps (e.g., standard biostatistical dose–response modeling) were being taken to ensure its exclusion from meaningful assessment, dialogue, and application within the scientific and regulatory communities.

Despite the above impediments to its acceptance, hormetic responses continued to be observed and published throughout the 20th century. However, until the book of Luckey in 1981 on ionizing radiation and hormesis, there was no major effort to summarize the published literature on the topic. Plans to develop a similar book on chemical hormesis as stated in Luckey's 1981 introduction never materialized. However, despite the efforts of Luckey (6), the acceptance of hormesis as a legitimate biological concept was quite limited. The principal reasons are somewhat speculative but most likely included enormous cultural opposition that was embedded with scientific and medical bodies, the politicization of the hormesis concept, the need for a sustained and substantial organization to nurture the hormesis hypothesis especially within a generally unsupportive and potentially hostile environment, and the difficulty in observing hormetic responses in experiments using high doses and too few doses.

Nonetheless, a breakthrough was made in the period 1990s to the present with a prolonged and detailed assessment of chemical and radiation hormesis. This development included an in-depth assessment of the history of chemical and radiation hormesis and why these concepts became marginalized (10–14), the creation of a relational retrieval database with now nearly 7000 dose responses with evidence of hormesis (29–31), the creation of a second hormesis database designed to provide an estimate of the frequency of hormesis in the toxicological literature (32), numerous pharmacologically oriented papers assessing the occurrence of hormesis and the underlying mechanisms of hormetic responses (33,34) with a focus on endogenous agonists such as adenosine (35), adrenergic compounds (36), prostaglandins (37), estrogens (38), androgens (39), nitric oxide (40), opioids (41), peptides (42), and other agents. Comprehensive assessments of immune-related hormetic responses (43) and hormetic responses of human tumor cell lines (56) have also been published. Other papers on radiation and chemical carcinogenesis (44), radiation and longevity (45), apoptosis (46), alcohol (47), chemotherapeutics (48), and inorganic agents (49) have also been published. These materials comprise a massive body of evidence that provides highly reliable and documented evidence of hormesis as published in leading scientific journals. In general, this evidence reveals that hormetic dose responses are common in the scientific literature, highly generalizable, being independent of the biological model, the end point measured and the chemical or physical stressor agent studied, having a frequency in the toxicological literature of 40% using highly rigorous a priori entry and evaluative criteria (32). Furthermore, in direct head-to-head comparison with the threshold model, the hormetic model was found to be far more common in the toxicological literature (48) and in comparison to human tumor cell lines (56).

These collective findings support the conclusion that the toxicological community made a profound error during its formative years by accepting the primacy of the threshold model as its most fundamental dose–response model and by the de facto rejection of the hormetic dose–response model. This is particularly notable because it is the dose–response relationship that is the most fundamental principle and pillar of toxicology. Such an error has had enormous implications for how toxicology is conducted, how its studies are designed, how risks are assessed, how costly various types of remedial activities and the implementation of exposure standards would be, and how toxicologists are taught to think and frame questions.

Some may doubt that multiple generations of a highly technical and informed scientific field such as toxicology, as well as all other biological subdisciplines using the dose–response, could be wrong on this most central core belief. Nonetheless, not only do the data support this conclusion but also a historical and technical evaluation provides cogent explanations that most likely account for this collective interdisciplinary failing. On the technical side, the field of toxicology has been dominated by a regulatory perspective to protect worker and community health. In the risk assessment process, this starts with hazard assessment on which toxicology operates. In this realm of hazard assessment, toxicology has long been a high dose–few doses discipline with its target being NOAEL or LOAEL derivation. Use of only two to three doses placed intellectual blinders on the field, focusing its entire attention on threshold and above-threshold responses. Because hormetic responses are in the below-NOAEL domain, it was not likely that hormetic responses would be regularly encountered or expected, and when they were observed they would tend to be dismissed as normal background variation, especially because the maximum hormetic response is typically only 30% to 60% greater than the response in controls. Furthermore, Calabrese and Baldwin (50) demonstrated that a high proportion (70%) of NOAEL values in vertebrate toxicological studies are below the control value with a reasonable likelihood that a modest degree of toxicity may be present even though such responses were not statistically different from the control value at the 95% level. It was also shown that the responses of doses immediately below the NOAEL dose also tended to be below the control value, suggesting residual toxicity there as well. These observations indicate that the high dose–few doses toxicological paradigm that has guided toxicology for nearly a century has prevented, via its limited study designs, an evaluation of the entire dose–response continuum, that is, the below-NOAEL domain.

These findings indicate that the intellectual playing field of the discipline of toxicology has been co-opted by the assumptions and needs of the regulatory community in essentially all countries and in the United States by agencies such as EPA, Food and Drug Administration, and Occupational Health and Safety Administration. This co-opting of the field of toxicology has been reinforced by the powerful impact of agency–mission oriented research funding to toxicologists who require such resources to conduct research, publish findings, and achieve professional success. Likewise, independent national toxicological leadership by the U.S. NAS is principally undertaken only after it receives funding and its mission statement from U.S. federal agencies (e.g., EPA), further reasserting continued regulatory agency control over the directions of scientific thought and developments in the field of toxicology.

Although there is convincing evidence to implicate high dose–few doses toxicological testing in the marginalization of the hormesis concept, a broader historical analysis reveals that the seeds of the rejection of the hormesis concept preceded the reinforcing actions of the regulatory agencies. In fact, this rejection was most likely due to the fact that Schulz made hormesis the explanatory factor of homeopathy, thereby placing it in the bulls-eye of traditional medicine. As a result, the hormesis concept was not only placed on the defensive since its scientific birth, but it also had the ever-powerful biomedical community being trained and eager to dismiss it. For example, the major text by Clark, *Handbook of Pharmacology* (1937), devoted considerable space to the refutation of this concept. Clark's text was a dominant one for about 40 years. In the forword to his book *Toward Understanding Receptors*,

Robison (51) referred to this text of Clark's as the "now classic monograph on General Pharmacology, a book that had great influence on a number of individuals." Most importantly, it was used to educate several generations of pharmacologists, some of whom became the first generation of toxicologists, who created the governmental regulatory framework that exists today. These individuals who came from a culture that could not even consider, let alone accept, hormesis, accepted the threshold model as dominant perspective and excluded the possibility of the hormetic effects even when experimental observations were revealed.

WHAT DOES THIS MEAN TODAY?

Acceptance of the existence and dominance of hormetic dose responses has the potential to markedly change the fields of toxicology, risk assessment, and risk communication. Most significant is the fact that it refutes the LNT model, establishing that carcinogens can act via thresholds. If this were accepted and implemented, it would have a major impact on a large number of environmental and occupational health standards. Likewise, acceptance of hormesis would force a reeducation of government risk communication personnel and restructuring of programs for educating future toxicologists, the media, and the public.

As lofty as such statements are the process of hazard assessment is still designed to provide a NOAEL to regulatory agencies for use in their risk assessment activities. These activities are still high dose–few doses schemes. This means that the plan of the agencies, as currently structured, excludes the hormesis concept unless the hormetic model was accepted as the default risk assessment model (43). Furthermore, a 2004 document of the EPA has stated that beneficial effects should not be incorporated into the risk assessment process (52), a conclusion that is designed to continue the marginalization of hormesis.

The future of toxicology must come to grips with its historical error of the rejection of the hormetic dose–response model. Its entire history has been one of victimization by its early intellectual imprinting that relegated the hormesis concept to a negligible status, permitted itself to be dominated by regulatory agencies, and never adequately questioned its past. In fact, if it were not for the costly implications of the LNT model for risk assessment of carcinogens, the concept of hormesis would not have received the opportunity for rebirth at this time. Although the toxicological revolution as led by the hormesis hypothesis is far from complete, it is clear that many in the toxicological community are now giving hormesis serious attention as evidenced by its inclusion in the most recent edition of leading textbooks (53,54), presence at national [U.S. Society of Toxicology (SOT), 2004; Canadian SOT, 2003] societal meetings, publication in the most prestigious journals (50,55), and now consideration by the U.S. NAS. These are hopeful signs of a toxicological awakening that will improve the nature of the science and the way chemicals and drugs are tested and risks estimated.

REFERENCES

  1. Calabrese EJ, Baldwin LA, Mehendale H. G2 subpopulation in rat liver induced into mitosis by low level exposure to CCl4: an adaptive response. Toxicol Appl Pharm 1993; 121(1):1–7.

2. Muller HJ. Artificial transmutation of the gene. Science 1928; 66:84–87.
3. National Academy of Sciences (NAS). Drinking Water and Health. Washington, D.C.: National Academy of Sciences, 1977:939.
4. Calabrese EJ. Methodological Approaches to Deriving Environmental and Occupational Health Standards. New York, NY: Wiley-Interscience Publication, 1978:402.
5. Calabrese EJ, Kenyon E. Air Toxics and Risk Assessment. Lewis Publishers, 1991:650.
6. Luckey TD. Hormesis with Ionizing Radiation. Boca Raton, FL: CRC Press Inc., 1980.
7. Schulz H. Zur Lehre von der Arzneiwirdung. Virchows Arch Pathol Anat Physiol Klin Med, 1887; 9:161–175.
8. Schulz H. Uber Hefegifte. Pflugers Arch Gesamte Physiol Menschen Tiere 1888; 42: 517–541.
9. Crump T. NIH-98-134: contemporary medicine as presented by its practitioners themselves Leipzig 1923:217–250. NIH Library Translation. Nonlinearity Biol Toxicol Med 2003; 1:295–318 (translated).
10. Calabrese EJ, Baldwin LA. Chemical hormesis: its historical foundations as a biological hypothesis. Hum Exp Toxicol 2000; 19:2–31.
11. Calabrese EJ, Baldwin LA. The marginalization of hormesis. Hum Exp Toxicol 2000; 19:32–40.
12. Calabrese EJ, Baldwin LA. Radiation hormesis: its historical foundations as a biological hypothesis. Hum Exp Toxicol 2000; 19:41–75.
13. Calabrese EJ, Baldwin LA. Radiation hormesis: the demise of legitimate hypothesis. Hum Exp Toxicol 2000; 19:76–84.
14. Calabrese EJ, Baldwin LA. Tales of two similar hypotheses: the risk and fall of chemical and radiation hormesis. Hum Exp Toxicol 2000; 19:85–97.
15. Brenchley WE. Inorganic Plant Poisons and Stimulants. London: Cambridge University Press, 1914.
16. Brenchley WE. Inorganic Plant Poisons and Stimulants. 2nd ed. London: Cambridge University Press, 1927.
17. Pfeffer W. The physiology of plants. In: Ewart AJ, ed. A Treatise upon the Metabolism and Sources of Energy in Plants. 2nd ed. Oxford: Clarendon Press, 1900.
18. Lamanna C, Mallette MF. Basic Bacteriology its Biological and Chemical Background. 3rd ed. Baltimore, MD: The Williams & Wilkins Co., 1965.
19. Clifton CE. In: Introduction to Bacterial Physiology. New York: McGraw-Hill Book Co. Inc., 1957:317–338.
20. Salle AJ. Fundamental Principles of Bacteriology. New York: McGraw-Hill Book Co. Inc., 1939:166–197.
21. Townsend CO. The correlation of growth under the influence of injuries. Ann Bot 1897; 11:509–532.
22. Branham SE. The effects of certain chemical compounds upon the course of gas production by baker's yeast. J Bacteriol 1929; 18:247–264.
23. Jensen GH. Toxic limits and stimulation effects of some salts and poisons on wheat. Bot Gaz 1907; 43:11–44.
24. Hotchkiss M. The influence of various salts upon the growth of bacterium communnis. Ph.D. dissertation, Yale University, 1923.
25. Hotchkiss M. Studies on salt actions. VI. The stimulating and inhibitive effect of certain cations upon bacteria growth. J Bacteriol 1923; 8:141–162.
26. Clark AJ. Handbook of Experimental Pharmacology. Springer, Berlin: Verlig Von Julius, 1937.
27. Calabrese EJ. Paradigm lost paradigm found: the re-emergence of hormesis as a fundamental dose response model in the toxicological sciences. Environ Poll 2005; 138(3): 378–411.
28. Bliss CI. When toxic turns to treatment. Chem Ind 1935:10–11.
29. Calabrese EJ, Baldwin LA. The dose determines the stimulation (and poison). Int J Toxicol 1997; 16:545–559.

30. Calabrese EJ, Baldwin LA, Holland C. Hormesis: a highly generalizable and reproducible phenomenon with important implications for risk assessment. Risk Anal 1999; 19:261–281.
31. Calabrese EJ, Blain R. The hormesis database: an overview. Toxicol Appl Pharmacol 2005; 202(3):289–301.
32. Calabrese EJ, Baldwin LA. The frequency of U-shaped dose responses in the toxicological literature. Toxicol Sci 2001; 62:330–338.
33. Calabrese EJ, Baldwin LA. Special issue: Hormesis: environmental and biomedical perspectives. Crit Rev Toxicol 2003; 33(3–4):213–467.
34. Calabrese EJ, Baldwin LA. Special issue: scientific foundations of hormesis. Crit Rev Toxicol 2001; 31(4–5):351–695.
35. Calabrese EJ. Adenosine: biphasic dose responses. Crit Rev Toxicol 2001; 31:539–552.
36. Calabrese EJ. Adrenergic receptors: biphasic dose responses. Crit Rev Toxicol 2001; 31:523–538.
37. Calabrese EJ. Prostaglandins: biphasic dose responses. Crit Rev Toxicol 2001; 31: 475–488.
38. Calabrese EJ. Estrogen and related compounds: biphasic dose responses. Crit Rev Toxicol 2001; 31:503–516.
39. Calabrese EJ. Androgens: biphasic dose responses. Crit Rev Toxicol 2001; 31:517–522.
40. Calabrese EJ. Nitric oxide: biphasic dose responses. Crit Rev Toxicol 2001; 31:489–502.
41. Calabrese EJ. Opiates: biphasic dose responses. Crit Rev Toxicol 2001; 31:585–604.
42. Calabrese EJ, Baldwin LA. Peptides and hormesis. Crit Rev Toxicol 2003; 33:355–405.
43. Calabrese EJ. Hormesis: from marginalization to mainstream. A case for hormesis as the default dose-response model in risk assessment. Toxicol Appl Pharmacol 2004; 197(2): 125–136.
44. Calabrese EJ, Baldwin LA. Radiation hormesis and cancer. Hum Ecolog Risk Assmnt 2002; 8:327–353.
45. Calabrese EJ, Baldwin LA. The effects of gamma-rays on longevity. Biogerontology 2000; 1:309–319.
46. Calabrese EJ. Apoptosis: biphasic dose responses. Crit Rev Toxicol 2001; 31:607–614.
47. Calabrese EJ, Baldwin LA. Ethanol and hormesis. Crit Rev Toxicol 2003; 33:407–424.
48. Calabrese EJ, Baldwin LA. The hormetic dose response model is more common than the threshold model in toxicology. Toxicol Sci 2003; 71:246–250.
49. Calabrese EJ, Baldwin LA. Inorganics and hormesis. Crit Rev Toxicol 2003; 33:215–304.
50. Calabrese EJ, Baldwin LA. Toxicology rethinks its central belief–hormesis demands a reappraisal of the way risks are assessed. Nature 2003; 421(6924):691–692.
51. Robison GA. Forward. In: Lamble JW, ed. Towards Understanding Receptors. New York: Elsevier/North-Holland, 1981:V–X.
52. U.S. Environmental Protection Agency (EPA). An examination of EPA risk assessment principles and practices (EPA/100/B-04/001). Prepared by members of the Risk Assessment Task Force. Office of the Science Advisor, Washington, D.C., 2004:181.
53. Eaton DL, Klaassen CD. Principles of toxicology. In: Klaassen CD, Watkins JB, eds. Casarett and Doull's Essentials of Toxicology. New York: McGraw-Hill, 2003:6–20.
54. Beck B, Slayton TM, Calabrese EJ, Baldwin LA, Rudel R. The use of toxicology in the regulatory process. In: Hayes AW, ed. Principles and Methods of Toxicology. 4th ed. Raven Press, 2000:77–136.
55. Kaiser J. Sipping from a poisoned chalice. Science 2003; 302:376–379.
56. Calabrese J. Cancer biology and hormesis: human tumor cell lines commonly display hormetic (biphasic) dose responses. Crit Rev Toxicol 2005; 35:1–120.

7

Chemical Risk Assessment

Ronald E. Baynes and Jennifer L. Buur
Center for Chemical Toxicology Research and Pharmacokinetics,
College of Veterinary Medicine, North Carolina State University, Raleigh,
North Carolina, U.S.A.

INTRODUCTION

Risk assessment (RA) can be defined as a systematic iterative process that helps to characterize potential adverse health effects in humans following exposure to chemical substances in their home or work environment. The risk assessor is often interested in first identifying the hazardous substance and then describing and quantifying the risk based on evidence from sound human and/or animal toxicology studies. As often stated, "it's the dose that makes the poison," the primary focus in any chemical RA exercise is to be able to provide a quantitative relationship between the dose and adverse health effect and then be able to determine at what exposure level the human population is least likely to be at risk of experiencing adverse health effects from the chemical in question. Ultimately, the risk assessor is focused on arriving at the dose metric or some expression of the toxicologically relevant dose in the target organ. The simultaneous growth in microcomputer capability and software, the mechanistic data from advances in molecular biology, and the biologically based dose–response models in the last decade have allowed the risk assessor to better quantify the human health risk associated with exposure to potentially hazardous chemicals. The reader should be aware that although there have been significant advances in the art and science of the process, there are still many hurdles to overcome in our attempt to reduce uncertainty in the RA process.

The source of much of this uncertainty is often associated with the quality of the available toxicology data, and default options may be required to complete a scientifically sound RA. For example, many times the risk assessor has limited scientific information about the true shape of the dose–response curve (1) and whether there is linear or nonlinear relationship. The available experimental doses may be way in excess of any human exposure, and extrapolation with the help of various curve fitting techniques may be required to assess the risk at low dose exposures. Various other extrapolations techniques may be required to account for differences within and between species and differences between male and female test animals, and exposed dose. Routine toxicology experiments are performed in laboratory animals that are biochemically and physiologically homogenous. However, human populations are heterogenous in this regard, and the measured variability from the animal studies may provide limited insights into the heterogeneity of the toxicant response in human populations (2). The burden therefore is to reduce the uncertainty in extrapolation process and reducing the reliance on default uncertainty factors (UFs), which oftentimes has no little or no physiological relevance. In reality, not all uncertainties will be resolved, but the use of defaults in deriving risk estimates is intended to be health-protective as well as scientifically defensible. However, it must be emphasized that risk characterizations are expected to detail key data uncertainties and a description of range or distribution of plausible risk estimates in a specified population. The intent here is to make the RA process clearer and more transparent and to improve communication of the strengths and weaknesses of the available data used in the risk characterization.

The purpose of this chapter is to review some of the more traditional or classical RA methods that have been used and then to highlight several of the new modeling strategies available to risk assessors. The latter has been gaining popular attention not only in academic circles but also in federal regulatory agencies where some day these methods may become part of routine chemical RA.

HAZARD IDENTIFICATION

Hazard identification is the initial step in RA, and it provides more of a qualitative assessment of available data that are associated with the chemical. The question often asked at this point of the assessment is whether the chemical can cause an increase in incidence of adverse health effects. This can include a weight-of-evidence (WOE) summary of the relationship between the substance and toxic effects. Ideally, human epidemiological and animal toxicology data should be identified to assess both cancer and noncancer endpoints. With little or no toxicology data structure–activity relationships (SARs) and short-term assays can also be included at this initial phase of the assessment.

Data from well-designed human epidemiological studies are ideal for RA as human effects are directly measured without need for animal to human extrapolations. Epidemiological studies can be designed as: (i) Cross-sectional studies that identify risk factors (exposure) and disease, but not useful for establishing cause–effect relationships; (ii) Cohort studies (also called prospective) that target individuals exposed and unexposed to chemical agent, and they are monitored for development of disease; and (iii) Case–control studies or retrospective studies where diseased individuals are matched with disease-free individuals. In general, epidemiological studies often reflect the realistic exposure conditions that may be difficult to simulate in the laboratory, and they describe the full range of human susceptibility to the chemical. However, these studies are often hindered with poor definition of exposure and related confounding factors and endpoints are often relatively crude (e.g., mortality).

Route specific effects are often overlooked as has been the case with assessments of hexavalent chromium [Cr(VI)] compounds. Epidemiological studies of occupational exposures by inhalation have demonstrated that this chemical is a human carcinogen (3). However, it is very unlikely to cause cancer by the oral route because the gastrointestinal tract reduces Cr(VI) to Cr(III) which is nontoxic (4,5). Epidemiological studies are also limited by the need to demonstrate a relatively large increase in disease incidence (e.g., twofold) depending on the sample size of the population (6).

In addition to being costly and time consuming, the very few controlled epidemiological studies are limited to short-term effects that may be reversible.

SARs can be used to help set priorities for testing the potential toxicant. Unique molecular structures or structural alerts such as *n*-nitroso or aromatic amine groups can be used for prioritizing chemical agents for further testing. In most SARs, regression analyses are performed to link physicochemical parameters of a series or subset of related chemicals with various toxicological endpoints. The SAR is used to predict information such as lethality, mutagenicity, and carcinogenicity. There are numerous SAR models that have been generated to predict numerous toxicity endpoints such as skin irritation endpoints (7) and toxicokinetic endpoints such as chemical partitioning behavior and permeability (8–10).

Quantitative structure–permeability relationship (QSPR) models have been used to relate physicochemical parameters such as lipophilicity and molecular weight (MW) to dermal permeability to predict dermal absorption of potential human toxicants. This is critically important for dermal RA which is not as developed as oral and inhalation routes of exposure. Many of the more recent QSPR models have been based on permeability (K_p) data compiled by Flynn (11) for 94 compounds from numerous sources and experimental protocols that can however be described as being more heterogeneous than other chemical clusters or series previously analyzed for QSPR. This data was utilized to generate the now widely cited Potts and Guy (12) model [Eq. (1)] that however reported a poor fit ($r^2 = 0.67$), and there was no thorough statistical analyses of the variance.

$$\log K_p = 0.71 \log K_{o/w} - 0.0061 \text{MW} - 6.3 \tag{1}$$

The U.S. Environmental Protection Agency (EPA) has refined this model by excluding several experimental data points, and it has recommended that this refined model be utilized in predicting permeability (K_p) values. It should be recognized that it is based on small hydrocarbons and pharmaceutical drugs that bear little resemblance to hazardous toxicants the workers are exposed to in the chemical industry. The more recent QSPR approaches now utilize such physicochemical descriptors as hydrophobicity (e.g., $\log K_{o/w}$), electronic properties (e.g., H-bonding), and steric properties (e.g., MW, molar volume) that are really solvation energy descriptors. However, these QSPR models have not been adapted by U.S. EPA in their dermal RAs.

In vitro short-term assays are known to rapidly identify potential human toxicants, and because they are inexpensive compared to in vivo test, they are widely used. Such classical examples include bacterial mutation assays (e.g., Ames test) to *identify* carcinogens, and there are many other short-term tests that can be used to identify neurotoxicity, developmental effects, or immunotoxicity (13,14). Unfortunately, any of these in vitro studies can be associated with some false positives and false negatives.

Quantitative SARs (QSARs) have been attempted to estimate noncarcinogenic toxicity. In one recent study, only about 55% of the 234 compounds with diverse structures and chemical classes estimated the lowest observed adverse effect level (LOAEL) within a factor of 2 for the observed LOAEL while more than 93% were within a factor of five (14). QSARs have gained more recognition with identifying mutagenicity and carcinogenicity using several computerized systems to predict these endpoints (15,16). Others have reported mixed or limited success with QSAR modeling to link physicochemical or structural properties of toxicants to the toxicological endpoints (17). Basak et al. (18) have utilized a hierarchical QSAR approach which entails use of more complex and computationally demanding molecular descriptors of the toxicant of interest. The hierarchy of descriptors consists of topostructural, topochemical, geometrical, and quantum chemical indices. For example, if the toxicity of a chemical is dependent on high-level quantum chemical indices, then this highlights the critical role of stereo-electronic factors as opposed to general shape, etc. In summary, QSARs will continue to be used as a tool in RA to help prioritize toxicologically important chemicals from a large sample, however, these evaluations are by no means conclusive of a toxicological endpoint or related mechanism of toxicity but can be seen as complimentary to relevant in vivo data (19).

In vivo animal studies are usually route specific, and usually involves two species, both sexes, 50 animals per dose group, and near-lifetime exposures, and doses are usually at the maximum tolerated dose (MTD), 50% MTD, and 25% MTD, as well as unexposed control. The MTD is obtained from subchronic studies as it is not a nontoxic dose, and is expected to produce some level of acceptable toxicity thus increasing sensitivity of the animal assay while using limited animal numbers. There is continuous debate as to whether the use of MTD in carcinogenicity bioassays discriminates between genotoxic and nongenotoxic agents and that carcinogenicity may be associated with large dose–induced toxicity (20–22). However, these rodent bioassays may not always be predictive of human carcinogenicity. For example, the widely used glycol ether, 2-butoxyethanol, causes hemangiosarcoma of the liver in male mice, forestomach squamous cell papilloma or carcinoma in female mice, and abundant changes in blood parameters in mice and rats (23). However, humans have no organ analogous to the rodent forestomach, and although the human esophagus is histologically similar, food contact time in the esophagus is less when compared to the rodent forestomach, and thus the risk to humans for esophageal tumors is minimal. Liver tumors in male mice are secondary to 2-butoxyethanol–induced hemolysis. However, because human red blood cells are less sensitive to this chemical (24), liver tumors are less likely to occur in humans exposed to this chemical. There is also the well-documented situation of renal tumors in male rats being associated with $\alpha_{2\mu}$-globulin–chemical binding and accumulation leading to neoplasia; however, this may be irrelevant to human RA as $\alpha_{2\mu}$-globulin is not found in humans, mice, or monkeys. On the other hand, the National Toxicology Program (NTP)–type two-year cancer bioassays in mouse and rat strains may not be sensitive to hormonally regulated cancers such as breast cancer in humans (25). Furthermore, humans are as sensitive as rats to aflatoxin B_1–induced liver tumors, while mice bioassays may not be predictive of aflatoxin B_1–induced tumors in humans. These and numerous other examples stress the importance of critically assessing rodent bioassays in terms of their human relevance to hazard identification of toxicants.

WOE

Carcinogens

The WOE approach is to qualitatively characterize the extent to which the available data support the hypothesis that an agent causes adverse effects in humans. In cancer RA, there are numerous carcinogen classification schemes developed by the U.S. EPA, American Conference of Governmental Industrial Hygienists (ACGIH), NTP, and by numerous other international organizations, such as the International Agency for Research on Cancer (IARC). Needless to say that many of these schemes do not always concur as some schemes may base a chemical's classification considering only positive evidence while other schemes consider a combination of positive and negative evidence along with other data from human and animal studies. These conflicts are usually associated with the quality of the individual study, consistency across studies, demonstration of cancer at same organ site in multiple studies, and absence of confounding exposures. Recent advances in our understanding of molecular mechanisms associated with chemical-induced carcinogenesis have provided significant bodies of mechanistic data that have been included in some WOE schemes, but not considered in other classifications. The U.S. EPA classification scheme of 1986 has been revised several times up until 2005 (26) to reflect much of this mechanistic information. Please note that the recent WOE guidelines have discarded the groups A through E as described in the 1986 guidelines and reported in Integrated Risk Management System (IRIS) summaries in favor of placing emphasis on a narrative description of the carcinogenic potential of a chemical under specific conditions of exposure. The 2005 final guidelines (Table 1) for carcinogen RA document the descriptors and narrative for summarizing WOE of carcinogens. One can only expect revisions of these guidelines to continue so as to make greater use of increasing scientific understanding of carcinogenic mechanisms.

Chemical Mixtures

A WOE approach has also been proposed for assessing toxicological interactions in binary chemical mixtures (27). This is an alphanumeric classification scheme that can be used for qualitative RA and potentially in quantitative assessments. It is based on all toxicological evidence from bioassays and pharmacokinetic studies, and it takes into account factors such as relevance of route, sequence, and duration of exposure, toxicological significance of the interactions, and the quality of the in vivo and in vitro data. This WOE approach has been used to estimate qualitatively and quantitatively the toxicity of binary chemical mixtures through modeling of chemical interactions (28). This system provides scores for (i) direction of interaction, (ii) mechanistic understanding, (iii) toxicological significance of the interaction, and (iv) modifiers. While this WOE approach correctly adjusted for the observed interactions between

Table 1 WOE Designation Based on U.S. EPA Guidelines

Carcinogenic to humans: This descriptor is appropriate with convincing epidemiologic evidence demonstrating causality or when there is strong epidemiological evidence, extensive animal evidence, knowledge of the mode of action, and information that the mode of action is anticipated to occur in humans and progress to tumors

Likely to be carcinogenic to humans: This descriptor is appropriate when the available tumor effects and other key data are adequate to demonstrate carcinogenic potential to humans, but does not reach the WOE for the descriptor "carcinogenic to humans"

Suggestive evidence of carcinogenic potential: This descriptor is appropriate when the evidence from human or animal data is suggestive of carcinogenicity, which raises a concern for carcinogenic effects but is judged not sufficient for a stronger conclusion

Inadequate information to assess carcinogenic potential: This descriptor is used when available data are judged inadequate to perform an assessment

Not likely to be carcinogenic to humans: This descriptor is used when available data are considered robust for deciding that there is no basis for human hazard concern

Abbreviations: U.S. EPA, U.S. Environmental Protection Agency; WOE, weight-of-evidence.
Source: From Ref. 26.

Table 2 Modified WOE Classification from U.S. EPA with Positive Values Indicating
Synergism and Negative Values Indicating Antagonism

Category	Description	Greater than additive	Less than additive
I	Interaction has been shown to be relevant to human health effects	1.0	−1.0
II	Interaction has been demonstrated in vivo in animal model and relevance to potential human health effects is likely	0.75	−0.5
III	Interaction is plausible, but the evidence supporting the interaction and its relevance to human health effects is weak	0.5	0.0
IV	Insufficient evidence to determine whether any interaction would occur or adequate evidence that no toxicologic interaction between/among the compounds is plausible	0.0	0.0

Abbreviations: U.S. EPA, U.S. Environmental Protection Agency; WOE, weight-of-evidence.
Source: From Ref. 29.

chemicals with similar modes of action, this was not so for dissimilar binary mixtures.
The EPA (29) has modified this Mumtaz and Durkin (27) scheme to one described in
Table 2 with Roman numeral I through IV.

Developmental Effects

A WOE scheme has also been developed for assessing developmental toxicity that
includes any detrimental effect produced by exposures during embryonic development. Adverse effects include death, structural abnormalities, altered growth, and
functional deficiencies; maternal toxicity is also included in this scheme. The evidence is assessed and assigned a WOE designation as follows: Category A, Category
B, Category C, and Category D.

   The scheme takes into account the ratio of minimum maternotoxic dose to
minimum teratogenic dose, the incidence of malformations and thus the shape of
the dose–response curve or dose relatedness of the each malformation, and types
of malformations at low doses. A range of UFs are also utilized according to designated category as follows: Category A = 1 to 400; Category B = 1 to 300; Category
C = 1 to 250; and Category D = 1 to 100. Developmental reference doses (Rf Ds) are
unique in that they are based on a short duration of exposure and therefore cannot
be applied to lifetime exposure.

DOSE–RESPONSE ASSESSMENT

Dose–response primarily involves characterizing the relationship between chemical
potency and incidence of adverse health effect. There are numerous approaches that
have been used to characterize dose–response relationships, and these include effect

levels such as LD50, LC50, ED50, no observed adverse effect levels (NOAELs), margins of safety, and therapeutic index. The first step is to utilize all available datasets from human and/or animal studies to arrive at a suitable point of departure (POD), and the second step will inevitably require extrapolation of the dose–response relationship from this POD to a relevant human exposure. This is often an extrapolation from relatively high dose usually associated with experimental exposure in laboratory animals to significantly lower doses that are characteristic of environmental exposure in humans. Therefore the focus, in quantitative RA, is in attempting to accurately model the dose–response relationship that will clearly influence the shape of the dose–response function below the experimentally observable range and therefore the range of inference (Fig. 1). These relationships are not always linear in this low dose range, and as numerous modeling approaches are available, care must be taken in choosing the appropriate model(s) during the two phases of dose–response assessment.

Carcinogens

The traditional approach for cancer RA is that no dose is thought to be risk free. A threshold for an adverse effect does not exist with most individual chemicals, but it is assumed that a small number of molecular events can lead to uncontrolled cellular proliferation and eventually to a clinical state of disease. Dose–response assessments are generally performed for human carcinogens and likely human carcinogens in two broad steps: (i) derive a POD from within the range of experimental observation and (ii) use the POD to extrapolate to lower doses. In the first step, biologically based models can be used if the carcinogenic process is fully understood or empirical curve fitting procedures can be used to fit data such as tumor incidence datasets. The linearized multistage model is often used as the default method to arrive at the POD. An estimated dose associated with a 10% increase in tumor or nontumor response [lower limit on a effective dose (LED_{10})] is often used as a POD. The derived POD (e.g., LED_1 or LED_{10}) is not the central estimate (e.g., ED_1 or ED_{10}) but the 95% confidence limit on the lowest level. The use of the lower limit instead of the central estimate allows for variability in sampling error, etc. The new guidance (26) suggests situations where linear or nonlinear approaches can be used to extrapolate from the observed region of the dose–response curve to lower doses of the inference region that are more relevant to human exposure in the environment. The linear extrapolation approach is used when the mode of action information

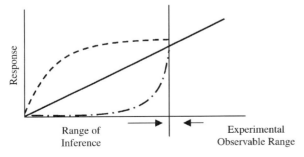

Figure 1 Dose–response curve, with emphasis on the shape of the dose response below the observable range. The range of inference is often where people are realistically exposed.

suggest that the dose–response curve is linear in the low dose region. It can also be used as a default approach in the absence of information on modes of action of the carcinogen as the linear approach is generally considered to be health-protective for addressing uncertainty. Other low dose extrapolation approaches include the use of probit, gamma multi hit, Weilbull, multistage, and logit models. Again, biologically based dose–response models with sufficient data represent the most ideal approach to extrapolate below the observed dose range. The nonlinear approach should be used only when there is sufficient evidence that the mode of action is not linear at low doses. For linear extrapolations, a line is drawn from the derived POD to the origin, and then the slope of this line, slope factor, is derived. This slope factor is described by U.S. EPA as an upper bound estimate of risk per increment of dose that can be used to estimate risk probabilities for different exposure levels. The slope factor is expressed as q_1^*:

$$\text{slope factor} = \text{risk per unit dose (risk per mg/kg/day)} \tag{2}$$

The slope factor can therefore be used to calculate the upper bound estimate on risk (R)

$$\text{Risk} = q_1^*[\text{risk} \times (\text{mg/kg/day})^{-1}] \times \text{exposure (mg/kg/day)} \tag{3}$$

The above risk is a unit less probability (e.g., 2×10^{-5}) of an individual developing cancer following chronic daily exposure averaged over 70 years. This can be estimated if we can determine the slope factor and human exposure at the waste site or occupational site. The U.S. EPA usually sets a goal of limiting lifetime cancer risks in the range of 10^{-6} to 10^{-4} for chemical exposures, while the Food and Drug Administration (FDA) typically aims for risks below 10^{-6} for general population exposure.

If a nonlinear dose–response extrapolations at low doses can be justified for some carcinogens, then an acceptable daily intake (ADI) for a carcinogen can be derived by dividing the RfD or benchmark dose (BMD) by UFs (Gaylor, 2005). These two estimates will be discussed in further detail in the next section of this chapter. Finally, it is well recognized that carcinogens can act by different mechanisms, and therefore different types of dose–response relationships and the same carcinogen can cause cancer at different body sites via different mechanisms. For example, the potent carcinogen, 2-acetylaminofluorene, exhibits a linear dose–response relationship for liver cancer, but a highly nonlinear dose–response relationship for bladder cancer (30).

Noncarcinogens

As with the carcinogens, there is a two-step approach for dose–response assessment for noncarcinogens. This section will focus on two approaches that have been used to derive the POD for noncarcinogens. These are (i) the NOAEL approach and (ii) the BMD approach.

There are some situations where there is limited scientific information about the true shape of the dose–response curve. In the case of exposure to noncarcinogens, the "NOAEL" approach is considered. This approach assumes that a *threshold* exists for many noncarcinogenic effects and that prior to achieving this threshold, there are protective biochemical and physiological mechanisms that must be overcome before one observes an adverse effect. It is assumed that the NOAEL is the exposure level at which there are no statistically or biologically significant increases in frequency and its appropriate control. In essence, the NOAEL will be the highest dose from a given data set that did not cause a significant change in effect. There are many examples where there are insufficient test dosages, and a NOAEL cannot be derived, and the LOAEL is derived from the study data.

The NOAEL or LOAEL is then used to calculate RfDs for chronic oral exposures and reference concentrations (RfCs) for chronic inhalation exposures. Other agencies such as the Agency for Toxic Substances and Disease Registry (ATSDR) use the above methodology to calculate minimum risk levels (MRLs). However, MRLs are derived for acute, subchronic, and chronic exposure for oral and inhalation routes. The World Health Organization (WHO) usually derives ADIs.

These RfDs, MRLs, and ADIs are intended to represent an estimate of an exposure dose that is least likely to cause harmful effects even in the most sensitive human population following a lifetime of exposure. To provide such conservative estimates, the NOAELs or LOAELs are divided by UFs, which will be further discussed in a subsequent section of this chapter:

$$RfD = NOAEL/(UF) \tag{4}$$

The reader should however be aware that there are several problems associated with using the NOAEL approach to estimate the above RfDs and RfCs. The first obvious constraint is that the NOAEL must by definition be one of the experimental doses tested. Once this dose is identified, the rest of the dose–response curve is ignored. In some experimental designs where there is no identifiable NOAEL but LOAEL, the dose–response curve is again ignored, and the NOAEL is derived by application of UFs as described earlier. This NOAEL approach does not account for the variability in the estimate of the dose–response, and furthermore, experiments that test fewer animals result in larger NOAELs and thus larger RfDs and RfCs.

Because of the above concerns with the NOAEL approach, an alternative approach known as the BMD approach was developed by Crump (31) as a more quantitative alternative to the first step in the dose–response assessment in arriving at a POD that can be used to estimate RfDs and RfCs. The BMD approach is not constrained by experimental design as the NOAEL approach, and incorporates information on the sample size and shape of the dose–response curve. In fact, this approach can be used for both threshold and nonthreshold adverse effects as well as dichotomous (quantal), categorical, and continuous data sets. Calculation of the BMD requires determination of the benchmark response (BMR). For quantal data, an excess risk of 10% is the default BMR, because the 10% response is at or near the limit of sensitivity in most bioassays (29). The BMR is usually specified as a 1% to 10% response; that is, it corresponds to a dose associated with a low level of risk, e.g., 1% to 10%. In the absence of any idea of what level of response is deemed adverse, a change in the mean equal to one control SD from the control mean is used as the BMR.

The next step is to model the dose–response data, and this requires use of U.S. EPA's BMD Software to fit a model to the dose–response data with the lower confidence bound for a dose at a specified BMR level of 1% to 10%. Several models are used in this exercise, the goodness-of-fit of the model needs to be ascertained, and the Akaike Information Criterion is often used to compare and select the model for BMD computation. Figure 2 shows how an effective dose that corresponds to a specific change of effect or response (e.g., 10%) over background and a 95% lower confidence bound on the dose is calculated. The latter is often referred to as the BMD level (BMDL) or LBMD, as opposed to the BMD, which does not have this confidence limit, associated with it.

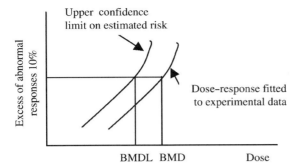

Figure 2 BMD determination from dose–response relationship with the BMDL corresponding to the lower end of a one-sided 95% confidence interval for the BMD. *Abbreviations*:
BMD, benchmark dose; BMDL, benchmark dose level.

The BMDL represents a statistical lower limit, and larger experiments will tend
on average to give larger benchmarks, thus rewarding good experimentation. This is
not the case with NOAELs, as there is an inverse relationship between NOAEL and
size of experiments. For example, poorer experiments possessing less sensitivity for
detecting statistically significant increases in risk inappropriately result in higher
NOAELs and RfDs, which may have an unknown unacceptable level of risk. The
BMD approach is not constrained to one of the experiments as in the NOAEL
approach, as the doses and slope of the curve influences the calculations, the variability of the data is considered, and the BMD is less variable between experiments
compared to NOAEL. The BMDL accounts for the uncertainty in the estimate of
the dose–response associated with the experimental design. In the BMD approach,
quantitative toxicological data such as continuous data (e.g., organ weights and
serum levels) and quantal or incidence data (e.g., pathology findings and genetic
anomalies) are fitted to numerous dose–response models described in the literature.
Dose–response modeling of continuous data is more difficult than other datasets as
there is no natural probability scale to characterize risk. However, these continuous
dataset can be dichotomized and then modeled as any other quantal dataset, or a
hybrid modeling approach can also be used (32,33). In the case of the latter, this
approach fits continuous models to continuous data, and presuming a distribution
of data, it determined the BMD in terms of fraction affected. Ultimately, the desired
POD is the derived BMDL, which can be used to calculate RfDs and RfCs using
appropriate UFs that are described in further detail below. It should be stressed here
that although the BMDL is the desired POD, there are chemicals for which dose–
response datasets are limited and insufficient for deriving a BMDL. In such circumstances, the NOAEL approach is acceptable for deriving a default POD.

Default Uncertainty and Modifying Factors

The previous section described how a POD can be calculated for a noncarcinogen,
however, the next series of steps requires a series of extrapolations to ensure that
exposure to these chemicals does not cause adverse health effects in humans. As
many of the NOAELs and BMDLs are derived from animal studies, extrapolations
from animal experimental data in the RA process require the utilization of various
UFs, because we are not certain how to extrapolate across species, within species
for the most sensitive population, and across duration. For this reason, an UF of

10 is intended to protect sensitive subpopulations. The value of 10 is derived from a 3.16-fold factor for differences in toxicokinetics and a 3.16-fold factor for toxicodynamics. An additional UF of 10 is used to extrapolate from animals to humans, that is, to account for interspecies variability between humans and other mammals. As with intraspecies extrapolations, this tenfold factor is assumed to be associated with uncertainty in toxicodynamics and toxicokinetics. Sometimes, there are no available chronic studies, and therefore an UF of 10 is used by U.S. EPA (not ATSDR) when a NOAEL is derived from a subchronic study instead of a chronic study. ATSDR does not perform this extrapolation as they derive chronic and subchronic MRLs. Sometimes, only a LOAEL can be derived from the available dataset, and in this case an UF of 10 is used when deriving a RfD or MRL from a LOAEL. A modifying factor ranging from 1 to 10 is included by EPA only to reflect a qualitative professional assessment of additional uncertainties in the critical study and in the entire data base for the chemical not explicitly addressed by preceding UFs. We know from pharmacokinetic studies with some human pharmaceuticals that drug elimination is slower in infants up to six months of age than in adults, and therefore potential for greater tissue concentrations and vulnerability for neonatal and postnatal effects exists. Based on these observations, the U.S. EPA as proposed by the 1996 Food Quality Protection Act (FQPA) supports a default safety factor greater or less than 10 unless the U.S. EPA can show that an adequate margin of safety is assured with out it (34). In summary, the BMDL or NOAEL can be adjusted by a factor of as much as 10,000-fold which in many RA circles can be regarded as overly conservative estimates.

With improved dose–response datasets and sound modeling approaches, relevant biochemical and mechanistic information is becoming increasingly available, and in these circumstances, the RfD or RfC can be refined by minimizing the use of several of the above tenfold default UFs (35). For example, the EPA has included dosimetry modeling in RfC calculations, and the resulting dosimetric adjustment factor (DAF) used in determining the RfC is dependent on physiochemical properties of the inhaled toxicant as well as type of dosimetry model ranging from rudimentary to optimal model structures. In essence, the use of the DAF can reduce the default UF for interspecies extrapolation from 10 to 3.16. Much of the research efforts in RA are therefore aimed at reducing the need to use these default UFs, although the risk assessor is limited by data quality of the chemical of interest. With sufficient data and the advent of sophisticated and validated physiologically based pharmacokinetic (PBPK) models and biologically based dose–response models (36), these default values can be replaced with science-based factors. In some instances, there may be sufficient data to be able to obtain distributions rather than point estimates.

QUANTIFYING RISK FOR NONCARCINOGENIC EFFECTS: HAZARD QUOTIENT

The measure used to describe the potential for noncarcinogenic toxicity to occur is not expressed as the probability. A probabilistic approach is used in cancer RA. For noncancer RA, the potential for noncarcinogenic effects is evaluated by comparing an exposure level (E) over a specified time period with an RfD. This ratio is called a hazard quotient (HQ):

$$HQ = E/RfD \qquad (5)$$

In general, the greater the value of E/RfD, i.e., exceeds unity, the greater the level of concern. Note that this is a ratio and is not to be interpreted as a statistical probability.

CHEMICAL MIXTURES

Humans are more likely to be exposed simultaneously or sequentially to a mixture of chemicals rather than one single chemical. The RA can involve evaluation of the whole mixtures and/or component data. Whole-mixture RfDs and slope factors can be derived depending on the availability of data generated from exposure to identical or reasonably similar mixtures. For component-based mixtures approach, dose-additive models or response-additive models are recommended default models by U.S. EPA (29) when there is no adequate interaction information. Toxicological interactions have been defined by mixture data showing statistically or toxicologically significant deviations from the no-interaction prediction for the mixture.

As stated earlier, calculation procedures differ for carcinogenic and noncarcinogenic effects, but both sets of procedures assume dose additivity in the absence of toxicological interaction information on mixtures, and risk from exposure to these mixtures can be calculated as follows:

$$\text{Cancer risk equation for mixtures: } \text{risk}_t = \sum \text{risk}_I \tag{6}$$

$$\text{Noncancer hazard index (HI)}$$
$$= E_1/RfD_1 + E_2/RfD_2 + \cdots + E_i/RfD_I \tag{7}$$

This HI approach as well as others [e.g., relative potency factor (RPF) approach] is applied for mixture components that induce the same toxic effect by identical mechanism of action. The HI is in essence a sum of individual HQs (E/RfD) for corresponding chemical components in a mixture.

Standard default approaches to mixture RA consider doses and responses of the mixture components to be additive. Dose-additive models (also referred to as concentration addition) are based on components in these mixtures that act on similar biological systems and elicit a common response, while response addition models are based on mixture components that act on different systems or produce effects that do not influence each other. The mixture response (p^{MIX}) for two chemicals exhibiting dose addition can be mathematically expressed in terms of an equivalent dose and dose–response function for one chemical as follows:

$$p^{MIX} = f(d_1 + t \times d_2) \tag{8}$$

where d_1 and d_2 represent the doses for the two chemicals and t represents the proportionality constant reflecting the relative effectiveness of chemical 2 to chemical 1. In essence, dose addition assumes that the chemicals in the mixture behave as if they are dilutions or concentrations of each other. Thus, the dose–response slopes for individual chemicals are identical. Isoboles for fixed response levels (e.g., 10% response level) can be used to graphically determine dose addition. Stated differently, all points along the isobole (linear slope) correspond to various combinations of different doses of two chemicals that produced the same response level. In this situation, two chemical components in this mixture are termed dose-additive, and various statistical methods can be used to decide which points near the isobole depart from dose additivity. It is plausible to assume that any departure from the isobole reflects antagonistic or synergistic interactions.

The toxicity equivalency factors (TEFs), which were developed explicitly for risk estimation of chlorinated dioxin and chlorinated dibenzofurans, are a modification of the above dose-additive method and can be used for three component mixtures. In this method, the exposure levels are added after being multiplied by a scaling factor that accounts for differences in toxicologic potency. The TEF is usually specific for the defined mixture, is applied to all health effects, all routes and duration of exposure, and it implies use of a large database and greater certainty about the mode of action. RPF has been developed for characterizing mixtures of chemicals with similar modes of action, but these RPFs are less stringent than the TEFs and require an evaluation of mixtures of *related* chemical compounds that are assumed to be toxicologically similar. The RPF approach relies on dose–response data for at least one chemical (also called the index compound) in the mixture and scientific judgment as to the toxicity of the other chemicals in the mixture. The RPF of a mixture component, which is really a proportionality constant or scaling factor, is based on an evaluation of the small sets of available toxicology data, and this RPF therefore represents the relative toxicity of this mixture component with respect to the *index* chemical described above. For example, if one chemical component in a mixture is one-tenth the toxicity of the index chemical the RPF is 0.1. The RPFs of each mixture component can be used to calculate a mixture exposure (C_m) in terms of an equivalent exposure to the index compound as follows:

$$C_m = \sum C_k \times RPF_k \tag{9}$$

where RPF_k represents the proportionality constant for toxicity of the kth mixture component relative to the toxicity of the index chemical. It must be emphasized that these RPFs (or TEFs) are estimates of toxicity and are meant to be an interim measure pending development of mixture-specific data. The resulting approach would be the HI, which is based on dose addition or response addition as is common for cancer RAs.

With response addition (also called independent joint action), chemicals behave independently of one another, and the response is measured by the percentage of exposed animals that show toxicity. This definition is most applicable to the individual, but not necessarily to a population, as one has to determine whether the chemicals cause toxicity to the same proportion of the population. For example, if the tolerance distributions of two chemicals in a mixture are perfectly correlated ($r=1$), then the ordering of individual sensitivities is the same for both chemicals, that is, the individual most sensitive to chemical 1 is also most sensitive to chemical 2. However, when the correlation is zero ($r=0$) the two chemicals are said to act independently on the population. Using the same notation as in Eq. (8), response addition for a binary mixture can be mathematically expressed according to statistical law of independence as follows:

$$p^{MIX} = 1 - (1 - p_1) \times (1 - p_2) \tag{10}$$

also rewritten as

$$p^{MIX} = p_1 + p_2 - (p_1 \times p_2) \tag{11}$$

It should be noted that these and related formulae have limited use in RA as it works best with binary mixtures.

Toxicological interactions can be adequately described as departures from dose addition as described above. WOE procedures described (27) in previous sections of this chapter are used as a qualitative judgment of the impact or *strength* of the potential interactions. The strength of this WOE is assigned a numerical binary WOE (BINWOE) which is scaled to reflect the relative importance of the component exposure levels. This BINWOE or WOE_N is then used to modify the dose-additive HI to an adjusted HI_{INT} as follows:

$$HI_{INT} = HI \times UF_I^{WOE_N} \tag{12}$$

where UF_I is the UF of the toxicological interaction with a default value of 10.

The problem with this approach as described by others (37,38) is that there is no guidance on selecting the UF for interactions, and this relatively simple equation is not flexible. The recent U.S. EPA guidance (29) describes algorithms that are more flexible than those proposed by Mumtaz and Durkin (27). For example, the HI (now termed HI_{INT}) can now be modified if chemical specific information is available. This procedure does not use UFs, but incorporates the estimated magnitude of each pairwise interaction (M_{ij}), WOE scores (B_{ij}), normalizing function (f_{ij}), and another weighing factor (θ_{ij}) that reflects the degree to which components are equitoxic. These estimates are used to modify the HQ for each chemical in the mixture as follows:

$$HI_{INT} = \sum_{i=1}^{n} \left(HQ_i \times \sum_{j \neq i}^{n} f_{ij} \times M_{ij}^{B_{ij}\theta_{ij}} \right) \tag{13}$$

The interaction magnitude (M_{ij}) is usually derived from qualitative interactions such as increase in severity of the histopathology or a x-fold change in lethal dose. Ideally, this M-value is recommended to reflect a change in effective dose, ED_x. The U.S. EPA recommends a default value for M to be five, although the M-value is sometimes not expected to remain constant over a dose range. However, when relevant information is available this data should be used instead of the default value.

The approaches described in the previous paragraphs have been recognized by scientist as interim as they do not really facilitate the use of data on interaction mechanisms in a quantitative manner. Mixture interactions can complicate tissue dosimetry at the route of entry (e.g., gastrointestinal tract and skin surface) and clearance/metabolic mechanisms. Furthermore, these and other considerable uncertainties can complicate extrapolation of toxicological effects from single or binary chemical exposure to multiple chemical mixture exposures. Numerous investigators have used SARs and PBPK models to quantify these mixture effects (39,40). These PBPK models take into account multiple pharmacokinetic interactions amongst mixture constituents by interconnecting the PBPK models for the individual chemicals at the level of the tissue compartment where the interaction is hypothesized or shown to occur. These interaction-based PBPK models can quantify change in tissue dose metrics of chemicals during exposure to mixtures and thus improve the mechanistic basis of mixture RA. While many of these PBPK models have looked at binary-level interactions, there has also been the development of PBPK models for more complex mixtures with all component models being interconnected at the binary level (41). A final cautionary note here is that there are still significant obstacles for the regular use of PBPK models for RA of mixtures. These include the fact that not all of the parameters in the model will be independently estimated and then there is the significant cost of validation studies. The reader is advised to consult recent excellent reviews in the subject and the U.S. EPA (29) guidance for more details on many of the RA procedures described above.

DERMAL RA

Dermal RA is focused on chronic health effects resulting from low-dose and long-term exposure. The U.S. EPA (42) guidance describes the process for water pathway and soil pathway, and in the case of the former pathway, the dermal route is significant if this route contributes at least 10% of the exposure derived from the oral pathway. Current dermal RAs are plagued by the many conflicting experimental protocols and internal dose calculations that limit characterization of chemical absorption in skin. Calculating chemical absorption has been based on (i) percent dose absorbed into the systemic circulation and/or (ii) steady state flux that can be used to calculate solute permeability. The latter is more useful primarily because chemical permeability is concentration independent, whereas expressing penetration as a fraction of applied dose may cause large errors associated with variations in external dosing and exposure times. Permeability is therefore preferred for extrapolating across dose in dermal RA and is also better suited for assessing and ultimately extrapolating across formulation and mixture effects. Assuming that solutes obey Fick's first law of diffusion as they diffuse across the human epidermal membrane, skin permeability can be defined by the equation:

$$\text{Permeability } (K_p)(\text{cm/hr}) = J_{ss}/C_v \tag{14}$$

where J_{ss} represents the solute steady state flux and C_v represents the solute dosing concentration. Solute permeability is dependent on solute diffusivity, D (cm^2/hr), in the membrane and its ability to partition from the dosing solution to the stratum corneum layer of skin. The latter is referred to as the stratum corneum–vehicle partition coefficient, $K_{s/v}$, and is often correlated to octanol–water partition coefficient, $K_{o/w}$. Permeability can therefore be redefined by Eq. (15), where $l =$ membrane thickness

$$K_p = (D \times K_{s/v})/l \tag{15}$$

The permeability of chemicals in water is estimated by an empirical correlation as a function of octanol–water partition coefficients ($K_{o/w}$) and MW as described by the EPA modified Potts and Guy (1992) model below:

$$\log K_p = -2.80 + 0.66 \log K_{o/w} - 0.0056 \, \text{MW} \quad (r^2 = 0.66) \tag{16}$$

This permeability value is then used to calculate the absorbed dose event (mg/cm^2-event) once the concentration in water and duration of the event exposure are known. The absorbed dose event can now be used to calculate the dermally absorbed dose (DAD, mg/kg day) for a 70-kg adult spanning exposure duration and frequency and available skin surface contact. The dermal HQ can now be defined as DAD/RfD_{ABS}, where RfD_{ABS} is the RfD adjusted for fraction oral bioavailability. Similarly, the dermal cancer risk can be calculated by multiplying the DAD by the absorbed cancer slope factor with the latter being the product of adjusting the oral slope factor by the oral bioavailability.

Clearly, the K_p value either derived or predicted from the above model has a significant effect on the final dermal cancer risk or HI characterizations for humans exposed to chemicals by the dermal route. This permeability has been identified as contributing significantly to the uncertainty of the RA. For example, there is a significant measurement error associated with the in vitro determination of permeability values and $K_{o/w}$ values. The above model was based on the Flynn (11) dataset of permeability values that consisted of only 90 chemicals that bear little physicochemical resemblance to environmental contaminants found at most superfund sites. This is especially applicable to environmental chemicals that have very large and very small $K_{o/w}$ values. Many of the latter have been classified by U.S. EPA as being outside of the effective prediction domain (EPD). The boundaries of the EPD were developed from statistical analysis of the above model for permeability and provide the ranges of the $K_{o/w}$ and MW where extrapolation of the K_p correlation will be valid. In spite of these issues and related uncertainties, the above model can be used as a starting point for estimating dermal permeability of environmental contaminants especially when there is little or no available permeability data that can be used to assess risk of contaminants from exposure to the dermal route.

PBPK MODELING

PBPK modeling has been used in RA to make scientifically based extrapolations, and at the same time it helps to explore and reduce inherent uncertainties. These PBPK models mathematically describe the absorption, distribution, metabolism, and excretion of xenobiotics in a biological system based on physiologic and mechanistic processes. Unlike traditional methods of RA, the use of these models spans species, life stage, dose, and route of exposure allowing us to describe more accurately and to predict ultimately the safe levels of exposure under a variety of circumstances. These mathematical modeling approaches also help identify areas of potential scientific research that could improve the human health assessment. These and other applications have been extensively reviewed (43), and ultimately, these analyses reduce the need for UFs in adjustments of POD (44).

Model Construction

A PBPK model is simply a series of tissue blocks linked together by a fluid plasma block. The tissue blocks represent tissues according to anatomical and physiological properties. Tissues can be combined into fewer blocks or separated into more blocks depending on the need of the researcher. For example, Figure 3 depicts the simplest model that includes a single high flow tissue block, a single low flow tissue block, and a single excretory tissue block. However, if a researcher was interested in the toxicological effects in the brain, then a central nervous system block could be added. Ultimately, the number and characteristics of the tissue blocks are dependent upon the sensitivity and specificity needed in the model itself. Each tissue block (Fig. 4) can then be subdivided into subcompartments that represent the vascular space of blood flow through the tissue, extracellular space, and finally the intracellular space. These subcompartments, like the blocks themselves, can be combined and simplified if needed.

Tissue blocks are then further categorized into either flow limited or membrane limited depending on the rate-limiting characteristic of that specific tissue. Flow limited tissues are based on the assumptions of a well-mixed model, that is, all partitioning of the xenobiotic from the blood into the tissue takes place instantaneously and in a homogenous manner. Thus the rate-limiting step is the blood flow. This is

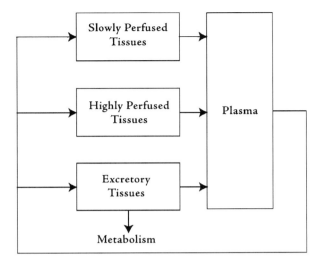

Figure 3 Generic PBPK model showing simplified tissue blocks linked together through a single plasma compartment. *Abbreviation*: PBPK, physiologically based pharmacokinetic.

normally the case for compounds that have a small MW and are lipophilic. Organ systems that are relatively small in volume are also normally treated in this manner. Membrane limited tissues, on the other hand, are described when the diffusion across membranes is the rate-limiting step in partitioning. This occurs with higher MW compounds that are polar in nature and in large volume organs.

Once the number of tissue blocks and their rate-limiting features have been defined, the model is constructed by writing a series of differential equations based on the idea of mass balance. Mass balance is the mathematical concept that the total mass of a xenobiotic in a system is constant and can be accounted for. The differential equations describe the rate of change of concentration in the tissue block per unit time. For a flow limited tissue block, the simple form of this equation is

$$V_t \, dC_t/dt = Q_t(C_a - C_v) \tag{17}$$

where Q_t, V_t, and C_t are the blood flow, anatomic volume, and concentration of the xenobiotic in tissue t, respectively, and C_a and C_v are the concentrations of

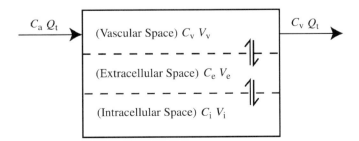

Figure 4 Tissue block with subcompartments for vascular space, extracellular space, and intracellular space. Q_t is blood flow to the tissue. C represents xenobiotic concentration, and V represents volume. Subscripts a, v, e, and i represent arterial blood, venous blood, extracellular, and intracellular subcompartments, respectively.

the xenobiotic in the arterial and venous circulations perfusing tissue t, respectively. If you assume that C_v is in equilibrium with the vascular space, C_v can be further defined as

$$C_v = C_t/P_t \qquad (18)$$

where P_t is the tissue-to-blood partition coefficient for tissue t. Combining Eqs. (17) and (18) yields

$$V_t \, dC_t/dt = Q_t(C_a - C_t/P_t) \qquad (19)$$

The basic equation for a membrane limited tissue block, where the vascular space is assumed to be instantaneously in equilibrium with the extracellular space, is defined by rate of change in the extracellular space per unit of time and can be written as

$$V_e \, dC_e/dt = Q_t(C_a - C_e) - K_t(C_e - C_i) \qquad (20)$$

$$V_i \, dC_i/dt = K_t(C_e - C_t/P_t) \qquad (21)$$

where V_e and C_e are the anatomic volume and xenobiotic concentration, respectively, of the extracellular space in tissue t, V_i and C_i are the volume and concentration of xenobiotic in the intracellular space, respectively, and K_t is the membrane permeability coefficient for tissue t.

Tissue blocks that metabolize or excrete the xenobiotic require further modification of Eqs. (19) and (20). The addition of a mass removal term, R_{ex}, is then added to account for the further loss of xenobiotic. R_{ex} can be defined by any set of parameters ranging from a simple first order equation to detailed Michaelis–Menton equations for multiple enzymes. The resulting equation for a flow limited tissue block is

$$V_t \, dC_t/dt = Q_t(C_a - C_t/P_t) - R_{ex} \qquad (22)$$

If R_{ex} is a first order process, it could be described as

$$R_{ex} = K_{el} C_t \qquad (23)$$

Thus the final equation is

$$V_t dC_t/dt = Q_t(C_a - C_t/P_t) - K_{el} C_t \qquad (24)$$

In much the same manner, a researcher can further refine the model by adding terms to describe other processes including protein binding, tissue binding, active transport, biliary excretion, and enterohepatic metabolism.

The final step in writing differential equations is to write the equations that describe the plasma block. This block is again defined by mass balance where the inputs are the mass from each tissue block's venous return. The rate of change in the plasma space per unit of time is then

$$V_p dC_p/dt = \sum Q_t C_v - Q_p C_p \qquad (25)$$

where V_p, Q_p, and C_p are the anatomic volume, the total blood flow, and the concentration of the xenobiotic in the plasma compartment, respectively. C_v represents the venous concentration of the xenobiotic from tissue t.

Parameter Estimation

PBPK models include both physiologic and physiochemical parameters. The physiologic parameters include tissue volumes and blood flow rates. These are normally taken from the literature for the species of interest. The physiochemical parameters include partitioning coefficients for the xenobiotic as well as protein and tissue binding properties, Michalis–Menton metabolism constants, elimination rates, and absorption rates. These values can be found in the literature or, as is more often the case, be derived from in vivo and in vitro experiments. If parameters are unable to be derived or found in the literature, the parameter can be estimated using known data points. In this case, the model is fit to known data points until a "best fit" is achieved. There are several computation methods using maximum likelihood ratios that can produce these parameters. Then the resulting parameter values are used in subsequent simulations.

The accuracy of a PBPK model is directly related to the accuracy of the parameters used within the mass balance equations. If the parameters are not accurate, the final model will not reflect true in vivo concentrations of the xenobiotic. Thus one major weakness of PBPK modeling is its dependence upon the source literature. However, PBPK models also allow for the inclusion of improved parameters as more information and mechanisms are elucidated.

Model Simulation and Validation

Simulations are achieved by the simultaneous solving of the differential equations. Currently there are many software packages that are equipped to handle these types of computations. They can range from simple spreadsheet programs to more complex computer programming packages including Simusolv, ACSL, Cmatrix, MatLab, and other Fortran-based programs. Validation of PBPK models occurs by comparing simulated values from the model with external data sets. Correlation plots, residual plots, and simulation graphs are evaluated to look at the overall goodness-of-fit of the model. There is no standardized way to apply statistics in the evaluation of goodness-of-fit. However, the evaluation of these graphs does provide substantial information. The goal is to have a regression line with an R^2 of 1 and an intercept of 0 on the correlation plot and an even distribution around 0 for the residual plot and simulation graphs. Figure 5 shows hypothetical examples of these types of plots with the distributions seen in a perfectly fitted model.

At this time, one can also expand the flexibility of the model by the inclusion of Monte Carlo or bootstrapping techniques. These techniques define parameters either by a distribution (Monte Carlo) or as a set of distinct data points (bootstrapping). The simulations are then run using randomly generated values for those parameters defined by distributions. This generates a data set that encompasses the known variability within the target population. In other words, these techniques provide a mean value for a specific variable along with upper and lower bounds. Figure 5 also provides an example of a Monte Carlo analysis output graph with mean values of the simulations along with upper and lower bounds.

Applications

The application of PBPK models to RA includes the incorporation of pharmacodynamic principles of tissue dosimetry into the PBPK model itself. This allows us to

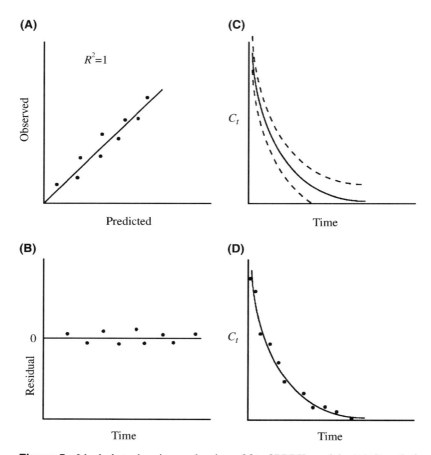

Figure 5 Ideal plots showing evaluation of fit of PBPK models. (**A**) Correlation between predicted and observed data points, (**B**) residuals between observed and predicted, (**C**) Monte Carlo analysis showing mean and upper and lower confidence bounds of the simulation for a single parameter, and (**D**) simulation where line is simulated data and points are observed data.

determine more accurately the minimum toxic dose of a xenobiotic in laboratory animals. This can then be scaled up to humans using the appropriate human PBPK model. For example, if you have a validated PBPK model you can calculate a dose-metric that corresponds to the NOAEL level found in routine animal studies. This dose metric is then modified by the same type of UFs applied to the RfD in classical studies. However, the UFs generally are significantly less because the pharmacokinetics, study design, and interspecies variations are generally accounted for in the PBPK models. The new dose metric is then used in the human PBPK model to calculate the dose associated with the adverse health effect. Thus, a more accurate RfD is calculated from a more scientifically derived dose metric.

An example of this process can be seen in the work done by Gentry et al. (44). Using validated PBPK models for acetone exposure in both the rat and the human, these investigators decreased the UF from 300 to 60 by eliminating uncertainty from pharmacokinetic, pharmacodynamic, and interspecies variability. The new UF contains a factor of 10 for intrahuman variability and a factor of six for chronic effects extrapolated from a subchronic study. The internal dose metric of area under the

Table 3 Summary of Xenobiotics with Validated PBPK Models for Humans Developed Since 1999

2-Methoxyethanal	Lidocaine
Acetone	Methoxyflurane
Acrylamide	Methyl chloride
Acrylic Acid	Methyl chloroform
Acrylonitrile	Methyl ethyl ketone
Arsenic	Methyl methacrylate
Benzene	Methyl *tert*-butyl ether
Caffeine	Methyl mercury
Carbon tetrachloride	Midazolam
Chloroform	Nitrous oxide
Chloropyrifos	Octamethylcyclotetrasiloxane
Dichloromethane	Perchlorate
Dichloromethane	Perchloroethylene
Dioxin	Pethidine
Enflurane	Propranolol
Epiroprim	Radon
Estradiol	Stryene oxide
Ethanol	Styrene
Ethyl acetate	TCDD
Ethylene glycol ethyl ether acetate	*term*-Amyl alcohol
Ethylene glycol ether	*term*-Amyl methyl ether
Ethylene glycol monomethyle ether (2-ME)	Tetrahydrofuran
Fentanyl	Theophylline
Glycyrrhzic acid	Toluene
Halogenatee hydrocarbons	Trichloroethylene
Halothane	Vinyl acetate
Iodide	Vinyl chloride
Isopronal	
Lead	

curve (AUC) was calculated using the rat PBPK model for a NOAEL dose of 900 mg/kg/day. The AUC of 10,440 mg hr/L was scaled using the UF of 60 to 626,400. The human PBPK model was run to find the dose associated with the new AUC. This new dose, 16 mg/kg/day, is thus the new RfD. For comparison, the proposed U.S. EPA RfD from classical RAs is 0.3 mg/kg/day using the same NOAEL. Table 3 provides a brief list of some of the xenobiotics for which validated human PBPK models have been developed since 1999.

Also, PBPK models can be applied in the area of food safety. Predicting drug residues within meat and milk is classically done by extrapolating plasma and tissue concentrations to extended time points based on classically derived pharmacokinetic parameters. This assumes that a linear relationship continues long after the final time point is measured. Unlike the descriptive classical studies, the predictive PBPK models are not dependent upon data derived from specific time points. This independence provides more realistic and accurate predictions at lower concentrations and at extended time points and are independent of dosage and route of administration. The addition of Monte Carlo analysis, in this case, accounts for population variability in the food animal population and further reduces uncertainty in the RA.

CONCLUSION

PBPK models reduce the uncertainty and variability inherent in classical RA. They are based on the concept of mass balance and designed around physiologic mechanisms. These modeling approaches can not only integrate data from a wide variety of pharmacokinetic and mechanistic studies, but they also are well suited for assessing the impact of biological variability. However, PBPK modeling is very dependent on the quantity and quality of the data used for parameterization. For example, while cytochrome P450 (CYP)–mediated metabolism may be the main metabolic pathway for many toxicants, wrong predictions can occur when in vivo hepatic clearance may involve active transport, direct conjugation, and non-CYP mediated metabolism (45). Also, PBPK models are time intensive to create and validate in comparison to classical pharmacokinetic or empirical curve fitting models. The information produced by these models is often complex and hard to correlate to classical kinetic parameters. This has made the incorporation of PBPK models into industry and regulatory agencies slow. In the pharmaceutical industry the test drug is eventually evaluated in humans, although this modeling tool could have a number of useful applications prior to human health safety trials as the company would have collected significant pharmacokinetic and dose–response data from preclinical animal studies. To date, the U.S. FDA acceptance of PBPK modeling is unknown. However, the U.S. EPA has recently incorporated PBPK models into the Integrated Risk Information System and will accept them as part of a total RA package. In addition, the United States Food Animal Residue Avoidance Databank is currently developing and validating PBPK models for use in estimating appropriate meat and milk withhold times in accordance with the Animal Medicinal Drug Use Clarification Act. As seen in the acetone example, once validated models are available, they can easily be used to predict more accurate RfDs. As more information becomes available these models will become further refined. Drug interactions, mixture effects, quantitative structural assessment relationships, and complex exposure scenarios will be incorporated. This will allow for the rapid screening of new substances and reduce the need for animal experiments and the cost and time associated with bringing a product to market while increasing the safety associated with human health RA. In the meantime, quantitative RAs such as these would ultimately provide scientifically sound information that will influence the risk management decision process.

REFERENCES

1. Conolly RB, Beck BD, Goodman JI. Stimulating research to improve the scientific basis of risk assessment. Toxicol Sci 1999; 49:1–4.
2. Aldridge JE, Gibbons JA, Flaherty MM, Kreider ML, Romano JA, Levin ED. Heterogeneity of toxicant response: sources of human variability. Toxicol Sci 2003; 76(1):3–20.
3. Crump C, Crump K, Hack E, et al. Dose-response and risk assessment of airborne hexavalent chromium and lung cancer mortality. Risk Anal 2003; 23(6):1147–1163.
4. Kuykendall JR, Kerger BD, Jarvi EJ, Corbett GE, Paustenbach DJ. Measurement of DNA-protein cross-links in human leukocytes following acute ingestion of chromium in drinking water. Carcinogenesis 1996; 17(9):1971–1977.
5. Fryzek JP, Mumma MT, McLaughlin JK, Henderson BE, Blot WJ. Cancer mortality in relation to environmental chromium exposure. J Occup Environ Med 2001; 43(7):635–640.
6. Enterline PE. Epidemiologic basis for the asbestos standard. Environ Health Perspect 1983; 52:93–97.

7. Kodithala K, Hopfinger AJ, Thompson ED, Robinson MK. Prediction of skin irritation from organic chemicals using membrane-interaction QSAR analysis. Toxicol Sci 2002; 66(2):336–346.

8. Smith JS, Macina OT, Sussman NB, Luster MI, Karol MH. A robust structure-activity relationship (SAR) model for esters that cause skin irritation in humans. Toxicol Sci 2000; 55(1):215–222.

9. Abraham MH, Martins F, Mitchell RC. Algorithms for skin permeability using hydrogen bond descriptors: the problem of steroids. J Pharm Pharmacol 1997; 49(9):858–865.

10. Abraham MH, Martins F. Human skin permeation and partition: general linear free-energy relationship analyses. J Pharm Sci 2004; 93(6):1508–1523.

11. Flynn GL. Physicochemical determinants of skin absorption. In: Getty TR, Henry CJ, eds. Principles of Route-to-Route Extrapolation for Risk Assessment. New York: Elsevier, 1990:93–127.

12. Potts RO, Guy RH. Predicting skin permeability. Pharm Res 1992; 9(5):663–669.

13. Enslein K. An overview of structure-activity relationships as an alternative to testing in animals for carcinogenicity, mutagenicity, dermal and eye irritation, and acute oral toxicity. Toxicol Ind Health 1988; 4(4):479–498.

14. Mumtaz MM, Knauf LA, Reisman DJ, et al. Assessment of effect levels of chemicals from quantitative structure-activity relationship (QSAR) models. I. Chronic lowest-observed-adverse-effect level (LOAEL). Toxicol Lett 1995; 79(1–3):131–143.

15. Blake BW, Enslein K, Gombar VK, Borgstedt HH. Salmonella mutagenicity and rodent carcinogenicity: quantitative structure-activity relationships. Mutat Res 1990; 241(3):261–271.

16. Patlewicz G, Rodford R, Walker JD. Quantitative structure-activity relationships for predicting mutagenicity and carcinogenicity. Environ Toxicol Chem 2003; 22(8): 1885–1893.

17. Benigni R. Chemical structure of mutagens and carcinogens and the relationship with biological activity. J Exp Clin Cancer Res 2004; 23(1):5–8.

18. Basak SC, Mills D, Gute BD, Hawkins DM. Predicting mutagenicity of congeneric and diverse sets of chemicals using computed molecular descriptors: a hierarchical approach. In: Benigni R, ed. Quantitative Structure–Activity Relationship (QSAR) Models of Mutagens and Carcinogens. New York, NY: CRC Press, 2003:208–234.

19. Benigni R, Giuliani A. Putting the predictive toxicology challenge into perspective: reflections on the results. Bioinformatics 2003; 19(10):1194–1200.

20. Haseman JK, Lockhart A. The relationship between use of the maximum tolerated dose and study sensitivity for detecting rodent carcinogenicity. Fundam Appl Toxicol 1994; 22(3):382–391.

21. Ames BN, Gold LS. The causes and prevention of cancer: the role of environment. Biotherapy 1998; 11(2–3):205–220.

22. Gaylor DW. Are tumor incidence rates from chronic bioassays telling us what we need to know about carcinogens? Regul Toxicol Pharmacol 2005; 41(2):128–133.

23. Boatman R, Corley R, Green T, Klaunig J, Udden M. Review of studies concerning the tumorigenicity of 2-butoxyethanol in B6C3F1 mice and its relevance for human risk assessment. J Toxicol Environ Health B Crit Rev 2004; 7(5):385–398.

24. Udden MM. In vitro sub-hemolytic effects of butoxyacetic acid on human and rat erythrocytes. Toxicol Sci 2002; 69(1):258–264.

25. Strauss HS. Sex biases in risk assessment. Ann NY Acad Sci 1994; 736:49–57.

26. EPA. Guidelines for Carcinogen Risk Assessment. Washington, D.C.: US Environmental Protection Agency, March 2005, EPA/630/P-03/001F.

27. Mumtaz MM, Durkin PR. A weight-of-evidence approach for assessing interactions in chemical mixtures. Toxicol Ind Health 1992; 8(6):377–406.

28. Mumtaz MM, De Rosa CT, Groten J, Feron VJ, Hansen H, Durkin PR. Estimation of toxicity of chemical mixtures through modeling of chemical interactions. Environ Health Perspect 1998; 106(suppl 6):1353–1360.

29. EPA. Supplementary Guidance for Conducting Health Risk Assessment of Chemical Mixtures. Washington, D.C.: US Environmental Protection Agency, August 2000, EPA/540/R/99005.
30. Cohen SM, Ellwein LB. Proliferative and genotoxic cellular effects in 2-acetylaminofluorene bladder and liver carcinogenesis: biological modeling of the ED01 study. Toxicol Appl Pharmacol 1990; 104(1):79–93.
31. Crump KS. A new method for determining allowable daily intakes. Fundam Appl Toxicol 1984; 4:854–871.
32. Crump K. Critical issues in benchmark calculations from continuous data. Crit Rev Toxicol 2002; 32(3):133–153.
33. Gaylor DW, Slikker W Jr. Risk assessment for neurotoxic effects. Neurotoxicology 1990; 11(2):211–218.
34. Scheuplein RJ. Pesticides and infant risk: is there a need for an additional safety margin? Regul Toxicol Pharmacol 2000; 31(3):267–279.
35. Jarabek AM. The application of dosimetry models to identify key processes and parameters for default dose–response assessment approaches. Toxicol Lett 1995; 79(1–3): 171–184.
36. Conolly RB, Butterworth BE. Biologically based dose response model for hepatic toxicity: a mechanistically based replacement for traditional estimates of noncancer risk. Toxicol Lett 1995; 82–83:901–906.
37. Hertzberg RC, MacDonell MM. Synergy and other ineffective mixture risk definitions. Sci Total Environ 2002; 288(1–2):31–42.
38. Hertzberg RC, Teuschler LK. Evaluating quantitative formulas for dose-response assessment of chemical mixtures. Environ Health Perspect 2002; 110(suppl 6):965–970.
39. Krishnan K, Clewell HJ, Andersen ME. Physiologically based pharmacokinetic analyses of simple mixtures. Environ Health Perspect 1994; 102(suppl 9):151–155.
40. Haddad S, Beliveau M, Tardif R, Krishnan K. A PBPK modeling-based approach to account for interactions in the health risk assessment of chemical mixtures. Toxicol Sci 2001; 63:125–131.
41. Krishnan K, Haddad S, Beliveau M, Tardif R. Physiological modeling and extrapolation of pharmacokinetic interactions from binary to more complex chemical mixtures. Environ Health Perspect 2002; 110(suppl 6):989–994.
42. EPA. Risk Assessment Guidance for Superfund. Volume I: Human Health Evaluation Manual (Part E, Supplemental Guidance for Dermal Risk Assessment). Washington, D.C.: US Environmental Protection Agency, July 2004, EPA/540/R/99/005.
43. Andersen ME. Toxicokinetic modeling and its applications in chemical risk assessment. Toxicol Lett 2003; 138(1–2):9–27.
44. Gentry PR, Covington TR, Clewell HJ III. Application of a physiologically based pharmacokinetic model for reference dose and reference concentration estimation for acetone. J Toxicol Environ Health 2003; 66:2209–2225.
45. Theil FP, Guentert TW, Haddad S, Poulin P. Utility of physiologically based pharmacokinetic models to drug development and rational drug discovery candidate selection. Toxicol Lett 2003; 138(1–2):29–49.

8

Toxicokinetics: Fundamentals and Applications in Drug Development and Safety Assessment

Rakesh Dixit

Toxicology Department, Johnson and Johnson Pharmaceutical Research and Development, L.L.C., San Diego, California, U.S.A.

INTRODUCTION

An understanding of the kinetics of absorption, distribution, metabolism, and excretion (ADME) in preclinical and clinical species is the key to risk assessment of preclinical toxicology data for humans. Given that doses used in preclinical toxicity are greater than pharmacological doses by orders of magnitude, it is likely that nonlinear kinetics may occur as a consequence of high concentrations mediated saturation of absorption and/or clearance process that may lead to unexpected dose and/or species-specific toxicities. Classical compartment-based blood/plasma toxicokinetic models are often sufficient to meet the objectives; however, they are too simplistic to relate biochemical and physiological processes to the kinetics of drug ADME. To meet the deficiencies of classical models, physiologically based toxicokinetic models have been developed. These models have found great utility in describing the kinetics of tissue distribution of drugs under a variety of exposure and disease conditions. It is to be emphasized that there are excellent reviews available on this topic, and readers are encouraged to consult these very detailed reviews on pharmacokinetics and toxicokinetics (1,2, and references therein).

PHARMACOKINETICS AND TOXICOKINETICS

When a drug is administered to the body, or when the body is exposed to an environmental chemical, the body's anatomy, physiology, and biochemistry act upon the drug or chemical entity for it to be absorbed into the systemic circulation, distributed to organs and tissues, metabolized to other active or inactive chemical species, and for its elimination out of the body. The rate and extent for each of these processes are dependent upon both the biology of the organism that is exposed and the physiochemical properties of the drug or chemical. The study of these mechanisms is

117

known as pharmacokinetics, and an appreciation for this area of study is important in understanding drug and chemical exposures in biological systems. Oftentimes, pharmacokinetic studies are referred to as ADME studies in consideration of the processes involved in a drug or chemical's disposition in the body.

The term *toxicokinetics* has evolved over recent years and is generally used to indicate exposure to high doses where toxicity is likely to occur, whereas pharmacokinetics is generally reserved for low, pharmacologically active doses. In the environmental industry, these terms are generally used interchangeably in ADME studies for chemical agents for which there are no pharmacologically relevant doses, probably because exposure to these agents generally occurs only through accidental contamination situations. However, in the pharmaceutical industry, toxicokinetics is reserved as a descriptive tool for exposure assessment in studies that involve safety evaluation, with less focus on the mechanisms of ADME.

The value of pharmacokinetics comes into play when trying to understand dose–response relationships to drug and chemical agents, especially when the response does not translate linearly with dose. The reasons why dose–response relationships may not be linear may be due to pharmacodynamic aspects; however, they may also be due to pharmacokinetic ones. Some of the pharmacokinetic factors that may affect drug response include differences in drug absorption at different doses (e.g., decreased solubility at high doses), in blood or tissue distribution (e.g., due to saturable protein binding, changes in tissue pathology), in metabolism (e.g., saturable enzyme kinetics), and in drug elimination. The best measure of tissue dose–response relationships or prediction of toxicity is through the quantitation of the time course of drug or chemical concentration at the site of action.

However, it is difficult to obtain relevant biological tissues to ascertain a chemical or chemical concentration in the body, and then relate that chemical concentration to a toxicological response. The least invasive and simplest method to gather information on absorption, distribution, metabolism, and elimination of a compound is by sampling blood or plasma over time. If one assumes that the concentration of a compound in blood or plasma is in equilibrium with concentrations in tissues, then changes in plasma chemical concentrations reflect changes in tissue chemical concentrations, and relatively simple pharmacokinetic calculations or models are necessary to describe the behavior of that chemical in the body.

The first thing that an investigator should do when presented with concentration versus time data is to plot the data on a log-linear scale. By quick observation of the plotted data, one can get a quick sense of the rate of absorption of the compound, of estimates of maximum drug concentration (C_{max}) and time to maximum concentration (T_{max}), whether or not concentrations are sustained for any period of time, and of the rate of elimination of the compound.

The simplest manner to quantitate drug pharmacokinetic behavior is through noncompartmental methods. Noncompartmental methods do not rely upon assumptions often inherent in compartmental models; however, they do rely on good study design for accurate assessment of pharmacokinetic parameters. Time point selection is especially critical when evaluating data by noncompartmental methods.

The most common pharmacokinetic parameters generally include plasma or tissue area under the concentration versus time curve (AUC), (C_{max}) achieved, T_{max}, apparent volume of distribution (V_d), systemic clearance (CL_s), and terminal half-life ($T_{1/2}$). The estimates for each of these parameters are described below. The derivation of each of these parameters has been described previously (3–5) but will be generally described here.

C_{max}

Using noncompartmental methods, the C_{max} achieved is often easily identified as the sample that contains the highest observed concentration.

T_{max}

The T_{max} is the time, or time point, at which C_{max} occurred.

AUC

One of the most important pharmacokinetic parameters is the drug or chemical AUC because it relates dose to exposure. AUC is the quantitative measure of the apparent amount of compound at the site from which samples were collected and concentrations measured, which in most cases is the systemic circulation. When sampling occurs from the systemic circulation, it is often an indication of systemic exposure.

The simplest manner to calculate AUC is by the linear trapezoidal rule (3). For this calculation, a concentration versus time curve is constructed as in (Fig. 1). The overall curve is then divided into a series of trapezoids, typically demarcated by observed time points and achieved concentrations. The area within each trapezoid is calculated, and the overall area under the curve is then the summation of all of the trapezoids.

The area under each trapezoid is calculated:

$$\text{Area}_{\text{trap}} = 0.5 \times (C_n + C_{n+1}) \times (t_{n+1} - t_n)$$

where C_n is the concentration at the earlier time, C_{n+1} is the concentration at the next latter time, t_{n+1} is the latter time, and t_n is the earlier time. The overall $\text{AUC}_{0-t_{\text{last}}}$ is then calculated by

$$\text{AUC}_{0-t_{\text{last}}} = \sum 0.5 \times (C_n + C_{n+1}) \times (t_{n+1} - t_n)$$

The area calculated above is $\text{AUC}_{0-t_{\text{last}}}$ when the concentration at time zero is the first concentration; t_{last} is the time of the last sample collected from which a concentration was measured. In toxicokinetics studies in the pharmaceutical industry, $\text{AUC}_{0-t_{\text{last}}}$ is often measured from time 0 to 24 hours. This is a measure of the daily systemic exposure for the investigational drug.

However, the most accurate assessment of total systemic exposure following a single dose is by measurement of AUC from time zero to time infinity. For this calculation, the remaining AUC from t_{last} to time infinity is calculated using the terminal elimination rate constant, and is added to $\text{AUC}_{0-t_{\text{last}}}$. $\text{AUC}_{0-\infty}$ is then estimated as follows:

$$\text{AUC}_{0-\infty} = \text{AUC}_{0-t_{\text{last}}} + \frac{C_{\text{last}}}{k_{\text{el}}}$$

where C_{last} is the last measured concentration, and k_{el} is the terminal elimination rate constant. Calculation of k_{el} is described below.

Terminal Elimination Rate Constant (k_{el}) and Half-life ($T_{1/2}$)

The terminal elimination rate constant (k_{el}) is a measure of the fraction of drug removed from the site of collection per unit of time, and has units of reciprocal time

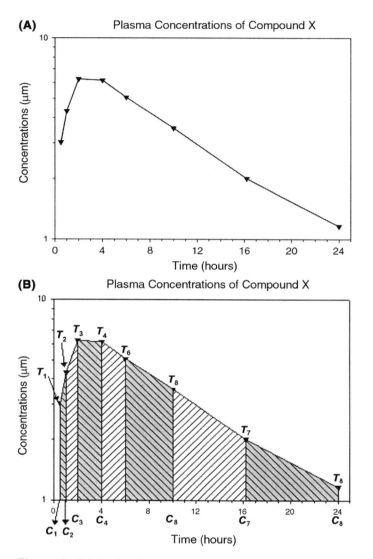

Figure 1 (**A**) Estimation of AUC from plasma concentration versus time data following extravascular administration. (**B**) A series of trapezoids is generated by dropping vertical lines from each concentration data point to the time axis. *Abbreviation*: AUC, area under concentration versus time curve.

(e.g., min^{-1} and hr^{-1}). It is determined from the slope of the straight-line portion of the terminal phase of the concentration versus time curve (Fig. 2) when the concentration data are log-transformed, as follows:

$$k_{el} = -2.303 \times slope$$

The multiplier 2.303 is a conversion factor from log units to natural log. The first-order elimination rate constant k_{el} is independent of dose.

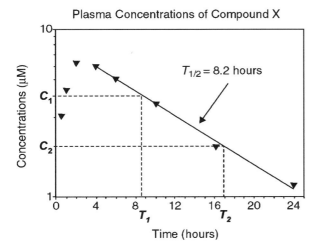

Figure 2 Concentration versus time curve of chemicals exhibiting behavior of a one-compartment pharmacokinetic model on a semilogarithmic scale. Half-life is the time required for blood or plasma chemical concentration to decrease by one-half.

Half-life ($T_{1/2}$) is the time required for the blood or plasma drug concentration to decrease by one-half and is determined from the terminal elimination rate constant by the following calculation:

$$T_{1/2} = \frac{0.693}{k_{el}}$$

The numerator 0.693 is the natural log of 2.

The accuracy of k_{el} and $T_{1/2}$ estimates is dependent upon how carefully time points are selected during the design of the pharmacokinetic study. It is especially important that time points be selected over an interval that covers at least 94% or more of total elimination of the compound. In other words, $T_{1/2}$ and k_{el} are more accurately estimated when the straight-line portion of the concentration versus time data is sampled over four half-life values. It should also be noted that both k_{el} and $T_{1/2}$ are dependent upon both volume of distribution and clearance by the relationship described below. Thus, care should be taken in analyses of data when relying upon $T_{1/2}$ as the sole determinant parameter in toxicokinetics studies.

$$T_{1/2} = \frac{0.693 \times V_d}{CL_s}$$

CL$_s$

Drugs and chemicals are cleared from the body from a variety of routes that may include fecal and urinary excretion, excretion in tears or sweat, metabolism in liver, kidneys, lungs, intestinal, or other tissues, or by exhalation. Clearance relates the volume of drug removed from the body per unit of time, and has the units of flow (mL/min). A CL_s value of 100 mL/min indicates that 100 mL of blood or plasma containing the drug is completely cleared each minute. This parameter may also be normalized to body weight, and thus values are often reported in mL/min/kg units. Total body clearance, or CL_s, describes clearance as measured from the systemic circulation. CL_s can only be calculated from concentration versus time data collected following intravenous (IV) bolus or infusion dose administration because 100% of the drug is bioavailable following IV administration. Following extravascular routes, estimates of clearance are generally normalized to bioavailability. In these instances, actual clearance can be determined if bioavailability is known.

$$CL_s = \frac{F \times \text{dose}}{\text{AUC}_{0-\infty}}$$

In the preceding equation, F is the fraction of the drug dose that entered the systemic circulation following extravascular administration. This equation is often used following bolus administration. Following IV infusion to steady state, CL_s may be calculated as a function of the infusion rate and the achieved steady-state concentration by the following relationship:

$$CL_s = \frac{k_0}{C_{ss}}$$

where k_0 is the rate of IV infusion, and C_{ss} is the steady-state concentration.

Total body clearance is also the sum of clearances by individual eliminating organs such that

$$CL_s = CL_r + CL_h + CL_i + \cdots$$

where CL_r describes renal, CL_h hepatic, and CL_i intestinal clearance. It is noteworthy that clearance of compounds from a particular organ cannot be higher than blood flow to that organ. For example, for a compound that is eliminated by hepatic biotransformation, hepatic clearance cannot exceed hepatic blood flow. This is true even if the rate of metabolism in the liver is more rapid than hepatic blood flow because the rate of overall hepatic clearance is limited by the delivery of the compound to the metabolic enzymes through the blood.

Apparent Volume of Distribution (V_d)

The apparent volume of distribution (V_d) is a proportionality constant that relates the total amount of drug in the body to the concentration of the drug in plasma. It is the apparent space or volume into which drug distributes in the body that results in an observed plasma concentration, and is an indicator of extravascular distribution. Because this parameter is a measure of volume, it has units of liters or liters per kilogram of body weight.

Pharmacokinetic volumes can be compared to the volume of a physical tank that holds some capacity of water. Let us say that we would like to know the exact volume of water in this tank, but cannot measure that volume directly. Another approach to estimate the volume of water in the tank might be by adding a known amount of dye or some other quantifiable substance into the water. After equilibrium is attained an aliquot of the water that contains the diluted dye can be sampled, and the concentration of dye in the aliquot can be analyzed. The volume of water in the tank can then be calculated by dividing the total amount of dye added to the tank by the concentration of dye sampled from the tank.

The apparent volume of distribution of a drug in the body is best determined after IV bolus administration, and is mathematically defined as the quotient of the amount of drug in the body to its plasma concentration. V_d is calculated as

$$V_d = \frac{F \times \text{dose}}{k_{el} \times \text{AUC}_{0-\infty}}$$

where F is, again, the fraction of dose that enters the systemic circulation. Following IV administration, F has a value of 1.

V_d is called the "apparent" volume of distribution because it has no actual physiological significance and usually does not refer to a real biological volume. The magnitude of the V_d term is drug specific and represents the extent of distribution of chemical out of the systemic circulation and into other body tissues. Thus, a drug with high affinity for tissues will have a large value for V_d. In fact, binding to tissues may be so extensive that the V_d of the compound is much larger than the actual volume of the body. The lowest V_d that can be achieved is the volume of the tissue that is sampled. For example, a compound that predominantly remains in the plasma will have a low V_d that approximates the volume of the plasma (e.g., 40 mL/kg in man) (6).

Once the V_d for a compound is known, it can be used to estimate the amount of chemical remaining in the body at any time given the relationship that $X_{drug} = V_d \times C_p$, where X_{drug} is the amount of drug in the body, and C_p is the plasma drug concentration.

Bioavailability

For most chemicals in toxicology, and drugs in the pharmaceutical industry, exposure occurs by extravascular routes (e.g., inhalation, dermal, or oral), and absorption into the systemic circulation is incomplete. The fraction of dose that ends up in the systemic circulation is termed "bioavailability" (F), and it is an exceedingly important concept in drug discovery and development, as it is the amount of a substance that enters the systemic circulation and is made available to the target site that is the most critical factor influencing efficacy and toxicity. Several factors can greatly affect bioavailability, and these include limited absorption after oral dosing, intestinal first-pass effect, hepatic first-pass effect, formulation, and dissolution rate.

Bioavailability is determined experimentally by comparing the plasma systemic exposure ($\text{AUC}_{0-\infty}$) values following IV and extravascular dosing. Bioavailability can be determined by using different doses, provided that the compound displays first-order kinetics by each of the routes of administration. For the determination of absolute bioavailability, pharmacokinetic data following IV administration are used as the reference to compare extravascular absorption because all chemical is delivered (i.e., 100% bioavailable) to the systemic circulation. For example, bioavailability following an oral exposure is determined as

$$F\% = \frac{\text{dose}_{i.v.} \times \text{AUC}_{0-\infty,ev}}{\text{dose}_{ev} \times \text{AUC}_{0-\infty,i.v.}} \times 100$$

where $\text{dose}_{i.v.}$ and dose_{ev} are the respective doses for IV and extravascular dose administration, and $\text{AUC}_{0-\infty,i.v.}$ and $\text{AUC}_{0-\infty,ev}$ are the areas under the concentration versus time curves for the IV and extravascular doses, respectively. Bioavailabilities for various chemicals range in values between 0% and 100%. Complete absorption of the chemical is demonstrated when $F = 100\%$. When F is less than 100%, incomplete absorption of chemical is demonstrated.

Relative bioavailabilities are often investigated in the pharmaceutical industry to help determine if there is any impact of different dose forms (e.g., particle size, solubility, dissolution, vehicle delivery, etc.) on the ability to achieve similar or different systemic exposures. In this instance, IV data is not essential, as one extravascular dose form can be compared with another extravascular dose form, where one of the dose forms may be used as the reference material.

First-Order Pharmacokinetics

When compounds exhibit first-order pharmacokinetics, the rate at which a chemical is eliminated at any time is directly proportional to the amount of that chemical in the body at that time. In addition, the rates and extent of absorption, distribution, metabolism, and elimination are predictable over a wide range of doses, and extrapolation across dose is simple. For such compounds, the rates of clearance, absorption, and half-life are independent of dose, as are T_{max} and the volume of distribution. However, the magnitude of change in AUC and C_{max} changes proportionately with dose. Thus, if one administers a compound at dose X and at $2X$, then AUC and C_{max} will increase twofold, while CL, F, k_a, k_{el} remain the same. Figure 3 demonstrates pharmacokinetic behavior of parameters for compounds that display first-order pharmacokinetics.

Saturation Toxicokinetics

For some compounds, as the dose of a compound increases, its volume of distribution or its rate of elimination may change, as shown in Figure 3. An example of a compound that displays saturation kinetics is ethanol (7). This is usually referred to as saturation kinetics and is often observed at high doses. Biotransformation, active transport processes, and protein binding have finite capacities and can be saturated. When the concentration of a chemical in the body is higher than the k_M (chemical concentration at one-half the maximum capacity or V_{max}), the rate of elimination is no longer proportional to the dose. The transition from first-order to saturation kinetics is important in toxicology because it can lead to prolonged residency time of a compound in the body or increased concentration at the target site of action, which can result in increased toxicity.

AUC and C_{max} are often normalized for dose, then plotted versus dose to assess dose proportionality of exposure (Fig. 4). The ratio of AUC/dose and C_{max}/dose will remain the same across a dose range for data that are dose proportional. However, when the ratios change with dose, the pharmacokinetics are often nonlinear. The slope of the line on a dose-normalized AUC or C_{max} versus dose plot indicates whether systemic exposure increases in a greater than, or less than, dose-proportional manner.

Some of the conditions for which nonlinear toxicokinetics may be displayed are as follows: (i) the decline in the levels of the chemical in the body is not exponential, (ii) AUC is not proportional to the dose, (iii) V_d, CL, k_{el}, or $T_{1/2}$ change with increasing dose, (iv) the metabolite profile changes with dose, (v) competitive inhibition by other chemicals that are biotransformed or actively transported by the same enzyme system occurs, and (vi) dose–response curves show a disproportionate

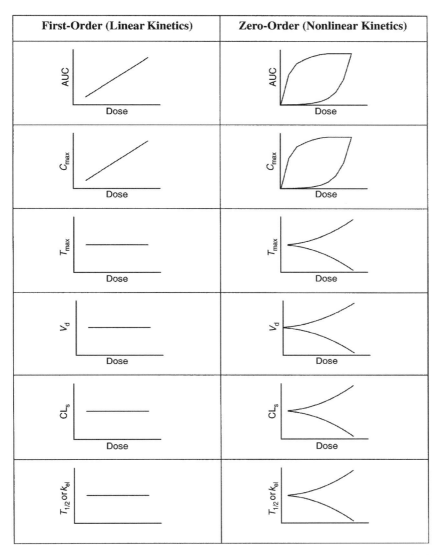

First-Order (Linear Kinetics)	**Zero-Order (Nonlinear Kinetics)**

Figure 3 AUC, C_{max}, T_{max}, V_d, CL_s, and $T_{1/2}$ following first-order toxicokinetics (*left panels*) and changes following saturable toxicokinetics (*right panels*). Pharmacokinetic parameters for compounds that follow first-order toxicokinetics are independent of dose. *Abbreviation*: AUC, area under the concentration versus time curve.

change in response with an increasing dose, starting at the dose level at which saturation effects become evident.

The elimination of some chemicals from the body is readily saturated. These compounds follow zero-order kinetics. Ethanol is an example of a chemical for which elimination follows zero-order kinetics and for which biotransformation is the rate-limiting step (7). Because ethanol elimination is zero order, a constant amount rather than a constant proportion of ethanol is biotransformed per unit of time, regardless of the amount of ethanol present in the body. Important characteristics of zero-order processes are as follows: (i) an arithmetic plot of plasma

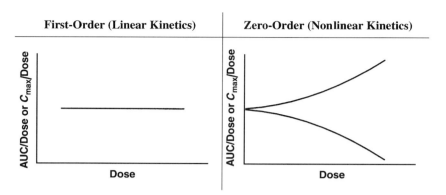

Figure 4 Dose-normalized AUC and C_{max} values following first-order toxicokinetics (*left panel*) and changes following nonlinear toxicokinetics (*right panel*). Changes may occur due to saturation of distributional processes, inhibition of metabolism or transporters, or induction of metabolism or transporters. *Abbreviation*: AUC, area under the concentration versus time curve.

concentration versus time becomes a straight line, (ii) the rate or amount of chemical eliminated at any time is constant and is independent of the amount of chemical in the body, and (iii) a true $T_{1/2}$ or k_{el} does not exist but differs depending upon ethanol dose.

Multiple Dosing and Accumulation

Upon repeated dosing with a compound, the concentration versus time profile may change compared to a single dose regimen. Accumulation (R) is often observed when the elimination rate for a compound is significantly large compared with the dosing interval. The extent of accumulation can be predicted if the terminal elimination rate constant (or half-life) and dosing interval are known by the following equation:

$$R = \frac{1}{1 - e^{-(k_{el} \times \tau)}}$$

Or alternatively

$$R = \frac{1}{1 - e^{-(0.693/T_{1/2}) \times \tau}}$$

where τ is the dosing interval in units of time. From this equation, one can easily see that if the dosing interval is significantly greater than the half-life of the compound, then accumulation is insignificant. This is best visualized in (Fig. 5), when the ratio of the dosing interval to half-life is plotted against accumulation. However, as τ begins to approximate or becomes less than the half-life, significant accumulation may be observed. It needs to be acknowledged at this point that the predicted accumulation is based upon the pharmacokinetics not changing with repeated dosing. However, if there is a change in biochemistry, for example, due to protein binding, induction or inhibition of metabolism, or toxicity that affects the pharmacokinetics of the compound, then the extent of accumulation may not be easily predicted.

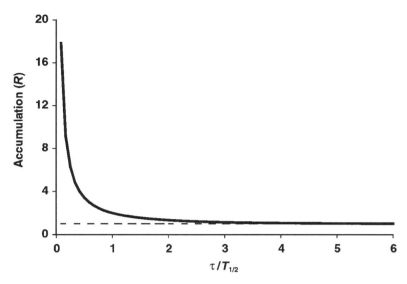

Figure 5 Effect of dosing interval relative to half-life on drug accumulation. If no accumulation is expected, $R = 1$. When $R > 1$, accumulation may be observed with repeated dosing.

Pharmacokinetic Models

Mathematical models can be developed that help the user with assessing the pharmacokinetics of the compound under study. In general, models are generated to provide nonlinear fits to concentration versus time data, and pharmacokinetic parameter estimates are calculated based upon the fitted or predicted data as opposed to the observed data. It should then be appreciated that the accuracy of the pharmacokinetic parameter estimates depends upon the accuracy of the fitted curve through the observed data points. The goodness of fit is often determined by visual inspection of the fitted concentration versus time curve to the observed data and with the use of statistical analyses.

There are a wide variety of models that one can choose, and choosing a model is dependent upon the needs of the investigator to help answer specific questions. For example, a simple pharmacokinetic model may be selected for descriptive purposes, while a more sophisticated pharmacokinetic model may be preferred for extrapolative purposes or to aid in evaluating nonlinearities in, or to discern mechanisms of, pharmacokinetic behavior.

Classical Models

The simplest of the pharmacokinetic models are the classical pharmacokinetic models. These models typically consist of a central compartment representing plasma and tissues into which drug rapidly equilibrates. The central compartment may be connected to one or more peripheral compartments that represent tissues that more slowly equilibrate with drug (Fig. 6). In the multiple compartmental models, drug is administered into the central compartment and distributes between the central and peripheral compartments. The rates of distribution between central and peripheral compartments are generally of first-order. Chemical elimination occurs from the

One-Compartment Model

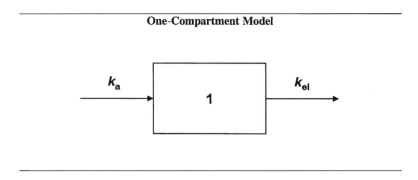

Two-Compartment Model

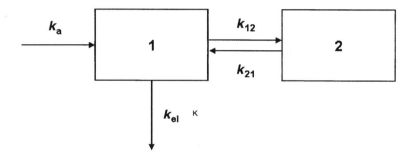

Figure 6 Compartmental pharmacokinetic models where k_a is the first-order extravascular absorption rate constant into the central compartment (**1**), k_{el} is the first-order elimination rate constant from the central compartment, and k_{12} and k_{21} are the first-order rate constants for distribution of chemical into and out of the peripheral compartment (**2**) in a two-compartment model.

central compartment, which is assumed to contain rapidly perfused drug eliminating tissues (e.g., kidneys and liver).

Advantages of the classical compartmental pharmacokinetic models are that they require no information on tissue physiology or anatomical structure. These models are valuable in describing and predicting the time course of drug concentrations in the systemic circulation at different doses, the extent of chemical accumulation with multiple doses, and aid in selecting effective dose and dose regimens in efficacy and toxicity studies to achieve specific exposures.

A disadvantage of classical models is that because compartments do not reflect the biology of specific tissues, and the rates of movement between compartments and of elimination are not physiologic, but rather are simple mathematic solutions for goodness of fit, the time course and exposure information in a specific tissue are often difficult to assess. Additionally, drug disposition is often described by first-order processes, thereby leading to difficulties when saturation kinetics apply. Oftentimes, models with different solutions for the model parameter estimates are generated to describe the pharmacokinetics of drugs across a range of doses when nonlinearity in pharmacokinetics is observed.

One-Compartment Models

The simplest toxicokinetic analysis entails measurement of the plasma concentrations of a drug at several time points after the administration of a bolus IV injection. If the data obtained yield a straight line when they are plotted as the logarithms of plasma concentrations versus time, the kinetics of the drug can be described with a one-compartment model (Fig. 6). Compounds whose toxicokinetics can be described with a one-compartment model rapidly equilibrate, or mix uniformly, between blood and the various tissues relative to the rate of elimination. The one-compartment model depicts the body as a single homogeneous unit. This does not mean that the concentration of a compound is the same throughout the body, but it does assume that the changes that occur in the plasma concentration reflect proportional changes in tissue chemical concentrations.

In the simplest case, a curve of this type can be described by the algebraic expression

$$C = C_0 \, e^{-k_{el} \times t}$$

where C is the blood or plasma chemical concentration over time t, C_0 is the initial blood concentration at time $t = 0$, and k_{el} is the first-order elimination rate constant with dimensions of reciprocal time.

Two-Compartment Models

After the rapid IV administration of some drugs, the semilogarithmic plot of plasma concentration versus time does not yield a straight line but a curve that implies more than one dispositional phase. In these instances, the drug requires a longer time for its concentration in tissues to reach equilibrium with the concentration in plasma, and a multicompartmental analysis of the results is necessary (Fig. 6). A multiexponential mathematical equation then best characterizes the elimination of the drug from plasma.

Generally, a curve of this type can be resolved into two monoexponential terms (a two-compartment model) and is described by

$$C = A e^{-\alpha t} + B e^{-\beta t}$$

where A and B are proportionality constants and α and β are the first-order distribution and elimination rate constants, respectively. During the distribution (α) phase, drug concentrations in plasma decrease more rapidly than they do in the post-distributional elimination (β) phase. The distribution phase may last for only a few minutes or for hours or days. Whether the distribution phase becomes apparent depends on the time of obtaining the first plasma samples, and on the relative difference in the rate of distribution relative to the rate of elimination. If the rate of distribution is considerably rapid relative to elimination, the timing of blood sampling becomes critical in the ability to distinguish a distribution phase. The equivalent of k_{el} in a one-compartment model is β in a two-compartment model.

Occasionally, the plasma concentration profile of many compounds cannot be described satisfactorily by an equation with two exponential terms, for example, if the chemical has an exceptionally slow distribution into and redistribution out of a deep peripheral compartment or tissue. Sometimes three or four exponential terms are needed to fit a curve to the log C versus time plot. Such compounds are viewed as displaying characteristics of three- or four-compartmental models.

Physiologically Based Pharmacokinetic Models

Physiologically based pharmacokinetic (PBPK) models are important biologically based tools that aid in quantitating exposure to compounds and/or their metabolites either in the systemic circulation or in a putative target tissue or active site. An advantage of PBPK models is that they provide quantitative information about tissues that cannot be readily collected, or from which determination of concentration is difficult (e.g., due to low doses and limitations on bioanalytical methods). Moreover, the models provide important information that helps to relate tissue dosimetry with pharmacological or toxic effects at the active site.

For extrapolative purposes, PBPK models provide greater flexibility due to the fact that compartments in the models reflect true tissue or organ systems, or groups of tissues and organs, with delivery of drug to a tissue governed by blood flow to that tissue. Ordinary differential equations are used to solve for the time course of drug concentrations throughout the tissues, and these account for the anatomy (tissue volume), biochemistry (e.g., protein binding, active transport systems, and metabolic enzymes), and physiology (tissue blood flows) of the tissue and of the organism. Another important feature of PBPK models is that mathematical equations can be incorporated into the structure to describe the physicochemical properties of the compound (e.g., solubility in tissues and stability).

Physiological parameters are typically obtained from the literature (6,8), but can also be obtained experimentally, and are not dependent on the compound under study, unless the compound is shown to alter physiological parameters through pharmacological or toxicological action. Chemical-specific biochemical parameters are experimentally determined for the compound. For example, tissue: blood partition coefficients are determined in the laboratory (9,10), as are metabolic rate constants (e.g., k_M and V_{max}) (9) for saturable metabolism. Metabolic rate constants can be specific to the enzyme or group of enzymes responsible for the metabolism of the compound. Administration routes and excretion mechanisms of the particular compound are also included in the mathematical description of the model.

Figure 7 illustrates a schematic of a PBPK model. In this particular model, the body is divided into physiologically and toxicologically relevant compartments (i.e., fat, kidney, liver, and skin). Other tissues are lumped into the rapidly perfused or slowly perfused compartments based upon the percentage of the total cardiac output that is delivered to these tissues. Exposure occurs by IV injection. Once in the blood, the compound circulates through the body and is delivered to the tissues where it may leave the blood and enter the cells of the tissue itself, or it may simply pass through the tissue. Distribution of compound into tissue is a function of its solubility in that tissue. In certain tissues such as the liver and, in this case, the kidneys, the compound may be metabolized. In addition to metabolism, this model contains a description of elimination of the compound from the body by exhalation.

The basic mathematical description for the rate of change in amount of compound in each tissue with respect to time may be written as follows:

$$\frac{\mathrm{d}A_i}{\mathrm{d}t} = Q_i \times (C_{art} - CV_i)$$

where i represents the tissue, Q_i is the rate of blood flow to the tissue, C_{art} is the concentration of compound in the arterial blood supply, and CV_i is the concentration of

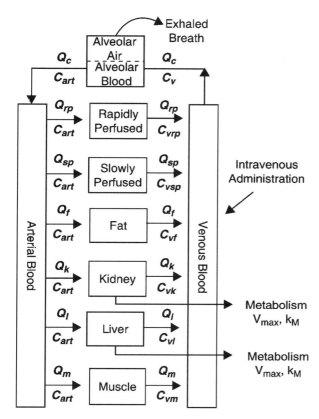

Figure 7 PBPK model structure. Q_c is total cardiac output. Q_{rp}, Q_{sp}, Q_f, Q_k, Q_l, and Q_m are blood flows to lumped rapidly perfused tissues, lumped slowly perfused tissues, fat, kidneys, liver, and muscle tissue groups, respectively. C_{art} and C_v are concentrations of compound in the arterial blood and venous blood, respectively. C_{vrp}, C_{vsp}, C_{vf}, C_{vk}, C_{vl}, and C_{vm} are concentrations of the compound in venous blood leaving the lumped rapidly perfused tissues, lumped slowly perfused tissues, fat, kidneys, liver, and muscle tissue groups, respectively. k_M and V_{max} are the metabolic rate constants in both liver and kidneys. Arrows indicate direction of flow. *Abbreviation*: PBPK, physiologically based pharmacokinetic.

compound that exits the tissue and enters the central venous blood supply. CV_i may be approximated as follows:

$$CV_i = \frac{A_i}{V_i \times P_i}$$

where A_i is the amount of the compound in the tissue, V_i is the tissue volume, and P_i is the respective chemical-specific tissue:blood partition coefficient.

For a metabolizing tissue, for example the liver, the rate equation may be written to include Michaelis–Menten saturable metabolism (11)

$$\frac{dL_i}{dt} = Q_L \times (C_{art} - CV_L) - \frac{V_{max} \times CV_L}{k_M + CV_L}$$

where L represents the liver, V_{max} is the maximum rate of metabolism, and k_M is the Michaelis constant. After a PBPK model is generated and validated against laboratory-obtained data, they offer greater flexibility for extrapolative purposes (12–14), for example, across dose, dose regimen, sex, routes of administration, and within species and across species (e.g., animal to man) (7,15–17). For this reason, they are widely used as tools in human health risk assessment, where they are advantageous in reducing the uncertainty in estimating human risk from exposure to compounds.

Modeling and knowledge of toxicokinetics data can be used in deciding on what dose or doses of chemical to use in the planning of toxicology studies (e.g., if specific blood concentrations are desired), in evaluating dose regimens (e.g., intravascular vs. extravascular, bolus injection vs. infusion, or single dosing vs. repeated doses), in choosing appropriate sampling times, and in aiding in the evaluation of toxicology data (what blood or plasma concentrations were achieved to produce a specific response, effects of repeated dosing on accumulation of chemical in the body, etc.).

Classical Plasma/Tissue Toxicokinetics: Applications in Drug Discovery and Drug Development

Prior to conducting efficacy and safety evaluations in humans, physicians and government regulators need to know both short- and long-term effects of a drug in at least one rodent and one nonrodent species, and whether or not the drug is genotoxic. Additionally, prior to exposing young women of child-bearing age to new drug entities, it is also critical to know any potential reproductive and developmental toxicity liabilities. Carcinogenic potential in two rodent species after a two-year exposure must also be assessed for those drugs that may be used on a chronic basis (more than six months continuously or intermittently). Nonclinical safety assessment studies in laboratory animal species and/or appropriate in vitro models are conducted and provide much needed information regarding the safety and potential target organ toxicity of an investigational new chemical entity. Species differences in ADME processes as well as differences in pharmacodynamics are key limiting factors in utilizing nonclinical animal safety assessment data to predict human safety. During the last 15 years, with a better integration of metabolism and toxicokinetics data and mechanisms of biological activity, it has been possible to greatly reduce the uncertainty regarding the extrapolation of animal toxicology data to predict human safety. Preclinical toxicokinetics studies combined with toxicity data can provide information on systemic exposure and its relationship to dose levels and toxicity. In addition, the toxicokinetics data are also used to justify the choice of species/strain, study design, and dose selection for subsequent studies.

Applications of ADME and Toxicokinetics in Drug Discovery and Development

Drug discovery and development processes are key to bringing a new chemical entity for gaining marketing approval. The drug discovery often starts with a hypothesis that a specific disease may involve an upregulated target enzyme or receptor, and its blockage may lead to effective treatments. Drug discovery basic researches create suitable in vitro/in vivo models to test the proof of mechanisms in laboratory settings and develop plans for screening potential new drug molecules from a vast library of chemicals. During this process, thousands of molecules are screened to seek molecules with desired biological activity at low-nM range and is followed by confirmation of efficacy at low in vivo doses in animal models. Once a drug is optimized for efficacy and adequate pharmacokinetics and metabolism characteristics are attained, a potential clinical candidate passes through the early development stage. During development stages efforts are focused on evaluation of the toxicity in preclinical species followed by a detailed evaluation of efficacy and safety in late stage clinical trials.

Most screening assays involve in vitro receptors or enzyme activity models. The extrapolation of data from in vitro to complicated in vivo systems cannot be made without a knowledge on ADME and toxicokinetics. This is attained by early pharmacokinetic and metabolic evaluation, which helps to ascertain that the selected molecule has druglike properties. Detailed information on the metabolic processes of a new drug is necessary to fully understand if there are active metabolites that may have the desired pharmacological activities or may have safety liabilities. It is interesting to note that many of the currently available antihistamines came into existence owing to the discovery of their active metabolites which had favorable safety and pharmacological activity profile. Early drug metabolism and pharmacokinetics studies are key to gaining a good understanding of the basic mechanisms of the events involved in ADME, metabolic induction/inhibition with a possibility of drug–drug interaction, sources of interindividual variability in pharmacokinetic and pharmacodynamics, and their impact on potential drug toxicities and efficacy. Failures due to undesirable pharmacokinetic properties, including short $T_{1/2}$, poor absorption, and extensive first-pass metabolism have been substantially reduced in recent years when compared to about 15 years ago (18). However drug–drug interactions and polymorphisms continue to pose significant problems. Figure 8 lists properties for ADME optimization of a potential clinical candidate and a brief discussion follows.

Absorption

Oral route is the major desirable route of administration of pharmaceuticals. A large number of physiological factors as well as many physicochemical factors impact the absorption. Some of the important biological factors that impact drug absorption (extent as well as the rate of absorption) include the stomach emptying rate with an effect on drug absorption and gastrointestinal (GI) residence time, lipid solubility, surfactants, etc. (19). Most drugs are absorbed using a passive absorption process through the blood stream; however, a very small fraction of drug molecules (i.e., large molecules, steroids, etc.) is absorbed through lymphatics. Passive process lipid solubility and dissolution are critical for a good drug absorption. The GI epithelial membrane is composed of tightly packed phospholipids interspersed with proteins, and the transcellular passage of drugs depends on their lipid solubility and permeability characteristics to cross the lipid bilayer of the epithelial cell membrane. The classical study of barbiturates (20) demonstrated that the absorption of barbiturates is greatly increased by increasing lipophilicity as a consequence of increased membrane permeability. Peptide and protein drugs are transported across the intestinal epithelium by energy-dependent carrier-mediated peptide transporters. Some of the examples are beta-lactam antibiotics, angiotensin converting enzyme inhibitors, encephalins, etc. In addition to absorption, presystemic first-pass metabolism occurs first in intestines and then in the liver. This is one of the many reasons why a given drug with high metabolic clearance generally shows low systemic availability. The lipophilicity of a drug not only affects the membrane permeability, but the metabolic activity as well. It is just not a coincidence that first-pass metabolic clearance appears

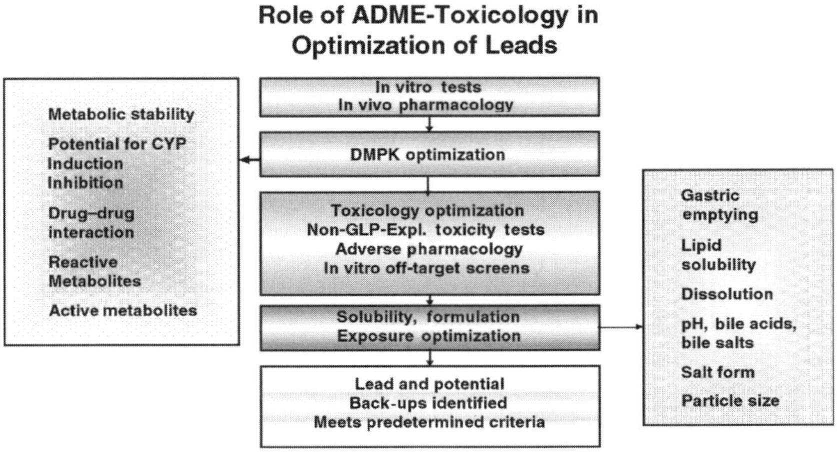

Role of ADME-Toxicology in Optimization of Leads

Figure 8 A schematic presentation of the role of ADME in identification of lead candidates is presented. Once a candidate drug is optimized by in vitro tests, it passes through in vivo pharmacology testing to get a proof of desired pharmacology in animal models of indicated human diseases. This is followed by a process of DMPK optimization. In this process DMPK scientists conduct in vitro–in vivo studies to show the metabolic stability of a candidate compound in a variety of drug-metabolizing systems (microsomes, hepatocytes, liver slices, etc.), potential for concentration and time-dependent induction and inhibition of CYPs to predict drug–drug interaction, reactive metabolites through the trapping of reactive metabolites by glutathione or cyanide, and quantitative comparison of metabolites across animal toxicology species. Once the drug has been optimized for DMPK properties, it goes through ancillary pharmacology tests to assess the potential to cause significant cardiovascular, renal, and pulmonary adverse pharmacological effects. In vitro off-target activity on key relevant pharmacological targets is often assessed to predict toxicological liabilities. Given that many synthetic compounds are poorly absorbed, extensive formulation optimization studies are often needed to improve the systemic bioavailability. Limited in vitro (e.g., genotoxicity and cytotoxicity/ apoptosis) and in vivo toxicity testing (acute to one week) may be conducted to assess early toxicity potential of a candidate compound before nominating for a full GLP toxicity testing. *Abbreviations*: ADME, absorption, distribution, metabolism, and excretion; CYP, cytochrome P-450; DMPK, drug metabolism–pharmacokinetics; GLP, good laboratory practice.

to be higher for lipophilic drugs due to their high membrane permeability and availability to microsomal enzymes (21,22).

In addition to high membrane permeability, lipophilic drugs also tend to show increased affinity for drug-metabolizing enzymes. In vitro studies by Martin and Hansch (23) demonstrated that variations in maximum velocity (V_{max}) values for a series of lipophilic compounds with chemical structure had the Michaelis constant (k_M) values, which varied by approximately 1000-fold. Overall, the k_M values were found to correlate significantly with their lipophilicity. The higher lipophilicity contributed to lower k_M values (higher enzyme affinities), and kinetic studies have consistently showed a positive correlation between metabolic clearance and lipophilicity for dihydropyridine calcium channel blockers (24).

In addition to lipid solubility, aqueous solubility plays an important role in drug absorption because a drug must be reasonably soluble in the aqueous environment to be available to cell membranes. The inability of poorly soluble drugs to show good absorption is very well known. The discovery of HIV protease inhibitors is an example that illustrates the concept of enhancing drug solubility to enhance exposures. Initial efforts to make serine protease inhibitors such as HIV protease inhibitors sufficiently bioavailable were unsuccessful partly due to their high lipophilicity, high first-pass clearance, and low aqueous solubility. Systemic availability was enhanced by incorporating a basic amine into the backbone of this series which eventually led to the discovery of indinavir (crixivan) (25,26). It is interesting to note that the solubility of indinavir increased from 0.07 mg/mL at pH 7.4 to 60 mg/mL at pH 3.5 due to the protonation of the pyridine nitrogen ($pK_a = 3.7$).

Enhancement of Systemic Exposure in Safety Studies

Because of the ethical constraints in directly evaluating high dose toxicities in humans, preclinical safety assessments at fairly high doses must be extensively studied in laboratory animals. The safety of a drug candidate requires very thorough evaluations and must follow mandated Good Laboratory Practice (GLP) regulations. It is critical that a new chemical entity undergoing toxicity testing must have adequate systemic exposure to induce target organ toxicity. Additionally, this provides assurance to regulatory agencies that preclinical toxicity evaluations were carried out at exposures exceeding potential human exposures. A poorly absorbed drug makes neither a good preclinical candidate nor a good clinical candidate. The balance between aqueous and lipid solubility becomes critical with increasing doses of even a well-absorbed drug because absorption can become limited due to inadequate aqueous solubility, poor dissolution, or precipitation of drug at high doses.

It is worth mentioning that there has been an increasing trend to have new molecular entities that have high molecular weight and are lipophilic and poorly water soluble. These drugs fall in Biopharmaceutics Classification System class 2 drugs (27). The modern high throughput screening is unintentionally biased toward finding high-molecular-weight compounds for biological activity. Lipinski's rule-of-five (28) has helped to provide guidance about properties and structural features of drug-like molecules. The rule-of-five (derives its name from the fact that the relevant cutoffs are all multiples of five) predicts that an orally administered compound is more likely to be poorly absorbed or poorly permeable when it possesses the following properties:

- molecular mass greater than 500 Da,
- high lipophilicity (expressed as $\log P > 5$),
- more than five hydrogen bond donors,
- more than 10 hydrogen bond acceptors.

As a general rule, drugs with $\log D_{7.4}$ values (octanol/aqueous buffer partition coefficients at pH 7.4 uncorrected for degree of ionization) between –0.5 and 2.0 are expected to be readily absorbed (29) and can be easily administered in aqueous formulations as solutions or suspensions. These types of drugs tend to be acidic/neutral, and generally have water solubility greater than 0.5 mg/mL; however, deviations from these norms have been observed due to decreasing solubility at higher doses, as well as high presystemic metabolism. (Table 1) lists some of the vehicles for maximizing exposures for nonclinical toxicology studies.

Solutions tend to absorb better because they can easily pass through the GI tract intestinal lumen into the blood stream. When compared to a decade ago, newer molecules tend to be bulky and poorly soluble and, therefore, are often poorly absorbed by animal species at higher doses. To enhance solubility, early studies

Table 1 Exposure Maximization (Absorption Enhancement) Vehicles for Nonclinical Toxicology Studies

---

Potential aqueous vehicles
0.5% or 1% methylcellulose (up to 10 mL/kg) with and without 0.15% SDS
Acidified (0.1 N HCl) 0.5% or 1% methylcellulose (up to 10 mL/kg)
3% hydroxypropylcellulose, 15% sucrose, and 0.15% SDS (up to 10 mL/kg)
10% Tween 80 in 0.5% methylcellulose at 5 mL/kg
Nonaqueous vehicles
PEG400 at 2 mL/kg
Imwitor 742/Tween 80 (50/50): 1 mL/kg for rodents, monkeys;
    0.5 mL/kg for dogs
Other potential emerging vehicles
Miglyol 810 or 812
Labrafil N2125CS or M1944CS
Imwitor 742
Soybean oil/Tween 80/span 80 (50/25/25)
Labrafil N2125CS/Tween 80 (70/30)
Cremophor EL/Tween 80 (70/30)
Vegetable oil/span 80 (70/30)
Vegetable oil/Tween 80 (70/30)
Appropriate Gattefosse lipid systems
Other vehicles from FDA inactive ingredients guide
Oils (vegetable, corn, sesame, soybean, olive)
Lipids (corn glycerides, glyceryl caprylates, glyceryl oleate, caprylic/capric triglycerides,
    medium chain triglycerides)

Abbreviations: SDS, sodium dodecyl sulfate; FDA, food and drug administration.

may involve creation of suitable salt forms, a more soluble amorphous and crystal form. Additionally many solubilizing vehicles such as polyethylene glycols (PEGs) 300/400, imwitor, propylene glycol, sorbitol, Tween (polysorbate 80), acidified carboxymethyl cellulose, hydroxypropylcellulose/sucrose/sodium lauryl sulfate, cremophor, cyclodextrin, and span (sorbitan monoester) have been successfully used in enhancing exposures to poorly soluble drugs. For safety assessment studies, it is also critical to assure that lipophilic vehicles are well tolerated by the animal species involved, and the toxicity of the formulated drug is not enhanced or reduced by the chosen formulation other than what is expected from the increase in exposures. Of all new vehicles, cyclodextrins provide an interesting approach to solubilize small molecule drugs. Cyclodextrins are cyclic glucose polymers with hydrophilic "outer surface" and hydrophobic "inner surface" cavities, and with these unique properties masses up to 500 Da can be easily accommodated. With cyclodextrins, large doses of fairly insoluble drug can reside as stable complexes, and solubility of cyclodextrins in GI medium essentially drives the drug absorption. Nomier et al. (30) reported up to a fourfold increase in systemic exposures with a hydroxypropyl-β-cyclodextrin complex than with standard methylcellulose suspensions.

The utility of various lipophilic formulations in improving solubility-mediated absorption is best illustrated by the example presented in Figure 9. An investigational drug X being developed to treat multiple neurological diseases, presented significant challenges to safety assessment studies. The drug had a water solubility of <0.05 μg/mL. In early studies, the systemic drug bioavailability from allstandard aqueous formulations was very poor in both rodent and nonrodent species.

Absorption Enhancement by Lipophilic Vehicles of a Compound X (Water Solubility <0.1 µg/mL)

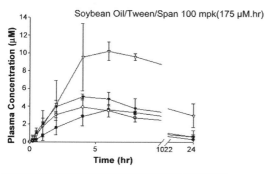

	Dose in mg/kg	AUC (µM.hr)
HPC/SDS	150	70
Imwitor/Tween	150	54
PEG 400	150	50
Soybean oil/Tween/ Span	100	175

Figure 9 Toxicokinetics data to demonstrate absorption enhancement of a poorly water soluble compound is presented. Compound X, a lipophilic compound, was being developed to treat CNS diseases. Given its poor water solubility of $< 0.1\,\mu g/mL$, initial attempts to create a soluble salt form were unsuccessful. Several lipophilic vehicles are as follows: HPC/SDS, imwitor/ Tween 80, and PEG400 and soybean oil/Tween 80 and span. Of all the lipophilic vehicles, the soybean oil/Tween 80/span vehicle was superior to other vehicles in improving absorption (Table 1). The absorption enhancement was likely due to the permeability enhancing ability of span and Tween 80. The top line in the graph is related to soybean oil/Tween span vehicle. The other three lines are related to HPC/SDS, imwitor/Tween and PEG400. *Abbreviations*: CNS, central nervous system; HPC/SDS, hydroxycellulose/sodium dodecyl sulfate; PEG, polyethylene glycol.

Extraordinary efforts were made to enhance solubility and improve oral exposures by formulating the drug in a variety of lipophilic vehicles, including PEG400, imwitor: Tween 80, and combinations of soybean oil, Tween, and span. When compared to other lipophilic vehicles, the combination of soybean oil, Tween, and span was found to give highest exposures in rats. This was likely related to enhanced lipid solubility attained by soybean oil and Tween and the emulsifying action of span.

For compounds that show low solubility, the oral systemic availability can be greatly enhanced by maximizing the rate of dissolution. According to the Noyes–Whitney equation (31), the dissolution rate can be described as follows:

$$\frac{\mathrm{d}W}{\mathrm{d}T} = \frac{DA(C_\mathrm{s} - C)}{L}$$

where $\mathrm{d}W/\mathrm{d}T$ is the dissolution rate, A, surface area, C, the concentration of drug in the dissolution medium, C_s, the concentration in the saturable solution, D, diffusion coefficient, L is the thickness of the diffusion layer.

For most drugs absorbed through a simple diffusion process, the membrane absorption can be considered generally very rapid ($C = 0$); therefore, increasing the A (surface area) and C_s (the concentration in the dissolution medium) will likely increase the rate of dissolution, the rate of passage from solid to solvated state, and hence the absorption.

Reduction in particle size can greatly enhance the surface area. Micronization of particles has been shown to enhance absorption. Drugs with a water solubility of $< 1 \text{mg/mL}$ can greatly benefit from the micronization process; however, a caution is warranted because small particles can agglomerate and may disperse poorly. Dispersants and stabilizers need to be added to prevent agglomeratization. Additionally, surfactants need to be added to provide drug in a form that can disperse on wetting. Figure 10 shows the utility of nanoparticles and formulation enhancers in improving the absorption of a drug X.

Permeability enhancers are being used to enhance the absorption of hydrophilic molecules (32). Generally, hydrophilic molecules have a tendency to remain in the aqueous medium and not partition into lipophilic enterocyte membranes. Most permeation enhancers are surfactants that increase permeability by solubilizing membranes and modifying trans/paracellular absorption. Some examples of permeation enhancers include span 80 (sorbitan ester), polyoxyethylene ether, dodecylmaltoside, sodium caprate, medium chain glycerides, and N-acetyl amino acids. It must be emphasized that many permeability enhancers tend to be not well tolerated, and chronic safety data will be necessary prior to their use in safety studies.

The passage of a well-absorbed drug from enterocytes into systemic circulation can be greatly modified by both efflux and influx drug transporters. P-glycoprotein

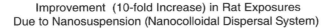
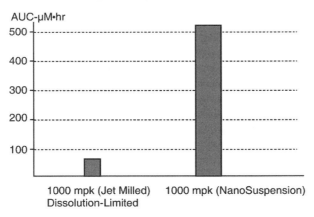

Figure 10 Absorption enhancement with a nanocolloidal dispersion system. Nanosuspensions are submicron colloidal dispersions of pure particles of drug, which are stabilized by surfactants. Nanosuspensions differ from nanoparticles, which are polymeric colloidal carriers of the drug and solid lipid nanoparticles, which are lipidic carrier of drug. Nanosuspension of a candidate drug was created in hydroxypropylcellulose, sucrose, and SDS. The candidate drug showed a solubility/dissolution-limited saturable absorption occurring at 1000mg/kg/ day in both aqueous and nonaqueous formulations. Its absorption was substantially (approximately 10-fold increase in systemic availability) improved with the use of a nanosuspension. *Abbreviation*: SDS, sodium dodecyl sulfate.

(P-Gp) is a major efflux transporter and serves as a major barrier to host defense mechanism to eject toxic molecules from cells. P-Gp is present in many important barrier systems such as enterocytes, blood–brain barrier, and placenta. The inhibitors of enterocytes and blood–brain barrier P-Gp have been exploited to enhance the systemic exposure and target tissue retention of drugs that are P-Gp substrates. Nucleoside and peptide transporters are good examples of influx transporters, and efforts are being made to utilize properties of influx transporters to enhance drug bioavailability of polar hydrophilic chemical moieties. Given that cytochrome P-450 (CYP) and phase II metabolizing enzymes are the major pathways of first-pass clearance, the inhibitors of these metabolizing enzymes are being considered in enhancing the drug bioavailability of important medical breakthrough drugs that are poorly bioavailable. The use of combination of metabolism inhibitors with a rapidly cleared drug must be carefully balanced against unwanted side effects related to metabolic inhibition, including potential toxicity and drug–drug interactions in clinical settings.

Other physiological variables such as diet, feeding (fed vs. fasted), intraluminal pH variations, and GI transit times have also been exploited to enhance drug absorption. Delays in gastric emptying generally increase the time of exposure of the entire GI tract to a poorly soluble drug which may be a slow and sustained absorption. Dressman (33) provided evidence that under the fed stage, the gastric residence is generally increased for drugs irrespective of whether it is delivered as aqueous solutions, suspensions, pellets, or tablets. It has also been shown that density and viscosity of the formulation have significant effect on gastric transit time. High density or thick formulations tend to enhance GI exposure time and total drug absorption.

Distribution, Metabolism, and Excretion

Distribution of systemically available drugs to various organs is critical to our understanding of pharmacokinetics and toxicity. Simple concepts regarding drug distribution are best illustrated by Welling (34). Drugs that can enter the intracellular water (60% of body weight) will be most bioavailable to tissues. In contrast, drugs that have difficulty in crossing cell membranes to reach intracellular sites will be largely distributed in extracellular volume (21% of body weight). Therefore, depending upon the distribution of drug in whole blood (7% body weight), extracellular water (21% body weight), intracellular water (39% body weight), and total body water (60% body weight), an eightfold difference in plasma concentrations can be expected. This simple distribution concept is complicated by binding of drugs to plasma and tissue proteins as well as to other components in tissues. Red blood cell (RBC) partitioning by certain drugs may greatly provide conflicting exposure data based on plasma concentrations versus whole blood concentrations. A good example is an anticancer drug tubercidin which concentrates (80–90%) exclusively in RBCs leading to underrepresentation of exposures when based on plasma only. Drugs that are highly lipophilic tend to give a volume of distribution that exceeds the volume of total body water. Pentothal is rapidly distributed in fatty tissues and then slowly redistributed through the general circulation. Digoxin shows an apparent volume of distribution of 500 L which exceeds the volume of total body water by 500-fold. This occurs because digoxin is not homogenously distributed and has a high extravascular penetration. Species differences in organ distribution was shown for gabapentin, an anticonvulsant drug which is predominantly distributed in rat pancreas but is poorly distributed in pancreas of monkeys and humans which can give rise to species differences in toxicities. Drug distribution into tissues could be perfusion or diffusion dependent. Blood perfusion rates (mL/min g tissue) for various tissues are as follows: lung (10), kidney (4), liver (0.8), heart (0.6), brain (0.5), muscle (0.3), and fat (0.03). If perfusion controls drug distribution, then muscle and fat will be slowest to equilibrate while lung, kidney, liver, brain, and heart will show a fairly rapid equilibration. This process tends to dominate for compounds that are highly lipophilic. Diffusion rate limited distribution is substantially more dominant for water soluble polar drugs than for nonpolar drugs, and best examples of diffusion-limited distribution are provided by central nervous system (CNS) active drugs that require crossing the blood–brain barrier.

Plasma protein binding is an important concept for evaluating drug distribution (34). Plasma albumin is the major drug binding moiety; however, alpha$_1$-acid-glycoprotein (AAG) also plays an important role. Basic drugs tend to favor AAG, whereas acidic drugs bind preferentially to albumin. Albumin with a molecular weight of 69,000 exists mostly in plasma (4 g/100 L); however, extravascular tissues (e.g., liver) possess a small amount of albumin. Generally the binding of a drug to albumin is a rapidly reversible process, and the process is dependent on the number of binding sites on albumin for a given drug and the concentration of drug and albumin. Intravascular plasma protein–bound drugs owing to their molecular size and polarity cannot cross the blood–brain barrier as well as most organs such as liver and kidney. For certain drugs failure to take into account the protein can result in misrepresentation of the volume of distribution and overall systemic exposure. When binding exceeds 99%, a saturable binding is likely to occur at higher concentrations, which can lead to a disproportionate increase in exposure to free drug. Plasma protein binding and tissue binding have an opposite effect on circulating drug levels. It is to be emphasized that a fine equilibrium exists between plasma and tissues to maintain an equilibrium between intravascular and extravascular compartments. As the concentration falls in one compartment, the other compartment releases more drug to establish a new equilibrium. Also, it is a misconception that highly protein-bound drugs are not available to intracellular targets. In fact, this is not true because many protein-bound drugs are very readily distributed extravascularly and very effective in reaching to intracellular target sites. Percent protein binding should not be taken at face value, but must be evaluated along with the rate of reversibility and saturability of protein binding as well as extravascular tissue binding.

Marked differences in metabolism are major contributors to species-related drug-induced toxicity, and these differences often complicate and limit the interpretation and risk assessment of the preclinical toxicology data (25,26). Species differences in the key CYP isozymes and phase II conjugation pathways greatly impact both the rate of drug metabolism and the metabolite patterns. The example below highlights the importance of qualitative and quantitative species differences in metabolism. Indinavir (MK-639) is a potent HIV protease inhibitor. The major metabolic pathways of indinavir include (a) glucuronidation at the pyridine nitrogen to yield a quaternized ammonium conjugate, (b) pyridine *N*-oxidation, (c) *para*-hydroxylation of the phenylmethyl group, (d) 3′-hydroxylation of the indan, and (e) *N*-depyridomethylation (35). While all oxidative metabolites were observed in safety species, rats, dogs, and monkeys, and in humans, *N*-glucuronide was found only in monkey and human urine (36). A unique metabolite *cis*-2′-3′-dihydroxyindan was formed only in monkeys, however, other species lacked it. In addition to species differences in qualitative metabolism, there were substantial quantitative species differences. The intrinsic clearance (CL$_{int}$) (V_{max}/k_M) of the oxidative metabolism of indinavir was the lowest in humans

(17/mL/min/kg). In preclinical species, the intrinsic clearance was as follows: dog (29 mL/min/kg), rat (157 mL/min/kg), monkey (162 mL/min/kg). Stevens et al. (37) reported marked species differences in phase I and phase II hepatic drug metabolism activities using human and monkey liver microsomes. Generally all CYP-dependent enzyme activities showed marked species differences; however, phase II enzymes were less different across species. Only 17α-ethynyl estradiol glucuronidation was significantly higher in the humans than in other species. When compared to other species, N-glucuronidation of some drugs, including quaternary amines, is uniquely specific to nonhuman primate species, including humans and monkeys (38).

Induction of metabolism has an important influence on systemic exposures, metabolism, and toxicities. Marked species differences have been observed in CYPs. Phenobarbital induces predominantly members of the CYP2B subfamily in rats, whereas in humans CYP3A subfamily of enzymes are the major inducible enzymes (39). CYP3A subfamily in rats are inducible by the steroidal agent, pregnenolone-16α-carbonitrile, though generally not inducible by the antibiotic rifampin. Inhibition of drug-metabolizing enzymes is often one of the main reasons for serious drug–drug interactions. Some of the well-characterized mechanisms of inhibition include competitive reversible inhibition, noncompetitive (allosteric) interaction with the enzyme, suicidal inactivation of the enzyme, and time-dependent inhibition. Competitive inhibition is one of the most important processes in the enzyme inhibition. When enzyme inhibition occurs involving competition between the substrate and the inhibitor for binding at the same site, the k_M value of the substrate and the dissociation constant of an inhibitor (K_i) as well as their concentrations at the site of enzymes (40) are the major determinants of enzyme inhibition. Quantitative differences in the k_M and K_i values between preclinical animal species and humans can result in differences in enzyme inhibition. Sexual dimorphism–related differences in drug metabolism are well known in all laboratory species and humans; however, sex differences in metabolism are substantially more common in laboratory rats than in other species, including humans (41,42). This dominant pattern of sexual dimorphism in rats has been shown to result from the differential expression of sex-dependent CYPs. A large body of evidence suggests that the sexual dimorphic secretion pattern of growth hormone directly regulates the expression of certain hepatic CYPs in male versus female rats (43–45). Generally male rats have higher activities of certain important most abundant CYP enzymes than females; however, female rats have higher activities of certain specific CYP enzymes than the males (42). CYP2A2, CYP3A2, and CYPC11 are male dominant; however, CYP2A1, CYP2C7, and CYP2C12 are generally female dominant (46–50). Given that male rats predominantly have higher activity of many CYP enzymes, it is not surprising that male rats with high CYP metabolism tend to have lower exposures to many drugs than female rats. The existence of sex-related differences in drug metabolism is not unique to the rat but is less frequent in other species, and slight differences in metabolism have been seen in mice (51), ferrets (52), dogs (53), and humans (54). Lin et al. (55) evaluated sexual dimorphism of indinavir in rats, dogs, and monkeys and demonstrated marked sex-related differences in metabolic clearance in rats and dogs. The metabolic clearance was twofold higher in male rats than in female rats while female dogs had approximately 1.5-fold greater clearance than male dogs (55). This was supported by additional data on differences in functional activity of CYP3A, measured as testosterone 6β-hydroxylation, and immunoblot analysis of the level of CYP3A proteins (55). Monkeys and humans demonstrated no substantial sex-related differences in indinavir metabolism. With a large interhuman variability confounded by polymorphic CYPs, sexual dimorphism is poorly diagnosed in humans. The sexual dimorphism in humans could be related to the phase of the menstrual cycle, sex hormones, and the use of oral contraceptives in humans. Stereoselective metabolism of enantiomers (isomeric molecules with nonsuperimposable mirror images) can result in differences in pharmacokinetic, pharmacodynamic, and toxicological properties (25,26). The drug albuterol (e.g., Proventil®) contains equal amounts of two enantiomers. Only one of them is effective, and the other may be responsible for the occasional unacceptable side effects associated with the drug, which is used to dilate the bronchi, e.g., during an attack of asthma. The active form can now be synthesized pure and—called levalbuterol (Xopenex®)—is available by prescription. It is important to decide whether to develop a compound as a separate enantiomer or as a racemic mixture, and early studies can be very helpful. The hypnotic drug, thalidomide, was removed from the market in Europe due to its extreme teratogenicity. Thalidomide was developed as a racemic mixture because both enantiomers are equally sedating; the stereoselective teratogenicity of thalidomide was species dependent. Blaschke et al. (56,57) reported that the S-enantiomer of thalidomide was teratogenic in rodents; however, the R-isomer was not teratogenic. In contrast to rodents, both enantiomers of thalidomide were equally teratogenic in rabbits and the racemic mixture was more teratogenic than individual enantiomers (58,59). Binding of MK-571, a potent leukotriene D_4 antagonist studied in 12 mammalian species (60). MK-571 enantiomers demonstrated extensive, stereoselective, and species-dependent protein binding. Depending upon the species, the S-enantiomer and the R-enantiomer showed great differences in plasma protein binding with about an eightfold species difference in unbound fraction. The elimination clearance of the enantiomers correlated greatly with the unbound fraction.

The metabolism of mephenytoin was stereoselective and species specific (61). In rabbits, dogs, and rats the rates of microsomal 4'-hydroxylation were two to six times higher for R-mephenytoin than for the S-enantiomer. In contrast, the rates of microsomal 4'-hydroxylation were 5 to 15 times higher for S-mephenytoin than for R-enantiomer in monkeys and humans. Phase II biotransformation reactions also show species differences in stereoselectivity. The glucuronidation of three racemic 2-arylprorionic acids, naproxen, ibuprofen, and benoxaprofen, was stereoselective and species dependent (62). Qualitative species differences in stereoselective metabolism have also been shown to occur. S-Enantiomer of warfarin, an oral anticoagulant, is much more potent than the R-enantiomer. While S-warfarin in humans is eliminated almost entirely as S-7-hydroxywarfarin, R-warfarin is metabolized principally as R-6-hydroxywarfarin (63). The enantiomer-specific hydroxylation reactions of warfarin are mediated by different CYP isoforms (64). Diisopyramide, a quinidine-like antiarrhythmic agent, contains a chiral center and its enantiomers show interesting differences in metabolism. Cook et al. (65) have shown that arylhydroxylation is the major metabolic pathway of racemic diisopyramide in rats, whereas N-dealkylation is the only pathway in dogs. Considerable interspecies variability exists with respect to the process of chiral inversion. Flurbiprofen, a 2-arylprorionic acid NSAID, is a racemate, and its anti-inflammatory activity is thought to be due to S-enantiomer only. Unidirectional inversion of the R- to the S-enantiomer has been demonstrated to occur in a species selective manner (66,67).

The excretion of drugs and their metabolites can occur by multiple routes, including urine, bile, GI tract/fecal, lung, milk, sweat, and saliva. Urinary and biliary excretion tend to be the major route of elimination. Toxic drugs and their products are removed by kidneys similarly to the processes used to remove intermediary metabolism products. Renal excretory mechanisms involve glomerular filtration (excludes proteins exceeding molecular weight of 60,000), renal secretion, renal reabsorption, etc. Protein binding also impacts urinary excretion because highly protein-bound drugs/metabolites are too large to be filtered through the glomerulus. Depending on acidity or alkalinity of urine, the excretion of basic and acidic compounds can be modulated. A good example is the enhanced excretion of acidic phenobarbital in poisoned patients by alkalinization of urine. Active secretion process is an important mechanism of the secretion of compounds or metabolites from tubular cells into urine. There are many active transporters (e.g., organic anion and organic cation) in kidneys, and all active processes can be saturated or competitively inhibited. Inhibition of penicillin secretion by probenicid (active organic acid transporter substrate) was exploited to enhance its half-life and antibacterial activity during the shortage of penicillin during World War II. Many lipophilic compounds and their metabolites can be reabsorbed from the glomerular filtrate through a passive diffusion process into proximal tubules (similar to that in the small intestines), and polar compounds or metabolites will be excreted owing to the difficulty in getting reabsorbed by tubular epithelium. The reabsorption of lipophilic toxicants can lead to excess accumulation within proximal tubules leading to renal toxicity. Excretion of compounds through bile is an important mechanism for removal of drugs and their metabolites through biliary-fecal excretion. Class B substances, including metabolites with bile to plasma ratio of 10 to 1000, are most preferred for biliary excretion. The class B substances include bile acids, bilirubin, lead, copper, arsenic, high-molecular-weight (>300) compounds, and large lipophilic molecules. There are a large number of active transporters which are involved in the biliary transport. Compounds (parent molecule and/or metabolites) excreted in bile can be reabsorbed through a process known as enterohepatic recirculation. Many polar glucuronide conjugates are not sufficiently lipophilic to get reabsorbed in the GI tract. When glucuronide and sulfate conjugates are excreted through bile into GI tract, bacterial microflora enzymes can hydrolyze these conjugates back to parent molecules which are sufficiently lipophilic to get reabsorbed into liver to bile to GI tract. This process often results in an increase in liver and systemic exposure to drugs and their metabolites. Biliary excretion can greatly modulate the toxicity of drugs. For example, biliary excretion of indomethacin contributes to its intestinal toxicity, and species sensitivity to toxicity is directly proportional to the amount of indomethacin in the bile. Indomethacin toxicity was prevented by bile duct ligation (68). Fecal excretion is an important route of drug and metabolite excretion. The fecal excretion can result from unabsorbed drug after oral administration, biliary excretion, and secretion of drug in saliva, intestinal, and pancreatic secretory fluid.

Applications of Toxicokinetics in Nonclinical Safety Assessment of Pharmaceuticals

The major objective of nonclinical toxicology and safety studies in laboratory rodents and nonrodents is to determine a dose that will produce adverse effects and a dose that will be safe and well tolerated. This information is necessary for initiation and continuation of clinical trials and to assure that prior to exposing humans to an investigation drug, adequate margins exist between the proposed does in humans and toxic doses in humans. Given the substantial differences in ADME between preclinical species and humans, the risk assessment of animal toxicology data cannot be fully made without a complete understanding of the toxicokinetic

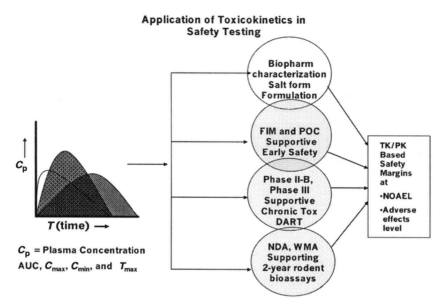

Figure 11 Toxicokinetic information has multiple applications in drug development. This includes toxicokinetic support for (i) biopharmaceutical optimization, including the selection of an appropriate salt form and formulation optimization to maximize drug exposure in toxicology studies as well as in clinical trials, (ii) assessment of toxicokinetics-based safety margins at NOAEL and toxic effect level at FIM and POC trials, (iii) toxicokinetics for chronic toxicology support for exploratory and confirmatory efficacy and safety studies in humans, and (iv) toxicokinetic support for two-year carcinogenicity studies. These studies are needed for filing for NDA and Worldwide Manufacturing Application. *Abbreviations*: DART, developmental and reproductive toxicity; FIM, first in man; NDA, new drug application; NOAEL, no observable adverse effect levels; POC, proof of concept; TK/PK, toxicokinetics/pharmacokinetics.

profile. This becomes more critical when doses used on mg/kg basis vary across species by a factor of 100 or more, and dose response in toxicokinetics is nonlinear. In this aspect, comparative understanding of dose response in toxicokinetics/pharmacokinetics can assist in prediction of possible initial safe doses and maximal doses or clinical trials. Toxicokinetics information for various stages in drug development vary in nature and scope. These include toxicokinetic support for early biopharmaceutical characterization, early toxicology studies to support phase I through proof of efficacy of phase II trials, key chronic toxicology studies to support phase III trials, and carcinogenicity studies to support the filing of new drug application. To support the inclusion of women of child-bearing potential in clinical trials and assure the reproductive, prenatal (lack of teratogenicity) and postnatal safety, toxicokinetics data–based risk assessment is needed for nonclinical developmental and reproductive toxicology studies. Figure 11 summarizes the application of toxicokinetics studies in drug development.

Biopharmaceutical Characterization to Support Early Toxicology Studies

Systemic availability of drug and its pharmacologically active metabolites are key to the success of a drug development program. Toxicology studies conducted with poorly bioavailable compounds do not provide any assurance of the drug safety and may create doubts about the adequacy of preclinical safety testing. At a minimum, a toxicologist needs to have a preliminary estimate of dose proportionality, potential for drug accumulation, and metabolic induction at proposed doses. This is critical to assure that appropriate safety margins based on AUC (in some cases for C_{max}) exist for the proposed starting clinical doses as well as the maximal clinical doses. Bio-pharmaceutical characterization of a drug candidate is critical to achieving the above objectives. In general, aqueous solutions are preferred over aqueous/nonaqueous suspensions and nonaqueous solutions. Ideally the safety of the proposed vehicles must be known prior to their use in toxicology studies; however, appropriate risk assessment of vehicles includes the safety of dose volumes and duration. Table 1 summarizes the proposed list of aqueous and nonaqueous vehicles. For additional discussion on biopharmaceutical characterization, please review discussions under absorption.

Formulation Optimization and Dose Proportionality (Exposure Maximization) Studies to Support Dose Selection

Formulation optimization and dose proportionality studies are valuable in selecting dose levels and are described below: Drugs that have poor aqueous solubility, poor absorption, and/or very fast first-pass clearance (e.g., high intestinal clearance, liver metabolism, and substrates for drug transporters) may not achieve desired exposure in animal studies. Every effort must be made to select safe vehicle formulations that can deliver high circulating drug levels. Dose proportionality studies are conducted to understand the relationship between dose and exposure (AUC, C_{max}, and trough levels) with increasing dose levels. Additionally, these studies provide important information on potential exposure margins, metabolic saturation, and dissolution-limited absorption. These studies should be conducted in relevant species (e.g., rats, dogs, or monkeys) at dose levels ranging from the expected low dose to the maximal feasible dose (based on solubility and practicality considerations in a selected vehicle for each species) or to a limit dose (typically 2000 mg/kg/day). In most cases single dose toxicokinetics studies in safety species are adequate; however, if there is possibility of metabolic autoinduction/autoinhibition, a 5- to 14-day (depending on the half-life of the drug) repeated dose study may be needed to predict steady-state exposure levels. If dissolution-limited absorption is limiting exposure, it will be important to establish that it is not due to drug-related increase in first-pass clearance. The levels of both parent drug and major metabolites should be measured to demonstrate that a plateau was obtained for the parent drug and the major metabolites.

Early First in Man and Proof of Concept Supporting Safety/Toxicology Studies

In the early toxicology studies, it is imperative to find doses (exposures) that would be safe to administer to healthy humans and sick patients to determine the pharmacokinetics and safety in phase I and the proof of concept/dynamics/mechanisms phase IIa trials. These early toxicology studies are usually from 2 to 13 weeks in duration, typically employ three treatment groups, and are conducted in one rodent (typically rat) and one nonrodent (typically dog or monkey).

The choice of toxicokinetic parameters and the selection of time points are discussed below. The choice of matrix is important, and appropriate species-specific analytical methodology is critical to assuring accurate determination of drug concentrations. In certain cases, a given drug can preferentially partition into blood cells, plasma, or serum; therefore, analytical methodology for quantifying drug in plasma and whole blood may be needed. Additionally, analytical methods must be specific to a given species/matrix because species differences in drug stability may exist for certain compounds. For example, amlodipine is degraded by an amine oxidase in monkey and dog plasma only; however, due to the absence of this pathway in mouse, rat, rabbit, and human plasma, it is very stable in plasma from these species (69).

Assessment of Toxicokinetics in Toxicology Studies

Given the animal-to-animal variability and the fact that experimental conditions can greatly influence toxicokinetics, it is often desirable to obtain toxicokinetics data under the condition of toxicity evaluation. It must be understood that detailed information (e.g., volume of distribution, clearance, and half-life) on pharmacokinetics following oral administration in toxicology studies cannot be easily obtained from the integrated toxicity–toxicokinetics study design. This is because of the small number of blood samples that can be taken from laboratory animals in integrated toxicity–toxicokinetics studies. Therefore, it is important to accurately assess, at a minimum, the following parameters that are critical for exposure assessment. These parameters include C_{max}, T_{max}, and AUC. The area under the plasma concentration versus time curve (AUC) is the most ideal parameter for assessment of exposure; however, in certain cases, C_{time} (e.g., C_{max} and trough levels) can be used to monitor drug exposures. C_{max} and T_{max} are useful in understanding the rate of absorption or rate of exposure. A ratio of C_{max} to AUC can be useful in determining the rate and the extent of absorption in many cases, though in many cases a descriptive absorption profile may be more useful. In the absence of half-life, clearance, and volume of distribution data, a comparison of C_{min} (trough concentration) and C_{max} can provide some general information on the rate of plasma drug elimination.

With the advances in quantitation of drugs or metabolites, a small blood volume is often appropriate for quantitation of drug levels; however, frequency of blood sampling is greatly impacted by species-specific restrictions on blood withdrawals. Also, blood samples are taken at regular intervals throughout the duration of toxicity studies to monitor clinical pathology (e.g., hematology, clinical biochemistry, and urinalysis) changes and excessive blood sampling for toxicokinetics may undermine toxicological evaluations. It is likely that high doses may show a very different plasma concentration profile than the low doses; therefore, special consideration must be given to the possibility of slow absorption, multiphasic elimination, and enterohepatic circulation at very high doses. The half-life of the drug as well as the variable absorption rate usually has significant impact on exposure assessment. As a general rule, two or three time points representing absorption, distribution, and early elimination and late (terminal) elimination phases are sufficient to estimate systemic exposure as well as the patterns of exposure. With a minimum of five to eight time points, desirable toxicokinetics information can be obtained in most toxicity studies. In the absence of very good pharmacokinetic data, certain default time points can be 0.5, 1, 2, 4, 8, 12, and 24 hours postdosing. When metabolite(s) are to be quantified, and their formation and elimination are different than the parent drug, additional time points may be necessary to quantify exposure to metabolite(s). In certain cases (e.g., dietary two-year carcinogenicity studies) where the drug may be given orally as dietary admixture, it is recommended that blood sampling for toxicokinetics follow the rate of oral intake of drug matching closely to the feeding patterns of rodents. Blood sampling for toxicokinetic parameters of interest may be as follows: every four hours during the light cycle (after offering the diet–drug mixture) and every two to four hours during the dark cycle (rodents consume most food during the night time). For continuous infusion studies to assess steady-state pharmacokinetics, one needs to know the clearance and half-life of the drug because the product of projected steady-state concentration and clearance provides the infusion rate for IV drug administration. Frequent sampling may be needed until steady state is attained followed by infrequent sampling to monitor steady state. Subcutaneous/intramuscular dosing also require less frequent sampling because the absorption is rather slow, and plasma concentrations are generally sustained over time.

Application of Toxicokinetics in Species and Dose Selection

Species Selection

Species selection is an important consideration prior to designing single or repeated dose studies. Scientific considerations may include (not limited to) the relevance of species to humans, the species sensitivity and metabolism or pharmacokinetics, and the availability of background data (e.g., clinical pathology, histopathology, tumors, etc.). Other considerations include the practicality, such as the size of the animal, the availability, and the length of gestation (e.g., reproductive and developmental toxicology studies). There are specific recommendations for selection of species to evaluate the safety of biological products (e.g., vaccines and monoclonal antibodies), and it is recommended to consult the International Conference on Harmonisation (ICH) S6 Guideline at the Center for Drug Evaluation and Research Food and Drug Administration (FDA) website (70).

Qualitative similarity in metabolism: Without mechanistic data, it is often difficult to know whether toxicity or lack of it is due to parent drug and/or metabolites; therefore, it is important to show that the selected toxicology species is as close to humans as possible with regard to exposure to all major drug-related moieties. The phrase *major metabolites* has been subject to numerous debates, and there have been several documents to define it precisely. Although there is a lack of consensus regarding quantitative aspects of the major metabolite, it is generally expected that a major metabolite is one that comprises at least 10% to 25% of the total circulating drug-related moieties. A recent FDA draft document (70) defines the major metabolite as that which forms 10% or more of the total circulating drug-related material or 10% of the dose. Over time, it is expected that a more precise assessment of metabolites to ensure drug safety will evolve. The quantitation of major metabolites in preclinical toxicology species also ensures that the safety of all major drug-related moieties, including all major metabolites that may form in humans, has been evaluated appropriately in animals.

Similarity in metabolism between species can be demonstrated by comparing in vitro metabolic profiles (qualitatively based on metabolite peaks in chromatograms or by quantification of identified metabolites) between selected species and humans (hepatocytes or microsomal incubations). In most cases, a comparative in vitro metabolic profile of humans, rats, dogs, and rhesus monkeys is sufficient. Under circumstances where in vitro profiles between selected animal species and humans differ, in vivo studies in selected species may be desirable to fully understand the metabolic profile across species and provide justification for the use of a species which may differ from humans with regard to major metabolites.

Pharmacokinetics and bioavailability considerations: Good pharmacokinetics and bioavailability are an important consideration in selection of nonclinical species. This helps to assure that adequate safety margins exist for potential adverse effects in humans. In this regard, whenever it is possible it is important to select species that have an adequate plasma half-life and/or clearance. If solubility-related absorption and metabolic autoinduction/autoinhibition following repeated dosing are not limiting bioavailability, generally drugs with a moderately sufficient half-life (e.g., 5–10 hours) will likely provide an adequate systemic exposure. It is critical to have these data at toxicologically relevant doses because drug bioavailability can change at high doses due to saturation of metabolic and elimination pathways. Biopharmaceutical optimization studies to maximize exposure at proposed doses can be very useful.

Desired pharmacology considerations: Pharmacodynamic relevance (pharmacodynamic response based on desired pharmacology end points) of selected species is an important scientific consideration to justify the relevance of selected nonclinical animal species. This helps to distinguish between pharmacological mechanism (on target)-based and nonpharmacological mechanism (off-target)-based toxicities. If toxic effects are largely off-target and safety margins exist, the human safety of low clinical doses can be assured on the basis of thresholds. When toxicity occurs at low pharmacological doses and toxic effects cannot be separated from desired pharmacological effects, it may not be advantageous to develop specific therapeutics. While pharmacological relevance of species is not a regulatory requirement for pharmaceuticals (e.g., small molecules), species relevance for desired pharmacology end points will provide a sound justification for the relevance of species for prediction of human safety at pharmacologically relevant dosages. For biological pharmaceuticals, species that have the desired pharmacology (e.g., appropriate antibodies and epitopes) is a specific regulatory recommendation. Pharmacodynamic relevance of the selected species can be demonstrated by comparing in vitro/in vivo pharmacology end points [e.g., obtained by assessing receptor binding activity, enzyme(s) inhibition, and functional pharmacodynamic activity].

Species sensitivity: All species have deficiencies with regard to their relevance to humans; however, scientifically, it is important to avoid animal species that show species-specific unique toxicities which may be irrelevant for human safety. For example, if a specific toxic effect is due to a unique preclinical species-specific metabolite or a toxicity mechanism unique to a specific species/strain, it is critical not to use that species for toxicity testing because toxicities observed in the selected species will mislead prediction of human toxicities or may lead to premature termination of a promising drug development.

Dose Selection for Repeated Dose Toxicity Studies

Selection of appropriate dosages to elicit target organ toxicity/maximum tolerated dose (MTD) and no observable adverse effect levels (NOAEL) is critical to the success of a nonclinical toxicology program. The discussion below is for only general toxicity studies. MTD is an important concept in toxicity evaluation because it ensures that doses are high enough to produce some overt toxicity (lower end of toxicity response) without excessive adverse effects (e.g., toxicities leading to morbidity and mortality). This concept of avoidance of excessive toxicity is important because maximal pharmacological dosages for most noncancer pharmaceuticals are often kept low relative to the doses exhibiting toxicity. For anticancer drugs, dose-limiting toxicities (DLTs) are important considerations because high clinical doses up to but slightly below DLTs are often used in clinical trials to attain maximal efficacy. DLTs are the maximal adverse effects that limit further dose escalation. Doses that cause lethality should be avoided because adverse effects associated with morbidity and mortality in animals will not be relevant for humans.

Exploratory MTD Range–Finding Studies

These are exploratory dose range–finding studies to assist in dose selection for detailed regulatory GLP toxicity studies. These studies should be conducted prior to GLP toxicology studies. These studies can be of up to two weeks in duration and up to four doses may be tested. The low dose can be close to a small multiple of projected pharmacological dose (exposure) and the high dose should be high enough to produce some toxicity or at least 10-fold excess of the projected human exposure. These studies can be conducted in a very small number of animals of one sex (if it can be justified based on lack of metabolic or pharmacological differences) with a limited amount of drug. Toxicity end points may involve histopathology, limited clinical pathology, and toxicokinetics investigations in major organs (typically liver, kidneys, heart, lung, brain, GI tissues, sex organs, and limited primary and secondary immune organs).

General principles of dose selection for toxicity studies are described in ICH S1C and S3A Guidelines (70).

High dose: The selection of the high dose should be based on toxicological considerations. It is regulatory expectation that the high dose should be sufficiently high enough to produce overt toxicity in at least one species. The demonstration of toxicity helps to better design phase I and II clinical trials with an appropriate adverse effects monitoring program. Additionally, it ensures that if humans (healthy and targeted sick) are more sensitive than animals to certain toxicities, appropriate safety margins exist to assure the safety of drugs in sensitive humans. Toxic effects can be defined by dose-limiting adverse pharmacological effects (e.g., unexpected adverse effects unrelated to the desired pharmacology such as increased bleeding with anticlotting drugs) and/or by target organ toxicity (principally assessed by histomorphological evaluation).

It is to be recognized that some drugs may have low toxicity potential in selected preclinical species. For potentially nontoxic drugs, it is important to show that either adequate exposure margins (e.g., at least 10-fold or greater) were attained or dissolution-limited drug absorption had limited further dose escalation. Dissolution-limited drug absorption can be demonstrated by establishing that systemic exposure to parent drug and major metabolites has plateaued (e.g., with a doubling of dose, the plasma AUC for the drug and each of its major metabolites may not increase by more than 20%; however, if interanimal variability is high, demonstration of statistical significance may be necessary).

For drugs that show poor absorption and low degree of toxicity, the high dose can be based on maximal feasible dose (based on solubility and practicality considerations) or the limit dose (generally not to exceed 1500 to 2000 mg/kg/day); however, consideration should be given to the dose that provides saturable drug absorption.

Maximum tolerated dose: Dose that causes minimal toxicities and/or overt pharmacological adverse effects is the maximum tolerated dose; however, these effects must be compatible with long-term conduct of a study and must not decrease the natural lifespan of rodents. Other considerations are as follows:

- sustained decrease in body weight gain, though not exceeding 10% when compared to controls;
- mild target organ toxicity (confirmed by histopathology).

Maximum absorbed dose (saturable absorption dose)

- dose that results in saturation of absorption that limits systemic exposure to drug and its major metabolites;
- as guidance, it is important to demonstrate a clear plateau between two/three doses (one above and one below the indicated saturable absorption dose).

Middle dose: The selection of middle dose is typically based on establishing a dose range between the overt toxicity and the no toxicity. It is recommended to consider the following additional points when selecting the middle dose levels:

1. nonlinear saturable toxicokinetics: the middle dose may be the threshold dose that results in nonlinear saturable toxicokinetics;
2. safety margins (e.g., at least 5× or greater than projected human exposures, when possible);
3. multiples of dose (exposure) that produce adverse pharmacodynamic effects in ancillary pharmacology/safety pharmacology studies;
4. machanistic toxicity considerations.

Low dose selection: This level should be expected to provide no toxicologically relevant adverse effects. The selection of the low dose is based on achieving NOAEL. Ideally this dose level should be set to provide some small multiples of safety margin (e.g., 1–2 × greater than human exposures to attain efficacy).

Dose Selection for Carcinogenicity Studies

Dose selection for a two-year carcinogenicity assay remains a major challenge because inappropriate selection of doses may result in significant mortality or severe toxicity which may compromise the validity of the bioassays. To guide the selection of dose levels, the ICH has provided guidelines on Dose Selection for Carcinogenicity Studies of Pharmaceuticals (70). ICH is an international body made up of regulatory and industrial representatives from three major geographical areas: the United States, Japan, and the European Union, responsible for generating mutually agreed guidance for pharmaceuticals for human use. Additionally, the U.S. FDA has provided a guidance document for selecting dose levels for carcinogenicity testing. This guidance document also provides details of the information that is needed to seek approval of dose selection prior to initiating carcinogenicity studies (70) (Pharmacology/Toxicology, Procedural, Carcinogenicity Study Protocol Submissions). Figure 12 describes recommended approaches to high-dose selection and a brief discussion follows below.

Toxicokinetics-Based Approaches to Top Dose Selection

For compounds that exhibit a low degree of toxicity and are nongenotoxic in a standard battery of genotoxic tests, it may be appropriate to use toxicokinetics-based end

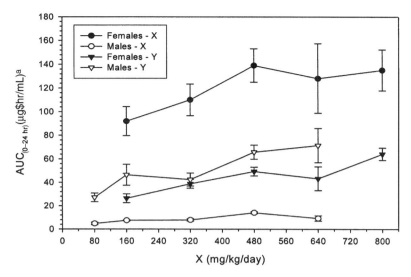

Figure 12 Saturable absorption toxicokinetics were used to justify the selection of a high dose for the two-year carcinogenicity study of compound X in rats. Male and female rats received drug X by oral gavage at doses ranging from 80 to 800 mg/kg/day. Exposure to parent drug (X) and its key metabolite (Y) plateaued at the dose of 160 mg/kg/day in males and at about 480 mg/kg/day in females. This provided a strong evidence for a dissolution/solubility limited drug absorption and indicated that doses above 160 mg/kg/day in males and above 480 mg/kg/day will not offer any advantage in improving the systemic exposure to drug.

points (70) for top dose selection. The following approaches have been recommended by the ICH Guidelines.

Twenty-Fivefold Multiples of Human Systemic Exposure

Based on the carcinogenicity databases available from the U.S. FDA, National Toxicology Program, and International Agency for Research on Cancer (IARC), the FDA scientists evaluated 35 compounds belonging to various pharmacological mechanisms. Their conclusion was that in the absence of an achievable MTD, a systemic exposure ratio of 25 between systemic exposure in rodents and humans (maximally recommended clinical dose) represents an adequate margin of safety for detecting a positive neoplastic response in animal studies for all IARC Class I (definitive human carcinogen) and 2A (possible human carcinogen) carcinogens. Overall, pharmaceuticals tested using a 25-fold or greater AUC ratio for the high dose will have exposure ratios greater than 75% of pharmaceuticals tested previously at the MTD in the carcinogenicity studies. In considering the dose that provides a 25-fold margin of systemic exposure, the following conditions must be met: (i) the drug must be nongenotoxic with low degree of toxicity (e.g., lack of target organ toxicity); (ii) the drug must be metabolized at least qualitatively similar in both rodents and humans; (iii) the systemic exposure should be corrected for protein binding especially when the plasma protein binding is significant (e.g., approximately >80%) and is greater in humans than in animals; (iv) the systemic exposure must be based on parent drug, parent drug plus major metabolite(s), or solely on metabolites; and (v) human systemic exposure must be determined at the maximum recommended human daily dose for clinical practice. There is a risk in setting the top dose based on 25-fold exposure multiples if the clinical daily dose (exposure) is not firmly established. Under this circumstance, it may be best to consider an additional safety factor of two to four to compensate for increases in efficacious clinical doses (exposure). Perhaps when faced with the uncertainty in clinical efficacious doses, a dose that provides an AUC margin of 50- to 100-fold over projected human exposures could be the top dose.

Saturation of Absorption

The solubility and the rate of dissolution are major limiting factors in drug absorption, because a prerequisite for absorption is that a drug remains soluble in GI fluids. Drugs that are poorly soluble in aqueous body fluids and show a poor dissolution characteristic (i.e., release from the dissolved GI fluid to the site of absorption) are poorly absorbed because they are unable to access the absorptive epithelial cells of the intestinal mucosa. Due to the limited aqueous solubility/dissolution rate, the poorly absorbed drugs will likely attain a saturation in systemic exposure (bioavailable dose) at fairly low doses, and higher oral doses will not offer any advantage for evaluating carcinogenicity (1). Another important point to consider is that chronic accumulation of large amounts of unabsorbed drug in the intestinal tract following chronic repeated oral dosing may cause adverse effects on GI homeostasis (e.g., nutrient absorption, GI emptying, changes in GI microflora, and chronic GI irritation and inflammation) and tumors may arise secondary to these noncancer local adverse effects.

For compounds which are poorly absorbed, the systemic exposure may reach a plateau due to saturable absorption of drug that is typically used to describe dose-limiting systemic exposure. It is to be emphasized that for the vast majority of compounds absorption occurs via a passive diffusion process and, therefore, in most cases the absorption is not a saturable process. Given the relatively small number of animals used coupled with the large interanimal variability in plasma drug concentrations that is often observed, it is important to evaluate dose versus systemic exposure using a wide range of doses, generally up to 2000 mg/kg/day or the maximal feasible dose whichever is lower. Based on the U.S. FDAs bioequivalence criteria that are used to qualify generic drugs, it is reasonable to define a plateau in exposure suggestive of dose-limiting absorption when there is a 20% or less increase in systemic exposure at the twofold higher dose. To demonstrate a plateau in systemic exposure, it is also important that the systemic exposure remains similar across multiple doses (at least one dose below and above the dose providing the maximal exposure). When there is a large variability in plasma drug concentrations due to variable absorption and/or elimination, an appropriate statistical test (e.g., comparison of mean values and confidence interval or trend test) may be necessary to define the plateau in systemic exposure over various doses. When the increase in systemic exposure is limited by absorption, the lowest dose level, which provides the maximum systemic exposure, should be considered as the top dose. It should also be established that limitations in systemic drug exposure are not related to the increased metabolic clearance of drug with increasing doses. In the vast majority of cases, the increased clearance of drug is related to increased metabolism. Therefore, it is important to show that limitations in systemic drug exposure are not related to increased metabolism resulting in increased exposure to major metabolites while limiting the exposure to the parent drug. Overall, the lowest dose that provides maximal systemic exposure to parent drug and major metabolites should be considered as the top dose.

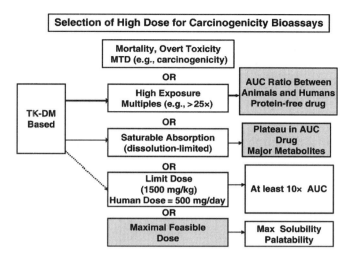

Figure 13 A stepwise approach is described to select dose levels for a two-year carcinogenicity study. As the first choice, drugs that show toxicity at relatively low doses will produce an MTD. MTD is the first choice and a preferred dose for high dose selection. For drugs that are generally well tolerated and nongenotoxic, toxicokinetics-based approaches such as 25× multiple of AUC and saturable absorption can be used for dose selection. *Abbreviations*: AUC, area under the concentration versus time curve; MTD, maximum tolerated dose; TK-DM, toxicokinetics-drug metabolism.

Figure 13 presents a case of drug X and the use of saturable absorption and exposure multiples for the top dose selection. Overall, in the absence of a clear MTD coupled with a low degree of DLT, the lowest doses that provided a trend toward a plateau (e.g., saturable absorption) in drug and metabolite exposures were between 160 and 480 mg/kg bw/day. When considering high systemic exposure multiples (>100-fold) relative to maximal human clinical exposures (see below), the saturable high doses were reduced. This reduction was necessary to avoid potential excessive mortality at higher doses; however, the selected high doses for the two-year carcinogenicity study in rats produced systemic exposure [$AUC_{(0-24\,hr)}$] margins of 888 (male rats) and 2192 (female rats) times the systemic exposure observed in humans receiving the dose of 1 mg (marketed for a specific therapeutic indication). For the 5-mg human dose (marketed for an additional therapeutic indication), these doses produced systemic exposure margins in rats of 110 (male rats) and 275 (female rats) times the systemic exposure observed in humans. The mean human $AUC_{(0-24\,hr)}$ values were 0.05 and 0.4 µg hr/mL for the 1 and 5 mg dosages, respectively.

Repeated Dose Toxicity—Toxicokinetics Study

The goal of toxicokinetics in repeated dose toxicity studies is to assess dose- and time–related (relative to the first dose) changes in systemic exposure to explain dose and time response in toxic effects. Whenever possible, toxicokinetics are usually conducted after a single dose as well as after repeated doses (toward the end of study). Comparison of toxicokinetics after repeated doses (near the end of study or at steady state) and after the first dose helps in determining if there is a metabolic induction or inhibition of systemic drug clearance. Additionally, the repeated dose exposure data may help to assess steady-state toxicokinetics and systemic exposure margins (relative to actual or expected human systemic exposure) at various dose levels. The systemic exposure data may also be useful in interpreting dose response in toxicity; however, in certain cases target organ exposure may be more valuable than systemic exposure in interpreting target organ toxicity.

Systemic Exposure Evaluation in Carcinogenicity Studies

The objective of toxicokinetics in carcinogenicity studies is to assure that the plasma drug concentration versus time profile in the carcinogenicity study is similar to previously determined toxicokinetic profiles in dose range–finding studies. This also ensures that carcinogenicity studies were appropriately conducted and animals were exposed to drug. Because a full profile and systemic exposure data are generally available at doses used in carcinogenicity studies, it may not be necessary to reassess exposure in carcinogenicity studies. Exposure can be generally monitored by sampling animals at two time points, including at previously known T_{max} (time to reach peak concentration) and T_{min} (time to reach trough concentration). If major metabolites show a very different plasma concentration versus time profile than the parent drug, additional time points may be necessary to monitor major metabolites. It is important to monitor toxicokinetics at regular intervals during the first six months of study only because rodents age rapidly, and the profile may change due to aging related factors and confound the interpretation of steady-state exposures in carcinogenicity studies. It is recommended that exposure monitoring be conducted in rats at the end of three months and six months while in mice the exposure monitoring may be conducted at the end of one month and six months of dosing. A similarity in limited plasma concentration versus time profiles between the two time periods would be indicative of steady-state toxicokinetics and consistency in daily systemic drug exposure in carcinogenicity bioassays.

Tissue Distribution of Drug

It is a normal practice to use plasma drug levels as surrogates for tissue(s) exposure; however, this is largely untrue for drugs that distribute in peripheral compartments, have long terminal half-life, and/or have high affinity for tissue macromolecules. Drugs with a large volume of distribution may get preferentially sequestered in certain tissues leading to delayed toxicities. The following conditions may require the conduction of tissue distribution studies as described below:

1. when the identified target tissue half-life exceeds the dosing interval by at least twofold (target tissue half-life of 48 hours for once-a-day drug) from single dose–tissue distribution studies;
2. when the half-life after repeated dosing is significantly longer than after a single dose indicating the possibility of tissue sequestration;
3. when target organ toxicity (histopathological lesions) correlates poorly with plasma drug concentration and the toxicity is suspected to result from the tissue sequestration of drug in target organs of toxicity;
4. when drugs are intended for tissue-specific delivery and tissue distribution.

Normally one to three weeks of tissue distribution study in a relevant rodent or nonrodent should be sufficient to monitor steady-state tissue drug concentration at critical time points (C_{max} and trough levels).

Toxicokinetics in Developmental and Reproductive Toxicity Studies

The evaluation of reproductive and development safety in humans relies exclusively rely on preclinical animal developmental and reproductive toxicity (DART) data. Similar to general toxicity studies, the information about similarity in metabolism between selected DART animal species and humans and adequacy of good systemic exposures are necessary to justify the choice of species, study design, and dosing schedules for DART studies. All DART studies may not require toxicokinetics evaluations; however, the lack of toxicity or low degree of toxicity may justify a very detailed assessment of exposure in reproductive toxicity studies. Exposure assessment can provide assurance that lack of reproductive toxicity and adverse effects on fetus are not due to lack of systemic exposure. Selection of DART species should greatly consider the relevance of the selected species for detecting reproductive and developmental toxicities for humans. Rats are preferred for fertility studies because of the practicality, and a large body of information is available as a reference for reproductive toxicants in rats. For teratology studies of pharmaceuticals, rats and rabbits are the preferred species because of the existence of a large amount of background data and the suitability of these species for laboratory teratology studies. Monkeys and dogs have been used for teratology testing on a case-by-case basis, and their use for teratology testing should be justified on the basis of specific objectives that cannot be met by standard teratology studies in rats and rabbits. It must be realized that all species have advantages and disadvantages for DART testing, and the species selected for testing should be based on specific objectives of the study and the data needed for human risk assessment. Although one rodent and one nonrodent species are required for DART evaluation, in some cases a single species may be sufficient if it can be shown that the selected species have relevance for humans on the basis of pharmacology, toxicokinetics or metabolism, and biology. Similar to standard toxicology testing, the selected DART species should have a metabolic profile similar to humans, and the drug bioavailability should be adequate to establish some margins of safety. An information on the placental transfer of drug and exposure of embryos and fetus to drug is often necessary to interpret developmental toxicity findings. Therefore, the selected rodent and nonrodent species must demonstrate fetal exposures to drug and its major metabolites. Selection of dose levels is mostly based on maternal toxicity findings from the dose range–finding studies; excessive doses with little or no relevance for human systemic exposure may be selected for compounds with low degree of toxicity.

Typically, high-dose selection is based on a demonstration of some maternal toxicity or MTD in pregnant animals. The middle- and low-dose selection may be based on providing 2- to 10-fold AUC margins for the clinically efficacious maximal AUC in humans. For drugs with low toxicity, toxicokinetics data may be helpful in selecting doses that may be based on either saturation of exposure of drug and its major metabolites or the limit dose of 1000 mg/kg. The route of administration should be generally identical to that intended to be used for human dosing. Alternate routes of administration can be used if it can be demonstrated, on the basis of toxicokinetic bioequivalence of other routes, to provide desired exposures without altering the metabolic profile.

The extent of toxicokinetics in pregnant animals is dependent on the type of information desired and the objectives of the study. If a compound has no biologically relevant teratological findings, it may be sufficient to assess maternal exposure (full profile) and placental transfer of drug by monitoring fetal exposure to drug at 1 to

3 time points generally at the end of the gestation period. Lack of developmental toxicity coupled with lack of placental transfer may pose special problems in risk assessment and in the validation of species/strain used for developmental toxicity. This problem is best illustrated with the example of indinavir, a HIV protease inhibitor. While pregnant rats showed a moderate placental transfer of indinavir, there was a lack of placental transfer of drug in rabbits accompanied by the absence of any significant maternal and developmental toxicity. This necessitated the evaluation of placental transfer of drug in other nonrodent species, including dogs and monkeys. Placental transfer studies in monkeys were complicated by poor maternal drug exposure. Pregnant dogs showed an approximately 50% transfer of drug, and this attribute was used to conduct developmental toxicity studies of indinavir in pregnant dogs.

Assessment of the secretion of drug in milk is needed to understand the role of milk in overall exposure of neonates to drug. Transfer of drug through milk may be studied by comparing milk and maternal plasma concentration in rats on lactation day 14 at a selected time point (typically $T_{\max}$). When there is significant neonatal toxicity in suckling animals, it may be appropriate to measure both milk secretion as well as neonatal systemic drug exposure.

Qualitative and Quantitative Assessment of Metabolites

The qualitative similarity in metabolic profiles in safety species and humans provides assurance regarding the validity of the selected species for human risk assessment. The qualitative similarity in metabolites can be demonstrated by comparing in vitro metabolic profile of radiolabeled drug in liver slices, hepatocytes, and hepatic microsomes from multiple species and humans; however, confirmation must be provided by comparing plasma metabolite profiles in preclinical species and humans. Unlike qualitative metabolism, species differences in quantitative metabolism are fairly common. The following points should be considered when deciding about the quantitation of metabolites to support the safety of drug-related substances:

1. when a prodrug is converted (nonenzymatically or enzymatically) to bioactive metabolites, both prodrug and active metabolite should be quantified because there may be species differences in conversion of prodrug to active metabolite.
2. when metabolites constitute the predominantly circulating drug-related moieties (e.g., 10% of total exposure), the systemic exposure to these metabolites should be assessed.

CONCLUSIONS

Integration of toxicokinetics into product safety assessment studies has greatly reduced the uncertainty in interpretation of preclinical toxicity data and its utility in human health risk assessment. The classical noncompartmental toxicokinetics-based data have been particularly useful in selecting the dose levels for short-term to long-term toxicity and carcinogenicity studies for compounds which show low degree of toxicity. The utility of toxicokinetics in designing and interpreting toxicity data is an important achievement and has allowed great improvements in the selection of relevant species, formulation, and doses for safety studies. The greatest utility of preclinical toxicokinetics data has been in the interspecies comparison of product toxicity. It is now widely accepted that toxic effects can be better extrapolated from animals to humans when these comparisons are based on toxicokinetics and disposition (ADME) data in preclinical species and humans. In this context, safety margin is based on a ratio of animal AUC at NOAEL and toxic effect dose levels to the human AUC, and is the key predictor of human toxicity risk. It is generally accepted that when the AUC ratio is large, the expected risk of toxicity in humans is low. Although model-independent or compartment-based plasma/blood toxicokinetics has served well as a practical means of assessing systemic exposure, it provides no information on the time course of exposure of target organs to drugs or metabolites. Overall, toxicokinetics has greatly enhanced our understanding of interspecies differences in toxicity and significance of safety margins based on toxicokinetically determined systemic exposure.

ACKNOWLEDGMENT

The author is grateful to invaluable contributions made by Dr. John Valentine in this book chapter.

REFERENCES

1. Dixit R, Riviere J, Krishnan K, Andersen ME. Toxicokinetics and physiologically based toxicokinetics in toxicology and risk assessment. J Toxicol Environ Health 2003; 6:1–40.
2. Medinsky MA, Valentine JL. Toxicokinetics. In: Klaassen CD, ed. Casarett and Doull's Toxicology. 6th ed. New York: McGraw-Hill, 2001.
3. Gibaldi M, Perrier D. Pharmacokinetics. 2nd. New York: Marcel Dekker, 1982.
4. Rowland M, Tozer TN. Clinical Pharmacokinetics. 3rd ed. Philadelphia: Lippincott Williams & Wilkins, 1995.
5. Shargel L, Yu ABC. Applied Biopharmaceutics and Pharmacokinetics. 3rd ed. Norwalk: Appleton & Lange, 1993.
6. Brown RP, Delp MD, Lindstedt SL, Rhomberg LR, Beliles RP. Physiological parameter values for physiologically based pharmacokinetic models. Toxicol Ind Health 1997; 13:407–484.
7. Pastino GM, Asgharian B, Roberts K, Medinsky MA, Bond JA. A comparison of physiologically based pharmacokinetic model predictions and experimental data for inhaled ethanol in male and female B6C3F1 mice, F344 rats, and humans. Toxicol Appl Pharmacol 1997; 145(1):147–157.
8. Arms AD, Travis CC. Reference physiological parameters in pharmacokinetic modeling. EPA/600/6-88/004, 1988.
9. Gargas ML, Burgess RJ, Voisard DE, Cason GH, Andersen ME. Partition coefficients of low-molecular weight volatile chemicals in various liquids and tissues. Toxicol Appl Pharmacol 1989; 98:87–99.
10. Jepson GW, Hoover DK, Black RK, McCafferty JD, Mahle DA, Gearhart JM. A partition coefficient determination method for nonvolatile chemicals in biological tissues. Fundam Appl Toxicol 1994; 22:519–524.
11. Andersen ME. A physiologically based toxicokinetic description of the metabolism of inhaled gases and vapors: analysis at steady state. Toxicol Appl Pharmacol 1981; 60:509–526.
12. Andersen ME, Clewell HJ III, Gargas ML, et al. Physiologically based pharmacokinetic modeling with dichloromethane, its metabolite, carbon monoxide, and blood carboxyhemoglobin in rats and humans. Toxicol Appl Pharmacol 1991; 108(1): 14–27.
13. Clewell HJ III, Andersen ME. Risk assessment extrapolations and physiological modeling. Toxicol Ind Health 1985; 1:111–131.

14. Wang X, Santostefano MJ, DeVito MJ, Birnbaum LS. Extrapolation of a PBPK model for dioxins across dosage regimen, gender, strain, and species. Toxicol Sci 2000; 56:49–60.

15. Corley RA, Mendrala AL, Smith FA, et al. Development of a physiologically based pharmacokinetic model for chloroform. Toxicol Appl Pharmacol 1990; 103:512–527.

16. Goodman JI. A rational approach to risk assessment requires the use of biological information: an analysis of the National Toxicology Program (NTP) final report of the advisory review by the NTP Board of Scientific Counselors. Regul Toxicol Pharmacol 1994; 19:51–59.

17. McClellan RO. Reducing uncertainty in risk assessment by using specific knowledge to replace default options. Drug Metab Rev 1996; 28:149–179.

18. Prentis RA, Lis Y, Walker SR. Pharmaceutical innovation by seven UK-owned pharmaceutical companies (1964–1985). Br J Clin Pharmacol 1988; 25:387–396.

19. Leahy DE, Lynch J, Taylor DD. Mechanisms of absorption of small molecules. In: Prescott LF, Nimmo WS, eds. Novel Drug Delivery. New York: John Wiley & Sons, 1989:33–44.

20. Schanker LS. On the mechanism of absorption from the gastrointestinal tract. J Med Pharm Chem 1960; 2:343–346.

21. Seydel JK, Schaper KJ. Quantitative structure-pharmacokinetic relationships and drug design. In: Rowland M, Tucker G, eds. Pharmacokinetics: Theory and Methodology. New York: Pergamon Press, 1986:311–366.

22. Toon S, Rowland M. Structure-pharmacokinetic relationships among the barbiturates in the rat. J Pharmacol Exp Ther 1983; 225:752–763.

23. Martin YC, Hansch C. Influence of hydrophobic character on the relative rate of oxidation of drugs by rat liver microsomes. J Med Chem 1971; 14:777–779.

24. Humphrey MJ. Pharmacokinetic studies in the selection of new drugs: a case history on dihydropyridine calcium channel blockers. In: Kato R, Estabrook RW, Gayen MN, eds. Xenobiotic Metabolism and Disposition. New York: Taylor & Francis, 1989:245–253.

25. Lin JH, Lu AYH. Role of pharmacokinetics and metabolism in drug discovery and development. Pharmacol Rev 1977; 49(4):403–449.

26. Lin J, Sahakian DC, de Morais SMF, Xu JJ, Polzer RJ, Winter SM. The role of absorption, distribution, metabolism, excretion and toxicity in drug discovery. Curr Top Med Chem 2003; 3:1125–1154.

27. Wu CY, Benet LZ. Predicting drug disposition via application of BCS: transport/absorption/elimination interplay and development of a biopharmaceutics drug disposition classification system. Pharmaceut Res 2005; 22(1):11–23.

28. Lipinski CA. Chris Lipinski discusses life and chemistry after the rule of five. Drug Discov Today 2003; 8:12–16.

29. Manners CN, Payling, DW, Smith DA. Distribution coefficient, a convenient term for the relation of predictable physico-chemical properties to metabolic processes. Xenobiotica 1998; 18:331–350.

30. Nomier AA, Kumari P, Hilbert MJ, et al. Pharmacokinetics of SCH 56592, a new azole broad-spectrum antifungal agent in mice, rats, rabbits, dogs and cynomolgus monkeys. Antimicrob Agents Chemother 2000; 44:727–731.

31. Curatolo W. Physical chemical properties of oral drug candidates in the discovery and exploratory development setting. Pharm Sci Technol Today 1998; 1:387–393.

32. Crowley PJ, Martini LG. Enhancing oral absorption in animals. Curr Opin Drug Discov Develop 2001; 4(1):73–80.

33. Dressman JB. Kinetics of drug absorption from the gut. In: Hardy JG, Davis SS, Wilson CD, eds. Drug Delivery to the Gastrointestinal Tract. Vol. 15. . Chichester, UK Ellis Harwood, 1989:195–221.

34. Welling PG. Pharmacokinetics. Processes, Mathematics, and Applications. 2nd ed. Washington, D.C.: American Chemical Society, 1997.

35. Chiba M, Hensleigh M, Nishime JA, Balani SK, Lin JH. Role of cytochrome P-450 3A4 in human metabolism of MK-639, a potent human immunodeficiency virus protease inhibitor [abstr]. Drug Metab Dispos 1996; 24:307–314.

36. Lin JH. Bisphosphonates: a review of their pharmacokinetic properties. Bone 1996; 18:75–85.

37. Stevens JC, Shipley LA, Cashman JR, Vandenbranden M, Wrighton SA. Comparison of human and Rhesus monkey in vitro phase I and phase II hepatic drug metabolism activities. Drug Metab Dispos 1993; 21:753–760.

38. Caldwell J, Weil A, Tanaka Y. Species differences in xenobiotic conjugation. In: Kato R, Estabrook RW, Cayen MN, eds. Xenobiotic Metabolism and Disposition. London: Taylor & Francis, 1989:217–224.

39. Rice JM, Diwan BA, Ward JM, Nimes RW, Lubet RA. Phenobarbital and related compounds: approaches to interspecies extrapolation. Prog Clin Biol Res 1992; 374:231–249.

40. Segel IH. Simple inhibition systems: competitive inhibition. In: Segel IH, ed. Enzyme Kinetics: Behavior and Analysis of Rapid Equilibrium and Steady-State Enzyme Systems. New York: John Wiley & Sons, 1975:100–112.

41. Shapiro BH, Agrawal AK, Pampori NA. Gender differences in drug metabolism regulated by growth hormone. Int J Biochem Cell Biol 1995; 27:9–20.

42. Skett P. Biochemical basis of sex differences in drug metabolism. Pharmacol Ther 1989; 38:269–304.

43. Legraverend C, Mode A, Westin S, et al. Transcriptional regulation of rat P-450 2C gene subfamily members by the sexually dimorphic pattern of growth hormone secretion. Mol Endocrin 1992; 6:259–266.

44. Waxman DJ. Regulation of liver specific steroid metabolizing cytochromes P-450: cholesterol 7α-hydroxylase, bile acid 6β-hydroxylase and growth hormone-responsive steroid hormone hydroxylase. J Steroid Biochem Mol Biol 1992; 43:1055–1072.

45. Kato R, Yamazoe Y. Sex-dependent regulation of cytochrome P-450 expression. In: Ruckpaul K, Rein H, eds. Principles, Mechanisms and Biological Consequences of Induction. New York: Taylor & Francis, 1990:82–112.

46. Kobliakov V, Popova N, Rossi L. Regulation of the expression of the sex-specific isoforms of cytochrome P-450 in rat liver. Eur J Biochem 1991; 195:585–591.

47. Bandiera S. Expression and catalysis of sex-specific cytochrome P-450 isozymes in rat liver. Can J Physiol Pharmacol 1990; 68:762–768.

48. Legraverend C, Mode A, Wells T, Robinson I, Gustafsson JA. Hepatic steroid hydroxylating enzymes are controlled by the sexually dimorphic pattern of growth hormone secretion in normal and dwarf rats. FASEB J 1992; 6:711–718.

49. Waxman DJ, Dannan GA, Guengerich FP. Regulation of rat hepatic cytochrome P-450: age-dependent expression, hormonal imprinting and xenobiotic inducibility of sex-specific isoenzymes. Biochemistry 1985; 24:4409–4417.

50. Waxman DJ, Ram PA, Notani G, et al. Pituitary regulation of the male-specific steroid 6β-hydroxylase P-450 2a (gene product IIIA2) in adult rat liver: suppressive influence of growth hormone and thyroxine acting at a pretranslational level. Mol Endocrinol 1990; 4:447–454.

51. Macleod JN, Sorensen MP, Shapiro BH. Strain independent elevation of hepatic monooxygenase enzyme in female mice. Xenobiotics 1987; 17:1095–1102.

52. Ioannides C, Sweatman B, Richards R, Parke DV. Drug metabolism in the ferret: effects of age, sex and strain. Gen Pharmacol 1977; 8:243–249.

53. Dogterom P. Development of a simple incubation system for metabolism studies with precision-cut liver slices. Drug Metab Dispos 1993; 21:699–704.

54. Hunt CM, Westerkam WR, Stave GM. Effect of age and gender on the activity of human hepatic CYP3A. Biochem Pharmacol 1992; 44:275–283.

55. Lin JH, Chiba M, Chen IW, Nishime JA, Vastag KJ. Sex-dependent pharmacokinetics of indinavir: in vivo and in vitro evidence. Drug Metab Dispos 1996; 24:1298–1306.

56. Blaschke TF, Kraft HP, Fickentscher K, Köhler F. Chromatographische racemattren-nung von thalidomid und teratogene wirkung der enantiomere. Drug Res 1979; 29:1640–1642.

57. Blaschke TF. Chromatographic resolution of racemates. Angew Chem Int Ed Engl 1980; 19:13–24.

58. Fabro S, Smith RL, Williams RT. Toxicity and teratogenicity of optical isomers of thalidomide (comment). Nature 1967; 215:296.

59. Simonyi M. On chiral drug action. Med Res Rev 1984; 4:359–413.

60. Lin JH, deLuna FA, Ulm EH, Tocco DJ. Species-dependent enantioselective plasma pro-tein binding of MK-571, a potent leukotriene D_4 antagonist. Drug Metab Dispos 1990; 18:484–487.

61. Yasumori T, Chen L, Nagata K, Yamazoe Y, Kato R. Species differences in stereoselec-tive metabolism of mephenytoin by cytochrome P-450 (CYP 2C and CYP 3A). J Phar-macol Exp Ther 1993; 264:89–94.

62. El Mouelhi M, Ruelius HW, Fenselau C, Dulik DM. Species-dependent enantioselective glucuronidation of three 2-arylpropionic acids: naproxen, ibuprofen and benoxaprofen. Drug Metab Dispos 1987; 15:767–772.

63. Lewis RJ, Trager WF, Chan KK, et al. Warfarin: stereochemical aspects of its metabo-lism and its interactions with phenylbutazone. J Clin Invest 1974; 53:1607–1617.

64. Kunze KL, Wienkers LC, Thummel KE, Trager WF. Warfarin-fluconazole: I—inhibi-tion of the human cytochrome P-450-dependent metabolism of warfarin by fluconazole: in vitro studies. Drug Metab Dispos 1996; 24:414–421.

65. Cook CS, Karim A, Sollman P. Stereoselectivity in the metabolism of diisopyramide enantiomers in rat and dog. Drug Metab Dispos 1982; 10:116–121.

66. Wechter WJ, Loughead DG, Reischer RJ, Van Giessen GJ, Kaiser DG. Enzymatic inver-sion at saturated carbon: nature and mechanism of the inversion of R(−)-p-isobutyl-hydratropic acid. Biochem Biophys Res Commun 1974; 61:833–837.

67. Hutt AJ, Caldwell J. The metabolic chiral inversion of 2-arylpropionic acids: a novel route with pharmacological consequences. J Pharm Pharmacol 1983; 35:693–704.

68. Duggan DE, Hooke KF, Noll RM, Kwan KC. Enterohepatic circulation of indometha-cin and its role in intestinal irritation. Biochem Pharmacol 1975; 24:1749–1754.

69. Smith DA, Humphrey MJ, Charuel C. Design of toxicokinetic studies. Xenobiotica 1990; 20(11):1187–1199.

70. http://www.fda.gov/cder/guidance/index.htm.

9

Validation of In Vitro Methods for Toxicology Studies

William S. Stokes
National Toxicology Program Interagency Center for the Evaluation of Alternative Toxicological Methods, National Institute of Environmental Health Sciences, National Institutes of Health, Department of Health and Human Services, Research Triangle Park, North Carolina, U.S.A.

INTRODUCTION

In vitro test methods are increasingly being used as components in an integrated approach to assess the safety or potential toxicity of various chemicals, medicines, and products (1,2). Their development has been stimulated by advances in new technologies and enhanced understanding of the molecular and cellular mechanisms of toxicity. Advances in tissue culture methods and development of genetically modified stable cell lines have contributed to improved in vitro model systems. New scientific tools such as toxicogenomics, proteomics, and metabonomics are facilitating the identification of more sensitive and earlier biomarkers of toxicity that will likely be incorporated into both in vitro and animal safety testing methods. The number and diversity of in vitro test systems incorporating these sensitive biomarkers will undoubtedly expand greatly in the coming years.

New in vitro test methods are developed for various reasons. From a public health perspective, regulatory agencies continually seek new test methods, including in vitro methods that will improve the accuracy of predictions of the safety or hazard of new chemicals and products, thereby promoting and providing for improved protection of human health. Improved safety information can then be used to implement risk management practices that will further reduce or avoid the potential for injury, illness, or death. New test methods are also sought to help address new toxicity endpoints and mechanisms of concern. One such example is the recent test method development program undertaken to identify potential endocrine disrupting chemicals (3). In vitro methods are also being developed to help reduce and replace the use of animals in research and testing, as well as to refine their use to further minimize or eliminate pain and distress (4,5). Finally, new methods are sought that can provide for improved efficiency in terms of time and expense.

Prior to using in vitro data for regulatory safety assessment decisions, the test methods used to generate such data must be determined to be scientifically valid and acceptable for their proposed use (6,7). Adequate validation is therefore a prerequisite for test methods to be considered for regulatory acceptance. Demonstration of scientific validity requires evidence of a test method's relevance and reliability, and is necessary to determine the usefulness and limitations of a test method for a specific intended purpose. Regulatory acceptance involves determining that a test method can be used to meet a specific regulatory need. This chapter will discuss key aspects of the process for conducting validation studies for in vitro methods and review established criteria for validation and regulatory acceptance of toxicological test methods.

THE CONCEPT OF TEST METHOD VALIDATION

In the context of toxicity testing, validation has been defined as the scientific process by which the relevance and reliability of a test method are determined for a specific purpose (6). Relevance is defined as the extent to which a test method correctly measures or predicts a biological or toxic effect of interest. Relevance incorporates consideration of the accuracy of a test method for a specific purpose and consideration of mechanistic and cross-species or other test system relationships. Reliability is a measure of the degree to which a test method can be performed reproducibly within and among laboratories over time. It is assessed by determining intralaboratory repeatability and intra- and interlaboratory reproducibility.

EVOLUTION PROCESS FOR TEST METHODS

A test method normally progresses through various stages as it evolves from concept to development and then through validation to regulatory acceptance and use (Fig. 1). The first stage involves determining the need for a new, revised, or alternative method. This should involve a review of the adequacy of existing hazard identification methods used to support the risk assessment process for a specific toxicity endpoint. If a need or opportunity for a new or improved test method is identified, then the next stage may involve research necessary to better understand the mechanisms and pathways

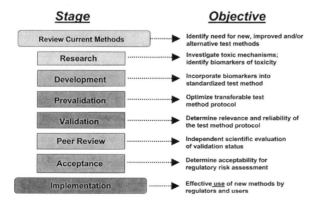

Figure 1 ICCVAM test method evolution process. *Abbreviation*: ICCVAM, Interagency Coordinating Committee on the Validation of Alternative Methods. *Source*: From Ref. 8.

involved in the toxicity of interest, and identification of mechanism-based biomarkers. If a potential predictive biomarker of toxicity can be identified, then the next stage involves the development of a test method that incorporates the biomarker into a standardized test method. This test method development stage involves initial investigations to determine if the test method is sufficiently promising to warrant validation studies for a specific proposed purpose.

If a decision is made to proceed with validation, this is normally accomplished in a phased process. The initial phases preceding a final formal validation phase are often referred to as prevalidation. The object of prevalidation is to identify any further protocol modifications necessary to maximize reproducibility and accuracy, which is often referred to as protocol optimization. The first of these phases should involve limited testing with one or more positive control substances and a vehicle control to develop a historical positive control range and to further define standardized test method protocol procedures necessary to obtain consistent intralaboratory repeatability and intra- and interlaboratory reproducibility. After this is accomplished, a second phase should be conducted using a limited number of coded substances of varying potencies and physical/chemical properties. It is often beneficial to conduct this second phase in two or more stages, with the first stage involving only a few chemicals. Following necessary protocol modifications, a second stage with an expanded number of chemicals can be used, with additional protocol modifications made as needed. If significant protocol modifications are made, it may be necessary to repeat the testing with the same coded substances to ensure that the revised test method protocol has sufficient accuracy and reproducibility. The desired final product of prevalidation is an optimized, standardized test method protocol that will be used for the formal interlaboratory validation study without further modification. However, it is important to recognize that the prevalidation process may determine that the accuracy and/or reproducibility are not adequate to justify a final formal validation phase. In this case, a decision might be made that further development work is necessary to improve reproducibility and/or accuracy. Criteria for adequate validation of a test method have been developed and are discussed in the next section (6,9).

Following completion of formal validation studies that provide specific assessments of accuracy and reproducibility, the complete package of data and information is normally submitted for consideration by an independent organization or regulatory authority. In the United States, new, revised, and alternative test methods with regulatory applicability can be submitted for evaluation by the Interagency Coordinating Committee on the Validation of Alternative Methods (ICCVAM). In Europe, the European Center for the Validation of Alternative Methods provides a similar function (5,10). Evaluation by these validation organizations typically includes an independent scientific peer review evaluation of the validation study results and proposed test method protocol. These organizations then forward recommendations on the scientific validity and demonstrated usefulness and limitations to appropriate regulatory authorities.

Regulatory acceptance decisions about proposed new, revised, and alternative test methods are made by regulatory authorities based on their statutorily mandated responsibilities. Accordingly, the proposed test method must have regulatory applicability for the specific testing needs of the agency that is considering the test method. Criteria that should be met by a test method so that it will be considered acceptable for regulatory purposes have been developed and will be discussed below (6,9).

Following regulatory acceptance, an implementation stage is usually necessary to ensure effective and appropriate use of the new method. This typically should involve workshops and training sessions to familiarize the toxicology and regulatory communities about the appropriate use, applications, interpretation, and limitations of the test method.

VALIDATION AND REGULATORY ACCEPTANCE CRITERIA

The criteria that have to be fulfilled by a new or revised test method, so that the method will be considered as being adequately validated for regulatory acceptance consideration, have been developed (6). Criteria that describe the general attributes that a test method should have, so that it would be considered as being acceptable for regulatory use, have also been developed (6). An ad hoc ICCVAM developed these validation and regulatory acceptance criteria in response to statutory mandates to the National Institutes of Health (NIH) (4,6). The criteria are summarized in Table 1. It is important to recognize that the extent to which each of these criteria should be met will vary with the method and its proposed use. The ICCVAM emphasized that there should be flexibility in assessing a test method given its purpose and the supporting database. Test methods can be designed and used for different purposes by different organizations and for different categories of substances. Accordingly, the determination by regulatory authorities as to whether a specific test method is adequately validated and useful for a specific purpose will be on a case-by-case basis. Regulatory acceptance of new test methods generally requires a determination that using the information from the test method will provide equivalent or improved protection of human or animal health or the environment, as appropriate for the proposed use. Further guidance on adequately addressing established validation criteria is provided in the following description of the validation process.

Table 1 Test Method Validation and Acceptance Criteria[a]

Validation criteria
Clear statement of proposed use
Biological basis/relationship to effect of interest provided
Formal detailed protocol
Reliability assessed
Relevance assessed
Limitations described
All data available for review
Data quality: ideally GLPs
Independent scientific peer review
Acceptance criteria
Fits into the regulatory testing structure
Adequately predicts the toxic endpoint of interest
Generates data useful for risk assessment
Adequate data available for specified uses
Robust and transferable
Time and cost effective
Adequate animal welfare consideration (3Rs)

[a]These are summary versions of the adopted criteria.
Abbreviation: GLPs, Good Laboratory Practices.
Source: From Ref. 6.

THE VALIDATION PROCESS

Defining Test Method Purpose and Rationale

The specific proposed purpose and scientific rationale for a test method should be carefully defined prior to the initiation of validation studies, because these considerations will impact many design aspects of validation studies. Test methods proposed for eventual regulatory applications must define how the outcomes of the proposed test method will be used in the regulatory decision-making process with regard to hazard identification. The design of validation studies must also take into consideration the scientific rationale for a test method. The mechanistic basis of the proposed method and the context in which it will be used to measure or predict the toxicological activity of a test substance should be clearly stated.

Regulatory Rationale and Applicability

Regulatory authorities have established numerous standardized test guidelines that can be used to meet regulatory safety and hazard assessment requirements for various toxicity endpoints (11–13). The purpose of the specific test methods incorporated in regulations and guidelines will vary. Most methods will serve as definitive test methods, while others may serve as screening tests, mechanistic adjunct tests, or components of a testing battery. A test method may be proposed as a complete replacement for an existing test method or as a substitute for an existing test method for certain testing situations, such as for test articles in specific well-defined product or chemical classes or those with specific physical and/or chemical properties.

Definitive test methods are those that provide sufficient data to characterize a specific hazard potential of a substance for hazard classification and labeling purposes without further testing. Examples include specific tests for skin irritation, eye irritation, allergic contact dermatitis, acute oral toxicity, and multigenerational reproductive toxicity, and the rodent carcinogenicity bioassay.

Screening test methods are those that may be used to determine if the test substance is hazardous in a tiered testing strategy or that may provide information helpful in making decisions on prioritizing chemicals for more definitive testing. For example, several in vitro tests have now been validated and accepted for determining if a substance has the potential to cause dermal corrosion. Positive results can be classified and labeled as corrosives, while negative results would undergo additional testing to identify any false negative corrosive substances and to determine dermal irritation potential. The use of information from screening tests to meet regulatory requirements must consider the precautionary principle and the need to avoid potential underclassification of hazard.

Mechanistic adjunct test methods are those that provide data that add to or help interpret the results of other assays and that provide information useful for the hazard assessment process. An example is the estrogen receptor–binding assay. A positive result in this assay indicates that a substance has the potential to bind to the estrogen receptor in an in vitro system. However, it does not definitively indicate that the substance will be active in vivo because it does not take into account absorption, distribution, metabolism, and excretion (ADME) factors. In conjunction with a positive rodent uterotrophic bioassay result, a positive result in this in vitro assay contributes mechanistic information for a weight-of-evidence decision supporting the likelihood that the in vivo bioassay response resulted from an estrogen active substance.

A testing battery is a series of test methods that are generally performed at the same time or in close proximity to reach a decision on hazard potential. In such cases, the component test methods of the proposed battery will need to undergo validation as individual test methods. For the test methods proposed for a test battery, it is essential that each individual test method validation study uses the same reference substances. This is necessary so that the accuracy of each of the possible combinations of component test methods can be calculated and the most predictive combination identified.

Test methods proposed to replace an existing definitive test method will require evidence from validation studies that the use of the proposed method will provide for a comparable or better level of protection than the currently used test method or approach. In some cases, there may be limitations of a new test method with regard to certain types of physical or chemical properties (e.g., solubility in an in vitro system) that do not allow for it to completely replace an existing test. In this case it may be determined as being an adequate *substitute* for the existing test method for many but not all test substances or testing circumstances.

Scientific Rationale

The scientific rationale for a new test method should consider the mechanistic basis and relationship of the biological model used in the test system compared to that for the species of interest for which the testing is being performed (e.g., humans for health-related testing). The extent to which the mechanisms and modes of action for the toxicity endpoint of interest are similar or different in the proposed test system compared to that in the species of interest must be considered. The potential role and impact of in vivo ADME on the toxicity of interest, as well as the extent to which each of these parameters is or is not addressed by the in vitro test system, and the impact of any ADME limitations of the in vitro test system must be considered. For in vitro systems, it is also important to consider what is known or not known about the similarities and differences in responses between the target tissues in the species of interest, the surrogate species used in the currently accepted test method, and the cells or tissues of the proposed in vitro test system.

Developing a Standardized Test Method Protocol

The outcome of test method development should be a detailed, standardized test method protocol that can be evaluated and optimized in the initial phases of the validation process, often referred to as prevalidation. The test method protocol should be sufficiently detailed such that it can be reproduced in other appropriately equipped laboratories that have trained personnel. Modifications are often necessary during the prevalidation phases to reduce sources of inter- and intralaboratory variation, and to optimize the accuracy of the test method to measure or predict the toxicity or biological activity of interest. The version of the test method protocol determined to be sufficiently accurate and reproducible during the last phase of prevalidation should be finalized for the formal validation phase. Because the objective of the formal validation phase is to determine the reproducibility and accuracy of this optimized and standardized test method protocol, no changes should be made to the protocol during this phase. Because most regulatory testing must be conducted in accordance with national and/or international Good Laboratory Practices (GLPs), the test method protocol should be prepared as a GLP compliant protocol.

Table 2 Selected In Vitro Test Method Protocol Components

Biological systems, materials, equipment, reagents, and supplies
Concentration selection procedures, e.g., defined limit concentration, range-finding studies,
  procedures for determining limit of solubility, highest noncytotoxic concentration
Test system endpoints measured
Duration of test article exposure, postexposure incubation
Positive, vehicle, and negative control substances; basis for their selection
Acceptable response ranges for positive, vehicle, and negative control substances, including
  historical control data and basis for acceptable ranges
Decision criteria for interpreting the outcome of a test result, basis for the decision criteria for
  classifying a chemical, accuracy characteristics of the selected decision criteria
Information and data to be included in the study report
Standard data collection and submission forms

The test method protocol should provide a detailed description for all aspects
of the proposed test method (Table 2). This will need to include a description of all
materials, equipment, and supplies. Detailed procedures for dose or concentration
selection should be provided, including procedures for dose-range–finding studies
and solubility testing to select appropriate solvents, as applicable. It is especially
important to define criteria for the highest concentration or dose that should be used.
For in vitro methods, this may be a defined limit concentration (e.g., 1 mM), the
highest noncytotoxic concentration, or the highest soluble concentration. The duration and basis for test substance exposure and postexposure incubation should be
provided. The nature of data to be collected and the methods and procedures for
data collection must be specified.

1. *Positive and negative controls*: Concurrent positive and negative controls
   should be designated and used for every test run. These are necessary to
   ensure that the test system is operating properly and capable of providing
   appropriate positive and negative responses. When a vehicle or solvent is
   used with the test article, a vehicle or solvent control should also be used.
   A positive control substance that is intermediate in the potential dynamic
   response range of the test system should normally be selected. An acceptable positive control response range should be developed for each laboratory participating in a validation study. Accordingly, a test result would
   not normally be considered acceptable when the positive control was outside of the established acceptable positive control range.

2. *Benchmark controls*: In some cases, it may be desirable to include substances for which potential toxicity has previously been established in
   human, animal, and/or in vitro test systems. These substances, commonly
   referred to as benchmark controls, could include substances that are in the
   expected response range of the test articles or that have similar chemical
   structure or physical–chemical properties as the test articles. Benchmark
   controls can be helpful in providing information about the relative toxicity
   of a test article compared to other well-characterized substances, and can
   also be used to ensure that the test system is functioning properly in specific
   areas of the response range.

3. *Decision criteria*: For test methods that determine the hazard classification
   category of a test substance, the test method protocol will need to specify

the decision criteria used to determine the classification category based on results from the test system. For methods that provide qualitative assessments of toxicity, these may be the criteria used to determine if something is positive, negative, or equivocal. For example, in Corrositex®, a test method for determining the corrosivity category of substances, the corrosivity hazard category is based on the time taken by the substance to penetrate a biobarrier membrane. For test methods used to predict a toxic effect currently determined by an existing method, a formula or algorithm, sometimes referred to as a "prediction model," is often used to convert test results into a prediction of the toxic effect (6). A test method prediction model contains four elements: a definition of the specific purpose of the test method; specifications of all possible results that may be obtained when using the test method; an algorithm that converts each study result into a prediction of the toxic effect of interest; and specification of the accuracy associated with the selected decision criteria used in the prediction model (i.e., sensitivity, specificity, false positive, and false negative rates) (9). Decision criteria should always be specified in the test method protocol that will be used for validation studies. However, decision criteria may need to be revised following a validation study to minimize false negative and/or false positive rates, as appropriate for the proposed regulatory use.

4. *Test system*: The basis for selection of the test system should be described in the test method protocol and should include a detailed description and specifications for cells, tissues, or other critical components used. Procedures for assuring the correct identity and critical parameters of cells and tissues should be provided in the protocol, including the basis for determining that the components are acceptable (14).

Selection of Reference Substances for Validation Studies

Reference substances are those for which the response of the substance is known in the existing reference test method and which are used to characterize the accuracy and reproducibility of the proposed test method. The selection of appropriate reference substances is a critical aspect of validation studies. The ideal reference chemicals are those for which high quality testing data are available from both the reference test method and the species of interest (e.g., humans). However, human testing data are rarely available for ethical reasons. Exceptions are for substances and endpoints that do not result in severe or irreversible effects such as allergic contact dermatitis and mild to moderate dermal irritation. These studies are usually limited to premarketing assessment of products that are intended for human contact, such as cosmetics and some mild consumer products. For test methods proposed for predicting human health effects, reference substances for which there are accidental human exposures and toxic effects should be considered.

The number and types of reference substances selected must adequately characterize the accuracy and reproducibility of a test method for its specific proposed use. Reference chemicals should represent the range of chemical classes, product classes, and physical and chemical properties (e.g., pH, solubility, color, solids, and liquids) for which the test method is expected or proposed to be applicable. Reference chemicals should also represent the range of expected responses proposed for the test method, including negatives and weak to strong positives.

Reference chemicals and formulations should ideally be of known purity and composition and should be readily available from commercial sources. Formulations should provide detailed information on the type, purity, and percentage of each ingredient. Unless justified, chemicals should not normally pose an extreme environmental or human health hazard or involve prohibitive disposal costs.

Coding and Distribution of Test Substances

Test substances should normally be coded during both the prevalidation and formal validation phases to exclude bias. This can be accomplished by the use of a chemical distributing facility not directly associated with the participating laboratories. Each substance should be uniquely coded for each different laboratory so that the identity is not readily available to laboratory personnel. However, provisions must be taken to ensure that the designated safety officer in each laboratory has the Safety Data Sheets available for each coded substance, in case there is a need to access the information. One approach is to provide participating laboratory testing staff with sealed packages containing all relevant health and safety data, including instructions for accidental exposures or other laboratory accidents. The envelopes can then be returned to the study sponsor at the end of study, with an explanation for any opened envelopes. Laboratories will need to ensure that all environmental, safety, disposal, and handling procedures are in compliance with regulatory requirements.

Assuring Test Method Quality

Ideally, all data supporting the validity of a test method should be obtained and reported in accordance with national and/or international GLPs (14–16). GLPs provide a formal quality assurance system for data collected in the study. If the studies are not conducted in accordance with GLPs, then aspects of data collection or auditing not performed according to GLPs should be documented. International guidance for the application of GLPs to in vitro testing is available (14). In any case, all laboratory notebooks and data should be retained and available for audit if requested by the reviewing authorities.

Selection of Laboratories for Validation Studies

Laboratories selected for validation studies should be adequately equipped and have personnel with appropriate training. For example, the validation of an in vitro test method that involves aseptic tissue culture should utilize laboratories that have demonstrated proficiency in successfully conducting tissue culture experiments or testing. The use of three laboratories has generally been found to be adequate for assessing the interlaboratory reproducibility of test methods during validation studies.

It is helpful to designate the laboratory most experienced with the test method as the lead laboratory during prevalidation studies to serve as a resource for technical issues that develop during the studies. In a recent in vitro validation study managed by the National Toxicology Program (NTP), dividing the prevalidation study into three phases was found to aid in efficiently optimizing the test method protocol (17). The first phase involved conducting a series of experiments with the positive control, with cycles of modifications and additions to the protocol until all laboratories were able to obtain similar and reproducible results. This phase also was used to establish acceptance criteria for the test system, including positive control acceptance values. The second phase tested three coded substances representing three different areas of the response range (low, moderate, and high toxicity), and was again followed by minor protocol revisions to minimize variation among and within the participating laboratories. The third phase tested nine coded substances, again representing the range of responses as well as range of solubility. Additional minor protocol revisions were made after this phase, and an optimized test method protocol was finalized for testing of the 60 remaining reference chemicals in the formal validation phase.

Assessment of Test Method Accuracy

Following the completion of a validation study, the accuracy of a test method should be calculated to determine the extent to which the test method can correctly predict or measure the biologic or toxic effect of interest (18). This should involve calculation of the sensitivity, specificity, false positive rate, and false negative rate of the proposed test method. This is accomplished by comparing the test results of the in vitro test method to the results from the accepted reference test method for the entire list of reference substances. Wherever appropriate, the accuracy for specific chemical classes and physical and chemical properties should be calculated. If sufficient human testing data are available, then this should be used to calculate and compare the accuracy of the reference test method and the proposed test method for predicting the human results. The basis for any discordance in results from the proposed test method and the reference test method should be discussed. Where human testing or other exposure/effect information is available, the basis for any discordance with this data should also be discussed. Finally, accuracy calculations should be used in conjunction with reliability data to define the usefulness and limitations of a test method.

Assessment of Test Method Reliability

Test method reliability involves determining the intralaboratory repeatability and intra- and interlaboratory reproducibility of a test method (9). The goal of this assessment is to determine if the test method protocol contains sufficient detail and procedures that will result in qualified laboratories obtaining similar and consistent results. It will also characterize the inherent variation in the measured biological responses of a test system. Interlaboratory reproducibility should be assessed using the same or a subset of the reference substances used to assess test method accuracy. Most importantly, reference substances representing the range of possible test outcomes, chemical/physical properties, and mechanisms of toxicity should be evaluated. However, this can sometimes be accomplished with a smaller number of reference chemicals than used to characterize accuracy. Interlaboratory reproducibility has typically been assessed using three qualified laboratories. The impact of the results of test method reliability assessments should always be considered and limitations identified for methods or situations where there is evidence of poor reproducibility.

ICCVAM ROLE IN VALIDATION AND REGULATORY ACCEPTANCE

The ICCVAM was first established as an ad hoc interagency committee in 1994 (19,20). It consisted of representatives from 15 Federal agencies and programs that require, generate, use, or disseminate toxicological testing information (Table 3). This committee was charged with developing validation and regulatory acceptance criteria and recommending a process for achieving the regulatory acceptance of

Table 3 Member Agencies

ICCVAM
Consumer Product Safety Commission
Department of Defense
Department of Energy
Department of Health and Human Services
  Agency for Toxic Substances and Disease Registry
  Food and Drug Administration
  National Institute for Occupational Safety and Health
  National Institutes of Health, Office of the Director
  National Cancer Institute
  National Institute of Environmental Health Sciences
  National Library of Medicine
Department of the Interior
Department of Labor
  Occupational Safety and Health Administration
Department of Transportation
  Research and Special Programs Administration
Department of Agriculture
Environmental Protection Agency

Abbreviation: ICCVAM, Interagency Coordinating Committee on the Validation of Alternative Methods.

scientifically valid alternative test methods (4). The principles embodied in the validation and regulatory acceptance criteria are based on good science and the need to ensure that the use of new test methods will provide for equivalent or better protection of human health and the environment than previous testing methods or strategies. The ICCVAM issued its report in 1997 (6). A summary listing the validation and regulatory acceptance criteria is provided in Table 1.

To implement a process for achieving regulatory acceptance of proposed new, revised, and alternative test methods with regulatory applicability, a standing ICCVAM was established to evaluate the scientific validity of these test methods. The National Institute of Environmental Health Sciences also established the NTP Interagency Center for the Evaluation of Alternative Toxicological Methods (NICEATM) to administer ICCVAM and to provide scientific and operational support. NICEATM collaborates with the ICCVAM to carry out scientific peer review and interagency consideration of new test methods of multiagency interest. The Center also performs other functions necessary to ensure compliance with provisions of the ICCVAM Authorization Act of 2000 (7) and conducts independent validation studies on promising new test methods.

The ICCVAM Authorization Act of 2000 formally established ICCVAM as a permanent interagency committee under NICEATM (7). The Act mandates specific purposes and duties of the ICCVAM (Tables 4 and 5). ICCVAM also continues to coordinate interagency issues on test method development, validation, regulatory acceptance, and national and international harmonization (7). The public health goal of NICEATM and ICCVAM is to promote the scientific validation and regulatory acceptance of new toxicity testing methods that are more predictive of human health, animal health, and ecological effects than currently available methods. Methods are emphasized that provide for improved toxicity characterization, savings in time and costs, and refinement, reduction, and replacement of animal use whenever feasible.

Table 4 The Purposes of the ICCVAM

Increase the efficiency and effectiveness of federal agency test method review
Eliminate unnecessary duplicative efforts and share experiences between federal regulatory
   agencies
Optimize utilization of scientific expertise outside the federal government
Ensure that new and revised test methods are validated to meet the needs of federal agencies
Reduce, refine, or replace the use of animals in testing where feasible

---

Abbreviation: ICCVAM, Interagency Coordinating Committee on the Validation of Alternative Methods.
Source: From Ref. 7.

ICCVAM Nomination and Submission Process

Any organization or individual can submit a test method for which adequate validation studies have been completed to ICCVAM for evaluation. ICCVAM has published guidelines for the information that should be submitted and has developed
an outline to organize the information and data supporting the scientific validity
of a proposed test method (9). Any organization or individual can also nominate test
methods for which adequate validation studies have not been completed to the ICC-
VAM for further study. Nominations are prioritized based on established ICCVAM
prioritization criteria (9). Specific activities, such as workshops and validation studies, are then conducted for those test methods with the highest priority and for
which resources are available.

 The ICCVAM has reviewed several in vitro and in vivo alternative test methods that have now been accepted by national and international authorities (9,20).
These methods have resulted in significant refinement, reduction, and partial replacement of animal use. These include four in vitro methods for identifying dermal corrosives (21–25), the local lymph node assay for assessing allergic contact dermatitis
(26–29), and the revised up-and-down procedure for determining acute oral toxicity
(30–33). Additional in vitro methods have been evaluated or are being developed for
a wide range of human health and ecological testing purposes.

Table 5 The Duties of the ICCVAM

Consider petitions from the public for review and evaluation of new and revised test methods
   for which there is evidence of scientific validity
Coordinate the technical review and evaluation of new and revised test methods of
   interagency interest
Submit ICCVAM test recommendations to each appropriate federal agency
Facilitate and provide guidance on validation criteria and processes
Facilitate interagency and international harmonization of test protocols that encourage the
      reduction, refinement, and replacement of animal test methods
   Acceptance of scientifically valid test methods and awareness of accepted methods
Make ICCVAM final test recommendations and agency responses available to the public
Prepare reports on the progress of the ICCVAM Authorization Act of 2000 and make these
   available to the public

Abbreviation: ICCVAM, Interagency Coordinating Committee on the Validation of Alternative Methods.
Source: From Ref. 7.

Performance Standards

ICCVAM recently developed the concept of performance standards for application to toxicological test methods (9). Performance standards are developed to communicate the basis on which a new proprietary or nonproprietary test method is determined to have sufficient accuracy and reliability for a specific testing purpose. Performance standards are based on a validated test method and provide a basis for evaluating the comparability of mechanistically and functionally similar test methods. ICCVAM develops and recommends performance standards as part of its technical evaluation of new, revised, and alterative test methods. Regulatory authorities, upon acceptance of a recommended test method, can adopt and reference performance standards in their regulatory guidelines and regulations.

Performance standards consist of three elements: essential test method components, a minimum list of reference chemicals, and accuracy and reliability values (9). Essential test method components are the essential structural, functional, and procedural elements of a validated test method that should be included in the protocol of a mechanistically and functionally similar proposed test method. These components include unique characteristics of the test method, critical procedural details, and quality control measures. Adherence to essential test method components helps assure that a proposed test method is based on the same concepts as the corresponding validated test method to which it is being compared. The minimum list of reference chemicals contain those chemicals that are used to assess the accuracy and reliability of a mechanistically and functionally similar proposed test method. These chemicals are a subset of those used to demonstrate the reliability and accuracy of the validated test method. Finally, accuracy and reliability values are provided for which the proposed test method should have comparable performance when evaluated using the minimum list of reference chemicals.

SUMMARY

Significant progress has been made in recent years regarding the scientific principles and processes for adequate validation of in vitro and other test methods proposed for regulatory applications. Regulatory authorities have also communicated the criteria that they will use as the basis for making decisions on the regulatory acceptability of new, revised, and alternative methods. The ICCVAM provides an efficient process for the interagency evaluation of new, revised, and alternative methods of multiagency interest. These established criteria and processes will facilitate the validation and regulatory acceptance of proposed test methods that incorporate new science and technology. Continued development, validation, and adoption of improved testing methods can be expected to support enhanced protection of public health, animal health, and the environment. Adoption of scientifically valid alternative methods will also benefit animal welfare by the reduction, replacement, and more humane use of laboratory animals.

ACKNOWLEDGMENTS

This work was supported by the Intramural Research Program of the NIH, National Institute of Environmental Health Sciences. The author acknowledges Ms. Debbie McCarley for her expert editorial assistance in preparing this manuscript.

REFERENCES

1. Stokes WS, Hill R. Validation and regulatory acceptance of alternatives. Cambridge Quart Healthcare Ethics 1999; 8:74–80.
2. Stokes WS, Marafante E. Alternative testing methodologies: the 13th meeting of the scientific group on methodologies for the safety evaluation of chemicals: introduction and summary. Environ Health Perspect 1998; 106(2):405–412.
3. EPA (Environmental Protection Agency). Health Effects Test Guidelines OPPTS 870.1100 Acute Oral Toxicity. Washington, D.C.: Office of Pesticides, Prevention, and Toxic Substances, 1999 (http://www.epa.gov/OPPTS_Harmonized/870_Health_Effects_Test_Guidelines/Series/870–1100.pdf) (accessed July 25, 2005).
4. USC (United States Code). National Institutes of Health Revitalization Act of 1993. Public Law 103–43. Washington, D.C.: U.S. Government Printing Office, 1993.
5. Stokes WS. The Interagency Coordinating Committee on the Validation of Alternative Methods (ICCVAM): recent progress in the evaluation of alternative toxicological testing methods. In: Salem H, Katz S, eds. Alternative Toxicological Methods. Washington, D.C.: CRC Press, 2003:15–30.
6. ICCVAM (Interagency Coordinating Committee on the Validation of Alternative Methods). Validation and regulatory acceptance of toxicological test methods: a report of the ad hoc Interagency Coordinating Committee on the Validation of Alternative Methods. NIH Publication No. 97–3981. Research Triangle Park, NC, U.S.A.: NIEHS, 1997 (http://iccvam.niehs.nih.gov/docs/docs.htm#general).
7. ICCVAM Authorization Act of 2000, Public Law No. 106–545, 114 Stat. 2721 (to be codified at 42 U.S.C. Sec. 285l-2), 2000.
8. ICCVAM Guidelines for the Nomination and Submission of New, Revised, and Alternative Test Methods. NIH Publication No. 03–4508. NIHS, 2003.
9. ICCVAM (Interagency Coordinating Committee on the Validation of Alternative Methods) Biennial Progress Report. NIH Publication No. 04–4509. Research Triangle Park, NC, U.S.A.: NIEHS, 2003 (http://iccvam.niehs.nih.gov/docs/docs.htm#general).
10. Schechtman LM, Stokes WS. ECVAM-ICCVAM: prospects for future collaboration. ATLA 2002; 30:227–236.
11. FDA. http://vm.cfsan.fda.gov/~redbook/red-toca.html (accessed July 25, 2005).
12. CPSC. http://www.cpsc.gov/businfo/notices.html (accessed July 25, 2005).
13. EPA. http://www.epa.gov/oppfead1/harmonization/ (accessed July 25, 2005).
14. OECD. http://www.oecd.org/document/63/0,2340,en_2649_34381_2346175_1_1_1_1,00.html (accessed July 25, 2005).
15. EPA-FIFRA Good Laboratory Practice Standards (GLPs) Enforcement Policy (http://www.epa.gov/compliance/resources/policies/civil/fifra/fifraglp).
16. FDA Good Laboratory Practices (http://www.fda.gov/ora/compliance_ref/bimo/glp/default.htm).
17. Paris M, Strickland J, Stokes W, et al. Protocol optimization for the evaluation of in vitro cytotoxicity assays for estimating rodent and human acute systemic toxicity. Toxicologist 2005; 84(S-1):332.
18. Stokes WS, Hill R. The role of the Interagency Coordinating Committee on the Validation of Alternative Methods in the evaluation of new toxicological testing methods. In: Balls M, van Zeller AM, Halder M, eds. Progress in the Reduction, Refinement, and Replacement of Animal Experimentation. 31A. Amsterdam: Elsevier Science, 2001: 385–394.
19. Purchase IFH, Botham PA, Bruner LH, Flint OP, Frazier JM, Stokes WS. Scientific and regulatory challenges for the reduction, refinement, and replacement of animals in toxicity testing. Toxicol Sci 1998; 43:86–101.
20. ICCVAM Biennial Progress Report. Research Triangle Park, NC, U.S.A.: Department of Health and Human Services, National Institutes of Health, National Institute of

Environmental Health Sciences, 2001 (http://iccvam.niehs.nih.gov/about/annrpt/annrpt.htm).

21. ICCVAM Biennial Progress Report. NIH Publication No. 04–4509. Research Triangle Park, NC, U.S.A.: Department of Health and Human Services, National Institutes of Health, National Institute of Environmental Health Sciences, 2003 (http://iccvam.niehs.nih.gov/docs/docs.htm#general).

22. ICCVAM (Interagency Coordinating Committee on the Validation of Alternative Methods). Corrositex®: An *in vitro* test method for assessing dermal corrosivity potential of chemicals. NIH Publication No. 99–4495. Research Triangle Park, NC, U.S.A.: Department of Health and Human Services, National Institutes of Health, National Institute of Environmental Health Sciences, 1999 (http://iccvam.niehs.nih.gov/docs/reports/corprrep.htm).

23. NIEHS (National Institute of Environmental Health Sciences). EPISKIN™, EpiDerm™, and Rat Skin Transcutaneous Electrical Resistance Methods: *in vitro* test methods proposed for assessing the dermal corrosivity potential of chemicals: notice of availability of a background review document and proposed ICCVAM test method recommendations and request for public comment. 66 FR 49685, September 28, 2001 (http://iccvam.niehs.nih.gov/docs/FR/frnotice.htm).

24. OECD. Test Guideline 430. In vitro skin corrosion: transcutaneous electrical resistance test (TER). Adopted April 13, 2004. In: OECD Guidelines for Testing of Chemicals. Paris: OECD, 2004 (http://www.oecd.org) (accessed July 25, 2005).

25. OECD. Test Guideline 431. In vitro skin corrosion: human skin model test. Adopted April 13, 2004. In: OECD Guidelines for Testing of Chemicals. Paris: OECD, 2004 (http://www.oecd.org) (accessed July 25, 2005).

26. OECD. Draft Test Guideline 435: In vitro membrane barrier test method for skin corrosion. Draft Version May 20, 2004. In: OECD Guidelines for Testing of Chemicals. Paris: OECD, 2004 (http://www.oecd.org) (accessed July 25, 2005).

27. ICCVAM (Interagency Coordinating Committee on the Validation of Alternative Methods). The murine local lymph node assay: a test method for assessing the allergic contact dermatitis potential of chemical/compounds. NIH Publication No. 99–4494. Research Triangle Park, NC, U.S.A.: Department of Health and Human Services, National Institutes of Health, National Institute of Environmental Health Sciences, 1999 (http://iccvam.niehs.nih.gov/llnarep.htm).

28. Sailstad D, Hattan D, Hill R, Stokes WS. Evaluation of the murine local lymph node assay (LLNA) I: the ICCVAM review process. Regul Toxicol Pharmacol 2001; 34(3):258–273.

29. Dean J, Twerdok L, Tice R, Sailstad D, Hattan D, Stokes W. Evaluation of the murine local lymph node assay (LLNA) II: conclusions and recommendations of an independent scientific peer review panel. Regul Toxicol Pharmacol 2001; 34(3):274–286.

30. Haneke K, Tice R, Carson B, Margolin B, Stokes WS. Evaluation of the murine local lymph node assay (LLNA): III. Data analyses completed by the national toxicology program (NTP) interagency center for the evaluation of alternative toxicological methods. Regul Toxicol Pharmacol 2001; 34(3):287–291.

31. ICCVAM (Interagency Coordinating Committee on the Validation of Alternative Methods). The revised up-and-down procedure: A test method for determining the acute oral toxicity of chemicals. NIH Publication No. 02–4501. Research Triangle Park, NC, U.S.A., 2001 (http://iccvam.niehs.nih.gov/docs/docs.htm#udp).

32. ICCVAM (Interagency Coordinating Committee on the Validation of Alternative Methods). Report of the international workshop on *in vitro* methods for assessing acute systemic toxicity, results of an international workshop organized by the Interagency Coordinating Committee on the Validation of Alternative Methods (ICCVAM) and the National Toxicology Program (NTP) Interagency Center for the Evaluation of Alternative Toxicological Methods (NICEATM). NIH Publication No. 01–4499. Research Triangle Park, NC, U.S.A.: Department of Health and Human Services, National

Institutes of Health, National Institute of Environmental Health Sciences, 2001 (http://iccvam.niehs.nih.gov/docs/docs.htm#udp).

33. ICCVAM (Interagency Coordinating Committee on the Validation of Alternative Methods). Guidance document on using *in vitro* data to estimate in vivo starting doses for acute toxicity, based on recommendations from an international workshop organized by the Interagency Coordinating Committee on the Validation of Alternative Methods and the National Toxicology Program (NTP) Interagency Center for the Evaluation of Alternative Toxicological Methods (NICEATM). NIH Publication No. 01–4500. Research Triangle Park, NC, U.S.A.: Department of Health and Human Services, National Institutes of Health, National Institute of Environmental Health Sciences, 2001 (http://iccvam.niehs.nih.gov/docs/docs.htm#udp).

10

Chemical Mixtures Risk Assessment and Technological Advances

M. M. Mumtaz, B. A. Fowler, and C. T. De Rosa
Division of Toxicology and Environmental Medicine, Agency for Toxic Substances and Disease Registry, Atlanta, Georgia, U.S.A.

P. Ruiz
Division of Toxicology and Environmental Medicine, Agency for Toxic Substances and Disease Registry, Atlanta, Georgia, and Oak Ridge Institute for Science and Education, Oak Ridge, Tennessee, U.S.A.

M. Whittaker
ToxServices, Washington, D.C., U.S.A.

J. Dennison
Colorado State University, Fort Collins, Colorado, U.S.A.

DISEASE CONDITIONS CAUSED BY CHEMICAL MIXTURES

Exposure to environmental contaminants or toxicants is one of many conditions or factors that compromise human quality of life. Evidence of illness associated with exposure to chemicals has been documented since the mid-19th century. A recent study has shown that toxic agents were responsible for 55,000 U.S. deaths in the year 2000, which is more than those caused by motor vehicle crashes (1). In this study, toxic agents were associated with increased mortality from cancer, respiratory, and cardio-vascular diseases. The characteristics and patterns of exposures from waste sites, unplanned releases, and other sources of pollution need to be understood clearly so as to prevent potential adverse human health effects and diminished quality of life. Often such exposures are to mixtures, but this is not well-documented in terms of the exact chemical composition and concentrations of the components of the mixture as a function of time and space. A majority of people are exposed to chemical mixtures at low environmental levels through multiple routes of exposure and environmental media such as air, water, and soil. Exposures to mixtures are often estimated based on water and air consumption and the scarce data on composition. However, several

177

default assumptions inherent to such estimations are influenced by life style, personal habits, nutritional status, and genetic factors. Humans are exposed to a wide range of chemical mixtures due to the astounding number of chemicals used in commerce: more than seven million chemicals have been identified, and of these chemicals, 650,000 are in current use with more than 1,000 chemicals added per year (2). Even though humans are exposed to chemical mixtures on a daily basis, approximately 95% of resources allocated to health effects research are devoted to assessing the health effects of single chemicals (3).

The Centers for Disease Control and Prevention (CDC) has established a bio-monitoring program to measure levels of environmental chemicals in the U.S. population through the National Health and Nutrition Examination Survey (NHANES) (4). Several commonly known environmental chemicals such as metals, pesticides, cotinine (a biomarker of tobacco smoke), and certain phthalates were measured in human biological matrices such as urine and blood (Fig. 1). This report provides geometric means, ranges and percentiles of chemicals found in minorities, children, women of childbearing age, and other potentially vulnerable groups (Fig. 2). The update of this survey, the third National Report on Human Exposure to Environmental Chemicals includes data on over 100 chemicals (Table 1) (5). This type of information has served dual purposes of evaluation of control measures and protection from accumulation of harmful chemicals in addition to providing a database of background data. Data show that current control and mitigation efforts have decreased levels of certain pollutants such as lead and dioxin by about 10-fold over the last 30 years (Fig. 3) (6). These decreases are not universal for all chemicals, and certain chemicals such as polybromi-nated biphenyl ethers have increased rapidly in human blood and breast milk, indicating possible health concerns if this trend continues (7).

Limited but several well-defined epidemiological studies have established a relationship between exposures to chemical mixtures and adverse health effects. Exposure to tobacco smoke and asbestos is the classical example of a synergistic health effect (8). The same is true of occupational exposure to tetrachloroethylene

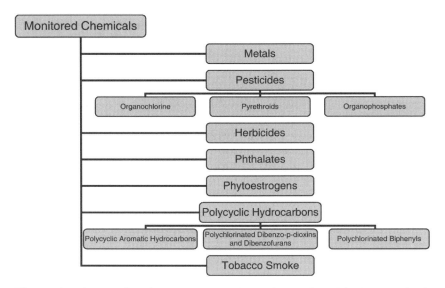

Figure 1 Classes of environmental chemicals bio-monitored in humans in the NHANES study. *Source*: From Ref. 4.

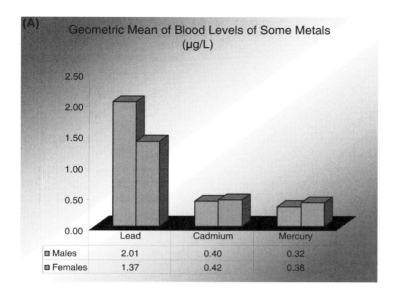

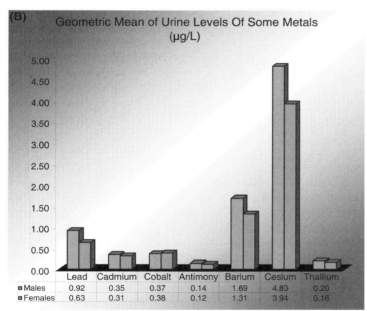

Figure 2   **(A,B)** Geometric mean of blood and urine levels of some metals. *Source*: From Ref. 4.

(TCE) followed by consumption of alcohol that causes "Degreaser's Flush" on the face, hands, and trunk (9). Occupational exposure to metals such as iron, copper, manganese, mercury, and lead has been documented to cause adverse effects such as neuropathy (10). In a case–control study, workers chronically exposed to individual metals did not show any signs of Parkinson's disease, in contrast to workers exposed to a simple mixture of metals. For workers exposed to lead and copper, the odds ratio (OR) of having Parkinson's disease was 5.24 (95% CI = 1.59, 17.21) while it was 2.83 (95% CI = 1.07, 7.50) for workers exposed to lead and iron (10).

Table 1 Environmental Chemicals Bio-monitored in Humans in the NHANES Study

Metals
Lead
Cadmium
Mercury
Cobalt
Uranium
Antimony
Barium
Beryllium
Cesium
Molybdenum
Platinum
Thallium
Tungsten
Herbicides
2,4,5-Trichlorophenoxyacetic acid
2,4-Dichlorophenoxyacetic acid
2,4-Dichlorophenol
Atrazine mercapturate
Acetochlor mercapturate
Metolachlor mercapturate
Organochlorine pesticides
Hexachlorobenzene
Beta-hexachlorocyclohexane
Gamma-hexachlorocyclohexane
Pentachlorophenol
2,4,5-Trichlorophenol
2,4,6-Trichlorophenol
p,p'-DDT
p,p'-DDE
o,p'-DDT
Oxychlordane
trans-Nonachlor
Heptachlor epoxide
Mirex
Aldrin
Dieldrin
Endrin
Pyrethroid pesticides
4-Fluoro-3-phenoxybenzoic acid
cis-3-(2,2-Dichlorovinyl)-2,2-dimethylcyclopropane carboxylic acid
trans-3-(2,2-Dichlorovinyl)-2,2-dimethylcyclopropane carboxylic acid
cis-3-(2,2-Dibromovinyl)-2,2-dimethylcyclopropane carboxylic acid
3-Phenoxybenzoic acid
Other pesticides
2-Isopropoxyphenol
Carbofuranphenol
N,N-diethyl-3-methylbenzamide
ortho-Phenylphenol
2,5-Dichlorophenol

(Continued)

Table 1 Environmental Chemicals Bio-monitored in Humans in the NHANES Study (*Continued*)

Organophosphate insecticides: dialkyl phosphate metabolites
Dimethylphosphate
Dimethylthiophosphate
Dimethyldithiophosphate
Diethylphosphate
Diethylthiophosphate
Diethyldithiophosphate
Organophosphate insecticides: specific metabolites
para-Nitrophenol
3,5,6-Trichloro-2-pyridinol
2-Isopropyl-4-methyl-6-hydroxypyrimidine
2-(Diethylamino)-6-methylpyrimidin-4-ol/one
3-Chloro-7-hydroxy-4-methyl-2H-chromen-2-one/ol
Phthalates
Mono-methyl phthalate
Mono-ethyl phthalate
Mono-n-butyl phthalate
Mono-isobutyl phthalate
Mono-benzyl phthalate
Mono-cyclohexyl phthalate
Mono-2-ethylhexyl phthalate
Mono-(2-ethyl-5-oxohexyl) phthalate
Mono-(2-ethyl-5-hydroxyhexyl) phthalate
Mono-(3-carboxypropyl) phthalate
Mono-n-octyl phthalate
Mono-isononyl phthalate
Phytoestrogens
Daidzein
Enterodiol
Enterolactone
Equol
Genistein
o-Desmethylangolensin
Polycyclic aromatic hydrocarbons
1-Hydroxybenz [a]anthracene
3-Hydroxybenz [a]anthracene and 9-hydroxybenz [a]anthracene
1-Hydroxybenzo [c]phenanthrene
2-Hydroxybenzo [c]phenanthrene
3-Hydroxybenzo [c]phenanthrene
1-Hydroxychrysene
2-Hydroxychrysene
3-Hydroxychrysene
4-Hydroxychrysene
6-Hydroxychrysene
2-Hydroxyfluorene
3-Hydroxyfluorene
9-Hydroxyfluorene
1-Hydroxyphenanthrene
2-Hydroxyphenanthrene

(*Continued*)

Table 1 Environmental Chemicals Bio-monitored in Humans in the NHANES Study (*Continued*)

3-Hydroxyphenanthrene
4-Hydroxyphenanthrene
9-Hydroxyphenanthrene
1-Hydroxypyrene
3-Hydroxybenzo [a]pyrene
1-Hydroxynapthalene
2-Hydroxynapthalene
Polychlorinated dibenzo-p-dioxins and dibenzofurans
1,2,3,4,6,7,8,9-Octachlorodibenzo-*p*-dioxin (OCDD)
1,2,3,4,6,7,8-Heptachlorodibenzo-*p*-dioxin (HpCDD)
1,2,3,4,7,8-Hexachlorodibenzo-*p*-dioxin (HxCDD)
1,2,3,6,7,8-Hexachlorodibenzo-*p*-dioxin (HxCDD)
1,2,3,7,8,9-Hexachlorodibenzo-*p*-dioxin (HxCDD)
1,2,3,7,8-Pentachlorodibenzo-*p*-dioxin (PeCDD)
2,3,7,8-Tetrachlorodibenzo-*p*-dioxin (TCDD)
1,2,3,4,6,7,8,9-Octachlorodibenzofuran (OCDF)
1,2,3,4,6,7,8-Heptachlorodibenzofuran (HpCDF)
1,2,3,4,7,8,9-Heptachlorodibenzofuran (HpCDF)
1,2,3,4,7,8-Hexachlorodibenzofuran (HxCDF)
1,2,3,6,7,8-Hexachlorodibenzofuran (HxCDF)
1,2,3,7,8,9-Hexachlorodibenzofuran (HxCDF)
1,2,3,7,8-Pentachlorodibenzofuran (PeCDF)
2,3,4,6,7,8-Hexachlorodibenzofuran (HxCDF)
2,3,4,7,8-Pentachlorodibenzofuran (PeCDF)
2,3,7,8-Tetrachlorodibenzofuran (TCDF)
Polychlorinated biphenyls
2,2′,5,5′-Tetrachlorobiphenyl (PCB 52)
2,3′,4,4′-Tetrachlorobiphenyl (PCB 66)
2,4,4′,5-Tetrachlorobiphenyl (PCB 74)
3,4,4′,5-Tetrachlorobiphenyl (PCB 81)
2,2′,3,4,5′-Pentachlorobiphenyl (PCB 87)
2,2′,4,4′,5-Pentachlorobiphenyl (PCB 99)
2,2′,4,5,5′-Pentachlorobiphenyl (PCB 101)
2,3,3′,4,4′-Pentachlorobiphenyl (PCB 105)
2,3,3′,4′,6-Pentachlorobiphenyl (PCB 110)
2,3′,4,4′,5-Pentachlorobiphenyl (PCB 118)
3,3′,4,4′,5-Pentachlorobiphenyl (PCB 126)
2,2′,3,3′,4,4′-Hexachlorobiphenyl (PCB 128)
2,2′,3,4,4′,5′ and 2,3,3′,4,4′,6-hexachlorobiphenyl (PCB 138 and 158)
2,2′,3,4′,5,5′-Hexachlorobiphenyl (PCB 146)
2,2′,3,4′,5′,6-Hexachlorobiphenyl (PCB 149)
2,2′,3,5,5′,6-Hexachlorobiphenyl (PCB 151)
2,2′,4,4′,5,5′-Hexachlorobiphenyl (PCB 153)
2,3,3′,4,4′,5-Hexachlorobiphenyl (PCB 156)
2,3,3′,4,4′,5′-Hexachlorobiphenyl (PCB 157)
2,3′,4,4′,5,5′-Hexachlorobiphenyl (PCB 167)
3,3′,4,4′,5,5′-Hexachlorobiphenyl (PCB 169)
2,2′,3,3′,4,4′,5-Heptachlorobiphenyl (PCB 170)
2,2′,3,3′,4,5,5′-Heptachlorobiphenyl (PCB 172)

(*Continued*)

Table 1 Environmental Chemicals Bio-monitored in Humans in the NHANES Study (*Continued*)

2,2′,3,3′,4,5′,6′-Heptachlorobiphenyl (PCB 177)
2,2′,3,3′,5,5′,6-Heptachlorobiphenyl (PCB 178)
2,2′,3,4,4′,5,5′-Heptachlorobiphenyl (PCB 180)
2,2′,3,4,4′,5′,6-Heptachlorobiphenyl (PCB 183)
2,2′,3,4′,5,5′,6-Heptachlorobiphenyl (PCB 187)
2,3,3′,4,4′,5,5′-Heptachlorobiphenyl (PCB 189)
2,2′,3,3′,4,4′,5,5′-Octachlorobiphenyl (PCB 194)
2,2′,3,3′,4,4′,5,6-Octachlorobiphenyl (PCB 195)
2,2′,3,3′,4,4′,5,6′ and 2,2′,3,4,4′,5,5′,6-octachlorobiphenyl (PCB196 and 203)
2,2′,3,3′,4,5,5′,6′-Octachlorobiphenyl (PCB 199)
2,2′,3,3′,4,4′,5,5′,6-Nonachlorobiphenyl (PCB 206)
Tobacco smoke
Cotinine

In another population-based case–control study, occupational exposure to mixtures of herbicides or insecticides was associated with elevated ORs for developing Parkinson's disease. Among these two groups, ORs were 4.1 (95% CI = 1.37, 12.24) and 3.55 (95% CI = 1.75, 7.18), respectively. This elevated risk could not be accounted for by pesticide exposure alone (11). From the foregoing, it is clear that chemicals must be evaluated as mixtures and not as single chemicals.

This chapter summarizes and highlights the methods in use and the development of new methods and their application to the toxicity evaluation of chemical mixtures. It is specifically written for students of human and environmental risk assessment, and those who are interested in public health. It provides a historical perspective of the methods development process as applied to chemical mixtures and sheds insight into research needs and future developments needed to strengthen chemical mixtures risk assessment.

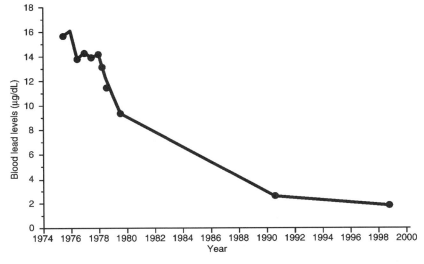

Figure 3 Change in blood lead levels as a function of time. *Source*: From Ref. 6.

ASSESSMENT APPROACHES

A chemical mixture is defined as "Any set of multiple chemical substances that may or may not be identifiable, regardless of their sources, that may jointly contribute to toxicity in the target population" (Table 2) (12,14). Chemical mixtures are generally classified into two distinct groups: simple and complex (12,14,15). Simple mixtures consist of fewer than 10 chemicals, while complex mixtures are made up of tens, hundreds, or even thousands of chemicals (15,16).

Table 2 Types of Mixtures

Chemical mixture
Any set of multiple chemical substances that may or may not be identifiable, regardless of their sources, that may jointly contribute to toxicity in the target population. May also be referred to as a "whole mixture" or as the "mixture of concern."
Components
Single chemicals that make up a chemical mixture that may be further classified as systemic toxicants, carcinogens, or both.
Simple mixture
A mixture containing two or more identifiable components, but few enough that the mixture toxicity can be adequately characterized by a combination of the components' toxicities and the components' interactions.
Complex mixture
A mixture containing so many components that any estimation of its toxicity based on its components' toxicities contains too much uncertainty and error to be useful. The chemical composition may vary over time or with different conditions under which the mixture is produced. Complex mixture components may be generated simultaneously as by-products from a single source or process, intentionally produced as a commercial product, or may coexist because of disposal practices. Risk assessments of complex mixtures are preferably based on toxicity and exposure data on the complete mixture. Gasoline is an example.
Similar components
Single chemicals that cause the same biologic activity or are expected to cause a type of biologic activity based on chemical structure. Evidence of similarity may include similar shaped dose–response curves, or parallel log dose–probit response curves for quantal data on the number of animals (people) responding, and same mechanism of action or toxic endpoint. These components are expected to have comparable characteristics for fate, transport, physiologic processes, and toxicity.
Similar mixtures
Mixtures that are slightly different, but are expected to have comparable characteristics for fate, transport, physiologic processes, and toxicity. These mixtures may have the same components but in slightly different proportions, or have most components in nearly the same proportions with only a few different (more or fewer) components. Similar mixtures cause the same biologic activity or are expected to cause the same type of biologic activity due to chemical composition. Similar mixtures act by the same mechanism of action or affect the same toxic endpoint. Diesel exhausts from different engines are an example.
Chemical classes
Groups of components that are similar in chemical structure and biologic activity, and that frequently occur together in environmental samples, usually because they are generated by the same commercial process. The composition of these mixtures is often well controlled, so that the mixture can be treated as a single chemical. Dibenzo-dioxins are an example.

Source: From Ref. 12, 13.

Although the importance of chemical mixture research and methods development is readily acknowledged by government bodies, national, and international organizations (12,14,15,17,18), relatively few controlled studies have been conducted to evaluate their health effects. Thus, there are very few methods available for their accurate toxicity assessment. The development of methodologies to assess the public health impact of chemical mixtures presents a unique challenge because human exposures to chemical mixtures involve multiple chemicals through complex exposure scenarios (19). In this chapter, three different approaches that are available for the assessment of the toxicity and health effects of diverse mixtures are described (Fig. 4) (14,15). These approaches are riddled with varying uncertainties and assumptions. Selection of an appropriate assessment method depends on known toxic effects of chemicals comprising a mixture, availability of toxicity data, and quality of available exposure data. Ideally, the assessment for each of the mixtures must be done by more than one method and the conclusions compared for consistency.

The Mixture of Concern

The first approach is the "mixture of concern" approach, in which risk is estimated using toxicity data from the same mixture. It is the most direct and simplest method and entails the fewest uncertainties. This approach, however, is the least frequently applied because it requires availability of adequate toxicity data on the specific chemical mixture of concern. These data are then used to derive an integrated defined allowable level such as a minimal risk level (MRL) for the mixture. There are very few mixtures that have been studied adequately for this type of assessment, however. Fuel oils, jet fuels, mixtures of polychlorinated biphenyls (PCBs), and polybrominated biphenyls (PBBs) are some mixtures of concern for which MRLs could be derived (Table 3) (20). Similarly, few occupational exposure limits (OELs) for complex mixtures have been established. Examples of OELs established by organizations such as the American Conference of Governmental Industrial Hygienists (21), the National Institute of Occupational Safety and Health, and the Occupational Safety and Health Administration (22) include OELs for asphalt fumes (0.5 mg/m^3), cotton dust (0.2 mg/m^3), coal dust (0.4 mg/m^3), and Stoddard solvent (100 ppm).

The Similar Mixture Approach

The second approach used is the "similar mixture" approach. As shown in Figure 4, this approach can be applied on a case-by-case basis to a candidate mixture or groups of mixtures that could act similarly (23). It is used when adequate information is not

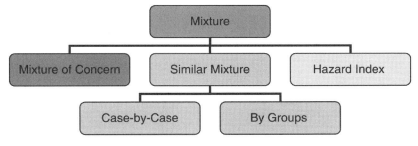

Figure 4 Risk assessment approaches.

Table 3 MRLs for Whole Mixtures

Mixture	MRL value	Duration	Route
Fuel oil	$0.02\,mg/m^3$	Acute	Inhalation
Jet fuel			
JP-4	$9\,mg/m^3$	Intermediate	Inhalation
JP-7	$0.3\,mg/m^3$	Chronic	Inhalation
JP-8	$3\,mg/m^3$	Intermediate	Inhalation
PBBs	$0.01\,mg/kg/day$	Acute	Oral
PCBs	$0.03\,\mu g/kg/day$	Intermediate	Oral
Arochlor 1254	$0.02\,\mu g/kg/day$	Chronic	Oral

Abbreviations: MRLs, minimal risk levels; PBBs, polybrominated biphenyls; PCBs, polychlorinated biphenyls.

available for the mixture of concern, and is often applied to complex mixtures that have been extensively investigated, such as coke oven emissions, diesel exhaust, and wood stove emissions. However, a minimum of information should be available to consider a mixture sufficiently similar to the mixture of concern. For example, if a risk assessment is needed for gasoline contamination of groundwater and information is available on the chronic toxic effects of gasoline, it may be possible to use the available information to assess risks from the contaminated groundwater. However, there are no set criteria to help decide when a mixture is sufficiently similar. Hence, the health assessor is left to determine whether the two chemical mixtures are sufficiently similar, and whether this similarity justifies the use of surrogate risk or toxicity data.

The Relative Potency Factor

One method that can be helpful in deciding the similarity is the relative potency factor (RPF) method (24–26). This method allows the use of information from short-term bioassays to determine sufficient similarity of a mixture. The relative potency method is based on the hypothesis that the "relative potencies" of chemical mixtures are consistent across various bioassays used for toxicity testing of certain endpoints. When validated, this approach offers a method for using short-term assays of complex mixtures as a surrogate for long-term in vivo assays. The approach first normalizes the results among each class of bioassays relative to some standard bioassay; thus the term "relative potency" method. This method has only been specifically applied to complex mixtures that cause cancer and for which the dose–response functions can be described by a simple linear dose–response model. The RPF method has been used for polycyclic aromatic hydrocarbons (PAHs), where RPFs were estimated for PAHs based on the cancer potency factor of benzo(a)pyrene (27). This method has also been used to estimate the cancer risk of dioxins and PCBs (28,29). The mixture of concern and the similar mixture approach are used for those mixtures that have been experimentally tested as a whole to some extent.

The Hazard Index Approach

The third approach, the hazard index (HI) approach, is the method most often used for chemical mixtures risk assessment. This approach integrates the exposure level and the related toxicity into a single value employing potency–weighted dose or response addition. The goal of the HI approach is to approximate the toxicity index that would have been determined had the mixture itself been tested (30). Initially, the potential health hazard from exposure to each chemical is estimated by calculating its individual hazard quotient (HQ). The HQ is derived by dividing a chemical's actual exposure level (E) through an environmental medium by its acceptable exposure (AE) level such as a MRL or a reference dose (RfD). The HI of the mixture is then calculated by adding together all the component HQs, as illustrated below for three chemicals in a mixture:

$$HI = \frac{chem.exposure_1}{AE_1} + \frac{chem.exposure_2}{AE_2} + \frac{chem.exposure_3}{AE_3}$$

In a manner analogous to the HI approach for noncarcinogens, a HI for carcinogens can be estimated by dividing chemical exposure levels by doses associated with a set level of cancer risk also known as risk-specific dose (RSD) (12,14):

$$HI = \frac{chem.exposure_1}{RSD_1} + \frac{chem.exposure_2}{RSD_2} + \frac{chem.exposure_3}{RSD_3}$$

In terms of estimating risk, the HI values obtained using the HI approach should be interpreted carefully. For example, if chemical mixture "X" yields an HI value of 4, it need not be interpreted as twice as toxic as mixture "Y" that yields a value of 2. However, it can be said that mixture "X" is more toxic than mixture "Y." Thus the HI approach can be used for priority setting of mixtures. As the value of the HI increases toward unity, the concern for the potential hazard of a mixture increases. The potential health effects of a mixture are further analyzed and investigated if the HI value is equal to or greater than one. The HI approach assumes that all components have similar joint action, that is their uptake, pharmacokinetics, and dose–response curves have similar shape (12,31). For carcinogens, the above equation assumes that each carcinogen has a linear dose–response curve and that each carcinogen is acting independently (14).

The Toxic Equivalency Factor Method

In addition to its use in the HI approach, the similar joint action principle also serves as the basis for the toxic equivalency factor (TEF) method, which assumes dose additivity. The TEF method is used for a mixture containing the same class of chemicals, and complete toxicity data exists for only one chemical of the class (Table 4) (14,32). In the first step, all the available data on the components are used to estimate the relative potencies in terms of an index chemical, thus converting the potency estimates of various components into equivalent doses of the index chemical. The TEF approach has been used to assess the health risks of chlorinated dibenzo-*p*-dioxins (CDDs) and coplanar PCBs for chlorinated and brominated dioxins and dibenzofurans (12,14,33). In the case of CDD's, TEFs for individual congeners were estimated based on a comparison of in vitro or acute in vivo data for specific congeners such as 2,3,7,8-TCDD. 2,3,7,8-TCDD is assigned a TEF of one, while the other TEFs are usually less than one (34). The concentration of each component of the mixture is multiplied by its TEF value, and values for each component are then summed into a value that is defined as the total toxic equivalents (TEQs) of a mixture. When using the TEF approach, the hazard of a mixture can be evaluated by comparing the TEQ with

Table 4 TEFs for the PCDD and PCDF Congeners

Congener	TEFs proposed by		
	Safe	International A	Nordic
PCDDs			
2,3,7,8-TetraCDD	1.0	1.0	1.0
1,2,3,7,8-PentaCDD	0.5	0.5	0.5
1,2,3,6,7,8-HexaCDD	0.1	0.1	0.1
1,2,3,7,8,9-HexaCDD	0.1	0.1	0.1
1,2,3,4,7,8-HexaCDD	0.1	0.1	0.1
1,2,3,4,6,7,8-HeptaCDD	0.01	0.01	0.01
OctaCDD	0.001	0.001	0.001
PCDFs			
2,3,7,8-TetraCDF	0.1	0.1	0.1
2,3,4,7,8-PentaCDF	0.5	0.5	0.5
1,2,3,7,8-PentaCDF	0.1	0.05	0.01
1,2,3,4,7,8-HexaCDF	0.1	0.1	0.1
2,3,4,6,7,8-HexaCDF	0.1	0.1	0.1
1,2,3,6,7,8-HexaCDF	0.1	0.1	0.1
1,2,3,7,8,9-HexaCDF	0.1	0.1	0.1
1,2,3,4,6,7,8-HeptaCDF	0.1	0.01	0.01
1,2,3,4,7,8,9-HeptaCDF	0.1	0.01	0.00
OctaCDF	0.001		

Abbreviations: PCDD, polychlorinated dibenzo-*p*-dioxins; PCDF, polychlorinated dibenzofurans; CDD, chlorinated dibenzo-*p*-dioxins; CDF, chlorinated dibenzofurans.
Source: From Ref. 32.

an MRL or other health-based criteria derived for the surrogate chemical representing the class of chemicals (such as 2,3,7,8-TCDD). Like the relative potency method, the TEF method allows incorporation of information from short-term bioassays or other forms of toxicologic data that might not otherwise be directly useful. Both the RFP and the TEF methods should only be used when there is no better alternative method, because there are many uncertainties associated with these methods (13).

The Target-organ Toxicity Dose

In terms of estimating risk, it is important that the estimates be realistic. The use of AEs (MRLs or RfDs) that are based on a critical effect to assess secondary effects could lead to overestimation of risk (35). To circumvent this problem, target-organ toxicity dose (TTDs) can be developed and employed (36). TTDs, in essence, are effect or organ specific MRLs and are calculated using the same methodology and process. Thus for a given chemical, there could be an MRL for hepatotoxicity and a series of TTDs for nephrotoxicity, neurotoxicity and reproductive toxicity (29,33). The TTD method is a simple modification of the HI approach and yields a series of HIs for various toxic effects. The values of endpoint-specific hazard indices are treated the same as that of a HI of a mixture (37).

One of the uncertainties inherent in the HI approach for both carcinogens and noncarcinogens is that the assumption of additivity does not consider potential interactions between chemicals. Specifically, dose additivity assumes that chemicals comprising a mixture do not influence each other's toxicity (15). The HI approach is simple but limited with respect to the influence of chemical interactions on the overall toxicity and estimation of adverse effects. It assumes no interaction among chemicals and also assumes additivity and as a result, may either underpredict or overpredict risk estimates of mixtures of industrial, occupational, and environmental chemicals, if synergistic and antagonistic interactions occur (11,12).

Each chemical component of a mixture has the potential to influence the toxicity of other mixture components. However, the magnitude or the capacity of this potential to interact is frequently unknown. Equally important is that the risk assessment values arrived at using the HI approach should also express the available information on chemical interactions. There is very little guidance on how the interactions should be evaluated or incorporated into the overall risk assessment. Even though there might be different types of interactions, in chemical mixtures toxicity, the term "interaction" implies that one agent affects the biological action of another agent (Table 5).

Chemical interaction is the combined effect of two chemicals resulting in a stronger (synergistic, potentiation, supra-additive) or a weaker (antagonistic, inhibitive, subaddittive, or infra-additive) effect than expected based on the basis of additivity (15). The National Academy of Sciences has proposed the use of additional safety factors if synergistic interactions are of concern (38).

A Weight-of-Evidence Method

To provide further guidance on this issue, a weight-of-evidence (WOE) method was developed (39,40). The WOE method yields a composite representation of all the toxicologic interaction evidence from animal bioassay data and human studies data; relevance of route, duration, and sequence; and the significance of interactions. The method consists of a classification scheme used to provide a qualitative and if needed a quantitative estimation of the effect of interactions on the aggregate toxicity of a mixture (Table 6).

The first two components of the scheme are major ranking factors for the quality of the mechanistic information, which supports the assessment and the toxicologic significance of the information. The last three components of the WOE are modifiers which express how well the available data correspond to the conditions of the specific risk assessment in terms of the duration, sequence, routes of exposure as well as the animal models. This method evaluates data relevant to joint action for each possible pair of components of a mixture, and as such requires mechanistic information and direct observation of toxicologically significant interactions. Initially, for each pair of component chemicals of a mixture, two binary WOE (BIN-WOE) determinations are made to estimate the effect of the first chemical on the second chemical's toxicity, and also to estimate the effect of the second chemical on the toxicity of the first chemical. Once all of the qualitative WOE determinations have been made for each pair of compounds in the mixture, these are arrayed in a qualitative WOE matrix (Table 7).

This matrix lists each potential binary classification along both axes. The diagonal line running from the upper left hand corner to the lower right hand corner corresponds to chemical identities. These are, by definition, dose additive and are left blank in the interaction matrix. The column headings indicate the chemicals that are affected by the compounds listed in the row headings. For example, the classification for the effect of selenium on the toxicity of chloroform is given in the third column (chloroform) of the fourth row (selenium). Similarly, the classification for

Table 5 Types of Toxicological Interactions

Additivity

When the "effect" of the combination is estimated by the sum of the exposure levels or the effects of the individual chemicals. The terms "effect" and "sum" must be explicitly defined. Effect may refer to the measured response or the incidence of adversely affected animals. The sum may be a weighted sum (see "dose addition") or a conditional sum (see "response addition").

Antagonism

When the effect of the combination is less than that suggested by the component toxic effects. Antagonism must be defined in the context of the definition of "no interaction," which is usually dose or response addition.

Chemical antagonism

When a reaction has occurred between the chemicals and a new chemical is formed. The toxic effect produced is less than that suggested by the component toxic effects.

Chemical synergism

When a reaction has occurred between the chemicals and a different chemical is formed. The toxic effect produced is greater than that suggested by the component toxic effects, and may be different from effects produced by either chemical by itself.

Complex interaction

When three or more compounds combined produce an interaction that cannot be assessed according to the other interaction definitions.

Dose additivity

When each chemical behaves as a concentration or dilution of every other chemical in the mixture. The response of the combination is the response expected from the equivalent dose of an index chemical. The equivalent dose is the sum of component doses scaled by their toxic potency relative to the index chemical.

Index chemical

The chemical selected as the basis for standardization of toxicity of components in a mixture. The index chemical must have a clearly defined dose–response relationship.

Inhibition

When one substance does not have a toxic effect on a certain organ system, but when added to a toxic chemical, it makes the latter less toxic.

Masking

When the compounds produce opposite or functionally competing effects at the same site or sites, so that the effects produced by the combination are less than suggested by the component toxic effects.

No apparent influence

When one substance does not have a toxic effect on a certain organ or system, and when added to a toxic chemical, it has no influence, positive or negative, on the toxicity of the latter chemical.

No observed interaction

When neither compound by itself produces an effect, and no effect is seen when they are administered together.

Potentiation

When one substance does not have a toxic effect on a certain organ or system, but when added to a toxic chemical, it makes the latter more toxic.

Response additivity

When the toxic response (rate, incidence, risk, or probability of effects) from the combination is equal to the conditional sum of component responses as defined by the formula for the sum of independent event probabilities. For two chemical mixtures, the body's response to the first chemical is the same whether or not the second chemical is present.

(Continued)

Table 5 Types of Toxicological Interactions (*Continued*)

Synergism

When the effect of the combination is greater than that suggested by the component toxic effects. Synergism must be defined in the context of the definition of "no interaction," which is usually dose or response addition.

Unable to assess

Effect cannot be placed in one of the above classifications. Common reasons include lack of proper control groups, lack of statistical significance, and poor, inconsistent, or inconclusive data.

Abbreviations: MRLs, minimal risk levels; PBBs, polybrominated biphenyls; PCBs, polychlorinated biphenyls.

the effect of carbon tetrachloride on the toxicity of chloroform is given in row two (carbon tetrachloride) column three (chloroform). Binary classifications starting with an "=" indicate an additivity, those starting with a ">" indicate a greater than additive interaction and those starting with a "<" indicate a less than additive interaction. The qualitative WOE matrix may then be used as a tool to quantify the risk assessment for a particular site. The qualitative BINWOE can have an absolute value

Table 6 WOE Scheme

WOE scheme for the qualitative assessment of chemical interactions

Determine if the interaction of the mixture is additive (=), greater than additive (>), or less than additive (<).

Classification of Mechanistic Understanding

I. Direct and Unambiguous Mechanistic Data:

    The mechanism(s) by which the interactions could occur has been well characterized and this leads to an unambiguous interpretation of the direction of the interaction.

II. Mechanistic Data on Related Compounds:

    The mechanism(s) by which the interactions could occur are not well characterized for the compounds of concern but structure/activity relationships, either quantitative or informal, can be used to infer the likely mechanisms and the direction of the interaction.

III. Inadequate or Ambiguous Mechanistic Data:

    The mechanism(s) by which the interactions could occur have not been well characterized or information on the mechanism(s) do not clearly indicate the direction that the interaction will have.

Classification of Toxicologic Significance

A. The toxicologic significance of the interaction has been directly demonstrated.

B. The toxicologic significance of the interaction can be inferred or has been demonstrated in related compounds

C. The toxicologic significance of the interaction is unclear

Modifiers

  1. Anticipated exposure duration and sequence

  2. A different exposure duration or sequence

    a. In vivo data

    b. In vitro data

      i. The anticipated route of exposure

      ii. A different route of exposure

Abbreviation: WOE, weight of evidence.

Source: From Ref. 39.

Table 7 A Qualitative Interaction Matrix

		Affected by			
		1. Cd	2. CCl$_4$	3. CHCl$_3$	4. Se
Affects	1. Cadmium		< I. A.2.	< I. B.2.	< III. C.1.
	2. CCl$_4$	?		> III. A.2.	?
	3. CHCl$_3$	?	> III. A.2.		?
	4. Selenium	< II. A.1.	< I. B.1.	< I. B.1.	

of zero to one, with one indicating the highest degree of confidence in the assessment. Ultimately, this WOE method can be used to modify the HI so as to include the role of chemical interactions (33,39,40). Improvements in assessing human health risks from chemical mixtures can only come about by reducing inherent uncertainties in risk assessment methods. This will require an understanding of how chemicals behave in biological systems as well as elucidating their collective mechanisms of action.

NEW DEVELOPMENTS

The Basic Principles of Physiologically Based Pharmacokinetic/ Pharmacodynamic (PBPK/PD) Modeling

The approaches and methods discussed above have dealt with media-specific exposures and risk assessments. In reality, complex exposure to multiple chemicals occurs by multiple routes through multiple media. Hence, this medium by medium evaluation is artificial and will have shortcomings. Irrespective of how chemical exposure occurs, chemicals absorbed into the body influence bodily processes and cause deviations from normalcy. A more recent approach, computational toxicology, which has shown a potential to integrate complex exposures takes advantage of the latest understanding of toxicological mechanisms, advances in computer technology, and mathematical tools to predict toxic effects. PBPK/PD modeling is one such tool that combines physiology with molecular biology and information technology to focus on dose in target organs and health effects in biological systems. PBPK/PD models are useful tools for estimation of internal dose, thereby improving the interpretation of information on external exposure and biological response. In biological systems, interpretation of dose–response is extremely complex. PBPK/PD models incorporate biological "realism" by transforming the dose to which an animal is exposed to the actual "delivered" target organ dose. This is done by considering the body as being composed of "compartments," and modeling the movement of a chemical(s) through the system of compartments with differential equations. The compartments and flows are described by a series of differential equations that are solved using mathematical modeling software. These differential equations incorporate such processes as uptake, distribution, metabolism, and elimination. Thus, the body is represented as a large number of compartments, each with a physiological identity. The parameters representing such things as volumes, flow rates, and biochemical parameters are determined experimentally for each of the compartments. There is obvious advantage to this level of specificity because this allows high dose to low dose as well as species to species extrapolation. In a sense, such models can be viewed as a means of transforming an administered dose to a delivered dose. These models can be tested with laboratory animals, after which physiological and biochemical parameters for humans are substituted in the model to imitate the human body (41).

PBPK/PD modeling has been applied for the toxicity and risk assessment of individual chemicals (42–44). More recently, it is being extended to chemical mixtures. Mixtures modeling parallels the individual component or whole mixture assessment approaches wherein single-chemical models are first developed and linked together to predict mixtures toxicity or whole mixture modeling is carried out for a given exposure scenario. These types of models can be used in the future to direct targeted toxicology experiments and reduce time and monetary resources spent on tests, as well as reduce the number of laboratory animals used in testing.

The bottom-up and top-down approaches are two mixtures modeling approaches that are used to evaluate joint toxicity and more specifically, interactive toxicity of chemical mixtures. Interactive toxicology studies have been carried out by toxicologists to study the influence of one chemical on the behavior of a second chemical in biological systems (45).

The Bottom-Up Approach

This approach starts with the simplest mixture, the binary mixture, and as mechanisms are understood, the models are extended to ternary and higher-order mixtures. In the bottom-up approach, ultimately the model is so developed that the overall toxicity and behavior of the mixture of interest in a biological system can be predicted by the model. Experimental studies are conducted to confirm the predictions of the model before it is applied for the evaluation of real world exposure scenarios. Where inconsequential interactions occur between chemicals in a biological system, an interaction threshold can be defined as the point where they can be statistically identified and become toxicologically significant.

Sometimes binary interactions have been modeled to assess joint toxicity of specific scenarios such as exposure among backpack applicators to specific pesticides or solvents (46–52). A rat PBPK model was developed to identify the interaction threshold for two pesticides using chlorpyrifos and parathion (53). These two pesticides are found together in the environment, especially around the farming communities (54). There are adequate data to show that both of these chemicals act through common mechanisms, and that they are activated by P450, and have the same target enzyme and inhibit acetylcholine esterase. The overall PBPK model consisted of four submodels: two for the parent chemicals and two for their respective active metabolites, chlorpyrifos-oxon and paraxon (Fig. 5). Each metabolite submodel was linked to its parent chemical via the liver compartment, where it is formed. Estimated levels of metabolites were then linked to a model for acetylcholinesterase kinetics describing enzyme synthesis, degradation, binding to the metabolites, and aging after binding. All the physiological, metabolic, and biochemical parameters needed were obtained from the literature. The model output was verified by historical data from literature. So as to explore the utility of the model in directing targeted research at low-dose ranges, the model was run at very low doses, and it was found that the threshold of interaction was approximately 0.08 mg/kg for each of the two components. Because experimental research at such low-level exposures is costly and resource intensive, this kind of modeling helps experimental scientists use model-predicted information to design and conduct studies that are narrowed down to doses that might be significant for each of the components of simple

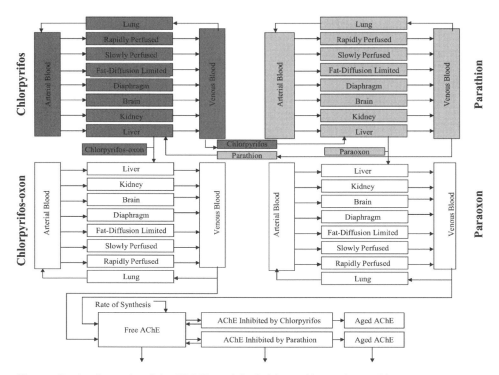

Figure 5 A schematic of the PBPK model of chlorpyrifos and parathion.

mixtures. Extrapolation of this rat model to humans just requires the modification or substitution of values for various parameters to account for the differences between rats and humans.

The direct utility of binary mixtures modeling has been generally limited because exposures are often to higher-order mixtures, i.e., those having more than two components. But this type of modeling can be used as the building block for modeling of higher-order mixtures as long as the quantitative information for each interacting pair is available. Again, models are developed for all the components of the mixture and are linked at the binary level through the tissue compartment where the interactions occur. Then the binary models are interconnected and the overall simulation for the kinetics of all the mixture components is carried out that accounts for the interactions that are occurring at various levels in the complex mixture. A PBPK model, based on this building block concept, for a mixture of m-xylene, toluene, ethyl benzene, dichloromethane, and benzene was shown to predict the interactions through inhalation route in rats (55). This ability to predict the interactions and joint toxicity of complex mixtures by accounting for binary interactions within a PBPK model can contribute enormously toward interaction-based risk assessment of chemical mixtures. This type of modeling also allows addition or substitution of chemicals to an existing mixture model wherein the binary interaction between a pre-existing mixture and the new component is included in a revised model. An interactive rat PBPK model was built for three common volatile organic solvents, TCE (PERC), and 1,1,1-trichloroethane (TCA), under different dosing conditions where PERC and TCA are competitive inhibitors for TCE (56). The model was then used to extrapolate from high dose to low dose so as to study the toxicological interactions at occupational exposure levels (57). The approach showed the viability of adding components one at a time. The simulations allowed one to determine the interaction between the components, via metabolic inhibition that increased the concentration of chemical(s) in the blood and tissues, for any combination of chemicals at any level within the range of experimental validation of the model.

The Top-Down Approach

The top–down approach circumvents the complications of using simple mixture models and linking them to build complex mixture models. This approach models directly higher-order or complex chemical mixtures that consist of several component chemicals and can explain the pharmacokinetic behavior of a complex mixture (58,59).

For example, petroleum hydrocarbon mixtures such as gasoline, diesel fuel, aviation fuel, and asphalt liquids typically contain hundreds of compounds. Although advances have been made in this area recently, the stage of being able to model all the components individually in a complex chemical mixture has not been reached. Hence, a representative mixture is selected considering the characteristics of the components, mechanisms that could be involved, and how well the chemicals can be subgrouped (59). In the case of gasoline, several markers, which are major components and have toxicological significance, were selected. These consisted of benzene, toluene, ethylbenzene, o-xylene, n-hexane, and a lumped chemical group that was chemically not characterized (60,61). Data were collected through gas uptake experiments wherein rats were exposed in separate experiments to each of the above individual components or one of two blends of gasoline. A model was developed that consisted of four compartments, viz., fat tissue, rapidly perfused tissue, slowly perfused tissue, and the liver. Partition coefficients for each chemical were obtained from the literature, and n-hexane values were used for the lumped portion. Computer simulations of the mixture model explained the behavior of the mixture, including the interactions between the lumped chemical and each of the marker chemicals.

In the next generation of PBPK/PD models, it may be possible to link recent advances in computational technology and molecular biology using reaction network modeling (62) and advances in toxigenomic, proteomics, metabonomics, and Quantitative Structure–Activity Relationships (QSAR). (This volume chapters by Amin R, Witzman F, Griffin J, Basak S.)

RESEARCH ADVANCEMENTS IN JOINT TOXICITY OF METALS

Environmental exposure to chemical mixtures is usually in the form of exposure to complex chemical mixtures such as tobacco smoke, car exhaust, or petroleum. Absolute quantities of individual components comprising a chemical mixture are small, and the expected exposure often occurs over a lifetime.

Interactions between mixtures of metals and/or metalloids are of particular public health importance, because toxic trace elements are chemically stable and tend to persist in human tissues and the environment (63,64). As eruditely observed by Merrill et al. (65), it is critical to precisely quantify and characterize the toxicity of metals because exposure to metals cannot be banned—metals are present in any given sample of food, water, soil, or air. Government agencies acknowledge the potential health hazards and risks posed by toxic trace elements. In a recent Agency for Toxic Substances & Disease Registry (ATSDR) document, the joint toxic action of lead, cadmium, arsenic, and chromium was evaluated (66). To coordinate metals-related policies and regulations, the U.S. Environmental Protection Agency is developing an action plan and draft framework for metal risk assessment (16,67).

From a public health perspective, lead, cadmium, and arsenic warrant continued scrutiny because all three elements are found at 95% of hazardous waste sites. Lead, arsenic, and cadmium are ranked as three of the most common pollutants found at facilities on the National Priorities List otherwise known as "Superfund" sites (68). Besides being present at elevated concentrations at such sites, these metals are often present at elevated concentrations in the environment near mining and smelting sites (69). Experimental data demonstrate that interactions do occur as a result of concurrent exposure to trace elements, viz., lead, cadmium, and arsenic (70). Specifically, concurrent administration of Cd and Pb resulted in lower lead levels and fewer lead intranuclear inclusion bodies in the kidney, indicating that cadmium influenced the bioaccumulation of lead. Concurrent exposure to lead and arsenic resulted in an additive effect on urinary coproporphyrin excretion. Disturbances in specific heme pathway molecules such as coproporphyrin have been utilized for years as biomarkers that detect sublethal toxicity of metals, and are regarded as sensitive biomarkers of exposure (71).

A series of 30-, 90- and 180-day mixture studies in rats exposed via drinking water using lowest-observed-effect-levels (LOELs) showed nephrotoxicity, mediated by oxidative stress (72–74). These studies demonstrated that exposure to lead, cadmium, and arsenic, as well as mixtures of these trace elements at LOEL dose levels resulted in increased production of the pro-oxidant aminolevulinic acid (ALA), as well as upregulation of cellular defensive mechanisms, such as glutathione production. These findings are important because they demonstrate that low-level exposure to trace elements or trace element mixtures results in the accumulation of ALA, which has been shown to induce pathological conditions in humans and laboratory animals (72). It has also been determined that concurrent exposure to binary mixtures of Pb, Cd, or As, particularly Pb and Cd or Pb and As at 30-, 90- and 180-day exposure times appeared to attenuate the excretion of ALA. This underscores the importance of considering the biological effects that mixtures have upon living systems (73). Increased excretion of urinary ALA was accompanied by statistically significant increases in kidney glutathione levels with some of the greatest increases measured among the four binary combinations (Pb + Cd, Pb + As, Cd + As, Pb + Cd + As). Glutathione maintains a reducing environment within the intracellular milieu, and can suppress free radical chain reactions. Low-level exposure to lead, cadmium, and arsenic mixtures does result in measurable increases in oxidative stress, as well as upregulation of compensatory mechanisms associated with cellular defense, and provides insight into how living systems respond to oxidative stress induced by low-level exposure to trace elements and their mixtures (72).

Other articles from the published literature demonstrate that metals do interact to produce biologically significant effects. The combined effects of Pb and As on neurotransmitters and brain distribution showed a greater accumulation (−165%) of Pb in specific regions of the brain and a lower accumulation of arsenic (25–50%) in specific regions of the brain (hypothalamus, midbrain, striatum, and cortex) compared to the single elements. Norephinephrine concentration in the hippocampus was decreased in the Pb–As group compared to controls (38%) (75). DNA single strand break induction studies in red blood cells sampled from 78 German workers simultaneously exposed to airborne Cd, Co (cobalt), and Pb in 10 work facilities showed statistical evidence of interaction among the three heavy metals, with greater than additive effects upon the OR of DNA single strand breaks in mononuclear blood cells among workers coexposed to Pb, Cd, and Co (76).

Exact mechanisms by which individual or combined trace elements effect a toxic response are still being elucidated. There is a growing appreciation of the role that trace elements play in oxidative stress (77–83).

PERSPECTIVES AND FUTURE NEEDS

For hundreds of years, toxicologists have appreciated the fact that chemicals interact with each other, but until recently, scant research has been performed to elucidate these interactive effects. Sir Robert Christison, a well-respected English toxicologist and pioneer of nephrology, commented more than 160 years ago that effects from chemical mixtures (termed "compound poisoning") had been practically overlooked by toxicologists (84). Continued investigations into the health effects of chemical mixtures are required to understand how individual components of a mixture ultimately affect an organism's ability to adapt, or ultimately succumb, to the toxicity of a chemical mixture.

Ideally, data from epidemiological findings supported by animal studies to verify mechanisms leading to the toxicity of chemical mixtures would be the most appropriate data needed for risk assessment. Human and animal studies are costly and time consuming and sometimes lead to inconclusive results. The existing epidemiological studies, which have examined the health effects of mixtures are usually based on retrospective epidemiological data, where exposure duration and concentrations can only be approximated. Apart from this, the epidemiological studies suffer from confounding factors such as genetic susceptibility, nutritional status, and lifestyle factors.

One short-term testing approach that has limited capability but enormous potential is the in vitro approach. For certain endpoints of concern, it may be feasible to develop a screen of tests employing specific in vitro assays. The underlying assumption of such testing is that if biologic activity in these tests is well-correlated with in vivo toxic potency, then the interaction coefficients for mixtures measured using such screening tests may be similarly correlated. A specific in vitro approach that has recently gained enormous recognition is the field of genomics (85). This technique exploits the property of DNA to form duplex structures between two complementary strands, and depending upon the number of genes used, allows for the analysis of gene expression on a truly genome-wide scale. Gene array techniques are being used extensively by researchers to study the influence of chemicals on genes (86), (This volume chapters by Amin R, Witzman F, Griffin J). These studies have either confirmed or provided additional insights that multiple genes and interconnected pathways contribute to the physiological and pathological changes in the normal body leading to toxicity and disease conditions. The results of these studies can be used to identify unique patterns of up or downregulation of genes for various classes of chemical mixtures and thus can be applied for bio-monitoring as well as exposure analyses. Also, analyses of these patterns are also paving the way to targeted research to study specific gene changes associated with certain disease conditions. Hence, like every in vitro method, scientists must evaluate the in vitro techniques used in genomic studies and verify their findings in vivo.

Given the different types of mixtures for which health risk assessments are performed, as well as the many factors that impact the overall toxicity of such mixtures, no single approach is suitable to conduct every health risk assessment. An ideal risk assessment would be based on toxicologic and epidemiological data on the mixture of concern for which health effects have been well characterized, thus requiring minimal extrapolation. Also, appropriate monitoring information either alone or in combination with modeling information is sufficient to accurately characterize human exposure to the mixture. But such characterizations of risk are rare. The risk assessment process must encompass all available toxicologic data and scientific evidence on the plausible toxicities of chemical mixtures. It is also imperative that research continues to develop appropriate methods, with an emphasis on a "systems" approach that studies multiple endpoints rather than a specific endpoint. Health risk assessors should utilize every plausible approach (Fig. 4), compare the results, and then decide to use the approach that best suits the exposure scenario. The results of such multiple analyses may be useful in describing the uncertainty in a risk assessment that is under consideration. In practice, the use of all three approaches may not be possible because of lack of data, time, and other resources.

In the meantime, professional judgment that is gained by conducting such assessments must be carefully used to ensure adequate public health protection. The chemical mixtures risk assessment process will benefit by utilizing a team approach wherein experimental scientists, model developers, and health risk assessors participate toward the development of consensus.

The development of alternative risk assessment procedures and models is a complex data-intensive task; and paucity of data has frequently been the bottleneck to developing hazard-assessment methods or models of risk assessment of chemical mixtures. Experimental research needs to be funded to obtain data to elucidate toxicologic mechanisms, and to better understand the molecular toxicology of chemical mixtures, particularly their mechanisms of interaction with biological systems so as to establish biologically based models. From the foregoing, it can be seen that regulatory and public health agencies are acknowledging the usefulness of new methods, and several are adapting them through establishment of coordination centers and investment of resources. Without a program to ensure sufficient coordination between appropriate data generation and data use, intricate models may be constructed that are not validated by experimental research data. Improving our understanding of joint toxic effects at levels below the lowest-observed-adverse-effect-level or no-observed-adverse-effect-level should be one of the priorities of any future research. It is equally important that such studies be collaborative between scientists from various disciplines such as pharmacologists, epidemiologists, and ecologists (87,88). Inclusion of information from community and health surveys will broaden our awareness of routine exposure to a broad class of chemicals on a daily basis and conduct a realistic evaluation of the possible impact of actual chemical mixtures (89). When developing new methodologies, it is useful to consider the existing knowledge base, existent technology, funding, research and academic opportunities, and the government needs that are acting to shape them (90). For various reasons, there is real shift in the field of toxicology away from the single-chemical model and toward multiple chemicals exposure (91). Some recent regulatory requirements have gone a step further and recommend cumulative risk assessment from multiple media, because even risk assessment of chemical mixtures through environmental medium-by-medium evaluation is artificial in its approach (92). In silico toxicology through advanced computer simulation tools such as reaction network modeling will, to a great extent, replace mechanistic and animal experimental studies, which may be impractical or impossible to perform in animal laboratory for complex biological systems involving chemical mixtures and multiple stressors (93). These techniques, in combination, will equip the field of toxicology with the much needed speed and predictability and replace the classical time-consuming, less-predictive, testing methods. Creative thinking and a new mindset are needed to deal with the wealth of new information that is being generated (94). Data generation, processing, storing, sharing, and analyses procedures need to be standardized so as to maximize its utility and benefits. It is also important to compare the findings of different laboratories and understand the underlying principles operative in biological/ecological systems and humans (95). Lack of standardization presents considerable difficulty and risk in transcribing results or augmenting data from one study to another. Bioinformatics and data analysis will play a key role in the future of toxicology, a field which is largely driven by issues that relate to human exposure to chemical mixtures and their health consequences (96).

The lay public is now appreciating that "some chemical traces appear to have greater effects in combination than singly" (97). Today's advancements in analytical chemistry have made possible to measure one-millionth of the smallest traces, not even measurable three decades ago. The possible low-dose risks of common industrial chemicals in combination that disrupt the endocrine system were documented through the observation of feminized sex organs in certain species (98). It took the toxicologists 25 years and $2 billion to study 900 chemicals. But in the United States, some 80,000 commercial and industrial chemicals are now in use, of which, over 30,000 are produced or used in one part of the county, the Great Lakes region. Thus, the modern toxicology must address the concerns of the today's well-informed communities to low level environmental chemical exposures through breast milk and fish consumption (99,100). All chemicals and their combinations cannot be tested experimentally, even if resources are available, in an expedited manner. Hence modern toxicology must employ computational tools to evaluate toxicity and join toxicity of environmental chemicals. We must communicate realistically the chemical risk of contaminants of breast milk and fish while at the same time educate in their nutrition advantages.

REFERENCES

1. Mokdad AH, Marks JS, Stroup DF, Gerberding J. Actual causes of death in the United States, 2000. JAMA 2004; 291(10):1238–1245.
2. http://www.USA29.org/USA/threshold_limit_values.html (accessed June 2005).
3. Groten JP. Mixtures and interactions. Food Chem Toxicol 2000; 38:S65–S71.
4. http://www.cdc.gov/exposurereport/2nd/ (accessed June 2005).
5. http://www.cdc.gov/exposurereport/pdf/third_report_chemicals.pdf (accessed June 2005).
6. Aylward LL, Hayes SM. Temporal trends in human TCDD body burden: decreases over three decades and implications for exposure levels. Exp Anal Environ Epidemiol 2002; 12(5):19–28.
7. Hites RA. Polybrominated diphenyl ethers in the environment and in people: a meta-analysis of concentrations. Environ Sci Technol 2004; 38(4):945–956.
8. US department of health and human services. The health consequences of smoking: cancer and chronic lung disease in the workplace. A report of the surgeon general, Rockville, MD, 1985.
9. Toxicological profile of trichloroethylene http://www.atsdr.cdc.gov/toxprofiles/tp19.html (accessed June 2005).
10. Gorell JM, Johnson CC, Rybicki BA, et al. Occupational exposures to metals as risk factors for Parkinson's disease. Neurology 1997; 48(3):650–658.
11. Gorell JM, Johnson CC, Rybicki BA, Peterson EL, Richardson RJ. The risk of Parkinson's disease with exposure to pesticides, farming, well water, and rural living. Neurology 1998; 50:1346–1350.

12. United States Environmental Protection Agency (U.S. EPA). Supplementary guidance for conducting health risk assessment of chemical mixtures, 2000a, EPA/630/R-00/002. http://cfpub.epa.gov/ncea/raf/chem_mix.cfm.

13. United States Environmental Protection Agency (U.S. EPA). Methodology for deriving ambient water quality criteria for the protection of human health, 2000b. http://www.epa.gov/waterscience/humanhealth/method/complete.pdf (accessed June 2005).

14. United States Environmental Protection Agency (U.S. EPA). Guidelines for the health risk assessment of chemical mixtures. Fed Reg 1986; 51(185):34014–34025. http://cfpub.epa.gov/ncea/raf/rafguid.cfm.

15. Agency for Toxic Substances and Disease Registry (ATSDR). Guidance manual for the assessment of joint toxic action of chemical mixtures, 2001a. http://www.atsdr.cdc.gov/interactionprofiles/ipga.html.

16. Feron VJ, Groten JP. Toxicological evaluation of chemical mixtures. Food Chem Toxicol 2002; 40:825–839.

17. United States Environmental Protection Agency (U.S. EPA). Framework for inorganic metals risk assessment (external review draft). U.S. Environmental Protection Agency, Washington, D.C., 2004, EPA/630/P-04/068B. http://cfpub.epa.gov/ncea/raf/recordisplay.cfm?deid=88903.

18. World Health Organization. Health effects of combined of combined exposures in the workplace. Technical report number 662, 1981. http://whqlibdoc.who.int/trs/WHO_TRS_662.pdf.

19. Mumtaz MM, Lichtveld MY, Wheeler JS. An approach to define chemical mixtures of public health concern at hazardous waste sites. In: De Rosa CT, Holler JS, Mehlman MA, eds. Impact of Hazardous Chemicals on Public Health, Policy, and Service. New Jersey: International Toxicology Books, 2002a:391–402.

20. http://www.atsdr.cdc.gov/mrls.html (accessed June 2005).

21. http://www.acgih.org (accessed June 2005).

22. http://www.cdc.gov/niosh/homepage.html (accessed June 2005).

23. Durkin P, Hertzberg R, Stiteler W, Mumtaz MM. Identification and testing of interaction patterns. Tox Lett 1995; 79:251–264.

24. Lewtas J. Development of a comparative potency method for cancer risk assessment of complex mixtures using short-term in vivo and in vitro bioassays. Toxicol Ind Health 1985; 1(4):193–203.

25. Lewtas J. Emerging methodologies for assessment of complex mixtures: application of bioassays m the integrated air cancer project. Toxicol Ind Health 1989; 5(5):838–850.

26. Schoeny RS, Margosches E. Evaluating comparative potencies: developing approaches to risk assessment of chemical mixtures. Toxicol Ind Health 1989; 5(5):825–837.

27. United States Environmental Protection Agency (U.S. EPA). Provisional guidance for quantitative risk assessment of polycyclic aromatic hydrocarbons. Office of Health and Environmental Assessment, 1993, NTIS PB94–116571.

28. United States Environmental Protection Agency (U.S. EPA). Interim procedures for estimating risks associated with exposures to mixtures of chlorinated dibenzo-p-dioxins and dibenzofurans (CDDs and CDFs) and 1989 update. Risk Assessment Forum, 1989, EPA/625/3–89/016.

29. United States Environmental Protection Agency (U.S. EPA). Workshop report on toxicity equivalency factors for polychlorinated biphenyl congeners. Risk Assessment Forum, 1991, EPA/625/3–91/020.

30. Mumtaz MM, Poirier KA, Colman JT. Risk assessment for chemical mixtures: fine-tuning the hazard index approach. In: De Rosa CT, Holler JS, Mehlman MA, eds. Impact of Hazardous Chemicals on Public Health, Policy, and Service. New Jersey: International Toxicology Books, 2002b:417–432.

31. Teuschler LK, Hertzberg RC. Current and future risk assessment guidelines, policy, and methods development for chemical mixtures. Toxicology 1995; 105:137–144.

32. Safe S. Polychlorinated biphenyls (PCBs), dibenzo-p-dioxins (PCDDs), dibenzofurans (PCDFs) and related compounds: environmental and mechanistic considerations which support the development of toxic equivalent factors (TEFs). CRC Crit Rev Toxicol 1990; 21:51.
33. http://www.atsdr.cdc.gov/toxprofiles/tp17.html (accessed June 2005).
34. Agency for Toxic Substances and Disease Registry (ATSDR). Guidance manual for the joint toxic action of chemical mixtures, 2004a, http://www.atsdr.cdc.gov/interactionprofiles/ipga.html (accessed June 2005).
35. U.S. Environmental Protection Agency (U.S. EPA). Risk assessment guidance for superfund. Vol. I. Human Health Evaluation Manual (Part A). Office of Emergency and Remedial Response, Washington, D.C., 1989, EPA/540/i-89/001.
36. Mumtaz MM, Poirier KA, Hertzberg RC. Feasibility of developing target-organ toxicity doses (TTDs) for estimation of toxicity of chemical mixtures. Toxicologist 1993; 13(1):1728.
37. Mumtaz MM, Poirier KA, Colman JT. Risk assessment for chemical mixtures: fine-tuning the hazard index approach. J Clean Technol Environ Toxicol Occup Med 1997; 6(2):189–204.
38. National Academy of Sciences (NAS). Risk assessment of mixtures of systemic toxicants in drinking water. Drinking Water and Health. Vol. 9. Safe Drinking Water Committee, National Research Council, Washington, D.C.: National Academy Press, 1989:121–132.
39. Mumtaz M, Durkin PR. A weight of the evidence approach for assessing interactions in chemical mixtures. Toxicol Ind Health 1992; 8(6):377–406.
40. De Rosa CT, Hansen H, Wilbur SB, et al. Weight of evidence of assessment of chemical interactions: further guidance. In: De Rosa CT, Holler JS, Mehlman MA, eds. Impact of Hazardous Chemicals on Public Health, Policy, and Service. Princeton: International Toxicology Books, 2002:403–416.
41. Andersen ME. Physiological modeling of organic compounds. Ann Occup Hyg 1991; 35:309–321.
42. Andersen ME, Clewell HJ, Gargas ML, Smith FA, Reitz RA. Physiological based pharmacokinetics and the risk assessment process for methylene chloride. Toxicol Appl Pharmacol 1987; 87:185–205.
43. Reitz RH, Mendrala AL, Park CN, Andersen ME, Guengerich FP. Incorporation of in vitro enzymatic data into the physiological-based pharmacokinetic (PBPK) model for methylene chloride: implications for risk assessment. Toxicol Lett 1988; 43:97–116.
44. Clewell HJ, Andersen ME. Risk assessment extrapolations and physiological modeling. Toxicol Ind Health 1985; 1:111–131.
45. Elsisi AED, Hall P, Sim WL, Earnest DL, Sipes IG. Characterization of vitamin A potentiation of carbon tetrachloride-induced liver injury. Toxicol Appl Pharmacol 1993; 119(2):280–288.
46. Andersen ME, Gargas ML, Clewell HJ, Severyn KM. Quantitative evaluation of the metabolic interactions between trichloroethylene and 1,1–dichloroethylene in vivo using gas uptake methods. Toxicol Appl Pharmacol 1987; 89:149–157.
47. El-Masri HA, Constan AA, Ramsdell HS, Yang RSH. Physiologically-based pharmacodynamic modeling of an interaction threshold between trichloroethylene and 1,1–dichloroethylene in Fischer 344 rats. Toxicol Appl Pharmacol 1996; 141:124–132.
48. Purcell KJ, Cason GH, Gargas ML, Andersen ME, Travis CC. In vivo metabolic interactions of benzene and toluene. Toxicol Lett 1990; 52:141–152.
49. Thakore KN, Gargas ML, Andersen ME, Mehendale HM. PBPK derived metabolic constants, hepatotoxicity, and lethality of BRCCl, in rats pretreated with clordecone, phenobarbital or mirex. Toxicol Appl Pharmacol 1991; 109:514–528.
50. Pelekis ML, Krishnan K. Assessing the relevance of rodent data on chemical interactions for health risk assessment purposes: a case study with dichloromethane-toluene mixture. Reg Toxicol Pharmacol 1997; 25:79–86.
51. Tardif R, Laparé S, Brodeur J, Krishnan K. Physiologically-based pharmacokinetic modeling of a mixture of toluene and xylene in humans. Risk Anal 1995; 15:335–342.

52. Durkin P, Hertzberg R, Diamond G. Application of PBPK model for 2,4-D to estimates of risk in backpack applicators. Environ Toxicol Pharmacol 2004; 16:73–91.
53. El-Masri HA, Mumtaz MM, Yushak ML. Application of physiological-based pharmacokinetic modeling to investigate the toxicological interaction between chlorpyrifos and parathion in the rat. Toxicol Pharmacol 2004; 16:57–71.
54. Finke RA, Lu C, Barr D, Needham L. Children's exposure to chlorpyrifos and parathion in an agricultural community in central Washington State. Environ Health Perspect 2002; 110:549–556.
55. Krishnan K, Haddad S, Béliveau M, Tardif R. Physiological Modeling and extrapolation of pharmacokinetic interactions from binary to more complex chemical mixtures. Environ Toxicol Pharmacol 2002; 10(S6):989–994.
56. Dobrev I, Andersen ME, Yang RSH. Assessing interaction thresholds for trichloroethylene, tetrachloroethylene, and 1,1,1–trichloroethane using gas uptake studies and PBPK modeling. Arch Toxicol 2001; 75:134–144.
57. Dobrev I, Andersen ME, Yang RSH. In silico toxicology: simulating interaction thresholds for human exposure to mixtures of trichloroethylene, tetrachloroethylene, and 1,1,1–trichloroethane. Environ Health Perspect 2002; 110:1031–1039.
58. El-Masri HA, Thomas RS, Benjamin SA, Yang RSH. Physiologically based pharmacokinetic/pharmacodynamic modeling of chemical mixtures and possible applications in risk assessment. Toxicology 1995; 105:275–282.
59. Yang RSH, El-Masri HA, Thomas RS, Constan AA. The use of physiologically based pharmacokinetic/pharmacodynamic dosimetry models for chemical mixtures. Toxicol Lett 1995; 82:497–504.
60. Dennison JE, Andersen ME, Yang RSH. Characterization of the pharmacokinetics of gasoline using PBPK modeling with a complex mixtures chemical lumping approach. Inhal Toxicol 2003; 15(10):961–986.
61. Dennison JE, Andersen ME, Dobrev ID, Mumtaz MM, Yang, RSH. PBPK modeling of complex hydrocarbon mixtures: gasoline. Environ Toxicol Pharmacol 2004; 6: 107–119.
62. Mayeno AN, Yang RSH, Reisfeld B. Biochemical reaction network modeling: a new tool for predicting metabolism of chemical mixtures. Environ Sci Tech 2005; 39:5363–5371.
63. Hu H. Exposure to metals. Occup Environ Med 2000; 27(4):983–996.
64. Goyer RA, Golub M. Issue paper on the human health effects of metals. Submitted to the U.S. Environmental Protection Agency. Risk Assessment Forum, 2003. http://cfpub.epa.gov/ncea/raf/recordisplay.cfm?deid=86119 (accessed June 2005).
65. Merrill JC, Morton JJ, Soileau SD. Metals. In: Hayes AW, ed. Principles and Methods of Toxicology, 4th ed. Taylor and Francis, 2001:649–698.
66. Agency for Toxic Substances and Disease Registry (ATSDR). Interaction profile for arsenic, cadmium, chromium, and lead, 2004b. http://www.atsdr.cdc.gov/interactionprofiles/ip04.html (accessed June 2005).
67. United States Environmental Protection Agency (U.S. EPA). Draft action plan for the development of a framework for metals assessment and guidance for characterizing and ranking metals (external review draft), 2002, EPA/630/P-02/003A. http://cfpub.epa.gov/ncea/cfm/recordisplay.cfm?deid=51736 (accessed June 2005).
68. Agency for Toxic Substances and Disease Registry (ATSDR). CERCLA Priority List of Hazardous Substances, 2001b. http://www.atsdr.cdc.gov/clist.html (accessed June 2005).
69. Agency for Toxic Substances and Disease Registry (ATSDR). Interaction profile: arsenic, cadmium, chromium, and lead, 2001c. http://www.atsdr.cdc.gov/interactionprofiles/ip04.html (accessed June 2005).
70. Mahaffey KR, Capar SG, Gladen B, Fowler BA. Concurrent exposure to lead, cadmium and arsenic: effects on toxicity and tissue metal concentrations in the rat. J Lab Clin Med 1981; 98:463–481.

71. Flora SJ. Possible health hazards associated with the use of toxic metals in semiconductor industries. J Occup Health 2000; 42:105–110.

72. Fowler BA, Whittaker MH, Lipsky M, Wang G, Chen XQ. Oxidative stress induced by lead, cadmium and arsenic mixtures: 30-day, 90-day, and 180-day drinking water studies in rats: an overview. Biometals 2004; 17(5):567–568.

73. Whittaker MH, Lipsky M, Wang G, Chen X, Fowler BA. Oxidative stress induced by lead, cadmium and arsenic mixtures: 30-, 90-, and 180-day drinking water studies in rats. Society of Toxicology 2004 meeting abstract, number 1158. Toxicologist 2004; 78 (1-S):238.

74. Whittaker MH, Lipsky M, Wang G, Chen XQ, Fowler BA. Induction of oxidative stress in response to ingestion of lead, cadmium and arsenic mixtures. Society of Toxicology 2005 meeting abstract, number 119587. Toxicologist 2005; 84(suppl S-1):237–238.

75. Mejia JJ, Diaz-Barriga F, Calderon J, Rios C, Jimenez-Capdeville ME. Effects of lead-arsenic combined exposure on central monoaminergic systems. Neurotoxicol Teratol 1997; 19(6):489–497.

76. Hengstler JG, Bolm-Audorff U, Faldum A, et al. Occupational exposure to heavy metals: DNA damage induction and DNA repair inhibition prove co-exposures to cadmium cobalt and lead as more dangerous than hithero expected. Carcinogenesis 2003; 24(1):63–73.

77. Stohs SJ, Bagchi D. Oxidative mechanisms in the toxicity of metal ions. Free Rad Biol Med 1995; 18(2):321–336.

78. Kasprzak KS, Buzard GS. The role of metals in oxidative damage and redox cell-signaling derangement. In: Zalups RK, Koropatnick J, eds. Molecular Biology and Toxicology of Metals. New York: Taylor Francis, 2000:477–527.

79. Ercal N, Gurer-Orhan H, Aykin-Burns N. Toxic metals and oxidative stress part I: mechanisms involved in metal induced oxidative damage. Curr Topics Med Chem 2001; 1:529–539.

80. Galaris D, Evangelou A. The role of oxidative stress in mechanisms of metal-induced carcinogenesis. Crit Rev Oncol /Hematol 2002; 42:93–103.

81. Richter C. Free radical-mediated DNA oxidation. In: Wallace KB, ed. Free Radical Toxicology. Washington, D.C.: Taylor and Francis, 1997:89–113.

82. Ramanathan K, Balakumar BS, Panneerselvam C. Effects of ascorbic acid and alpha-tocopherol on arsenic-induced oxidative stress. Hum Exp Toxicol 2002; 21(12):675–680.

83. Ahmad S, Kitchin K, Cullen WR. Arsenic species that cause release of iron from ferritin and generation of activated oxygen. Arch Biochem Biophys 2000; 382:195–202.

84. Christison R. Of compound poisoning. Chapter 17. A Treatise on Poisons: In Relation to Medical Jurisprudence, Physiology, and the Practice of Physic. 4th ed. London: Stark and Company, 1845:970–974.

85. Pennie WD, Tugwood JD, Oliver GJA, Kimber I. The principles and practices of toxicogenomics: applications and opportunities. Tox Sci 2000; 54:277–283.

86. Thomas R, Rank D, Penn S, Zastrow G, et al. Application of genomics to toxicology research. Environ Health Perspect 2002; 110(suppl 6):919–923.

87. Squillace P, Moran MJ, Lapham WW, Price CV, Clawges RM, Zogorski JS. Volatile organic compounds in untreated ambient groundwater of the United States, 1985–1995. Environ Sci Technol 1999; 33:4176–4187.

88. Squillace P, Scott JC, Moran MJ, Nolan BT, Kolpin DW. VOCs, pesticides, nitrate, and their mixtures in groundwater used for drinking water in the United States. Environ Sci Technol 2002; 36:1923–1930.

89. http://healthyamericans.org/reports/files/transition.pdf (accessed July, 2005).

90. Monosson E. Chemical mixtures: considering the evolution of toxicology and chemical assessment. Environ Health Perspect 2002; 113:383–390.

91. http://209.183.221.234/ai/meet/mixtures_meeting.asp#agenda (accessed August 2005).

92. http://www.epa.gov/oppfead1/fqpa/backgrnd.htm (accessed August 2005).

93. Yang RSH, EL-Masri HA, Thomas RS, et al. Chemical mixture toxicology: from descriptive to mechanistic, and going on to in silico toxicology. ETAP 2004; 18:65–81.

94. Johnson DE. The future of biotechnology testing in the next decade: a perspective. Int J Toxicol 2001; 20:9–13.
95. Marx J. DNA arrays reveal cancer in its many forms. Science 2000; 289:1670–1672.
96. Schwetz BA. Toxicology at the food and drug administration: new century, new challenges. Int J Toxicol 2001; 20:3–8.
97. Waldman P. Levels of risk: common industrial chemicals in tiny doses raise health issue. Wall St J 2005.
98. Colburn T, von Saal FS, Soto AM. Developmental effects of endocrine-disrupting chemicals in wildlife and humans. Environ Health Perspect 1993; 101:378–384
99. Oken E, Wright RO, Kleinman KP, et al. Maternal fish consumption, hair mercury, and infant cognition in a U.S. cohort. Environ Health Perspect 2005; 113:1376–1380.
100. Crofton KM, Craft ES, Hedge JM, et al. Thyroid-hormone-disrupting chemicals: evidence for dose-dependent additivity or synergism. Environ Health Perspect 2005; 115:1549–1554.

11

Risk-Based Approach to Foods Derived from Genetically Engineered Animals

Larisa Rudenko, Kevin J. Greenlees, and John C. Matheson III
USFDA Center for Veterinary Medicine, Rockville, Maryland, U.S.A.

INTRODUCTION: THE NEED TO MOVE BEYOND TRADITIONAL DOSE-EXAGGERATION STUDIES

Traditional toxicological testing of additives in food has relied on dose exaggeration (relative to anticipated human exposure) via relevant routes of exposure to provide a qualitative characterization of the nature of adverse events that could arise as the result of exposure to chemical substances. This was often followed by a stepped-down, but still exaggerated multiple dose regimen suitable to the nature of the study (e.g., subacute through chronic oral testing, reproductive toxicity, developmental toxicity, and neurotoxicity) to attempt to characterize relationship between dose and toxicity, and identify exposures that do or do not cause adverse outcomes within the limits of detection of the assay (1–4). Dose-exaggeration studies provide an opportunity to evaluate the dose–response relationship, making it possible to estimate impacts outside the observable range of responses. In addition, the exaggerated dose can increase the frequency of observed toxicological effects and increase the likelihood of observing a rare event. The results of well-designed exaggerated dose studies on relatively few animals make it possible to predict potential toxicological effects in the general population. Over time, these studies have proven their utility for both qualitative and quantitative characterizations of toxicity and the estimate of risk to humans or other animals (5). The studies form the basis of the approach used by toxicologists and regulators in assessing the risk to humans (or other animal species) from unintended exposures, or for estimating safe exposure levels for intended or unavoidable exposures.

The test materials in the traditional dose-exaggeration toxicological testing described above tend to be well-characterized chemical entities. They are often xenobiotics, and are frequently relatively small molecules. From time to time, well-defined mixtures are tested according to the dose-exaggeration paradigm (e.g., polychlorinated biphenyls (PCBs), dioxins), especially when environmental exposures are to mixtures rather than single chemical entity (6).

The traditional paradigm for toxicity assessment, however, is less amenable for determining whether any toxicity, or adverse outcomes, may be associated with complex mixtures such as foods (meat, milk, or eggs). There are several reasons for this, including the inability to adequately define or characterize the test materials of interest. For example, if the toxicity of milk were to be evaluated, then "milk" would need to be defined. Milk, however, is an exceptionally complex mixture (7) that varies qualitatively and quantitatively according to the breed of animal producing it, stage in the lactation cycle, diet, and husbandry (8). The second difficulty is dose exaggeration. It is often difficult to provide enough of a food constituent to elicit toxicological responses in a limited test population during a study of limited duration without also interfering with the dietary requirements of the test animal. Feeding edible tissue such as milk or meat to toxicological test species such as rats or mice, and looking for adverse effects as a way to assure the safety of the edible tissue has limited toxicological value. The likelihood of a feeding study such as this showing an adverse effect in the absence of a specifically identified hazard is extremely low. In such studies, the dose to the test animal cannot be increased above 100% of the diet, and the risk of showing effects due to nutritional deficits resulting from dietary perturbations rather than true toxicities is present (9). In addition to the inability to provide an exaggerated dose, the power of this analysis is severely limited by the number of animals in the test. Although it is possible to improve the sensitivity of a toxicological test by increasing the duration of exposure (up to a lifetime), traditional toxicological tests also attempt to increase the sensitivity of the toxicological assay and the ability to observe a low frequency event by increasing the dose while still allowing normal nutrition in the diet.

Most regulatory agencies, including the U.S. Food and Drug Administration (USFDA), have long considered "food" as safe and only evaluate substances added to food that are not considered to be generally recognized as safe (10). With the advent of biotechnology and the development of genetically engineered (GE) plants, the issue of developing methods for identifying potential food consumption hazards began to be addressed. Although traditional dose-exaggeration protocols could be applied, they had to be modified and expanded to address the (often) proteinaceous nature of the gene product. The question of how to address the characterization of the potential hazards that could be introduced into the resulting GE plant posed a problem for the toxicologic and regulatory communities during the late 1980s through the 1990s. Various individual regulatory agencies such as the USFDA, U.S. Environmental Protection Agency, and the U.S. Department of Agriculture, working with affected stakeholders, developed methods to assess the safety of these microorganisms and plants (11,12). Over time, these approaches were harmonized until in the late 1990s, significant concordance was reached in the international communities (13–15).

The introduction of GE animals into this toxicological milieu poses some toxicological issues similar to those posed by GE plants while raising issues specific to the generation of GE animals (14). GE animals, however, provide some inherent safeguards and challenges to the animal scientist, toxicologist, risk assessor, and regulator. In the following chapter, we outline these challenges and safeguards.

GE Animals

GE animals are those that contain either heritable or nonheritable recombinant DNA (rDNA) constructs. The term *transgenic* is most often applied to those GE animals that contain heritable constructs.

The first reports of the production of transgenic animals were published in 1982, when Brinster et al. reported on the development of transgenic mice (16,17). Many other transgenic rodent lines have been developed since that time, providing significant insights into the function of genes and their role(s) in the etiology of human and animal diseases. They have also served as replacements for conventionally bred mice that are used to detect tumor-forming abilities of various chemical agents (18,19). This chapter will not address those animals, as they are neither a species traditionally consumed for food, nor are they intended to produce other end products for human use.

Three years later, Palmiter and Brinster, aided by other colleagues had produced the first GE rabbits and pigs (20), with a transgene stably incorporated in the germ line. Now, some 20 years later, many different species including those traditionally consumed for food, have been transformed with various gene constructs. It may be useful to consider GE animals based on their intended use; a partial list might include:

1. GE animals with enhanced agronomic traits and intended for use as food (e.g., increased growth rate, altered protein, or fat content);
2. GE animals intended for the production of pharmaceuticals in their milk, blood, or eggs for use in humans or animals. These may include animals traditionally used for food (e.g., cows, pigs, sheep, goats, chickens), or companion animals (e.g., dogs, cats, horses);
3. GE animals intended for the production of substances with industrial uses (e.g., goats producing spider silk for use in personal protection fabrics and sheep with altered fleece characteristics to aid in wool processing);
4. GE animals modified with traits intended for therapeutic effect in situ. These may include animals that are traditionally used or not used for food (e.g., cows with decreased mastitis incidence and horses treated with gene therapy to increase collagen production);
5. GE animals with modifications that serve as model systems for studying human disease (e.g., knock-out rodents and pigs serving as models of degenerative eye diseases);
6. GE animals with "knock-out" modifications to be used either as sources for xenotransplantation or to improve their own health (e.g., deletion of surface glycoproteins causing acute transplant rejection, or loss of binding sites for pathogenic microorganisms); and
7. GE animals modified for esthetic or consumer purposes (e.g., aquarium fish with fluorescent markings and hypoallergenic cats).

Although the technology is evolving rapidly, there are two fundamental steps that are common to the production of all GE animals. The first involves the production of the rDNA construct containing the trait of interest plus the other regulatory signals required to package the inserted material for efficient expression in the GE animal. The second is the production of the GE animal containing the construct. The latter step may involve microinjection of early embryos or production of transgenic cells that are subsequently used as donors for somatic cell nuclear transfer to make transgenic animals. For the production of GE animals with nonheritable traits, other methods may be used to introduce the construct into the animal, including electroporation or the use of viral vectors or sequences derived from viruses.

Irrespective of the methods used to produce the final GE animal, the fundamental risk questions are the same:

1. Does the product of the gene construct used to make the GE animal pose a
   direct food consumption risk?
2. Does the gene construct pose an indirect food consumption risk by obvious
   or subtle changes, thereby making it a less wholesome source of food?

These questions may be answered best by applying the hazard-risk paradigm
described below (13).

UNDERLYING RISK BASIS AND PARADIGM SETTING

The hazard-risk paradigm or, "risk assessment," is a framework by which data can
be generated and evaluated to determine whether a particular substance or activity
has the potential to pose an adverse health outcome. When performing a risk analysis, it is critically important to distinguish between a *hazard* and the potential
risk(s) that may result from exposure to that hazard. A *hazard* can be defined as
an act or phenomenon that has the potential to produce an adverse outcome, injury,
or some sort of loss or detriment. These are sometimes referred to as *harms*, and are
often identified under laboratory conditions designed to maximize the opportunity
to detect adverse outcomes. Thus, such observational summaries are often referred
to as *hazard identification* or *hazard characterization. Risk*, on the other hand, is
the conditional probability that estimates the probability of harm given that exposure has occurred.

The National Academies of Science (NAS) have attempted to describe risk analysis in different ways, depending on the nature of the hazards (Table 1). The 1983
NAS report "Risk Assessment in the Federal Government" (21) first attempted to
consolidate the risk assessment procedures practiced in the U.S. regulatory agencies
(primarily FDA's Bureau of Foods, which subsequently became the Center for Food
Safety and Applied Nutrition) into four coherent steps. At that time, these steps were
appropriate to the nature of the substances on which risk assessments were performed,
e.g., potential radiation and chemical carcinogens. Chief among the shared characteristics of these substances was the ability to describe dose in discrete units, allowing for
the relative precision of exposure and dose–response estimates.

Table 1 Risk Assessment Steps as Described by the NAS

1983 *Red Book*	2002 Animal Biotechnology: Science-Based Concerns
• Hazard identification • Exposure assessment • Dose–response evaluation • Risk characterization	• Identify potential harms • Identify potential hazards that might produce those harms • Define what exposure means and the likelihood of exposure • Quantify the likelihood of harm given that exposure has occurred

Abbreviation: NAS, National Academies of Science.

By the time of the publication of the NAS's 2002 report "Animal Biotechnology: Science-Based Concerns" (22) the risk assessment process had been adapted to the potential risks associated with animal biotechnology. The most important differences reflect the change of etiologic agents from radiation and exogenous chemicals to biological agents or processes. These differences are most obviously seen in the hazard assessment and dose–response steps of the process, where the range of potential adverse outcomes (harms) differs in kind from radiation and chemical damage, and the concept of dose must accommodate biological potential. Biological potential can be thought of as the substance or organism's ability to either grow, replicate, and die, or perform a catalytic function so that concentration no longer follows the traditional pharmacokinetic models developed for chemical entities.

Another important consideration is who experiences the risk. Human toxicology has reasonably focused on toxicity to humans; veterinary toxicology examines toxicity to the exposed animal. At its inception, risk assessment tended to be anthropocentric; all risks were evaluated in the human sphere, and were expressed in units of the individual, that is, the probability of a person being exposed to a hazard and experiencing a harm over a lifetime. That individual is defined as the *receptor*. Human risks could also be expressed at the population level, or the probability of x individuals in the population experiencing the harm. The same approach can be used for assessing the risk to animals exposed to a hazard (e.g., pesticides, metals). Although the focus of this section is on human receptors as consumers of foods derived from GE animals, it is important to remember that receptors can include the GE animal itself, the surrogate dam carrying a fetal GE animal, nontransgenic breeding partners, and humans or other animals coming into contact with (including consuming edible products from) the GE animals.

TOWARD A RISK-BASED APPROACH

The introduction of chemicals and chemical products into food can range from food additives to pesticides to new animal drugs to chemical contaminants. The food matrix containing the introduced chemical product is generally presumed to be safe for consumption, and generally not considered as part of the evaluation (1). Chemical contaminants such as those found in the environment are typically evaluated using a risk assessment approach (i.e., estimating the probability of harm per unit of exposure) in part, because there is little control over their introduction. The risk assessment can begin with identification of the hazard, followed by characterization of the hazard(s), their potential harms, and then with characterization of the exposure to the human consumer to come up with an estimate of potential risk. Commercial products intentionally introduced into food (e.g., food additives, pesticides, and new animal drugs) are more often evaluated based on a safety assessment (i.e., the characterization of safe levels of exposure). The evaluation includes identification of the possible hazard(s) for human consumption, characterization of the hazard(s), evaluation of the exposure to the human consumer, and setting maximum allowable exposure levels to limit risk. Additional hazards, such as residues of inactive ingredients, or bioactive metabolites, are also identified during the toxicological and residue evaluation. In each case, the toxicological approach is similar. For food consumption risks, the hazard (e.g., residues of the chemical entity for food additives, pesticides, and new animal drugs) is introduced via the oral route of exposure. The hazard(s) is typically characterized through in vivo and in vitro toxicology tests.

The results are typically used to develop, when possible, a reference dose (RfD)[a] or an acceptable daily intake (ADI)[b]. Exposure to the human consumer may be considered through consumption databases, as is typically done for a pesticide, or through default exposure assumptions, as is typically done for food additives and new animal drugs (25). The exposure assumptions may consider the relative susceptibility of the impacted consumer (child, adult, or geriatric), variety and quantity of foods consumed, and quantity of material of toxicological concern in the food. The RfDs or ADIs, when considered together with the exposure information, are maximum exposure levels that assure safety for human consumption. Only when the data are insufficient to derive an RfD or ADI are assessments generally performed to determine the risk of consumption of food containing these compounds.

HAZARD ANALYSIS FOR GE ANIMALS TRADITIONALLY CONSUMED FOR FOOD

By contrast with relatively simple chemical contaminants or food additives, identifying and characterizing food consumption hazards posed by GE animals are more complex because of the increased opportunities for the introduced genetic material to interact with the cell/animal relative to a small molecule. Potential hazards may arise from a component of the introduced genetic construct, its intended expression product(s), perturbations of the genome resulting from insertion of the construct into the genome (i.e., insertional mutagenesis) that include alterations in coding sequences at the site of insertion, or changes distal to the insertion site resulting from alterations of regulatory regions of the DNA.

These hazards may be classified as posing *direct* or *indirect* risks to two receptors of that risk: the animal being transformed by the construct, and humans consuming food from the resulting transgenic animals. Direct risks can be defined as those arising immediately from the interaction of the gene construct with the genome of the animal (e.g., insertional mutagenesis resulting in developmental effects) and the gene product(s) with the physiology of the animal (e.g., perturbations in the physiology of the animal resulting from the poisoning of a metabolic pathway). Direct risks to the human receptors consuming food from GE animals are mostly centered on interactions with the gene product(s) (e.g., potential allergenicity). Indirect effects can be defined as those outcomes that may ultimately be caused by the presence of the transgene or its product(s), but whose effects are spatially or temporally removed from the construct or its product(s). One hypothetical example of indirect effects caused by the insertion of a transgene into the genome of an animal could cause a change in the expression of metallothionein altering the disposition of metals such as copper, cadmium, zinc, etc.

Although discussions of the hypothetical ways in which such changes might occur have drawn the attention of practitioners of biotechnology risk assessments,

---

[a] RfD: An RfD is defined as the no-observable adverse effect level divided by an uncertainty factor (typically 100) and modification factor. The RfD, while similar to an ADI in assuming doses below the RfD that are not likely to pose a hazard, explicitly makes no assumptions regarding the absolute safety of concentrations above and below the RfD, and is considered more appropriate than the ADI for safety evaluations to a population (23).

[b] ADI: An estimate of the amount of a substance in food or drinking water, expressed on a body-weight basis, that can be ingested daily over a lifetime without appreciable risk (standard human = 60 kg). The ADI is listed in units of mg per kg of body weight (24).

it may be more instructive to have a practical example of how such a hazard characterization might actually be performed. Recently, Wall et al. (26) reported on the production of transgenic Jersey cows containing a construct intended to produce an antibacterial protein to protect the animal from mastitis (27). Mastitis is an economically important and relatively common disease in dairy animals resulting from bacterial infections of the mammary gland (udder). One of the predominant pathogens that has been identified in mastitic dairy cows is *Staphylococcus aureus*. At this time there are no known vaccines to *S. aureus*, and infections tend to be refractory to treatment, leading to chronic or recurrent infections (26). Milk from cows with mastitis is rejected from the food supply, as is milk from cows treated with antibiotics. Cows with mastitis may experience significant distress, and may need to be culled or euthanized. By introducing resistance to mastitis in dairy cows, it may be possible to protect the animal from developing infection, resulting in increased health status and animal welfare, as well providing the dairy farmer with savings resulting from decreased veterinary care and lost production.

To protect dairy cows from mastitis, Wall et al. (26) constructed a transgene containing the gene for lysostaphin, a protein expressing peptidoglycan hydrolase activity from *Staphylococcus simulans*. The expression of this gene, which is intended to be specific to the mammary gland, is driven by an ovine β-lactoglobulin promoter. The construct also contains two genes used to help identify cells that have taken up the construct. One encodes the selectable marker that imparts resistance to the antibiotic neomycin, and the other is a "reporter gene" that codes for a protein that fluoresces green at certain wavelengths of light. The former is driven by a promoter derived from the SV40 virus (for selecting bovine cells containing the construct), while the latter uses a human elongation factor (hEF) promoter. Transgenic animals were generated from transformed bovine fibroblast cells expressing lysostaphin via somatic cell nuclear transfer.

The consideration of potential hazards associated with this GE animal may begin with a preliminary step that involves determining whether the animal would be likely to enter the food supply—that is, determining whether the animal is apparently healthy, wholesome, and suitable for food. The living GE animal would be examined for anatomical, physiological, and developmental abnormalities using comprehensive veterinary examinations appropriate for dairy cows and unmodified animals of the same breed as comparators. The comprehensive veterinary examinations would incorporate general indices of health such as visual observations, clinical chemistries, hematology, and indices of growth and development. Particular attention would be paid to evidence of normal organ systems, behavior, and successful sexual reproduction. This holistic approach has the advantage of allowing the animal to serve as the first screen for obvious food consumption risks due to the relative comparability of mammalian physiologies. An obviously sick or unthrifty animal can serve as a sentinel suggesting the presence of perturbations caused by the introduction of transgenic constructs. These results can help direct the toxicologist to particular organs or tissues. To identify less obvious hazards, a more reductionist approach is required.

This approach begins with a complete characterization of the transgene, including its synthesis and the source of the sequences used in the DNA construct. Consideration would be given to the origin of the promoters (ovine β-lactalbumin, SV40, and hEF) and the potential for any of these sequences to affect gene expression in the region of incorporation, as well as the potential for any of these sequences to recombine with endogenous viruses to alter virulence or host range. Similarly, the source (*S. simulans*) of the gene of interest (lysostaphin) would be evaluated to determine whether there would be a potential for any adverse outcomes. The construct itself would be evaluated to determine whether there were any additional open reading frames present that could encode unanticipated expression products, or whether any unintended regulatory elements that could affect gene expression may be found.

To determine whether lysostaphin poses an allergenic risk, however, it would be prudent to determine whether *S. simulans* is a human pathogen, if the organism has ever caused an allergic response in humans, the degree of homology with known allergenic proteins via a search of the primary amino acid sequence, and the digestibility of the protein as proposed in Codex Alimentarius Assessment of Possible Allergenicity (28). The antibacterial protein may be evaluated for bioavailability using methods ranging from in vitro digestion assays to in vivo models. Proteins surviving digestion may be evaluated for allergic potential. Databases of known and suspect allergens are rapidly expanding and, combined with structural analysis, are improving the ability to predict potential allergenicity (29). New models of allergenicity offer new insights into the potential for an allergic response (30,31). Protein products showing potential to survive acid degradation in the stomach may also be evaluated using toxicological studies such as exaggerated dose repeat-dose oral toxicity studies. The concentration of the protein in edible tissues may be determined to provide information about potential exposure.

Because risk is a function of both the inherent toxicity (hazard) and the exposure, estimating likely exposure to humans is a critical part of the risk assessment. How exposure is assessed very much depends on the basis under which the risk assessment is performed. With the lysostaphin transgenic dairy cow example, exposure assessments would likely consider people with high proportions of milk in their diets (infants, children, and the elderly). Additional considerations could include, but would not be limited to, standardized assumptions of dietary consumption, focused consumption survey data, or residual analysis of the edible tissues. Exposure of the human consumer to biologically active lysostaphin may be influenced by the effects of pasteurization, and other postpasteurization processing such as cheese making. Likewise, it is important to estimate exposure to lysostaphin resulting from other edible tissues as the result of ectopic expression (i.e., outside the anticipated tissue or organ) to determine potential risk.

Determining the indirect toxicity of lysostaphin is less straightforward, and would rely on the ability of changes in the overall contribution of *S. aureus* to alter microbiological ecologies in and around the animal and human receptors.

Both the direct and the indirect risks posed by the neomycin phosphotransferase used as a selectable marker have been evaluated (32). In fact, neomycin phosphotransferase has been approved as a food additive in transgenic plants (33). Characterization of the number of copies of the *neo*[r] gene and the estimated foodborne exposure to its gene product would aid in determining its safety. Although the green fluorescent protein (GFP) used as a reporter gene also has not been the subject of regulatory approval, Wall et al. claim that there is little to no expression of the gene (personal communication), which could be empirically demonstrated by an evaluation of the edible tissues for the presence of the GFP protein. The same assumption and the demonstration apply to the antibiotic selectable marker.

Although the preceding discussion is not comprehensive, it is illustrative of the overall approach that may be used to perform a hazard/risk evaluation for food derived from a GE animal. Implied in the outline of potential hazards for GE animals is the assumption that individually identified hazards can be amenable to traditional approaches used to evaluate food consumption risks.

Unintended Effects

How can we be certain that we have been as comprehensive as possible in identifying unanticipated or unintended potential hazards and risks? Identifying these hazards and risks has been likened to searching for car keys in a dark, unlit parking lot and only looking under the lamp posts because that is the only place where the keys could be seen. The probability of finding the lost keys is greatly enhanced by arranging the lamp posts to ensure that as much of the parking lot is illuminated as possible. Similarly, the confidence that may be placed in an assessment of the hazards and risks posed by unintended effect is likely enhanced or attenuated based on the number and nature of the chosen techniques used to search for them.

One of those approaches is related to the first screening tool mentioned at the beginning of this discussion (comparison of the overall health status of the live animal with appropriate comparators) that uses the animal itself as an integrating tool to determine which physiological pathways have been perturbed. This approach evaluates the composition of the edible tissues in the GE animal against conventional comparators matched for age, gender, breed, species, and husbandry. If there are no material differences between the GE animal and its conventional counterpart (with the exception of the intended gene product(s)), then the likelihood of increased risk from the background "food" is low. This approach has, at different times, been called the "substantial equivalence" or "conventional comparator" approach. The results of the comparison would need to be evaluated carefully, to determine whether variability in the composition of meat and milk is a function of genetic and environmental factors or a perturbation of the animal's physiology. Further, even if detected differences are due to genetic engineering, the questions "whether these changes pose a hazard to the animal or human consumer, and what, if any, risk may result" still remain. This approach is currently used in the evaluation of GE plants, and as part of the method for evaluation of the potential food consumption of meat and milk from livestock clones (34).

The difficulty with this approach, of course, is again the issue of placing the lamp posts appropriately. In general, producers tend to analyze proximates, essential and key nutrients such as vitamins, minerals, amino acids, essential fatty acids, and other substances (e.g., cytokines) that may be of toxicological interest (34). In addition, determining the appropriate comparator can be a nontrivial task. Possible comparators for milk or meat from a GE dairy cow could include closely related nontransgenic siblings or other genetic relatives, milk obtained from the same breed from an unrelated commercial dairy, milk in retail sale, or historical, published references.

Combining all three approaches (i.e., the health of the animal, rigorous evaluation of the construct, its integration and expression, and compositional analysis), allows the strengths of one approach to mitigate the weaknesses of the other. Edible tissue from an apparently healthy GE animal, compositionally indistinguishable from tissue obtained from a nonengineered healthy animal resulting in exposures to gene products that pose no direct or indirect risks, offers increased confidence that there are no significant unintended, unidentified risks arising from food derived from these animals for human consumption.

LOOKING FORWARD

Toxicogenomics has been defined as "the study of the relationship between the structure and activity of the genome (the cellular complement of genes) and the adverse effects of exogenous agents" (35). Because some of the tools developing under this discipline of toxicology may provide much more complete assessments of the biochemical pathways and the genetic signals driving them (additional lamp posts), they offer valuable opportunities for the assessment of GE animals. These approaches already have begun to impact discovery toxicology (36) and have been proposed to detect and predict unanticipated adverse drug reactions (37). High throughput array technologies and sophisticated handling of hundreds to thousands of data endpoints generated by genomic, proteomic, and metabolomic approaches offer a much broader basis for the critical biological system evaluation of the GE animal. As toxicogenomics approaches develop and mature, methods will become more standardized, and the ability to process and interpret the massive amount of data that will be generated will improve. Interest in the use of tools such as these in development of biomarkers of toxicity has led to the creation of the National Center for Toxicogenomics with the National Institute of Environmental Health Sciences (38). Two specific goals are to facilitate the application of gene and protein expression technology and to identify useful biomarkers of disease and exposure to toxic substances. Developments in this field of toxicology are likely to have an increasing impact on traditional areas of investigation as well as new areas such as GE animals. Nonetheless, until such assays have been validated for specific circumstances (age, stage of life, species, and breed) and food matrices (e.g., meat, milk, and eggs) their utility in the evaluation of risks associated with GE animals will be of limited value.

The ability of toxicology to contribute to the identification and characterization of hazards resulting from GE animals is developing at a pace that rivals the evolution of the techniques to develop the animals themselves as knowledge gained in one area feeds development in another. Keeping pace with the changing face of both the technology and the toxicological methods to evaluate it presents the toxicological community with exciting, if daunting, challenges.

ACKNOWLEDGMENT/DISCLAIMER

Special thanks to Dr. Robert Wall for agreeing to the use of his lysostaphin transgenic dairy cow as an illustrating example in this chapter. The discussion of this transgenic animal does not represent an official position of the USFDA of any regulatory requirements for commercial products derived from this or any other bioengineered animals.

REFERENCES

1. U.S. Food and Drug Administration Center for Food Safety and Applied Nutrition Redbook 2000. Toxicological Principles for the Safety Assessment of Food Ingredients. July 2000, Updated October 2001, November 2003, and April 2004.
2. U.S. Food and Drug Administration Center for Veterinary Medicine. Guidance for Industry 3. General Principles for Evaluating the Safety of Compounds Used in Food Producing Animals, June 21, 2005.
3. U.S. Food and Drug Administration, Center for Veterinary, Veterinary International Commission on Harmonization (VICH). Medicine Guidance for Industry 149. Studies to Evaluate the Safety of Residues of Veterinary Drugs in Human Food: General Approach to Testing. VICH GL33, May 18, 2004.

4. World Health Organization. Environmental Health Criteria 70. Principles for the Safety Assessment of Food Additives and Contaminants in Food, 1987.

5. Olson H, Betton G, Robinson D, et al. Concordance of the toxicity of pharmaceuticals in humans and in animals. Reg Toxicol Pharmacol 2000; 32:56–67.

6. U.S. Environmental Protection Agency. EPA/630/P-03/002A. External Review Draft. Framework for Application of Toxicity equivalence Methodology for Polychlorinated Dioxins, Furans, and Biphenyls in Ecological Risk Assessment, June 2003.

7. Jenness R. Composition of milk. In: Wong NP, Jenness R, Keeney M, Marth EH, eds. Fundamentals of Dairy Chemistry. New York: Van Nostrand Reinhold, 1988:1–38.

8. White SL, Bertrand JA, Wade MR, Washburn SP, Green JT, Jenkins TC. Comparison of fatty acid content of milk from Jersey and Holstein cows consuming pasture or a total mixed ration. J Dairy Sci 2001; 84:2295–2301.

9. Slikker W, Anderson ME, Bogdanffy MS, et al. Dose-dependent transitions in mechanisms of toxicity. Toxicol Appl Pharmacol 2004; 201:203–225.

10. U.S. Food and Drug Administration Office of Food Additive Safety Guidance for Industry. Frequently Asked Questions About GRAS, November 2004.

11. U.S. Food and Drug Administration. Draft Guidance for Industry. Drugs Biologics, and Medical Devices Derived from Bioengineered Plants for Use in Humans and Animals, 2002.

12. U.S. Food and Drug Administration. Draft Guidance for Industry. Recommendations for the Early Food Safety Evaluation of New Non-Pesticidal Proteins Produced by New Plant Varieties Intended for Food Use. November 2004.

13. Codex Alimentarius Commission. CAC/GL44–2003. Principles for the Risk Analysis of Foods Derived From Modern Biotechnology, 2003.

14. Codex Alimentarius Commission. Draft Guideline for the Conduct of Food Safety Assessment of Foods Derived from Recombinant-DNA Plants. CAC/GL45–2003.

15. Codex Alimentarius Commission. Guideline for the Conduct of Food Safety Assessment of Foods Produced Using Recombinant-DNA Microorganisms. CAC/GL 46–2003.

16. Brinster RL, Chen HY, Warren R, Sarthy A, Palmiter RD. Regulation of metallothionein-thymidine kinase fusion plasmids injected into mouse eggs. Nature 1982; 296(5852):39–42.

17. Palmiter RD, Brinster RL, Hammer RE, et al. Dramatic growth of mice that develop from eggs microinjected with metallothionein-growth hormone fusion genes. Nature 1982; 300(5893):611–615.

18. Jacobson-Kram D, Sistaire FD, Jacobs AC. Use of transgenic mice in carcinogenicity hazard assessment. Toxicol Pathol 2004; 32(suppl 1):49–52.

19. MacDonald J, French JE, Gerson RJ, et al. The utility of genetically modified mouse assays for identifying human carcinogens: a basic understanding and path forward. Toxicol Sci 2004; 77:188–194.

20. Hammer RE, Pursel VG, Rexroad CE, et al. Nature 1985; 315(6021):680–683.

21. National Research Council. Risk Assessment in the Federal Government: Managing the Process. Washington, D.C.: National Academy Press, 1983.

22. National Research Council. Animal Biotechnology: Science-Based Concerns. Washington, D.C.: National Academy Press, 2002.

23. Barnes DG. Reference dose (RfD): description and use in health risk assessments. Reg Toxicol Pharmacol 1988; 8:471–486.

24. Food Additive Organization of the United Nations and World Health Organization. Summary evaluations performed by the Joint FAO/WHO Expert Committee on Food Additives (JECFA 1956–2003). Internet ed. Washington, D.C.: ILSI Press, 2004 (http://jecfa.ilsi.org).

25. Friedlander LG, Brynes SD, Fernandez AH. The human food safety evaluation of new animal drugs. Vet Clin N Am Food Anim Pract 1999; 15(1):1–11.

26. Wall R, Powell AM, Paape MJ, et al. Genetically enhanced cows resist intramammary *Staphylococcus aureus* infection. Nat Biotechnol 2005; 23(4):445–451.

27. Rainard P. Tackling mastitis in dairy cows. Nat Biotechnol 2005; 23(4):430–432.
28. Codex Alimentarius, Guideline for the Conduct of Food Safety Assessment of Foods Derived from Recombinant-DNA Plants (CAG1GL 45-2003): Annex. Assessment of Possible Allergenicity.
29. Brusic V, Petrovsky N, Gendel SM, Millot M, Gogonzac O, Stelman SJ. Computational tools for the study of allergens. Allergy 2003; 58:1083–1092.
30. Goodman RE, Hefle SL, Taylor SL, van Ree R. Assessing genetically modified crops to minimize the risk of increased food allergy: a review. Int Arch Allergy Immunol 2005; 137(2):151–152.
31. Helm RM, Burks AW. Animal models of food allergy. Curr Opin Allergy Clin Immunol 2002; 2(6):541–546.
32. Astwood JD, Leach JN, Fuchs RL. Stability of food allergens to digestion in-vitro. Nat Biotechnol 1996; 14(10):1269–1273.
33. US Food and Drug Administration, "Secondary Food Additives Permitted in Food for Human Consumption; Food Additives Permitted in Feed and Drinking Water of Animals; Aminoglycoside 3'-Phosphotransferase II; Final Rule. Federal Register 1994; 59:26,700–26,711.
34. Rudenko L, Matheson JC, Adams AL, Dubbin ES, Greenlees KJ. Food consumption risks associated with animal clones: what should be investigated? Cloning Stem Cells 2004; 6(2):79–93.
35. Aardema MJ, MacGregor JT. Toxicology and genetic toxicology in the new era of "toxicogenomics": impact of "-omics" technologies. Mutat Res 2002; 499:13–25.
36. Hamadeh HK, Bushel P, Paules R, Afshari CA. Discovery in toxicology: mediation by gene expression array technology. J Biochem Mol Toxicol 2001; 15(5):231–242.
37. Lühe A, Suter L, Ruepp S, Singer T, Weiser T, Albertini S. Toxicogenomics in the pharmaceutical industry: hollow promises or real benefit? Mutat Res 2005; 575:102–115.
38. Waters MD, Olden K, Tennant RW. Toxicogenomic approach for assessing toxicant related disease. Mutat Res 2003; 544:415–424.

12

Toxicology of Nanomaterials

Nancy A. Monteiro-Riviere and Jessica P. Ryman-Rasmussen
Center for Chemical Toxicology Research and Pharmacokinetics, College of Veterinary Medicine, North Carolina State University, Raleigh, North Carolina, U.S.A.

INTRODUCTION

There has been explosive growth in engineering disciplines based on nanomaterials. These range from applications in the fields of ceramics to microelectronics (1,2). The unique physical properties of nanomaterials (conductivity, reactivity, etc.) compared to larger microparticles enable these novel engineering applications. Consequently, the field of nanoscience has experienced unprecedented growth during the last few years and has received a great deal of attention. However, there are many challenges that must be overcome before applying nanotechnology to the field of nanomedicine and prior to conducting science-based occupational or environmental exposure risk assessments. Insufficient data have been collected so far to allow for full interpretation or thorough understanding of the toxicological implications of occupational exposure or potential environmental impact of nanomaterials.

WHAT ARE NANOMATERIALS?

Nanomaterials are structures with characteristic dimensions between 1 and 100 nm; when engineered appropriately, these materials exhibit a variety of unique and tunable chemical and physical properties (Fig. 1, Table 1). These characteristics have made engineered nanoparticles central components in an array of emerging technologies, and many new companies have emerged to commercialize products. Although they have widespread potential applications in material science, engineering, and medicine, the toxicology of these materials has not been thoroughly evaluated under likely environmental, occupational, and medicinal exposure scenarios. Nanosize materials are also found in nature arising from combustion processes such as volcanoes and forest fires. Other naturally occurring nanosized particles are viruses, biogenic magnetite, and ferritin. Anthropogenic nanomaterials have been produced by combustion and possibly during the manufacturing process of nanomaterials (3). Engineered nanoparticles are used in personal care products such as sunscreens and cosmetics, as drug delivery devices and contrast imaging agents (4). The focus of this chapter is on assessing the toxicology of manufactured nanomaterials currently known.

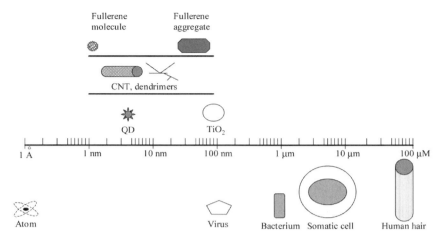

Figure 1 Size of nanomaterials relative to microscopic and macroscopic objects. *Abbreviations*: CNT, carbon nanotubes; QD, quantum dots; TiO_2, titanium dioxide.

NANOMATERIAL NOMENCLATURE

First, we should define nanotechnology-related terminology because without standardization of nomenclature, vague terms will cause confusion to regulators, journal editors, lawyers, and scientists of different disciplines. Due to the vast range in size, shape, and composition, the naming of nanomaterials is challenging (5).

Ambient air pollution particles have three size distributions. Combustion processes can generate ultrafine (UF) particles that are less than $0.1\,\mu m$ in diameter, but when they coalesce, they become mode particles that range from 0.1 to $1.0\,\mu m$. The third size distribution, the coarse mode particles can range from 1 to $100\,\mu m$ (6). Nanosized particles are referred to as *UF particles* by toxicologists *Aitkin mode* and *nucleation mode* particles by atmospheric scientists (7), and *engineered*

Table 1 Summary of Sizes and Uses of Nanomaterials

Nanomaterial	Size range	Current and proposed uses
TiO_2	~10 nm to 100 nm[a]	Sunscreens, cosmetics, pharmaceuticals, foods, plastics, paper products, pigments
Fullerenes (i.e. "bucky-balls")	~1 nm (molecule) to 100 nm[a] (e.g., nano-C_{60})	Materials science, electronics, optics, pharmaceutics
CNTs (MWCNT, SWCNT)	~1 nm to 100 nm	Materials science, electronics, pharmaceutics
Dendrimers	~10 nm to 100 nm	Materials science, electronics, pharmaceutics
Quantum dots	~1 nm to 10 nm	Biomedical research, diagnostics, pharmaceutics

[a]could be greater than 100 nm when agglomerated.
Abbreviations: CNTs, carbon nanotubes; MWCNT, multiwalled carbon nanotubes; SWCNT, single-walled carbon nanotubes; TiO_2, titanium dioxide.

nanostructured materials by materials scientists (7). Nanomaterials have structural features between 1 and 100 nm. Nanoparticles must have at least one dimension smaller than 100 nm. Engineered and manufactured nanoparticles have been engineered to exploit properties and functions associated with size. Nanostructured particles have a substructure greater than atomic/molecular dimensions but lesser than 100 nm and exhibit physical, chemical, and biological characteristics associated with nanostructures (Maynard, personal communication, 2005). The terms *agglomerate* and *aggregate* are commonly used interchangeably and refer to particles that are weakly or strongly held together. This terminology is commonly used by powder technologists to describe assemblages of powders in dry and liquid suspensions. In the pharmaceutical industry, the proposed term *agglomerate* is used to describe an assemblage of particles in a powder, whereas the term *aggregate* is confined to prenucleation structures. Thus, multiple terms are used in the nanotechnology field because scientists from many disciplines are collaborating in nanotechnology research. It is important that national and international standards organizations reach a consensus for a "nanovocabulary" before the imprecise terminology leads to universal confusion (8).

CHARACTERIZATION

The characterization of nanomaterials may differ from that of chemicals and bulk materials. Also, magnetic materials may require other types of characterization to reflect their magnetic properties. Therefore, several key physicochemical characteristics must be defined when reporting on the toxicity of nanomaterials. Contaminants present secondary to the manufacture of nanomaterials may influence the toxicity, so the core synthesis method should be referenced along with surface morphology and transmission electron microscopy (TEM) imaging, size, surface area, surface chemistry, surface charge, density, crystalline structure, and elemental composition. Providing most of these characteristics will enhance the understanding and mechanism of toxicity of the nanomaterial. To gain a better understanding of nanomaterials, the need for new detection techniques for localization within cells or tissues is needed, especially if one wants to quantitate the amount present. At the minimum, the basis of the analytical method used should be clearly stated. Easy access to radiolabeled nanomaterials would be helpful. It is extremely important that full characterization of the nanomaterials be conducted and described to facilitate comparisons between studies.

NANOMATERIALS AND TOXICITY

For the purposes of this chapter, we will focus on many nanomaterials that we consider likely to be candidates for industrial and biomedical applications. Also, we will discuss what is known about the relevant aspects of the toxicology of these materials. Titanium dioxide (TiO_2), fullerenes (C_{60}), single-walled nanotubes (SWNT) and multi-walled carbon nanotubes (MWCNT), dendrimers, and quantum dots (QD) will be discussed. The relative scale distributions of these materials are depicted in Figure 1 and summarized in Table 1.

TiO_2

TiO_2 is mined from mineral deposits and is widely used, with global production in 2004 estimated at 4.5 million metric tons (9). TiO_2 is utilized industrially in the manufacture of numerous consumer products, including paints, paper, and plastics and is also approved for use in cosmetics, sunscreens, foodstuffs, and pharmaceuticals. TiO_2 is available in several sizes, with the UF (less than 100 nm) being classifiable as a nanoscale material. An extensive review of the toxicology of TiO_2 is beyond the scope of this chapter. There are several studies, however, which suggest that mechanisms of toxic responses to $UF-TiO_2$ may prove relevant to the other nanomaterials described herein.

The majority of $UF-TiO_2$ studies have focused on the lung, where $UF-TiO_2$ has been shown to elicit a greater toxic response than larger TiO_2 particles (10–12). Toxicity is believed to be a consequence of inflammatory responses mediated by activated alveolar macrophages, which produce free radicals, proteases, and cytokines (13–16). The cellular mechanism by which phagocytosis of $UF-TiO_2$ leads to macrophage activation is not well understood, but is currently thought to result from the enhanced oxidizing activity of $UF-TiO_2$ and other UF particles, as a consequence of their large surface area (17). This idea of large surface area enhancing reactivity and toxicity may be applicable to other nanoscale materials. Some studies with nanosized preparations of TiO_2 have been conducted in nonpulmonary cell and tissue types. $UF-TiO_2$ has been shown to cause the production of micronuclei leading to apoptosis in hamster embryo fibroblasts. This study indicates that UF preparations of TiO_2 can be clastogenic in this cell type (18). Additionally, TiO_2 strongly absorbs UV radiation, which has led to its widespread use in sunscreens. These UV-absorbing properties of TiO_2 are shared by QD and fullerenes, the latter of which are proposed for use in sunscreens and cosmetics. Dermal absorption studies of micronised preparations of TiO_2 (ca. 20 nm) have indicated that this form of TiO_2 does not penetrate the stratum corneum to access the metabolically active and dividing cells of the epidermis and dermis (19). Thus, nanoscale preparations of TiO_2 are not expected to exert significant dermal toxicity due to the barrier function of the intact stratum corneum. The effect of altered barrier function has not been studied. The effects of other nanoscale materials on skin toxicity, as well as the effects of these nanomaterials or TiO_2 on skin with a compromised stratum corneum are unknown.

In summary, $UF-TiO_2$ is an important industrial and consumer product that is appropriately classifiable as a nanomaterial. The large surface area of $UF-TiO_2$ appears to be responsible for its enhanced toxicity relative to larger-sized TiO_2 particles. This enhanced surface area, and possibly, the mechanisms mediating the enhanced toxicity of $UF-TiO_2$ are shared by other nanoscale materials.

Fullerenes

Fullerenes or buckyballs are molecular structures made up of 60 or more carbon atoms. The fullerenes are used in materials science, superconductivity applications, electronic circuits, nonlinear optics, pharmaceuticals, and in everyday items such as clothing, tennis rackets, bowling balls, and numerous other applications. They have been added to the resin coatings of bowling balls to improve controllability. Producing fullerenes and to purchase them is quite expensive. They have been considered to be a commercial success for some companies and thus are being made in multiton quantities (20).

Fullerenes, although relatively new, have an interesting history. In 1985, scientists were studying "cluster-aggregates of atoms" in which they vaporized graphite with a laser in an atmosphere of helium gas to form carbon clusters. These clusters were very stable. It was discovered that only a geometric shape could combine 60 carbon atoms into a spherical structure of hexagons and pentagons. This combination of structure was the basis of a geodesic dome designed by Buckminster Fuller for the 1967 Montreal Exhibition. Because the newly discovered molecule resembled this architectural structure, they were termed buckminsterfullerene. It was later referred to as fullerene or C_{60} and is commonly known as buckyballs (21). In 1996, the scientists Curl, Kroto, and Smalley were awarded the Nobel Prize for their discovery. Since their discovery, fullerenes have been used in many biological applications, although little is known of their toxicity, potential carcinogenic effects, or overall health risk.

There is little information regarding the biodistribution and metabolism of C_{60}, probably due to the fact that C_{60} is insoluble in aqueous solutions coupled with a paucity of sensitive analytical techniques. Several in vivo studies, in contrast, have addressed this issue with water-soluble fullerenes. Yamago et al. (22) orally administered a water-soluble ^{14}C-labeled fullerene to rats. The fullerenes were not absorbed but were excreted in the feces. When given IV, the fullerenes were distributed to various tissues and remained in the body after one week. This ^{14}C-labeled fullerene even penetrated the blood–brain barrier, but acute toxicity proved to be low (22). One pharmacokinetic study used IV administration of bis(monosuccinimide) derivative of pp'-bis(2-amino-ethyl)-diphenyl-C_{60}, another water-soluble fullerene in rats. The fullerene plasma concentration–time profile was best described by a polyexponential model with a terminal half-life of 6.8 hours, a relatively large volume of distribution, and very low total body clearance. Protein binding was 99% and fullerenes were not detected in the urine at 24 hours. These data confirm extensive tissue distribution and minimal clearance from the body. A dose of 15 mg/kg of this fullerene derivative was well tolerated by the rats, but a dose of 25 mg/kg resulted in death within five minutes (23). Biodistribution of radiolabeled ^{99m}Tc-labeling of C_{60} (OH)$_x$ fullerenes in mice and rabbits indicated a wide distribution in all tissues, with a significant percentage in the kidneys, bone, spleen, and liver by 48 hours (24). Taken together, these studies indicate that water-soluble fullerenes are widely distributed after nonoral routes of administration and are slowly cleared from the body. Even less is known regarding fullerene metabolism, although one study has shown that fullerenol-1 can suppress levels of cytochrome P450-dependent monooxygenase in vivo in mice and mitochondrial oxidative phosphorylation in vitro in liver microsomes (25).

Uncoated fullerenes have also been shown to induce oxidative stress in juvenile largemouth bass at 0.5 ppm and cause a significant increase in lipid peroxidation of the brain and glutathione depletion in the gills after 48 hours (26). Some C_{60} fullerene derivatives can interact with the active site of HIV-1 protease, suggesting antiviral activity (27). Others have found that C_{60} protects quiescent human peripheral blood mononuclear cells from apoptosis induced by 2-deoxy-D-ribose or tumor necrosis factor alpha plus cycloheximide (28). The developmental toxicity and neurotoxicity of fullerenes have also been studied. Studies with C_{60} solubilized with polyvinyl pyrrolidone in water applied to the mouse midbrain cell differentiation system found inhibition of cell differentiation and proliferation. Harmful effects with C_{60} on embryos also occurred (27,29,30). Because C_{60} can convert oxygen from the triplet to the singlet state, there is some concern for potential health risks through this mechanism.

The dermal toxicity of fullerenes has also been the focus of some studies. Topical administration of 200 μg of fullerenes to mouse skin over a 72-hour period found no effect on either DNA synthesis or ornithine decarboxylase activity. The ability of fullerenes to act as a tumor promoter was also investigated. Repeated application to mouse skin after initiation with dimethlybenzanthracene for 24 weeks did not result in benign or malignant skin tumor formation, but promotion was observed with 12-O-tetradecanoylphorbol-13-acetate resulting in benign skin tumors (31).

In vitro studies using ^{14}C-labeled underivatized C_{60} exposed to immortalized human keratinocytes depicted cellular incorporation of the label uptake at various times. By six hours, approximately 50% of the radiolabel was taken up but it was unclear whether particles actually entered the cell or were associated with the cell surface. These investigators also found no effect of C_{60} on the proliferation of immortalized human keratinocytes and fibroblasts (32).

Four types of water-soluble fullerenes were assessed in human carcinoma cells and dermal fibroblasts for their toxicity in biomedical technologies. It was found that water-soluble functional groups on the surface of fullerenes can dramatically decrease the toxicity of pristine C_{60}. This least derivatized and most aggregated form of C_{60} was more toxic than the highly soluble derivatives such as C_3, Na^+_{2-3} $[C_{60}O_{7-9}(OH)_{12-15}]$, and $C_{60}(OH)_{24}$ (33). There are conflicting reports about the potential toxicity of fullerenes such as C_{60}. While C_{60} itself has essentially no solubility in water, it has been shown to aggregate with either organic solvent inclusion or partial hydrolysis to create water-soluble species, n-C_{60}. These aggregates have been shown to have exceptionally low mobility in aqueous solutions but have been proposed to have high cellular toxicity. A series of fullerene-substituted phenylalanine derivatives were prepared to compare with related functionalized fullerenes. The presence of the C_{60} substituent has been shown to alter the conformation of the native peptide (e.g., from a random coil to a β-sheet), making the conditions under which conversion to an alpha helix occurs important. Studies have shown that there is no apparent toxicity to the cells; however, early studies did not confirm that the peptide was incorporated into the cells (34).

These studies have stressed aspects of the toxicology of various fullerenes and it is important to emphasize, however, that fullerenes may have beneficial effects. Numerous studies have also indicated that specific functionalized fullerenes may be therapeutically useful in the treatment of a number of diseases.

Carbon Nanotubes

Carbon nanotubes (CNT), also known as buckytubes, are made up of seamless, cylindrical shells of graphitic carbon in the range of one to tens of nanometers in diameter and several micrometers in length. CNT can be either single-walled (SWCNT) or multi-walled (MWCNT), depending on whether they are comprised of one shell or two (or more) concentric shell layers. CNT have been utilized for their extraordinary electrical and mechanical properties. They have been used as tips for atomic force microscopy and scanning tunneling microscopy probes due to their nanoscale size, their ability to provide high resolution and to image deep narrow structures, and the fact that they are 100 times stronger than steel. They can be flexible, yet their hardness provides wear resistance (35,36).

One of the principal attributes of CNT and other nanoparticles that make their development such a breakthrough is their unique catalytic properties. For example, pure carbon buckytubes are referenced as being capable of reacting with many organic compounds due to their carbon chemistry base. Modifications including end-of-tube (e.g., via reaction with carboxyl groups at open tip ends of CNT) or sidewall derivatization would modify their physical properties and alter solubility or dispersion. When these attributes are deliberately modified, useful products and therapeutic approaches could be obtained.

SWCNT

There is limited data available on the toxicology and the biological effects of manufactured CNT (37). Acute lung toxicity studies were conducted by intratracheal instillation of SWCNT at high doses of 5 mg/kg in rats for 24 hours. While multifocal granulomas were observed, a mortality rate of 15% resulted from the mechanical blockage of the upper airways by the instillate due to a foreign body reaction and not due to the SWCNT particulate (38). Mice exposed to nanotubes manufactured by three different methods and catalysts depicted dose-dependent epithelioid granulomas formation (39). The primary finding of both the installation studies was the presence of multifocal granulomas and inflammation depending on the type of particle used. An important observation was that once CNT reached the lung tissue, they were more toxic than carbon black or quartz dust, two known pulmonary toxicants (39). This may be secondary to the particles' tendencies to self-aggregate when removed from controlled conditions. It must be stressed that inhalational exposure of particulate matter such as nanoparticles is fundamentally different from dermal or oral exposure, as the lung is designed to trap particulate matter.

Anecdotal reports of dermal irritation in humans (carbon fiber dermatitis and hyperkeratosis) suggest that particles may gain entry into the viable epidermis after topical exposure. Exposure to SWCNT in immortalized nontumorigenic human epidermal cells suggests that CNT may be toxic to epidermal keratinocyte cultures (40). These studies demonstrated significant cellular toxicity when unrefined SWCNT were exposed to cells for 18 hours. These investigators did not evaluate inflammatory markers of irritation/inflammation in this cell line. Previously, our group had demonstrated significant differences in the toxicological response of immortalized keratinocytes versus primary keratinocytes (41). Gene expression profiling was conducted on human epidermal keratinocytes exposed to 1.0 mg/mL of SWCNT that showed a similar profile to alpha-quartz or silica. Alpha-quartz is considered to be the main cause of silicosis in humans. Also, genes not previously associated with these particulates from the structural protein and cytokine families were significantly expressed (42).

In addition to toxicity, the use of SWCNT as therapeutic agents for drug delivery is also under investigation. Drug delivery can be enhanced by many types of chemical vehicles, including lipids, peptides, and polyethylene glycol (PEG) derivatives. Strategies using SWCNT and SWCNT–streptavidin conjugates as biocompatible transporters have shown to be localized within human promyelocytic leukemia (HL60) cells and human T (Jurkat) cells via endocytosis. Functionalized SWCNT exhibited little toxicity to the HL60 cells, but the SWCNT–biotin–streptavidin complex caused extensive cell death (43). Studies by Pantarotto et al. (44) have also demonstrated that functionalized, water-soluble SWCNT derivatives modified with a fluorescent probe can translocate across the cell membrane of human and murine fibroblasts without causing toxicity. The translocation pathway remains to be elucidated.

CNT can be filled with DNA or peptide molecules and can serve as a potential delivery system in gene or peptide delivery (45). These investigators have shown that SWCNT can inhibit cell proliferation and decrease cell adhesive ability in a dose- and time-dependent fashion using human embryonic kidney-293 cells (46). Mouse perito-neal macrophage-like cells can ingest SWCNT in a surfactant without a change in viability or population growth (47).

MWCNT

The dermal toxicity of MWCNT has also been addressed in a primary human ker-atinocyte model. Human neonatal epidermal keratinocytes exposed to 0.1, 0.2, and 0.4 mg/mL of MWCNT for 1, 2, 4, 8, 12, 24, and 48 hours depicted MWCNT within the cytoplasmic vacuoles of human epidermal keratinocytes (HEK). These are clas-sified as MWCNT because they exhibit a base mode growth, very little disordered carbon, and are well ordered and aligned. TEM depicted numerous vacuoles within the cytoplasm of HEK containing MWCNT of various sizes, up to 3.6 mm in length (Fig. 2A and B). At 24 hours, 59% of the keratinocytes contained MWCNT, com-pared to 84% at 48 hours at the 0.4 mg/mL dose. Viability decreased with an increase in MWCNT concentration, and IL-8, an early biomarker for irritation, increased with time and concentration (48). This data showed that MWCNT, neither derivatized nor optimized for biological applications, were capable of both localizing and initiating an irritation response in skin cells. These initial data are suggestive of a significant dermal hazard after topical exposure to select nanoparticles should they be capable of penetrating the stratum corneum barrier. Proteomic analysis con-ducted in human epidermal keratinocytes exposed to MWCNT showed both an increase and decrease in expression of many proteins relative to controls. These pro-tein alterations suggested dysregulation of intermediate filament expression, cell cycle inhibition, altered vesicular trafficking or exocytosis, and membrane scaffold protein downregulation (49,50). Nanosize carbon black has been recommended to serve as a negative control when conducting viability assays in cell culture. However, caution must be taken when utilizing carbon black because we have observed that carbon can adsorb the viability marker dyes such as neutral red and interfere with the absorption spectra causing false positives. In addition, the type of carbon black and its characterization and composition are extremely important. For instance, UF carbon black commonly used for in vivo inhalation studies with gross and micro-scopic endpoints may not be suitable for use in cell culture because it interferes with viability and cytokine assays.

In vivo studies with hat-stacked carbon nanofibers (resembles MWCNT char-acteristics) implanted in the subcutaneous tissue of rats depicted granulation and an inflammatory response that resembled foreign body granulomas. These fibers were present within the cytoplasm of macrophages. However, investigators reported no severe inflammation, necrosis, or degeneration of tissue (51).

Magnetic Nanomaterials

Iron magnetic nanoparticles are another form of nanomaterials that should be dis-cussed because they can accumulate in target tissues and are utilized in drug and gene therapy approaches. However, bare magnetic particles are insufficient for specific accumulations in some tissues. Modification of these particles could potentially enhance the biocompatibility and increase cellular uptake of site-specific delivery. Studies conducted in immortalized primary human fibroblasts exposed to magnetic particles of 40 to 50 nm in diameter, coated with PEG for 24 hours, depicted that PEG-coated particles localized within the vacuoles of the fibroblasts (52).

(A)

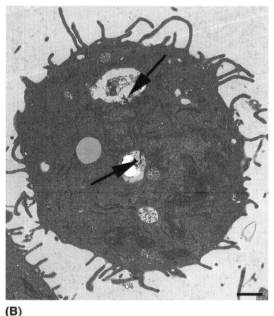

(B)

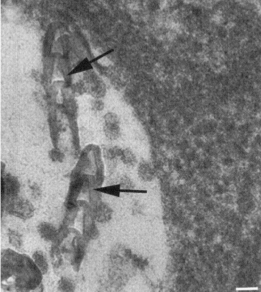

Figure 2   (**A**) TEM of a low magnification of a human epidermal keratinocyte. Arrows depict the location of the MWCNT within the cytoplasmic vacuoles. (**B**) TEM of a higher magnification of **A** to depict the localization of MWCNT within the cytoplasmic vacuole and to demonstrate that MWCNT retains their structure. *Abbreviations*: MWCNT, multi-walled carbon nanotubes; TEM, transmission electron micrograph. *Source*: From Ref. 48.

Dendrimers

Classes of nanomaterials known as dendrimers are planned to be used in gene therapy, drug delivery vehicles for oligonucleotides and antisense oligonucleotides, and imaging systems. They consist of dendritic polymers of polyamidoamine (PAMAM) that can serve a backbone for the attachment of biological materials. They have a well-defined chemical structure, globular shape, low polydispersity index, and controlled terminal groups.

The clusterlike architecture plays an important role on how they interact with lipid bilayers of a cell membrane. The mechanism of how PAMAM dendrimers can alter cells has been studied (53,54). Aqueous dendrimers can form holes 15 to 40 nm in diameter within intact lipid layers. It is thought that the dendrimers remove lipid molecules from the substrate and form aggregates of dendrimers surrounded by lipid molecules. Other investigators have studied the in vivo distribution of differently charged PAMAM dendrimers in B16 melanoma and DU145 human prostrate cancer mouse tumor model systems. Both positive and neutral surface charged dendrimers of 5 nm were localized within major organs and tumors with deposition peaking at one hour. Positive surface–charged dendrimers were higher than the neutral PAMAM in major organs such as the lungs, liver, and kidney, but were lowest in the brain (55).

QD

Semiconductor nanocrystals, or QD, are expected to have great utility as biomedical imaging and diagnostic agents. QD are usually spherical in shape and range in size from a few nanometers to over 10 nm. These heterogeneous nanoparticles consist of a colloidal core, commonly cadmium–selenide (CdSe) that is frequently surrounded by one or more coatings that serve to increase solubility in aqueous medium, reduce leaching of metals from the core, and facilitate customized surface chemistries such as the attachment of antibodies or receptor ligands (56). QD are highly fluorescent and excitable by UV and visible light. The resulting fluorescence emission in the visible or infrared is well defined and is dependent upon the size and chemical composition of the QD. Additionally, QD resist photobleaching and are not readily degradable by biological systems. These unique properties of QD have facilitated their development for a number of specialized detection and imaging applications, but their toxicology is not well understood. QD can be coupled to biomolecules to be used as ultrasensitive biological detectors (57). Compared to organic dyes, they are 20 times as bright and 100 times more stable against photobleaching. These nanometer size conjugates are water soluble and biocompatible. When labeled with the protein transferrin, they undergo a receptor-mediated endocytosis in cultured HeLa cells and dots labeled with immunomolecules recognize specific antibodies (58).

Water-soluble QD injected IV in mice were compared to conventional fluorescein isothiocyanate conjugated to dextran beads. For imaging, QD were more superior, and achieved depths using less power, especially in high scattering skin and adipose tissue. Mice exposed in this study showed no noticeable ill effects and were kept for long-term QD toxicity evaluation (59). Nanocrystals have also been encapsulated in phospholipid block polymer–copolymer micelles to increase their biocompatibility in *Xenopus* embryos. They were found to be stable and nontoxic and QD fluorescence could be followed through the tadpole stage allowing for tracing experiments in embryogenesis (60). A study by Jaiswal et al. (61) demonstrated that targeted CdSe/zinc sulfide (ZnS) QD could be taken up by HeLa cells and tracked in live cells for more than 10 days with no morphological signs of toxicity. The same study also showed that uptake of these QD by starved *Dictyostelium dicoideum* amoebae did not alter physiological responsiveness to cAMP. These observations fueled a great deal of optimism that QD were nontoxic at doses suitable for long-term imaging studies. Consequently, subsequent studies focused primarily on the biomedical applications of QD rather than more in depth assessment of potential toxic reactions.

A study by Derfus et al. (62) was the first designed to specifically investigate the cytotoxicity of QD at concentrations relevant to imaging applications. This study found that metabolic reduction of the tetrazolium dye, 3-(4,5-dimethylthiazol-2-yl)-2,5-diphenyl tetrazolium bromide (MTT), by primary human hepatocytes was significantly decreased after treatment with uncoated CdSe QD, but only when the QD were first UV irradiated or chemically oxidized. Another study revealed a dose-dependent increase in the cytotoxicity of mercaptoundecanoic acid–coated CdSe quantum of three different sizes in African green monkey kidney cells (Vero cells), HeLa cells, and primary human hepatocytes at the MTT reduction endpoint (63). Most recently, positively charged cadmium–tellerium (CdTe) QD coated with cysteineamine were shown to be cytotoxic at concentrations used for confocal imaging in N9 murine microglial and rat pheochromocytoma (PC12) cells at morphological and MTT reduction endpoints (64). Thus, some QD preparations have demonstrated cytotoxicity in some cell lines at doses relevant for biomedical applications.

Further study is clearly needed to evaluate the toxicology of QD in vitro and in vivo. Consideration of the specialized surface modifications, charge, size, and chemical composition of the QD of interest is a reasonable first step in study design. The surfaces of QD are often modified to bind antigens or receptors particular to specific cell types, which can be of great assistance in identifying target organ systems. A caveat is that targeting will not likely be absolute and may result in nonspecific binding, uptake, and toxicity at unintended sites. This idea is illustrated by a study showing that tumor-targeted QD were nonselectively taken up by the liver and spleen (65). Similarly, localization and retention of phosphine-capped QD in lymph nodes have also been reported (66). Additionally, cells undergoing rapid migration or those with high membrane turnover are potentially vulnerable to nonselective uptake of QD by association with the cell membrane (67).

Once vulnerable cell types are identified, the cellular mechanisms of QD cytotoxicity must be elucidated. Examples in the literature indicate that QD size, surface charge, and chemical composition, together with cell-dependent properties, will play a role in mechanisms of cytotoxicity. Studies have shown that in cell culture, QD with a cadmium–selenium core could be rendered nontoxic with additional coatings. In contrast, when exposed to air or UV radiation, they were toxic. Oxidized CdSe QD were shown to be cytotoxic to human hepatocytes at the MTT reduction endpoint (62). This cytotoxicity correlated with release of Cd^{2+} into the culture medium, which was particularly of interest because hepatocytes are highly cadmium sensitive. Cadmium release and cytotoxicity could be attenuated by coating QD with ZnS or bovine serum albumin prior to exposure to an oxidative environment. These observations illustrate that oxidative microenvironments, QD chemical composition, and cellular sensitivity can all contribute to QD cytotoxicity. This would also suggest that QD exposure to unintended physical or chemical stressors during occupational or environmental settings could modulate toxicity.

There are also indications that the size and charge of QD, together with their intracellular localization after uptake, may play a role in cytotoxicity mechanisms.

Positively charged CdTe-core, cysteineamine-capped QD were cytotoxic to PC12 and N9 cells (64). In both cases, the 2.3 nm QD were significantly more cytotoxic than the 5.2 nm QD. This cytotoxicity could be partially attenuated by N-acetylcysteine but not by α-tocopherol, suggesting a mechanism other than free radical generation resulting from Cd^{2+} release. Interestingly, the 5.2 nm QD localized to the cytoplasm of N9 cells, whereas the 2.3 nm QD localized to the nucleus. Nuclear localization of cationic QD has been reported previously in mouse fibroblast (3T3) cells (68). Whether or not this potentiated cytotoxicity of 2.3 nm QD is directly related to nuclear localization is unknown. There is one report in which CdSe/ZnS QD can cause nicking of supercoiled plasmid DNA (69).

In summary, QD offer great potential as therapeutic and diagnostic agents. Contrary to initial reports, QD can be cytotoxic at concentrations used in biomedical imaging applications in some cell types. QD size, charge, and chemical composition, together with cell-specific factors, will likely play a role in cytotoxicity. Investigations into the mechanisms by which toxicity is mediated will likely provide insight into the toxicology of other nanoscale materials, as well as indicate modifications for improved biocompatibility of QD.

EXPOSURE AND RISK ASSESSMENT

The health risk to engineered nanoparticles could be less significant by inhalation exposure than those observed for UF particulates or for larger micron-sized materials because they are produced in close gas phase reactors; therefore inhalation would have to occur in the solid state. Because many engineered nanomaterials are prepared and processed in liquids, there is a higher probability that dermal absorption and oral ingestion may be the more relevant exposure route during the manufacturing process or during accidental spills, shipping, and handling (4). Prevention of dermal exposure by the use of gloves is underscored by a study indicating that some workers had as much as 7 mg of nanotube material deposited on gloves in areas directly contacting nanotubes (70). This study showed that gloves offer dermal protection from nanotube material, but the permeability of different types of gloves to specific nanomaterials is unknown.

Many actual exposure conditions may further modify particle response. What happens when such modifications inadvertently occur such as in an occupational setting where particles may come into contact with harsh solvent containing mixtures or physical stressors (temperature, UV light, etc.)? To evaluate the potential health effects of nanoparticles, more research is needed to determine under what conditions dermal absorption and subsequent toxicity would be affected. Similarly, would nanoparticles that have not been cleaned of metal catalysts serve as an undesirable delivery system for these catalysts into cells? Is absorption altered when exposed to solvents and reactants known to modulate the absorption of other organic compounds and which, in this case, might also react with nanoparticles to alter their chemical or physical properties?

Skin is unique because it is a potential route for exposure to nanoparticles during their manufacture while also providing as a substrate within the avascular epidermis where particles could potentially lodge and not be susceptible to removal by phagocytosis. What are the toxicological consequences of "dirty" nanoparticles (catalyst residue) becoming lodged in the epidermis? In fact, it is this relative biological isolation in the lipid domains of the epidermis that has allowed for the delivery of drugs to the skin using lipid nanoparticles and liposomes. Larger particles of zinc and titanium oxide used in topical skin care products have been shown to penetrate the stratum corneum barrier of rabbit skin (71), with highest absorption occurring from water and oily vehicles. This should also apply to manufactured nanoparticles.

Before attempts are made to assess the environmental impact of nanomaterials, further research is needed to understand the release and distribution of nanoparticles in the environment. Can they accumulate, persist, degrade, or be transformed in water, soil, or air? Also, what is the potential health impact to all living organisms below humans in the food chain? The absorption, metabolism, and distribution or accumulation within target organs needs further investigation. Several studies have attempted to evaluate the toxic effect of nanomaterials and with these results we can attempt to set the hazards for risk assessment.

Block copolymer lipid micelles, which are of the same scale size as many nanoparticles, are known to have prolonged residence times in the body, an attribute that makes them desirable as drug delivery devices (72). Similarly, water-soluble fullerenes have been studied in rats (23) where extensive protein binding (greater than 99%) and body retention were demonstrated. Would these same properties result in accumulation or potentiate the toxicity of manufactured nanoparticles that inadvertently gained access to the systemic circulation? Nanoparticles (e.g., lipid based) are being designed to evade detection by the reticuloendothelial system by which phagocytosis normally removes particulate matter from the body (65). Although this is a desirable property when designed for drug delivery, what happens when nanoparticles from the manufacturing sector inadvertently gain access to the body? Can particles gain access to viable tissues and initiate adverse reactions? Recall that QD, designed for biological imaging applications, have some cytotoxic properties if not specifically formulated to avoid them. Would the extreme hydrophobicity of many of these particles allow for accumulation in lipid tissue once systemic exposure occur? Would such particles preferentially locate in the lipids within the stratum corneum after topical exposure (73)? These issues warrant further study.

ENVIRONMENTAL IMPACT

This emerging technology is producing significant quantities of nanomaterials that could be released into the environment and eventually migrate into porous media such as groundwater aquifers and water treatment plant filters. Deliberate release of nanomaterials has been used in environmental remediation to detoxify pollutants in the soil and groundwater. Some nanomaterials such as TiO_2 and zinc oxide can be activated by light, and could help to remove organic contaminants from various media. Nanomaterials may also be used to treat contaminated soils, sediments, and solid wastes. However, additional research is needed in this field before remediation can be recommended. Until there is sufficient evidence that there are no adverse health effects to humans and wildlife, caution must be exercised. Many ongoing studies are still in the testing phase. There is little information available on evaluating fullerene toxicity in aqueous systems. Monomeric C_{60} has extremely low solubility in water and is unlikely to have a significant effect in aqueous systems. However, nanomaterials can be modified or functionalized in the environment to exhibit different transport behaviors. Functionalized fullerenes that are used to facilitate dispersal in water have been shown to have the highest mobility and can migrate 10 m in unfractured sand aquifers (74). Studies have also shown that fullerol toxicity can increase in the presence of light (75).

CONCLUSION

Many of the studies discussed in this chapter raise concerns regarding the safety of nanomaterials in a variety of tissues and environments. There is a serious lack of information about human health and environmental implications of manufactured nanomaterials. This new field of nanotoxicology will continue to grow and emerge as new products are produced. The need for toxicology studies will increase for use in risk assessment. Knowledge of exposure and hazard is needed for understanding risks associated with nanomaterials. We will need to first understand the broad concepts that apply to pathways of dermal, oral, and inhalational exposures so that we can focus on hazard assessments. This chapter has explored the beginning threads of nanomaterial toxicology in a variety of systems. Nanomaterial toxicology will become a major challenge with regard to conducting a comprehensive safety evaluation of nanomaterials.

REFERENCES

1. Jortner J, Rao CNR. Nanostructured advanced materials. Perspectives and directions. Pure Appl Chem 2002; 74:1491–1506.
2. Zhan GD, Kuntz JD, Wan J, Mukherjee AK. Single-wall carbon nanotubes as attractive toughening agents in alumina-based nanocomposites. Nat Mater 2003; 2:38–42.
3. Oberdörster G, Oberdörster E, Oberdörster J. Nanotoxicology: an emerging discipline evolving from studies of ultrafine particles. Environ Health Perspect 2005. http://dx.doi.org/.
4. Colvin VL. The potential environmental impact of engineered nanomaterials. Nat Biotechnol 2003; 21:1166–1170.
5. Halford B. Nano dictionary. Chem Eng News 2005; 83:31.
6. Beckett WS, Chalupa DF, Pauly-Brown A, et al. Comparing inhaled ultrafine vs. fine zinc oxide particles in healthy adults: a human inhalation study. Am J Respir Crit Care Med 2005 (published ahead of print).
7. NNI. What is nanotechnology? 2004. http://www.nano.gov/html/facts/whatisnano.html.
8. Nichols G, Byard S, Bloxham MJ, et al. A review of the terms agglomerate and aggregate with a recommendation for nomenclature used in powder and particle characterization. J Pharm Sci 2002; 91:2103–2109.
9. United States Geological Survey. Mineral commodity summaries. 2005. http://minerals.usgs.gov/minerals/pubs/commodity/titanium/tidiomcs05.pdf.
10. Driscoll KE, Maurer JK, Lindenschmidt RC, Romberger D, Rennard SI, Crosby L. Respiratory tract responses to dust: relationships between dust burden lung injury alveolar macrophages fibronectin release and the development of pulmonary fibrosis. Toxicol Appl Pharmacol 1990; 106:88–101.
11. Oberdörster G, Ferin J, Lehnert BE. Correlation between particle size in vivo particle persistence and lung injury. Environ Health Perspect 1994; 102(suppl 5):173–179.
12. Oberdörster G. Pulmonary effects of inhaled ultra-fine particles. Int Arch Occup Environ Health 2000; 74(1):1–8.
13. Zhang Q, Kusaka Y, Sato K. Differences in the extent of inflammation caused by intratracheal exposure to three ultrafine metals: role of free radicals. J Toxicol Environ Health 1998; 53:423–438.

14. Churg A, Gilks B, Dai J. Induction of fibrogenic mediators by fine and ultra-fine titanium dioxide in rat tracheal explants. Am J Physiol 1999; 277:975–982.
15. Drumm J, Buhl R, Kienast K. Additional NO_2 exposure induces a decrease in cytokine specific mRNA expression and cytokine release of particle and fiber exposed human alveolar macrophages. Eur J Med Res 1999; 4(2):59–66.
16. Rahman Q, Narwood J, Hatch G. Evidence that exposure of particulate air pollutants to human and rat alveolar macrophages lead to different oxidative stress. Biochem Biophys Res Commun 1997; 240:669–672.
17. Donaldson K, Stone V. Current hypotheses on the mechanism of toxicity by ultrafine particles. Ann Super Sanita 2003; 39(3):405–410.
18. Rahman Q, Lohani M, Dopp E, et al. Evidence that ultrafine titanium dioxide induces micronuclei and apoptosis in Syrian hamster embryo fibroblasts. Envrion Health Perspect 2002; 110(8):797–800.
19. Schulz J, Hohenberg H, Pflucker F, et al. Distribution of sunscreens on skin. Adv Drug Del Rev 2002; 54(suppl 1):S157–S163.
20. Tremblay JF. Fullerenes by the ton. Chem Eng News 2003; 81:13–14.
21. Kroto HW, Heath JR, O'Brien SC, Curl RF, Smalley RE. C_{60}: buckminsterfullerene. Nature 1985; 318:162–163.
22. Yamago S, Tokuyama H, Nakamura E, et al. In-vivo biological behavior of a water-miscible fullerene-C-14 labeling absorption distribution excretion and acute toxicity. Chem Biol 1995; 2:385–389.
23. Rajagopalan P, Wudl F, Schinazi RF, Boudinot FD. Pharmacokinetics of a water-soluble fullerenes in rats. Antimicrob Agents Chemother 1996; 40:2262–2265.
24. Qingnuan L, Yan X, Xiaodong Z, et al. Preparation of ^{99m}Tc-$C_{60}(OH)_x$ and its biodistribution studies. Nucl Med Biol 2002; 29:707–710.
25. Ueng TH, Kang JJ, Wang HW, Chen YW, Chiang LY. Suppression of microsomal cytochrome P450-dependent monooxygenases and mitochondrial oxidative phosphorylation by fullerenol a polyhydroxylated fullerene C_{60}. Toxicol Lett 1997; 93:29–37.
26. Oberdörster E. Manufactured nanomaterials (fullerenes C_{60}) induce oxidative stress in the brain of juvenile largemouth bass. Environ Health Perspect 2004; 112:1058–1062.
27. Friedman SH, DeCamp DL, Sijbesma RP, Srdanov G, Wudl F, Kenyon GL. Inhibition of the HIV-1 protease by fullerene derivatives: model building studies and experimental verification. J Am Chem Soc 1993; 115:6506–6509.
28. Monti D, Moretti L, Salvioli S, et al. C_{60} carboxyfullerene exerts a protective activity against oxidative stress-induced apoptosis in human peripheral blood mononuclear cells. Biochem Biophys Res Commun 2000; 277:711–717.
29. Tsuchiya T, Yamakoshi YN, Miyata N. A novel promoting action of fullerene C_{60} on the chondrogenesis in rat embryonic limb bud cell culture system. Biochem Biophys Res Commun 1995; 206:885–894.
30. Tsuchiya T, Oguri I, Yamakoshi YN, Miyata N. Novel harmful effects of [60]fullerene on mouse embryos in vitro and in vivo. FEBS Lett 1996; 393:139–145.
31. Nelson MA, Frederick ED, Bowden GT, Hooser SB, Fernando Q, Carter DE. Effects of acute and subchronic exposure of topically applied fullerene extracts on the mouse skin. Toxicol Ind Health 1993; 9:623–630.
32. Scrivens WA, Tour JM. Synthesis of ^{14}C-labeled C_{60} its suspension in water and its uptake by human keratinocytes. J Am Chem Soc 1994; 116:4517–4518.
33. Sayes CM, Fortner JD, Guo W, et al. The differential cytotoxicity of water-soluble fullerenes. Nano Lett 2004; 4:1881–1887.
34. Yang J, Barron AR. A new route to fullerene substituted phenylalanine derivatives. Chem Commun 2004:2884–2885.
35. Dai H, Hafner JH, Rinzler AG, Colbert DT, Smalley RE. Nanotubes as nanoprobes in scanning probe microscopy. Nature 1996; 384:147–150.
36. Moloni K, Lal A, Lagally MG. Sharpened carbon nanotube probes. Proc SPIE 2000; 4098:76–83.

37. Dagani R. Nanomaterials: safe or unsafe? Chem Eng News 2003; 81:30–33.

38. Warheit DB, Laurence BR, Reed KL, Roach DH, Reynolds GAM, Webb TR. Comparative pulmonary toxicity assessment of single wall carbon nanotubes in rats. Toxicol Sci 2004; 77:117–125.

39. Lam CW, James JT, McCluskey R, Hunter RL. Pulmonary toxicity of single wall carbon nanotubes in mice 7 and 90 days after intratracheal instillation. Toxicol Sci 2004; 77: 126–134.

40. Shvedova AA, Castranova V, Kisin ER, et al. Exposure to carbon nanotube materials: assessment of nanotube cytotoxicity using human keratinocytes cells. J Toxicol Environ Health A 2003; 66:1909–1926.

41. Allen DG, Riviere JE, Monteiro-Riviere NA. Cytokine induction as a measure of cutaneous toxicity in primary and immortalized porcine keratinocytes exposed to jet fuels and their relationship to normal human epidermal keratinocytes. Toxicol Lett 2001; 119:209–217.

42. Cunningham MJ, Magnuson SR, Falduto MT. Gene expression profiling of nanoscale materials using a systems biology approach. Toxicol Sci 2005; 84(S-1):9.

43. Kam NWS, Jessop TC, Wender PA, Dai H. Nanotube molecular transporters: internalization of carbon nanotube-protein conjugates into mammalian cells. J Am Chem Soc 2004; 126:6850–6851.

44. Pantarotto D, Briand JP, Prato M, Bianco A. Translocation of bioactive peptides across cell membranes by carbon nanotubes. Chem Commun 2004:16–17.

45. Gao H, Kong Y, Cui D, Ozkan CS. Spontaneous insertion of DNA oligonucleotides into carbon nanotubes. Nano Lett 2003; 3:471–473.

46. Cui D, Tian F, Ozkan CS, Wang M, Gao H. Effect of single wall carbon nanotubes on human HEK293 cells. Toxicol Lett 2005; 155:73–85.

47. Cherukuri P, Bachilo SM, Litovsky SH, Weisman RB. Near-infrared fluorescence microscopy of single-walled carbon nanotubes in phagocytic cells. J Am Chem Soc 2004; 126:15638–15639.

48. Monteiro-Riviere NA, Nemanich RA, Inman AO, Wang YY, Riviere JE. Multi-walled carbon nanotube interactions with human epidermal keratinocytes. Toxicol Lett 2005; 155:377–384.

49. Monteiro-Riviere NA. Multi-walled carbon nanotube exposure in human epidermal keratinocytes: localization and proteomic analysis. Proceedings of the 229th American Chemical Society Division of Industrial and Engineering Chemistry, 2005.

50. Monteiro-Riviere NA, Wang YY, Hong SM, et al. Proteomic analysis of nanoparticle exposure in human keratinocyte cell culture. Toxicologist 2005; 84(S-1):447.

51. Yokoyama A, Sato Y, Nodasaka Y, et al. Biological behavior of hat-stacked carbon nanofibers in the subcutaneous tissue in rats. Nano Lett 2005; 5:157–161.

52. Gupta AK, Curtis ASG. Surface modified superparamagnetic nanoparticles for drug delivery: interaction studies with human fibroblasts in culture. J Mater Sci: Mater Med 2004; 15:493–496.

53. Mecke A, Uppuluri S, Sassanella TM, et al. Direct observation of lipid bilayer disruption by poly(amidoamine) dendrimers. Chem Phys Lipids 2004; 132:3–14.

54. Hong S, Bielinska AU, Mecke A, et al. Interaction of poly(amidoamine) dendrimers with supported lipid bilayers and cells: hole formation and the relation to transport. Bioconjugate Chem 2004; 15:774–782.

55. Nigavekar SS, Sung LY, Llanes M, et al. ^{3}H dendrimer nanoparticle organ/tumor distribution. Pharm Res 2004; 21:476–483.

56. Michalet X, Punaud FF, Bentolila LA, et al. Quantum dots for live cells in vivo imaging and diagnostics. Science 2005; 307:538–544.

57. Jovin TM. Quantum dots finally come of age. Nat Biotechnol 2003; 21:32–33.

58. Chan WCW, Nie S. Quantum dot bioconjugates for ultrasensitive nonisotopic detection. Science 1998; 281:2016–2018.

59. Larson DR, Zipfel WR, Williams RM, et al. Water-soluble quantum dots for multiphoton fluorescence imaging in vivo. Science 2003; 300:1434–1436.

60. Dubertret B, Skourides P, Norris DJ, Noireaux V, Brivanlou AH, Libchaber A. In vivo imaging of quantum dots encapsulated in phosopholipid micelles. Science 2002; 298:1759–1762.

61. Jaiswal JK, Mattoussi H, Mauro JM, Simon SM. Long-term multiple color imaging of live cells using quantum dot bioconjugates. Nat Biotechnol 2003; 21:47–51.

62. Derfus AM, Chan WCW, Bhatia S. Probing the cytotoxicity of semiconductor quantum dots. Nano Lett 2004; 4:11–18.

63. Shiosahara A, Hoshino A, Hanaki K, Suzuki K, Yamamoto K. On the cytotoxicity caused by quantum dots. Micobiol Immunol 2004; 48(9):669–675.

64. Lovrić J, Bazzi H, Cuie Y, Fortin GRA, Winnik FM, Maysinger D. Differences in subcellular distribution and toxicity of green and red emitting CdTe quantum dots. J Mol Med 2005 (published ahead of print).

65. Akerman ME, Chan WCW, Laakkonen P, Bhatia SN, Ruoslahti E. Nanocrystal targeting in vivo. Proc Natl Acad Sci 2002; 99(20):12617–12621.

66. Kim S, Lim YT, Soltesz EG, et al. Near-infrared fluorescent type II quantum dots for sentinel lymph node mapping. Nat Biotechnol 2004; 22(1):93–96.

67. Parak WJ, Boudreau R, Le Gros M, et al. Cell motility and metastatic potential studies based on quantum dot imaging of phagokinetic tracks. Adv Mater 2002; 14(12):882–885.

68. Bruchez M, Moronne M, Gin P, Weiss S, Alivisatos AP. Semiconductor nanocrystals as fluorescent biological labels. Science 1998; 281:2013–2016.

69. Green M, Howman E. Semiconductor quantum dots and free radical induced DNA nicking. Chem Commun 2005:121–123.

70. Baron PA, Maynard A, Foley M. Evaluation of aerosol release during the handling of unrefined single walled carbon nanotube material. NIOSH Dart-02 2002:191.

71. Lansdown ABG, Taylor A. Zinc and titanium oxides: promising UV-absorbers but what influence do they have on intact skin. Int J Cosmet Sci 1997; 19:167–172.

72. Kwon GS. Diblock copolymer nanoparticles for drug delivery. Crit Rev Ther Drug Carrier Syst 1998; 15(5):481–512.

73. Monteiro-Riviere NA. Comparative anatomy physiology and biochemistry of mammalian skin. In: Hobson DW, ed. Dermal and Ocular Toxicology. Boca Raton: CRC Press, 1991:1–71.

74. Lecoanet HF, Bottero JY, Wiesner MR. Laboratory assessment of the mobility of nanomaterials in porous media. Environ Sci Technol 2004; 38:5164–5169.

75. Pickering KD, Wiesner MR. Fullerol-sensitive production of reactive oxygen species in aqueous solution. Environ Sci Technol 2005; 39:1359–1365.

13

The Biological Basis of Experimental Toxicity of Jet Propulsion Fuel–8

Simon S. Wong and Mark L. Witten
Department of Pediatrics, University of Arizona Health Sciences Center, Tucson, Arizona, U.S.A.

INTRODUCTION

Jet propulsion fuel–8 (JP-8, also known as MIL-T-83133D, AVTUR, or F-34) is a kerosene-based hydrocarbon distillate fuel. It contains approximately 228 long/ short-chain aliphatic and aromatic hydrocarbons (C_6–C_{14}). The low volatility characteristics of JP-8 allow it to be a potential toxic irritant on the respiratory system. JP-8 has become the primary fuel source for the U.S. Armed Forces and North Atlantic Treaty Organization forces. JP-8 is an attractive alternative to JP-4 jet fuel due to its relatively low vapor pressure and high flash point temperature that reduces the risk of spontaneous explosions. It is estimated that worldwide utilizations are approximately five billion gallons (approximately 19×10^9 L) per year. Because of the volumes produced and the multipurpose nature of the fuel, there is potential for exposures to JP-8 in several forms—aerosol, vapor, or liquid. Occupational exposures to JP-8 may occur during fuel transport, aircraft fueling and defueling, aircraft maintenance, cold aircraft engine starts, maintenance of equipment and machinery, use of tent heaters, and cleaning or degreasing with fuel. The U.S. Navy Occupational Safety and Health Standards Board (NAVOSH) has proposed interim exposure limits of 350 and 1800 mg/m^3 as the eight-hour permissible exposure limit and the 15-minute short-term exposure limit, respectively. These limits are based on the U.S. National Institute of Occupational Safety and Health regulatory levels for more volatile petroleum distillates.

Most published data on toxicities of JP-8 have been critically reviewed by several publications, including the newly formulated toxicological assessment of JP-8 by the National Research Council. Recently, there is increasing evidence that exposures to JP-8 at the current limitations of NAVOSH can result in obvious adverse effects in experimental animals. With increased evidence of JP-8 toxicities, more attention has been paid to explore the potential cellular and molecular basis of JP-8 exposure. Therefore, this chapter especially summarizes current findings in experimental animal and cellular models with an emphasis on action mechanisms.

PULMONARY TOXICITY

In a simulated military flight-line exposure protocol, the effects of JP-8 exposure on lung toxicity in rats and mice have been demonstrated in the last decade. Seven-day exposures to aerosolized JP-8 in rats showed perivascular and interstitial edema, which was accompanied by increased respiratory permeability, leukocytic infiltration, and morphological alterations to the distal lung (1). A similar 28-day study in rats showed a jet fuel–induced effect on terminal bronchiolar airways that was accompanied by subendothelial edema (2). The observed effects were manifested by JP-8 exposures ranging from 500 to 1000 mg/m^3, which are above safety limits. An additional investigation of the acute one-hour exposure to NAVOSH permissible JP-8 aerosols also showed mild to moderate pulmonary toxicity (3). These adverse responses were characterized by a targeting of bronchiolar epithelium leading to an increased respiratory permeability, peribronchiolar edema cellular necrosis in airways, and type II alveolar epithelial injury. In vitro study further revealed that exposure to JP-8 induces increases of paracellular permeability in an SV-40 transformed human bronchial epithelial cell line (BEAS-2B) (4). Incubation of confluent BEAS-2B cells with concentrations of JP-8 or n-tetradecane (a primary constituent of JP-8) induced dose-dependent increases of paracellular flux. Following exposures of 0.17, 0.33, 0.50, or 0.67 mg/mL of JP-8, mannitol flux increased above vehicle controls by 10%, 14%, 29%, and 52%, respectively, during a two-hour incubation period. n-Tetradecane also caused higher mannitol flux increases of 37%, 42%, 63%, and 78%, respectively, following identical incubation concentrations. The transepithelial mannitol flux reached a maximum at 12 hours and spontaneously reversed to control values over a 48-hour recovery period, for both JP-8 and n-tetradecane exposure. These data indicate that exposures to JP-8 exert a noxious effect on bronchial epithelial barrier function that may preclude pathological lung injury. n-Tetradecane exposure could partially initiate JP-8–induced lung injury through a disruption in the airway epithelial barrier function.

To explore JP-8–induced inflammatory mechanisms, C57BL/6 mice (young, 3.5 months; adult, 12 months) were nose-only exposed to either room air or atmospheres of 1000 mg/m^3 JP-8 for 1 hr/day for seven days (5). This study showed that exposure of these mice to JP-8 resulted in multiple inflammatory mediator releases such as cytokines and arachidonic acid, which may be associated with JP-8–induced physiological, cellular, and morphological alterations. Data showed that bronchoalveolar lavage (BAL) fluid Prostaglandin E_2 (PGE_2) and 8-iso-$PGF_{2\alpha}$ levels were decreased in both the young and adult JP-8 jet fuel–exposed groups when compared to control values, suggesting that the metabolites of arachidonic acid may be important mediators in both models of JP-8 jet fuel exposure. These mediators are very potent regulators of cellular signaling and inflammation and can be further metabolized by lipoxygenase or cyclooxygenase enzymes to yield the family of leukotrienes or prostaglandins and thromboxanes.

A follow-up study (6) examined the effect(s) of JP-8 on cytokine secretion in a transformed rat alveolar type II epithelial cell (AIIE) lines (RLE-6TN) and primary alveolar macrophages (AMs, from Fischer 344 rats). The cell coculture study indicated that the balance of cytokine release could be regulated possibly by cross-communication of AIIEs and AMs, in close proximity to each other. JP-8 concentrations ranging from 0 to 0.8 μg/mL, which may actually be encountered in the alveolar spaces of lungs exposed in vivo, were placed in cell culture for 24 hours. Cultured AIIEs alone secreted interleukin (IL)-1β and IL-6 [below detectable limits for IL-10 and tumor necrosis factor-α (TNF-α)], whereas cultured AMs alone secreted IL-1β, IL-10, and TNF-α, in a concentration-dependent manner. These data suggest that the release of cytokines, not only from AMs but also from AIIEs, may contribute to the JP-8–induced inflammatory response in the lungs. However, the cocultures of AIIEs and AMs showed no significant changes in IL-1β, IL-6, and TNF-α at any JP-8 concentration when compared to controls. These cytokine levels in cocultures of AIIEs and AMs were inversely related to those of cultured AIIEs or AMs alone. Interestingly, IL-10 levels in the coculture system were increased, depending on the concentration, up to 1.058% at JP-8 concentration of 0.8 μg/mL, although under detectable limits in cultured AIIEs alone and no significant concentration change in cultured AMs alone. It is likely, we speculate, that AMs may possibly act via paracrine and/or autocrine pathways to signal AIIEs to regulate cytokine release. Further studies are required to identify the potential signaling mechanism(s) of cytokine expression and release between these two lung cells.

Proteomic analysis of lungs in male Swiss-Webster mice exposed 1 hr/day for seven days to JP-8 at concentrations of 1000 and 2500 mg/m^3 showed significant quantitative and qualitative changes in tissue cytosol proteins (7,8). Recent data have also shown that exposure of several human and murine cell lines, including rat lung AIIE (RLE-6TN), to JP-8 in vitro induces biochemical and morphological markers of apoptotic cell death, such as activation of caspase-3, cleavage of poly (ADP-ribose) polymerase, chromatin condensation, membrane blebbing, release of cytochrome c from mitochondria, and cleavage of genomic DNA. Generation of reactive oxygen species and depletion of intracellular reduced glutathione (GSH) were also shown to play important roles in the induction of programmed cell death by JP-8. At 250 mg/m^3 JP-8 concentration, 31 proteins exhibited increased expression, while 10 showed decreased expression (Table 1). At 1000 mg/m^3 exposure levels, 21 lung proteins exhibited increased expression and 99 demonstrated decreased expression. At 2500 mg/m^3, 30 exhibited increased expression, while 135 showed decreased expression. Several of the proteins were identified by peptide mass fingerprinting, and were found to relate to cell structure, cell proliferation, protein repair, and apoptosis. These data demonstrate the significant stress that JP-8 jet fuel puts on lung epithelium. Furthermore, there was a decrease in alpha1-antitrypsin expression suggesting that JP-8 jet fuel exposure may have implications for the development of pulmonary disorders.

Microarray analysis (11) has been utilized to characterize changes in the gene expression profile of lung tissue induced by exposure of rats to JP-8 at a concentration of 171 or 352 mg/m^3 for 1 hr/day for seven days, with the higher dose estimated to mimic the level of occupational exposure in humans. The expression of 56 genes was significantly affected by a factor of ≤ 0.6 or ≥ 1.5 by JP-8 at the lower dose. Eighty-six percent of these genes were downregulated by JP-8. The expression of 66 genes was similarly affected by JP-8 at the higher dose, with the expression of 42% of these genes being upregulated (Tables 2 and 3). Prominent among the latter genes was the centrosome-associated protein-synuclein, whose expression was consistently increased. The expression of various genes related to antioxidant responses and detoxification, including those for GSH S-transferases and cytochrome P450 proteins, were also upregulated. The microarray data were confirmed by quantitative reverse transcription-polymerase chain reaction (RT-PCR) analysis. This study may provide important insight into metabolism of the pulmonary response to occupational exposure to JP-8 in humans.

The adverse effects of JP-8 in the lungs were found to be conversely related to increased concentrations of neuropeptide substance P (SP) in bronchoalveolar

Table 1 Two Hundred Fifty Milligrams per Cubic Meter JP-8 Jet Fuel Aerosol–Mediated
Protein Expression Alterations in Identified Proteins

		Protein abundance		
Spot No.	Identity	Control	JP-8	Prob.
3314	*Actin*	*1038.3*	*1546.7*	*0.001*
2414	Actin, alpha	5348.4	5987.3	0.5
4404	Actin-related protein	400.6	491.5	0.4
6516	Aldehyde dehydrogenase	4277.9	4558.7	0.4
8311	Aldose reductase	1532.6	1708.8	0.6
1602	*AAT 1–4*	*2869.6*	*2296.2*	*0.02*
4304	Annexin III	2057.7	2414.2	0.06
1307	Annexin V	4054.4	5011.8	0.4
6107	Antioxidant protein 2	7325.2	7231.8	0.9
3102	Apolipoprotein A-l	5235.0	4492.5	0.2
2501	ATP synthase beta chain	2926.3	3275.9	0.2
5820	Brefeldin A–inhibited guanine nucleotide-exchange protein	140.1	161.9	0.5
712	Calreticulin	1710.9	1835.8	0.6
8409	Creatine kinase	2203.6	1889.5	0.6
4517	*CKA*	*1007.9*	*1365.2*	*0.03*
3501	*Desmin fragment*	*102.3*	*137.4*	*0.03*
6612	Dihydropyrimidinase related protein-2	1278.0	1681.3	0.04
7502	Enolase, alpha	3325.3	3117.5	0.8
6511	Enolase, alpha (charge variant)	1274.6	1387.0	0.4
5604	ER60	1534.8	1754.9	0.1
5720	Ezrin	991.7	1149.5	0.2
2523	*FAF1*	*349.9*	*503.6*	*0.05*
2701	grp78	2755.6	3009.0	0.1
1802	grp94 (endoplasmin)	1931.8	2513.2	0.3
8708	Hexokinase, type 1	1131.6	714.0	0.05
3706	hsc70	3634.9	3919.1	0.4
3607	*HSP60*	*1295.2*	*1753.1*	*0.007*
2506	*Keratin, type 1 cytoskeletal 9*	*190.4*	*266.7*	*0.02*
7707	Moesin	1233.5	1265.6	0.9
7715	Moesin	873.8	964.0	0.6
2105	Myosin light chain 1	385.6	108.7	0.1
3609	P21 activated kinase 1B	198.5	278.5	0.0004
4217	PA28, alpha subunit	950.8	1173.3	0.06
5513	Selenium binding protein 1	2335.5	2698.8	0.05
5503	Selenium binding protein 2	1346.0	1528.2	0.3
6002	Superoxide dismutase	3944.0	4635.8	0.1
5204	Thioether S-methyltransferase	5806.5	6883.5	0.08
3901	Ubiquitin carboxyl-terminal hydrolase 2	3048.0	2818.7	0.6
2611	Vimentin	1444.2	1217.4	0.3
5817	Vinculin	3413.0	3185.2	0.5
5811	Vinculin (charge variant)	571.6	632.3	0.4

Abbreviation: JP-8, jet propulsion fuel–8.
Source: From Ref. 7.

Table 2 Genes Whose Expression Was Shown to Be Upregulated by JP-8 in Jurkat Cells by Microarray Analysis

GeneBank no.[a]	Gene name	JP-8/V[b]	t-Test[c]
Cell cycle–related proteins			
L13698	Growth arrest specific protein 1	2.36	0.018
L35253	p38 mitogen-activated protein kinase	3.27	0.017
Apoptosis-related proteins			
U13737	Caspase-3	2.92	0.003
U56390	Caspase-9	3.67	0.016
X17620	Nucleoside diphosphate kinase A	2.46	0.002
X66362	Serine–threonine protein kinase PCTAIRE 2 (PCTK3)	2.0	0.026
Transcriptional activators or repressers			
L23959	E2F dimerization partner-1	2.11	0.017
X56134	E2F-1	2.32	0.015
U15642	E2F-5	2.66	0.012
Stress response-related proteins			
U90313	Glutathione transferase omega	2.51	0.012
Y00371	Heat shock 70-kDa protein	2.56	0.013
M34664	Heat shock 60-kDa protein	2.39	0.0029
M16660	Heat shock 90-kDa protein beta	6.01	0.016
Metabolic enzymes and DNA repair proteins			
U35835	DNA-dependent protein kinase	2.14	0.0016
K00065	Cytosolic superoxide dismutase 1	3.47	0.013
U51166	G/T mismatch-specific thymine DNA glycosylase	2.30	0.0025

[a] GeneBank is the NIH genetic sequence database, an annotated collection of all publicly available DNA sequences.
[b] JP-8/V, means the ratio of gene expression in cells treated with JP-8 vs control cells treated with vehicle (ethanol).
[c] t-test, Student's t-test are the statistical tests used to identify the altered genes. The level of significance considered was $P < 0.05$.
Abbreviation: JP-8, jet propulsion fuel–8.
Source: From Ref. 12.

lavage (BAL) fluid (13). SP has been demonstrated to play an important role, mediated through neurokinin-1 receptors on effector cells, in initiating and regulating the severity of inflammatory processes. Subsequent to its release from afferent nerve endings, SP increases substantial responses such as an increase of microvascular permeability, promotion of plasma extravasation, and priming of other inflammatory mediators. These effects are mostly modulated by neutral endopeptidase through degradative cleavage of SP. In addition, there is recent evidence that immunoinflammatory cells, such as macrophages, lymphocytes, and leukocytes, also express the SP gene and protein. To investigate the role of SP on JP-8–induced pulmotoxicity, B6.A.D. (Ahr d/Nats) mice received subchronic exposures to JP-8 at 50 mg/m^3 (14). Lung injury was assessed by the analysis of pulmonary physiology, BAL fluid, and morphology. Hydrocarbon exposure to target JP-8 concentration of 50 mg/m^3, with saline treatment, was characterized by enhanced respiratory permeability to 99mTc-labeled diethylenetriaminepentaacetic acid, AM toxicity, and bronchiolar epithelial damage. Mice administered with [Sar^9, met(O$_2$)^{14}] SP, SP's agonist, after each JP-8 exposure showed normal pulmonary values and tissue morphology.

Table 3 Genes Whose Expression Was Shown to Be Downregulated by JP-8 in Jurkat Cells by Microarray Analysis

GeneBank no.[a]	Gene name	JP-8V[b]	t-Test[c]
Cell cycle–related proteins			
X05360	Cell division control protein 2 homologue	0.49	0.0037
Apoptosis-related proteins			
X76104	Death-associated protein kinase 1	0.43	0.0026
X79389	Glutathione S-transferase theta 1	0.38	0.004
Stress response-related proteins			
M64673	Heat shock transcription factor 1	0.47	0.0048
U07550	Mitochondrial heat shock 10-kDa protein	0.35	0.005
M86752	Stress-induced phospho-protein	0.48	0.015
D43950	T-complex protein 1 epsilon subunit	0.38	0.018
U83843	T-complex protein 1 eta subunit	0.40	0.016
Metabolic enzymes and DNA repair proteins			
D49490	Protein dissulfide isomerase–related protein 1	0.45	0.014
Others			
X56134	Vimentin	0.39	0.005

[a] GeneBank is the NIH genetic sequence database, an annotated collection of all publicly available DNA sequences.
[b] JP-8/V, means the ratio of gene expression in cells treated with JP-8 vs control cells treated with vehicle (ethanol).
[c] t-test, Student's t-test are the statistical tests used to identify the altered genes. The level of significance considered was $P < 0.05$.
Note: Data are means of values from four independent experiment.
Abbreviation: JP-8, jet propulsion fuel–8.
Source: From Ref. 12.

In contrast, endogenous SP's receptor antagonism by CP-96345 administration exacerbated JP-8–enhanced permeability, AM toxicity, and bronchiolar epithelial injury. This study indicates that SP may have a protective role in preventing the development of JP-8–induced pulmotoxicity, possibly through the modulation of bronchiolar epithelial function.

The effects of combined JP-8 and SP on cell growth and survival, and cytokine and chemokine production, were studied using cocultures of rat AMs and AIIEs. These cell types were selected because the macrophages appear to communicate with the type II cells, which are the first to be affected by inhalation exposure to JP-8. A rat transformed type II cell line was allowed to form a monolayer, and primary pulmonary AMs were layered on top and allowed to settle on the epithelial monolayer. JP-8 was applied one-hour later, and the effects of [Sar^9,met(O$_2$)^{11}] SP on cell numbers and on responses of cellular factors were examined. Control cultures consisted of each individual cell type.

SP, alone, significantly depressed alveolar cell numbers when applied in the range of 10^{-4} to 10^{-9} M; however, the depression was not concentration related. Macrophage and coculture cell numbers were not affected by any SP concentration. All studies of cytokine or chemokine induction used 10^{-4} to 10^{-6} M SP. A concentration of 16 μg JP-8/mL was selected based on a preliminary study showing a nonsignificant increase in macrophage numbers, a significantly decreased survival of type II cells, and absence of an effect on the cell combination at this concentration. It was noted in the discussion that although most of the effects were observed at SP concentrations of 10^{-5} or 10^{-6} M, these levels are much higher than the 40 to 50 fmol/mL (40–50×10^{-15} M) of endogenous SP observed in untreated rat lungs. This value decreases to less than 5 fmol (below the level of detection) following JP-8 inhalation.

SP significantly increased the JP-8–induced release of IL-1α, IL-1β, and IL-18 by macrophages, alone, but there was no effect in the coculture. SP significantly decreased IL-6 production by the coculture, but not in the individual cell cultures. The results for the other cytokines and chemokines were less clear, and tended to be more variable. IL-10, MCP-1, and GRO-KC releases were significantly depressed in the coculture, but the responses were not clearly dose dependent. TNF-α release in the coculture was increased, but only at the high SP concentration. The results were interpreted to show the presence of cell-to-cell communication between the alveolar epithelial cells and macrophages, and that this type of coculture system could be used to measure the effects of SP on JP-8–induced lung injury. It was suggested that IL-12 also be measured because it is produced by macrophages under stress.

To examine the hypothesis that JP-8 inhalation potentiates influenza virus–induced inflammatory responses, we randomly divided female C57BL/6 mice (four-weeks old, weighing approximately 24.6 g) into the following groups: air control, JP-8 alone (1023 mg/m³ of JP-8 for 1 hr/day for seven days), A/Hong Kong/8/68 influenza virus (HKV) alone (a 10-μL aliquot of 2000 viral titer in the nasal passages), and a combination of JP-8 with HKV (JP-8 + HKV) (15). The HKV alone group exhibited significantly increased total cell number/granulocyte differential in BAL fluid compared to controls whereas the JP-8 alone group did not. The JP-8+HKV group further exacerbated the HKV alone–induced response. The pathological alterations followed the same trend as those of BAL cells (Fig. 1). However, increases in pulmonary microvascular permeability in JP-8 + HKV just matched the sum of JP-8 alone– and HKV alone–induced responses. Increases in BAL fluid SP in the JP-8 alone group and BAL fluid leukotriene B4 or total lung compliance in the HKV alone group were similar to the changes in the JP-8 + HKV group. These findings suggest that changes in the JP-8+HKV group may be attributed to either JP-8 inhalation or HKV treatment and indicate the different physiological responses to either JP-8 or HKV exposure. Taken together, this study provides initial evidence that JP-8 inhalation may partially synergize influenza virus–induced inflammatory responses.

DERMAL TOXICITY

Dermal contact to JP-8 jet fuel has been indicated to induce a broad spectrum of adverse health effects in the skin (9,10,16–24). These effects exhibit mild skin irritation or dermatitis with impaired epidermal protein synthesis, hyperemia, and cellular damage of the epidermis. Evidence has been provided that jet fuel contact under occluded dosing dramatically increased erythema and reduced hairless rat skin barrier function as assessed by transepidermal water loss (23). Increased epidermal thickness, intraepidermal and subcorneal microabscesses, and altered enzyme histochemistry patterns were also observed in vivo (22). The mild inflammation was also accompanied by formation of lipid droplets in various skin layers, mitochondrial and nuclear changes, cleft formation in the intercellular lipid lamellar bilayers, as well as disorganization in the stratum granulosum–stratum corneum interface

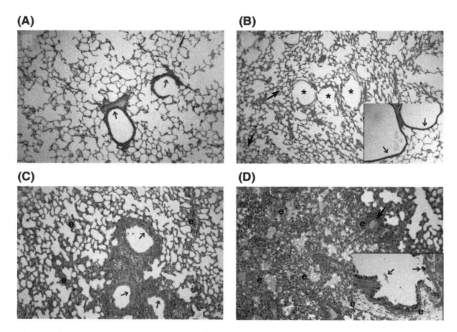

Figure 1 Representative micrographs of lungs illustrate JP-8 inhalation contributed to HKV-induced inflammatory response in C57BL/6 mice. Mice were exposed nose only to either air ($N = 6$) or 1023 mg/m^3 of JP-8 ($N = 6$), respectively, through an inhalational exposure chamber for 1 hr/day for seven days. On day 8, half of each group was then infected with a 10-μL vehicle or an aliquot of 2000 viral titer level of HKV, respectively, in the nasal passages and were examined at eight days after infection. (**A**) Air control, showing normal airways (*arrows*) and alveoli. (**B**) JP-8 alone, showing ectasia of respiratory bronchioles or alveoli (*) and inflammatory cells (*arrows*). (**C**) HKV alone, showing interstitial edema (e) and peribronchiolitis (*narrow arrow*). (**D**) JP-8 + HKV, showing edema (e) in alveoli or small airway (*thick arrow*) and airway epithelium sloping (*narrow arrow*). Hematoxylin and eosin stain; magnification, ×10. *Abbreviations*: JP-8, jet propulsion fuel-8; HKV, A/Hong Kong/8/68 influenza virus. *Source*: From Ref. 15.

(Figs. 2 and 3) (18). An increased number of Langerhans cells were also noted in jet fuel–treated skin. These changes suggest that the primary effect of JP-8 exposure is damaging to the stratum corneum barrier. However, underlying action mechanisms of JP-8 dermal toxicity remain unclear.

Proinflammatory cytokine or chemokine response in skin following JP-8 exposure was reported in many studies (18,24–26). Ullrich et al. (27,28) have conducted a series of studies (21) and demonstrated that low-dose (50 μL/day) repeated (five days) JP-8 exposure to C$_3$H/HeN mice suppressed contact and delayed hypersensitivity responses, including depressing the protective effect of prior vaccinations. They hypothesized that IL-10 and PGE$_2$ are produced by keratinocytes that distribute systemically and downregulated the cell-mediated immune response. JP-8 exposure for four hours has shown increases of proinflammatory cytokine TNF-α and IL-8 in human keratinocytes. This is the same time frame seen in rats, where IL-1α and inducible nitrous oxide synthetase (iNOS) expression were increased after JP-8 exposure (29). JP-8 jet fuel or acetone control (300 μL) was applied to the denuded skin of rats once a day for seven days (25). RT-PCR was performed utilizing skin total RNA to examine the expression of various inflammatory cytokines. The CXC chemokine

(A)

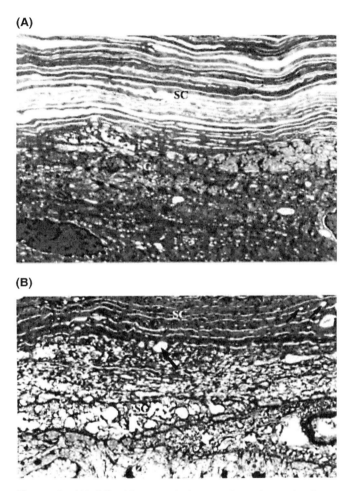

(B)

Figure 2 (A) SG with normal electron-dense keratohyalin granules and tightly packed filaments. Note the compact SC. Control; ×9600. (B) Loosely packed filaments in the SG layers. Lipid vacuoles (*arrow*) were frequently retained in the SC layers in four-day application of all jet fuels; ×7500. *Abbreviations*: SG, stratum granulosum; SC, stratum corneum. *Source*: From Ref. 18.

GROalpha was significantly upregulated at both time points, whereas GRObeta was only increased two hours after final exposure. The CC chemokines MCP-1, Mip-1alpha, and eotaxin were induced at 2 and 24 hours, whereas Mip-1β was induced only 24 hours after exposure. IL-1β and -6 mRNAs were significantly induced at both time points, while TNF-α was not significantly different from control. Enzyme-linked immunosorbent assay of skin protein confirmed that MCP-1, TNF-α, and IL-1β were modulated. However, skin IL-6 protein content was not increased two hours postexposure, whereas it was significantly upregulated by jet fuel after 24 hours. The increased expression of these cytokines and chemokines may lead to increased inflammatory infiltrate in exposed skin, resulting in JP-8–induced irritant dermatitis.

Using Hill Top Chambers, a study (29) showed that a one-hour exposure to JP-8 elevated IL-1α levels ranging from approximately 11% to 34% above controls

(A)

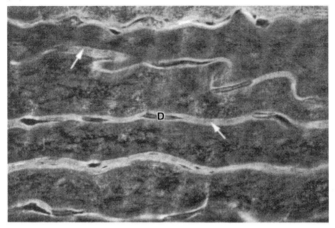

(B)

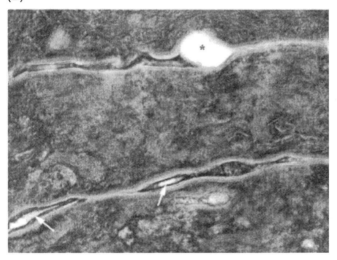

Figure 3 (**A**) Ruthenium tetroxide staining of the lipid bilayers (*arrows*) between the stratum corneum cell layers. Note the intercellular lamellae consisting of electron-dense and electron-lucent bands. Electron-dense desmosomes (D) between the stratum corneum layers were normal and spanned the entire width of the intercellular space. Control; ×86,600. (**B**) Jet fuel–treated skin depicting degradation of desmosomes and expanded intercellular space. The electron-dense desmosomes have separated from the central core (*arrows*) leaving a space within the desmosomes. Note the expansion of the intercellular space ∗ where the intercellular lipid lamellae appeared extracted. JP-8 + 100; ×117,000. *Abbreviation*: JP-8, jet propulsion fuel–8. *Source*: From Ref. 18.

(zero hour) over the six-hour period. Western blot analysis revealed significantly higher ($p < 0.05$) levels of inducible nitric oxide synthase (iNOS) at four and six hours compared to zero-hour samples. Increases in IL-1α and iNOS expression were also observed in the skin, immunohistochemically. In addition, increased numbers of granulocytes were observed to be infiltrating the skin at two hours and were more prominent by six hours. These data show that a one-hour exposure to JP-8 results in a local cytokine inflammation, which may be associated with histological changes.

A potential mechanism by which JP-8 causes keratinocyte cell necrosis was observed (30). Data revealed that exposure of keratinocytes to the toxic higher levels of JP-8 markedly downregulates the expression of the prosurvival members of the Bcl-2 family, Bcl-2 and Bcl-x(L), and upregulates the expression of antisurvival members of this family, including Bad and Bak. Bcl-2 and Bcl-x(L) have been shown to preserve mitochondrial integrity and suppress cell death. In contrast, both Bak and Bad promote cell death by alteration of the mitochondrial membrane potential, in part by heterodimerization with inactivation of Bcl-2 and Bcl-x(L), and either induce necrosis or activate a downstream caspase program. High intrinsic levels of Bcl-2 and Bcl-x(L) may prevent apoptotic death of keratinocytes at lower levels of JP-8, while perturbation of the balance between pro- and antiapoptotic Bcl-2 family members at higher levels may ultimately play a role in necrotic cell death in human keratinocytes.

cDNA microarray has also been utilized to identify the gene expression profile in normal human epidermal keratinocytes exposed to JP-8 for 24-hour and seven-day periods (11). Predictive neural networks were built using a multiplayer perception to carry out a proper classification task in microarray data in the untreated versus JP-8–treated samples. Increased transcription of genes involved in the stress response was observed in the time course study. High expressions of SERINE2, PLAU, and its receptor PLAUR mRNAs were present. Transcripts involved in metabolism, such as BACAH and PDHA, were upregulated and SDHC and ABFB showed reduced expression. Neither apoptotic genes nor the genes encoding for IL-1, IL-8, IL-10, COX-2, and iNOS reported by previous JP-8 studies in keratinocytes and in skin tissue (24,26,27,29) was changed. These differences may be attributed to the very low exposure level (equivalent to 0.01%) of JP-8 when compared to those used in other studies, such as in keratinocytes (0.1%) and in animals (undiluted fuel that was applied directly to the skin). Gene expression data in this study can be used to build accurate predictive models that separate different molecular profiles.

IMMUNE TOXICITY

Recent findings indicate that short-term exposure of animals to the high level of JP-8, either via inhalation or dermal contact, caused significant immune suppression, primarily cell-mediated immune effects (21,24,27,31–35). Such changes may have significant effects on the health of the exposed personnel. Harris et al. (34,35) reported that JP-8 nose-only inhalation to C57L/6 mice for 1 hr/day for seven days at a concentration of $100 \, mg/m^3$ caused a decrease in cellularity of the thymus, $500 \, mg/m^3$ led to decreased spleen weight and cellularity, and $1000 \, mg/m^3$ led to decreased ability of spleen cells to mediate several immune responses. Dermal exposure of mice to JP-8 at $50 \, \mu L/day$ for four to five days or in a single dose ($300 \, \mu L$) induced local and systemic immune reactions such as suppressed contact and delayed hypersensitivity responses. These studies raise concern about the potential mechanism of JP-8 to cause immunotoxicity.

More recently, a study examined the effects of JP-8 on humoral and cell-mediated and hematological parameters. Immunotoxicological effects were evaluated in adult female B6C3F1 mice gavaged with JP-8 (in an olive oil carrier) ranging from 250 to $2500 \, mg/kg/day$ for 14 days. One day following the last exposure, they found that thymic mass was decreased at exposure levels of $\geq$1500 mg/kg/day. Decreases in thymic cellularity, however, were only observed at exposure levels of $\geq$2000 mg/kg/day. Natural killer cell activity and T- and B-cell proliferation were not altered. Decreases in the plaque-forming cell response were dose responsive at levels of 500 mg/kg/day and greater, while, unexpectedly, serum levels of anti-SRBC immunoglobulin M were not altered. Alterations also were detected in thymic and splenic CD4/8 subpopulations, and proliferative responses of bone marrow progenitor cells were enhanced in mice exposed to 2000 mg/kg/day of JP-8. These findings indicate that humoral immune function is impaired with lesser exposure levels of JP-8 than the levels that are required to affect primary and secondary immune organ weights and cellularities, CD4/8 subpopulations, and hematological endpoints.

Most recently, studies to understand the mechanisms involved in JP-8 immune suppression have been conducted. Data have shown that JP-8–induced immune toxicity could be prevented or attenuated by blocking oxidative stress that induces platelet-activating factor (PAF), or blocking the binding of PAF to its receptors on target cells, or inhibiting the activity of the cyclooxygenase-2 thus suppressing PGE_2 secretion, or neutralizing the activity of IL-10, or injecting exogenous IL-12. In addition, SP also blocks dermal JP-8–induced immune suppression. It is so far unknown which pathway(s) plays a key role in pathogenesis of immune suppression. There is a hypothesis that the inflammatory phospholipid mediator, PAF, plays a role in immune suppression (36), because PAF upregulates cyclooxygenase-2 production and PGE_2 synthesis by keratinocytes. JP-8–induced PGE2 production was suppressed by treating the keratinocytes with specific PAF-receptor antagonists. Injecting mice with PAF, or treating the skin of the mice with JP-8, induced immune suppression. These observations suggest that PAF-receptor binding may be an early event in the induction of immune suppression by immunotoxic environmental agents that target the skin.

It was recently observed that inhalation of 1000 mg/m^3 by female C57Bl/6 mice for one day led to a significant decrease in thymus cellularity and a smaller decrease in the spleen (Dr. Harris). Immune cells in the spleen associated with apoptosis (CD4, DC8, Mac-1, and CD45R) were increased, Dec-205 was decreased, and CD-16 was unchanged at day 1 after exposure. CD8 and Mac-1 levels remained elevated at day 4, and no changes were seen at day 7. In the thymus, CD4 + CD8 were increased at day 1 rather than other time points. Mac-1 was increased at days 1 and 4, and was associated with necrosis. The cytokines IFNγ, IL-4, and IL-10 (associated with CD4, CD8, and Mac-1 cells) were increased at days 1 and 4.

On the basis of this brief review, it appears that the adverse effects of JP-8 on the biological system have been characterized. However, the effects and action mechanisms of JP-8 are not fully understood. Current data are not available or limited for human risk assessment. From a toxicological perspective, necessary studies from the basic to applied toxicology are not fulfilled in relation to the adverse effects of JP-8. For example, studies of exposure, absorption, and metabolism of JP-8 need to be explored in spite of the complex nature of the studies. In exposure analysis, ambient exposure and breathing-zone concentrations of JP-8 and its constituents (such as naphthalene and toluene) should be determined in relation to lung burden through assays of biological samples. The findings should be correlated with acute and chronic toxicities as well as symptoms or signs experienced by JP-8–exposed workers. Therefore, we will have to continue to rely on additional animal inhalation studies with appropriate cellular and molecular strategies that should help to understand some pathogenesis of the adverse effects associated with JP-8 exposure.

REFERENCES

1. Hays AM, Parliman G, Pfaff JK, et al. Changes in lung permeability correlate with lung histology in a chronic exposure model. Toxicol Ind Health 1995; 11(3):325–336.
2. Pfaff JK, Tollinger BJ, Lantz RC, Chen H, Hays AM, Witten ML. Neutral endopeptidase (NEP) and its role in pathological pulmonary change with inhalation exposure to JP-8 jet fuel. Toxicol Ind Health 1996; 12(1):93–103.
3. Robledo RF, Young RS, Lantz RC, Witten ML. Short-term pulmonary response to inhaled JP-8 jet fuel aerosol in mice. Toxicol Pathol 2000; 28(5):656–663.
4. Robledo RF, Barber DS, Witten ML. Modulation of bronchial epithelial cell barrier function by in vitro jet propulsion fuel 8 exposure. Toxicol Sci 1999; 51(1):119–125.
5. Wang S, Young RS, Witten ML. Age-related differences in pulmonary inflammatory responses to JP-8 jet fuel aerosol inhalation. Toxicol Ind Health 2001; 17(1):23–29.
6. Wang S, Young RS, Sun NN, Witten ML. In vitro cytokine release from rat type II pneumocytes and alveolar macrophages following exposure to JP-8 jet fuel in co-culture. Toxicology 2002; 173(3):211–219.
7. Drake MG, Witzmann FA, Hyde J, Witten ML. JP-8 jet fuel exposure alters protein expression in the lung. Toxicology 2003; 191(2–3):199–210.
8. Witzmann FA, Bauer MD, Fieno AM, et al. Proteomic analysis of simulated occupational jet fuel exposure in the lung. Electrophoresis 1999; 20(18):3659–3669.
9. Mattorano DA, Kupper LL, Nylander-French LA. Estimating dermal exposure to jet fuel (naphthalene) using adhesive tape strip samples. Ann Occup Hyg 2004; 48(2):139–146.
10. Singh S, Zhao K, Singh J. In vivo percutaneous absorption, skin barrier perturbation, and irritation from JP-8 jet fuel components. Drug Chem Toxicol 2003; 26(2):135–146.
11. Espinoza LA, Valikhani M, Cossio MJ, et al. Altered expression of {gamma}-synuclein and detoxification-related genes in lungs of rats exposed to JP-8. Am J Respir Cell Mol Biol 2005; 32(2):192–200.
12. Espinoza LA, Smulson ME. Macroarray analysis of the effects of JP-8 jet fuel on gene expression in Jurkat cells. Toxicology 2003; 189(3):181–190 .
13. Pfaff J, Parton K, Lantz RC, Chen H, Hays AM, Witten ML. Inhalation exposure to JP-8 jet fuel alters pulmonary function and substance P levels in Fischer 344 rats. J Appl Toxicol 1995; 15(4):249–256.
14. Robledo RF, Witten ML. NK1-receptor activation prevents hydrocarbon-induced lung injury in mice. Am J Physiol 1999; 276(2 Pt 1):L229–L238.
15. Wong SS, Hyde J, Sun NN, Lantz RC, Witten ML. Inflammatory responses in mice sequentially exposed to JP-8 jet fuel and influenza virus. Toxicology 2004; 197(2):139–147.
16. Serdar B, Egeghy PP, Gibson R, Rappaport SM. Dose-dependent production of urinary naphthols among workers exposed to jet fuel (JP-8). Am J Ind Med 2004; 46(3):234–244.
17. McDougal JN, Rogers JV. Local and systemic toxicity of JP-8 from cutaneous exposures. Toxicol Lett 2004; 149(1–3):301–308.
18. Monteiro-Riviere NA, Inman AO, Riviere JE. Skin toxicity of jet fuels: ultrastructural studies and the effects of substance P. Toxicol Appl Pharmacol 2004; 195(3):339–347.
19. Kanikkannan N, Locke BR, Singh M. Effect of jet fuels on the skin morphology and irritation in hairless rats. Toxicology 2002; 175(1–3):35–47.
20. McDougal JN, Robinson PJ. Assessment of dermal absorption and penetration of components of a fuel mixture (JP-8). Sci Total Environ 2002; 288(1–2):23–30.
21. Ramos G, Nghiem DX, Walterscheid JP, Ullrich SE. Dermal application of jet fuel suppresses secondary immune reactions. Toxicol Appl Pharmacol 2002; 180(2):136–144.
22. Monteiro-Riviere N, Inman A, Riviere J. Effects of short-term high-dose and low-dose dermal exposure to Jet A, JP-8 and JP-8 + 100 jet fuels. J Appl Toxicol 2001; 21(6):485–494.

23. Kanikkannan N, Burton S, Patel R, Jackson T, Shaik MS, Singh M. Percutaneous permeation and skin irritation of JP-8 + 100 jet fuel in a porcine model. Toxicol Lett 2001; 119(2):133–142.

24. Allen DG, Riviere JE, Monteiro-Riviere NA. Identification of early biomarkers of inflammation produced by keratinocytes exposed to jet fuels jet A, JP-8, and JP-8(100). J Biochem Mol Toxicol 2000; 14(5):231–237.

25. Gallucci RM, O'Dell SK, Rabe D, Fechter LD. JP-8 jet fuel exposure induces inflammatory cytokines in rat skin. Int Immunopharmacol 2004; 4(9):1159–1169.

26. Allen DG, Riviere JE, Monteiro-Riviere NA. Cytokine induction as a measure of cutaneous toxicity in primary and immortalized porcine keratinocytes exposed to jet fuels, and their relationship to normal human epidermal keratinocytes. Toxicol Lett 2001; 119(3):209–217.

27. Ullrich SE. Dermal application of JP-8 jet fuel induces immune suppression. Toxicol Sci 1999; 52(1):61–67.

28. Ullrich SE, Lyons HJ. Mechanisms involved in the immunotoxicity induced by dermal application of JP-8 jet fuel. Toxicol Sci 2000; 58(2):290–298.

29. Kabbur MB, Rogers JV, Gunasekar PG, et al. Effect of JP-8 jet fuel on molecular and histological parameters related to acute skin irritation. Toxicol Appl Pharmacol 2001; 175(1):83–88.

30. Rosenthal DS, Simbulan-Rosenthal CM, Liu WF, Stoica BA, Smulson ME. Mechanisms of JP-8 jet fuel cell toxicity. II. Induction of necrosis in skin fibroblasts and keratinocytes and modulation of levels of Bcl-2 family members. Toxicol Appl Pharmacol 2001; 171(2):107–116.

31. Keil D, Dudley A, EuDaly J, et al. Immunological and hematological effects observed in B6C3F1 mice exposed to JP-8 jet fuel for 14 days. J Toxicol Environ Health A 2004; 67(14):1109–1129.

32. Harris DT, Sakiestewa D, Robledo RF, Young RS, Witten M. Effects of short-term JP-8 jet fuel exposure on cell-mediated immunity. Toxicol Ind Health 2000; 16(2):78–84.

33. Harris DT, Sakiestewa D, Robledo RF, Witten M. Protection from JP-8 jet fuel induced immunotoxicity by administration of aerosolized substance P. Toxicol Ind Health 1997; 13(5):571–588.

34. Harris DT, Sakiestewa D, Robledo RF, Witten M. Short-term exposure to JP-8 jet fuel results in long-term immunotoxicity. Toxicol Ind Health 1997; 13(5):559–570.

35. Harris DT, Sakiestewa D, Robledo RF, Witten M. Immunotoxicological effects of JP-8 jet fuel exposure. Toxicol Ind Health 1997; 13(1):43–55.

36. Ramos G, Kazimi N, Nghiem DX, Walterscheid JP, Ullrich SE. Platelet activating factor receptor binding plays a critical role in jet fuel-induced immune suppression. Toxicol Appl Pharmacol 2004; 195(3):331–338.

14
Drug Safety Toxicology

Joseph P. Hanig and Robert E. Osterberg
Food and Drug Administration, Center for Drug Evaluation and Research, Silver Spring, Maryland, U.S.A.

INTRODUCTION

This chapter will define and provide an overview of the drug safety toxicology process. It will describe its evolution, development, and reliance on the interface between the cutting edge, validated and adopted, widely accepted methodologies and the application of sound principles of drug evaluation and review that allow these methods to be used as effective tools to predict toxicity and maximize drug safety. The ensuing discussion will describe the path that a new method or scientific approach should take to successfully evolve into what is normally considered to be a "gold standard" tool in the drug safety armamentarium. It will also include a description of the various stages at which completely new methods moving into ascendancy are, and contrast these with those methods that are no longer widely accepted or routinely used and the reasons for their decline. Finally, this chapter will highlight several widely accepted methods, and some of the most promising recent approaches to drug safety toxicology, as well as some of the newest frontiers and their potential for innovative changes to drug toxicology. The effective utilization of drug safety toxicology is the bridge between resource and time intensive drug development and a better quality of life for those who are the beneficiaries of approved, safe, and effective therapeutic agents.

BACKGROUND OF DRUG SAFETY REGULATION

The development of federal toxicological testing regulations began in the decade of the Mexican–American War of 1848. Many drugs in use at that time were either adulterated with toxicants or degraded because of improper storage or transportation conditions in shipments of drugs imported from Europe. During the Mexican–American War, more soldiers were disabled or died from impure drug product use than the total killed as a result of the armed conflict. As a result, Congress passed the Drugs and Medicine Act in 1848 that recognized the newly created United States Pharmacopeial Convention (USP) as the scientific body to determine the standards for pure drugs. Several decades later, in response to Upton

Sinclair's book *The Jungle* (1) among other political and social pressures, Congress passed the Pure Food and Drugs Act (1906). This Act mandated, among other things, that drug manufacturers abide by the standards set by the USP and that making false or misleading statements on labels constituted misbranding. However, in 1911, a setback to drug safety occurred when Congress passed the Sherley Amendment, which placed the burden of proof on the government to show fraud. At this point in time, there was virtually no testing for drug safety and a very high incidence of deaths occurred from "quack medicines" purported to be safe drugs (2).

Following the tragedy involving the Elixir of Sulfanilamide with many deaths of children from ingestion of the excipient diethylene glycol, Congress passed the Federal Food, Drug and Cosmetic Act in 1938. This Act required industry to prove the safety of drugs before marketing and to show on the product labeling the known hazards associated with the drug (3). In 1962, following the thalidomide-induced birth defect catastrophe, Congress passed the Kefauver-Harris Act. This Act increased the toxicity-testing burden on the drug sponsor, mandated efficacy testing for drugs, and therefore provided for a risk-benefit determination (3).

U.S. FOOD AND DRUG ADMINISTRATION TOXICITY TESTING DOCUMENTS

One of the earliest documents (1959) published by the Food and Drug Administration (FDA) that discussed the nonclinical toxicity testing of drugs was titled "Appraisal of the Safety of Chemicals in Foods, Drugs and Cosmetics" (4), which provided guidance and protocols for various toxicity tests. Approximately nine years later, a publication by Goldenthal appeared in the FDA Papers (May 1968) titled "Current Views on Safety Evaluation of Drugs" (5), which provided general guidelines for drug testing. In that same time period, an FDA document titled "Guidelines for Reproduction Studies for Safety Evaluation of Drugs for Human Use" (6) was written by W. D'Aguanno in which more specific information regarding teratological testing was published.

In the late 1970s, a team of pharmacologists and toxicologists from the Environmental Protection Agency (EPA), Consumer Product Safety Commission, National Institute of Occupational Safety and Health, and the FDA formed a subcommittee on toxicology within the Interagency Regulatory Liaison Group and began writing acute and chronic toxicity testing guidelines. These guidelines were eventually discussed at an open meeting with representatives of other government agencies, academia, and food, chemical, and pharmaceutical companies. The guidelines were subsequently incorporated into several drafts of FDA's book of toxicology guidelines, *Toxicological Principles for the Safety Assessment of Direct Food Additives and Color Additives Used in Food* (1982) also known as the *Red Book* (7).

For the past several years, FDA pharmacologists and toxicologists have participated in international activities such as Organization for Economic Cooperation and Development (OECD) guideline development, Interagency Coordinating Committee on the Validation of Alternative Methods (ICCVAM) or European Center for the Validation of Alternative Methods (ECVAM) evaluations, and International Conference on Harmonisation of Technical Requirements for Registration of Pharmaceuticals for Human Use (ICH) guidance development in efforts to modify existing toxicology protocols or to evaluate in vitro replacement toxicology methods. Such methods can be used within the drug development area to generate acceptable noncli-nical data to support human drug trials and accurate labeling statements. Because these are guidelines and not regulations, they can be modified as needed to reflect advances in the state-of-the-art of toxicology.

While considering the above information, one might ask the question, "How does a new toxicology test method and resultant data become accepted for use by a regulatory agency?" To answer that question, one should become familiar with the test method development process.

MODERN APPROACHES TO THE DEVELOPMENT OF METHODS FOR DRUG TESTING AND SAFETY EVALUATION

The very first questions that must be considered before we embark on a discussion of method development and validation are "What is drug safety?" What are the criteria for declaring that a drug is safe for use or at least presents a level of acceptable risk in comparison to the maximal level of therapeutic benefit to be derived? Drug safety can be defined in terms of the validity of a test in predicting, defining, or capturing essential toxicologic information that describes the deleterious effect of a drug on a biological system as well as the acceptability of these data to a reviewer who must make the ultimate decision on safety. The test or method defines the effect, but it is the reviewer who must place this information in the proper context and interpret its applicability. Many times there are "all or none" absolute signals that cannot be ignored (i.e., carcinogenicity and teratology). With this in mind, it should be recognized that graded responses within the context of the exposure require interpretation. Interpretation of these graded responses requires a judgment call by a trained safety reviewer, assuming that an acceptable method has been used to generate valid data. With this short definition or qualification, we can begin our discussion.

Philosophy of Test Method Development

Development of any potentially useful assay system originates with a requirement for specific data to answer a focused question. Either a new test is created or an existing test is modified in an effort to develop the needed information. Usually, the first step to be accomplished is the development and validation of a technique or methodology that has evolved from scientific basic research. It must be capable of being practically applied and sufficient in producing useful information as well as being cost effective. Useful methodology should be refined to be very efficient; the initial validation protocol should be written and then subjected to extensive evaluation. Reference chemical substances from the same source and lot number should be used during the evaluation of this test to better define the specificity and sensitivity of the defined or proposed test parameters. Issues of applicability, robustness, sensitivity, specificity, and reproducibility are usually addressed (Table 1). Highly competent statistical support is critical in terms of evaluating the validity of results and interpretations based on the data. Validation of the test method must be successful for the test to be recognized as useful in generating data acceptable for regulatory purposes. Whatever test protocols are developed and/or used by the drug sponsor, the sponsor must be mindful of the welfare of the test subjects and thus follow the considerations contained in the U.S. Department of Agriculture's, Animal Welfare Act

Table 1 Criteria for Acceptance of New Toxicology Methods for Drug Safety Evaluation

Relevent end points must correlate well with human end points using appropriate species of
   animals
Tests must be easy to conduct
The tests must be relatively inexpensive
The tests must be reliable
The tests must be justifiable in cost and timeliness
The tests must produce meaningful data
The data must be sufficient to convince CDER and other
   drug-regulatory bodies
The test must be easy to replicate in other laboratories
The data must be predictable for human safety and toxicity
The tests must have a wide consensus of utility
The tests must have a low rate of false positives and negatives
The tests must be useful
The test should be recommended by groups such as OECD, ICH,
   ICCVAM/ECCVAM, etc.

Abbreviations: ECVAM, European Center for the Validation of Alternative Methods; ICCVAM, Interagency Coordinating Committee on the Validation of Alternative Methods; ICH, International Conference
on Harmonisation; OECD, Organization of Economic Cooperation and Development; CDER, Center for
Drug Evaluation and Research.

and institutional animal care and use committee rules (8) regarding laboratory ani-
mals, or Department of Health and Human Services (DHHS) Office of Human
Research Protections policy, as well as the general concepts in the Declaration of
Helsinki regarding human drug trials (9).

   Validation requires extensive testing with standard reference compounds that
will establish the reliability of the procedures. Using the new test method, many chemically related, as well as unrelated substances, should be tested to obtain an extensive database upon which data reliability can be established. Furthermore,
collaborative studies should be encouraged using several accredited testing facilities
and laboratories to further establish the utility and validity of the test method. The
method must be useful, reliable, and produce comparable results in the hands of other
scientists. A test that can be performed successfully only in a single or very few laboratories will probably not be useful for routine testing of substances with which to
obtain data for regulatory use. However, exceptions to the lack of wide usage due
to cost of equipment and/or specialized test procedures are noted. New tests should
be sufficiently validated for estimating safety or potential hazards so that data interpretation has meaning. Successful tests must also be justifiable in terms of cost and
time spent in conducting them. Following successful collaboration, the final procedure or methodology should be independently and generally acknowledged, deemed
appropriate, and accepted in other laboratories for routine use (Table 1). Finally,
implementation of the new assay by regulators and/or drug developers is critical to
produce the best attainable data for scientific and regulatory purposes.

   With respect to the drug approval process within the FDA, the agency
encourages drug sponsors to develop specific and sensitive study protocols that
are appropriate for the particular substance with respect to its intended therapeutic
use. This statement applies to both the safety and the efficacy of the drug substance
in the test subjects. With a few exceptions, the FDA does not require routine, standardized test protocols, but may suggest certain guidelines and guidances that can be

followed and/or modified by the drug developer or testing laboratory when asked. FDA believes that guidelines promote flexibility in design rather than regulations that may prescribe rigid protocols. Many useful guidelines and guidances have been developed within the FDA (10), or other government agencies, or negotiated among scientific and regulatory organizations within the national and international community as previously mentioned.

Validation of a Method

Validation of methods can occur within any of the three major sectors of scientific activity (government, academe, or industry) or any combination of these groups. Ultimately, the results of these method-validation efforts come before a highly recognized international organization (Table 2) for evaluation and possible acceptance. Examples of these are as follows: The International Life Sciences Institute (ILSI) has a global branch known as the Health and Environmental Sciences Institute (HESI), a public nonprofit scientific foundation with branches throughout the world. Their stated mission is "to provide an international forum to advance the state of sciences on current issues of interest to the public, scientific community, government agencies and industry related to human health, toxicology, risk assessment and the environment. In particular, emphasis is placed on enhancing science-based safety assessment in order to make sound decisions concerning the use of chemicals to benefit people and the environment" (11). ILSI-HESI has a number of emerging issues committees that allows them to evaluate and publish reports on relatively young technologies and their associated methodologies. They also have a wide variety of technical committees that examine the more mature scientific areas in which the issue of validation of methods has a very important impact on their acceptance by

Table 2 Examples of International Regulatory Agencies and Organizations that Evaluate and/or Validate Methodology for Drug Safety Approval

CPMP—Committee for Proprietary Medicinal Products
ECVAM—European Center for the Validation of Alternative Methods
EEC—European Economic Community
EMEA—European Agency for the Evaluation of Medicinal Products
EPA—Environmental Protection Agency
FDA-BVM—Bureau of Veterinary Medicine (now Center for Veterinary Medicine)
FDA-CFSAN—Center for Food Safety and Applied Nutrition
FDA-CBER—Center for Biologics Evaluation and Research
FDA-CDER—Center for Drug Evaluation and Research
FDA-CDRH—Center for Devices and Radiological Health
FDA-PTCC—Pharmacology/Toxicology Coordinating Committee (CDER)
HESI—Health and Environmental Sciences Institute
ICCVAM—Interagency Coordinating Committee on the Validation of Alternative Methods
ILSI—International Life Sciences Institute
ICH—International Conference on Harmonisation of Technical Requirements for
　　　Registration of Pharmaceuticals for Human Use
JPMA—Japanese Pharmaceutical Manufacturers Association
MHLW—Ministry of Health, Labor and Welfare (Japan)
OECD—Organization of Economic Cooperation and Development
USP—United States Pharmacopeia

regulatory agencies that will recommend their use by those who submit studies in support of new drug approval.

The active technical committees of ILSI-HESI include those on: (a) Development and Application of Biomarkers of Toxicity, (b) Application of Genomics to Mechanism-Based Risk Assessment, (c) Developmental and Reproductive Toxicology, (d) Immunotoxicology, (e) Nonclinical Safety Issues, and others. Each of the above committees has published reports on collaborative, validation efforts, databases, and white papers for submission to various journals (11).

The OECD has participated in the publication of a variety of validation documents. Most recently, the OECD validated the Herschberger Assay: Phase 2 dose–response of methyl testosterone, vinclozoline, and *pp*-DDE (12). This assay allows for in vivo screening to detect androgen agonists or antagonists by measuring the response of five sex assessory organs and tissues in the rat. OECD has also conducted a program to validate the rat uterotrophic bioassay response to screen for new compounds that exhibit in vivo estrogenic activity (13).

The ICCVAM is a permanent body created by the ICCVAM Authorization Act of 2000 (Public Law 106-545). Its mission is to "establish criteria and processes for the validation of regulatory acceptance of toxicological test methods of interest to Federal Agencies, including alternative methods that replace, refine, or reduce the use of animals for research and testing purposes" (14). Fifteen U.S. federal agencies participate in efforts to establish standardization, validation, acceptance, regulatory implementation, international harmonization, and adoption of such test methods. Once a validation effort is complete, ICCVAM forwards recommendations to various agencies. FDAs response consists of the conclusions reached by each of its regulatory components with regard to practical applicability of the method to the product that the FDA Center regulates and the feasibility of the implementation of an accepted method to these regulated products. Recently, as part of its evaluation and validation process, ICCVAM published "The Revised Up and Down Procedure: A Test Method for Determining the Acute Oral Toxicity of a Chemical" (15). This revised up and down procedure has been proposed as an alternative to the existing conventional lethal dosage (LD-50) test. ICCVAM also published its "Guidance on Using In Vitro Data to Estimate In Vivo doses for Acute Toxicity" (16). Finally, ICCVAM evaluated three alternative in vitro test methods for assessing dermal corrosivity potential of chemicals and published "ICCVAM Evaluation of EPISKIN^{TM} (epi-200) (L'Oreal Products, Inc., New York, U.S.A.) and the rat skin Transcutaneous Electrical Resistance Assay: In Vitro Test Methods for Assessing the Dermal Corrositivity Potential of Chemicals" (17,18). These tests would be used as part of a strategy to evaluate corrositivity/irritation, thereby reducing and refining the use of animals.

The ECVAM subserves a very similar purpose to that of ICCVAM in that it also evaluates a variety of alternative methods. In 1997, ECVAM issued the following statement supporting the validity of the 3T3 neutral red uptake phototoxicity (NRU PT) test (an in vitro test for phototoxic potential). "The Committee therefore agrees that the conclusion from the formal validation study (of nine laboratories that correlated in vivo with in vitro studies) that the 3T3 NRU PT is a scientifically validated test which is ready to be considered for regulatory acceptance" (19). In 1998, ECVAM endorsed EPISKIN (http://www.medex.fr/Episkin.htm) as "scientifically validated for use as a replacement for the animal test and that it is ready to be considered for regulatory acceptance" (20). They gave a similar endorsement to EPIDERM^{TM} (MatTek Corporation, 200 Homer Avenue, Ashland, MA 01721) in 2000 for use as a model for human skin for corrositivity testing (21).

The European Medicine Agency and its Committee for Proprietary Medicinal Products has a vast network in Europe for "mobilizing scientific resources through the European Union to provide high quality medicinal products, to advise on research and development programs..." (22). Similarly the Japan Pharmaceutical Manufacturers Association has 11 committees that work very closely with the International Federation of Pharmaceutical Manufacturers Association to provide the evaluative expertise that is needed to evaluate and endorse the validations and collaborative studies on methods that are put before them (23).

The overall consensus that is developed and provided by the global evaluative bodies, described above, represents the cornerstone or foundation for the very broad agreements and guidelines for batteries of tests that have been issued in recent years by the ICH. These have allowed the United States, the European Union, and Japan to essentially accept a universal set of standards for the performance of toxicology testing that is the basis for regulatory decisions on drug safety.

The Johns Hopkins University's Center for Alternatives to Animal Testing is very active in providing grants to scientists who are developing nonanimal tests that could replace whole animal test methods. It also provides workshops on alternative methods and several types of publications to all interested scientists. Furthermore, it manages Altweb, an international online clearinghouse of alternatives resources (24).

Adoption of a Method

The final acceptance of a method for evaluation of drug toxicity is preceded by a process that goes through various stages (Fig. 1). Initially, the method is usually proposed and first reported at an appropriate scientific meeting (possibly the Society of Toxicology or the American College of Toxicology) as an abstract for poster or platform presentation. This is generally followed by the publication of the full method containing extensive supporting data in a widely read peer-reviewed journal. One can easily see within a year or so the impact of the method by the number of citations generated and the quality of the publications that these appear in as well as the nature of the collateral data that the use of this method generated. Unless the new method is unique, a process of comparison will occur, wherein the relative merits and disadvantages of the procedure will be compared with similar ones that have preceded it or even those published afterward.

It is at this point that the utility of the method really begins to emerge. If the disadvantages are overwhelming, then the method could be rapidly discarded unless subsequent modifications can be made. In contrast, if the elements of sensitivity and specificity are truly impressive and if there are no serious issues of false positives or false negatives, blind spots, incoherencies, or inconsistencies that are often a barrier to validation, then the method begins to acquire various levels of acceptance. It should be recognized that a different burden of proof is needed when a new test is to be used to replace an existing and accepted test rather than just adding a new test to the collection. Generally, if the method has the potential to play a crucial role in the production of data for regulatory decisions, it will undergo very serious scrutiny and peer review within government circles as well as the industrial sector. This does not, of course, preclude the often very rigorous participation of the academic community, which may utilize the method as a tool for mechanistic studies or in a consultative capacity in the resolution of regulatory issues.

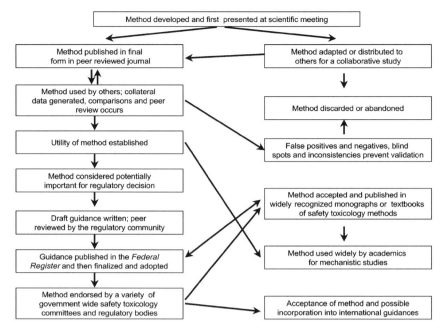

Figure 1 Decision tree—acceptance of a safety toxicology method.

Once a reasonable level of acceptance has been achieved, often the draft of a guidance will be generated that may suggest that the method be used when submitting data to a regulatory agency. This draft will be circulated within government to the various coordinating committees (having been originated by one of them) for critical review and optimization. At the same time, input is solicited from the industrial and academic sectors and the contribution of these partners are also incorporated. Finally, the document may be published in the *Federal Register* for comment by the public, and then after having incorporated all of the useful suggestions, it could be adopted in final form.

In an alternate pathway (although not mutually exclusive), and often in an earlier era, methods would achieve a very high level of acceptance by being incorporated into highly respected monographs such as the FDA *Red Book* (25), the *U.S. Pharmacopeia*, or *Current Protocols in Toxicology* (26). A method that has received broad acceptance might also be published in widely acclaimed textbooks such as *Principles and Methods of Toxicology* by A. Wallace Hayes (27) or others. This type of acceptance would only usually occur after the method was endorsed by a variety of government-wide or independent scientific committees such as OECD, ILSE, and HESI or, for example, was published in EPA guidelines (28) from the Office of Pesticide Programs and Toxic Substances. This type of recognition and acceptance would convey a very considerable amount of regulatory importance.

In today's era of globalization, the drug industry and others that generate toxicological data want harmonization of methodologies and policies in national and international markets. One costly drug-toxicology study should serve all. Thus the current harmonization efforts between the United States, the European Union, and Japan have resulted in ICH documents containing policies and methods that allow for uniformity and standardization of data requirements. ICH guidances have evolved and taken precedence in terms of acceptance over local methods and approaches. These documents provide a universal guide that confers the ultimate legitimacy on a particular method and its generated data to be used in drug development and regulation.

HISTORY OF REGULATORY ASSAY REVIEW AND/OR ACCEPTANCE

The Ames Salmonella Reversion Test in the Regulatory Setting

The Ames test started as a unique method for evaluating genotoxicity in prokaryotic organisms. In 1975, a paper appeared in the literature written by Ames et al. (29) in which a simple and inexpensive bacterial assay system using *Salmonella typhimurium* was shown to detect chemically induced reverse mutations. This assay also allowed the incorporation of an exogenous rodent liver microsomal enzyme system to produce reactive metabolites of a test compound (30).

The Department of Health, Education and Welfare (DHEW now DHHS) had established a Committee to Coordinate Toxicology and Related Programs in the early 1970s. In 1974, it also established a Subcommittee on Environmental Mutagenesis that comprised DHEW agencies, including the FDA and some observer agencies (i.e., Department of Defense). Its mission was "to draft a background document dealing with test procedures and approaches to testing chemicals for mutagenic activity to aid officials of regulatory agencies who have the responsibility for deciding: (i) advisability of promulgating test requirements for mutagenicity at the present time under any of their current legislative authorities; (ii) the appropriateness of mutagenicity tests for a wide range of product use and exposure categories; and (iii) the reliability and interpretation of data from mutagenicity tests developed on substances of commerce within their regulatory purview in spite of the absence of formal testing requirements." In 1977, it published a document entitled "Approaches to Determining the Mutagenic Properties of Chemicals; Risk to Future Generations" (31).

In 1977, a paper was presented in Montreal, Canada, by FDA's A.C. Kolbye, in which he declared that "The FDA has not required tests for mutagenic activity. However, it has suggested that the sponsors test for mutagenicity early in the study of a compound to provide some guidance in planning additional toxicity study programs. It is hoped that in the not too distant future, sufficient information will become available to support the use of a battery of mutagenicity tests" (32). Thus the thought process of the incorporation of mutagenicity data into FDA's regulatory concern began. In 1982, the FDA's Bureau of Veterinary Medicine released, for comment, its Guideline for Chemical Compounds in Food-Producing Animals, in which it mentioned the use of mammalian and nonmammalian test systems to determine the genetic toxicology of potential animal drugs used in food-producing animals (33). These data would be a part of the total assessment to determine if a compound was a candidate for two-year bioassays for carcinogenicity. The European Economic Community (1987), the FDA (Bureau of Veterinary Medicine, 1981, and Center for Food Safety and Applied Nutrition,1993), and the Ministry of Health, Labor and Welfare (1989) had guidances in regulatory use, and the harmonization was deemed to be a useful activity for global drug review. In an international activity to harmonize the genetic toxicity testing of pharmaceuticals, the ICH created an Expert Working Group in the early 1990s. The ICH S2A and S2B guidances (1996, 1997) developed by an ICH Expert Working Group incorporated the Ames test within its recommended battery of genetic toxicity tests (34,35). Throughout the 1970s, 1980s, and 1990s, the Ames test was widely used and extensive databases were assembled. As a result, national and international regulatory agencies accepted the data generated by the Ames test for the regulation of drugs and other chemicals.

Assays for Biomarkers in Drug Safety Toxicology

The idea of a biomarker is certainly not new when you consider that clinical chemistry and clinical pathology profiles have been used for disease diagnosis and drug safety monitoring in humans for decades. Problems arise, however, when one realizes that specificity often leaves much to be desired and the biomarkers used are not always predictive of a life threatening toxicity as occurred with some of the drugs, i.e., troglitazone that produced serious hepatotoxicity in the absence of aspartate aminotransferases, alanine aminotransferases, and bilirubin changes. The use of biomarkers to predict the onset of toxicity from drug exposure has become a very useful tool in preclinical research as their specificity has narrowed (for example, uses of troponin T to document cardiac toxicity).

Most recently, the three "omic" technologies, genomics, proteomics, and metabonomics/metabolomics, have gained prominence as being very promising for predictive purposes. Of these, genomics appears, at this point (December 2005), to be the most advanced in terms of presenting a complex signal that may be indexed or characterized as representative of a particular type of toxicity within a given target organ or tissue. A putative genomic biomarker may be either a gene expression pattern that is predictive or reflective of a given toxicity or a genetic polymorphism that is linked to interindividual differences in drug toxicity (36). The complexity of a gene expression signature presents an extremely large collection of data on many genes that absolutely requires great expertise in the use of informatics technologies. These data files may then be coupled with advanced statistical systems to sort out and segregate information obtained from the large number of genes that are expressed or remain silent. Further analysis and interpretation of the gene expression signal can lead to the isolation of a possible or probable valid biomarker, which is still a very important step away from a so-called valid biomarker that will be discussed below.

On November 4, 2003 the FDA issued a draft "Guidance for Industry on Pharmacogenomic Data Submission" (37) in which the agency encouraged the voluntary submission of genomic data, especially those being generated from animals used in toxicology studies conducted under Good Laboratory Practices for Investigative New Drug Applications. In addition, the draft guidance also acknowledged the existing difficulty of interpreting the exact impact of these data for safety evaluation and regulatory decision making. It is possible at this point to generate gene expression signatures for specific histopathological conditions. Obvious questions now arise. Are these reliable and reproducible regardless of the new drug used, and do they truly reflect the organ-specific toxicity, or are they nonspecific for such things as systemic inflammation, apoptosis, or cellular necrosis regardless of location? In other words, do these tools stand alone to always identify a specific toxicity for an organ or tissue such that conclusions can be made about the agent of exposure?

Clearly, the transition from gene expression signal to gene expression signature to probable biomarker and finally to a valid biomarker must rely on extensive experimental efforts and refinement of the approach to attain the analogous reliability of a single component biomarker such as troponin T (38). The same criteria apply to proteomic and metabolomic techniques. In the case of the former, the process of annotation of data from two-dimensional (2-D) gels may give protein signatures that have toxicological correlates, but once again the issues of specificity and reliability are of concern. In the case of proteomics, however, many would argue that the ultimate potential may be greater because the expressed proteins represent the tools that are necessary to empower action or are the result of an accomplished adverse action rather than merely a call to action as with gene expression in genomics. Clearly, the development of proteomic systems is not as far advanced as that of genomics. But it is no less promising, especially because it is to be found in a mobile phase within the body (in serum and urine) rather than being largely tissue bound. FDA has yet to issue guidance in this area.

Finally, metabonomics/metabolomics has begun to make an impact on predictive toxicology but in a somewhat different fashion. The complex metabolic patterns that are seen in urine and blood as a result of drug exposure, when analyzed, lend themselves to principal component analysis. The distinct divergent patterns that are generated allow one to sort out and distinguish between the various treatment and control groups. Correlation of these patterns with specific toxicities or localization of effects is yet to be achieved, but may occur in the near future as specific tissue loci and other fluids are sampled and subjected to comparison. This approach involving animal experiments would initially be quite invasive, but might confirm specificity that would allow the use of urine and serum samples only in later studies. Metabonomic and metabolomic techniques are clearly not ready, at this point (late 2005), to be used in clinical trials for safety monitoring. However, the potential in the near future is quite promising, especially because these signals, like those of the proteome, are quite mobile and reflective of rapid metabolic changes.

In contrast to the complex signal patterns generated by the "omic" technologies, troponin T is a single protein biomarker for cardiac necrosis which is highly specific for myocardial damage, and is now in the process of being accepted as a gold standard biomarker for clinical uses (39). It is being adopted as a clinical test for heart damage, whether it originates from drug exposure or the natural disease state. It is so sensitive that it has on occasion detected mild cardiac damage in normal individuals who have run 26 km marathons. In a similar context, but still in the experimental stage, biomarkers such as IL-6, tissue inhibitor of metalloproteinases-1, and haptoglobin are being explored as specific biomarkers for vasculitis, a relatively silent disease whose potential for life-threatening sequelae is very great (40).

In conclusion, the basic objective of these activities is to obtain valid biomarkers that are so specific, reliable, and characteristic for a given toxicity that they can be translated into the clinic to monitor human trials for early onset toxicity. It is hoped that in the preclinical arena, the use of genomic data would ultimately, after further refinement, be utilized to select lead compounds during drug development. More importantly, use of genomic data might help to "kill" those compounds that expressed the type of toxicity that often jeopardize the clinical trial subjects and/or approval of the drug during or after long and extremely costly phase II and III trials. This truncation of expensive human trials for drugs that should not have been marketed, while economically important, is certainly second to the goal of protecting the population from drugs that must ultimately be withdrawn from use because of lethal or serious adverse effects.

Development of Photo Cocarcinogenicity Testing

Among the oldest of written records, one can find references to the damaging effects of sunlight in skin. Even the knowledge that specific plant materials could greatly exaggerate the sensitivity of skin to sunlight was clearly documented in ancient Egyptian writings. Only in relatively recent history has it been appreciated that those and other chemical compounds could exaggerate the long-term effects of sun exposure as well. In current terminology, the chemical enhancement of light-induced cancer is "photo cocarcinogenesis." Such an enhancement is one of the recognized risks of certain classes of pharmaceuticals. The background of photo cocarcinogenesis as a useful safety test is as follows.

In 1928, Findlay (41) provided the first demonstration that UV radiation could induce skin tumors in animals. The discovery of UV-induced squamous cell tumors in genetically hairless mice by Winkelman et al. (42) in 1960 was a pivotal laboratory finding. In general, laboratory findings may be reported several decades before they become the subject of standardized preclinical tests. When a concern is raised for human safety, such tests and their validation could be requested or required by regulatory agencies.

Almost as early as the first demonstration of photo carcinogenesis in animals, investigators such as Blum attempted to assay the interactive influence of industrial, cosmetic, and pharmaceutical chemicals on the UV-induced neoplastic process (43). Epstein was probably the first investigator to test and report on a significant number of pharmaceuticals using a uniform set of procedures (in his case), involving repeated drug administration and UV light (44,45).

In 1975 and 1990, Forbes et al. (46,47) showed that the carcinogenic ability of UV light could be enhanced in the skin of hairless mice by certain chemicals such as 8-methoxypsoralen (8-MOP) (the active ingredient used in psoriasis phototherapy), subsequently confirmed by Stern et al. and then by Morison et al. (48,49) in large clinical trials.

In the late 1980s and early 1990s, the Center for Drug Evaluation and Research (CDER) became concerned about the potential photo cocarcinogenicity of retinoids that were being promoted for topical use on the skin. Experimental data of Forbes et al. (50) had shown that retinoids decreased the latency period to the formation of skin papillomas in mice. Photosensitivity reactions in humans had been reported for several members of drug classes including retinoids. Regulators suggested that drug sponsors provide data on their drugs under development that absorbed UV light in the range of 280 to 700 nm of the electromagnetic spectrum. At that time, there were few laboratories that had the capability to produce in vivo data. The two laboratories capable of generating photo cocarcinogenicity data were the Argus Research Laboratory in Horsham, Pennsylvania, U.S.A. and the A.D. Little Laboratory in Cambridge, Massachusetts, U.S.A., both having incorporated the methodology developed at Temple University. Both laboratories were asked to present seminars to the FDA/CDER regarding their testing capabilities in this area. CDER was also aware of literature data indicating that drugs from certain chemical classes, when given orally, could accelerate the rate at which papillomas/squamous cell carcinomas were induced by UV exposures.

Following these seminars, numerous scientific meetings on the subject of phototoxicity and photo cocarcinogenicity occurred in which CDER participated. A CDER advisory committee meeting in March of 1993 was held following a report to the CDER that oral fluoroquinolone antibacterials and UVA produced skin tumors in hairless mice with latency periods shorter than that seen in tumors produced by UVA alone (51). During the next 10 years, several meetings and symposia, both national and international, occurred that addressed the topics of phototoxicity and photo cocarcinogenicity. FDA published its final guidance, "Guidance for

Industry-Photosafety Testing," in May 2003 (52). This guidance is intended to assist pharmaceutical companies in deciding if tests for photosensitization are needed and to assess the potential human risk for photo cocarcinogenicity of their topical and/or systemically administered drugs during the clinical development process. While several in vitro assays have been proposed for use in providing relevant safety data, the hairless mouse protocol developed at Temple University is still a useful regulatory tool for detecting a drug's potential for enhancing UV-induced skin cancer (53).

It is important to recognize that the photo cocarcinogenicity test in the hairless mouse has, at present, not been validated in accord with procedures that are considered ideal by U.S. regulatory agencies. Such procedures call for replication of positive or negative photo cocarcinogenic findings, depending upon the test substance, in laboratories, in addition to those mentioned above using identical test procedures and protocols (Table 1). To our knowledge, this has not occurred at the present time. The reasons for this could be many and it is recognized that the cost of equipping and appropriately staffing a test facility are significant factors. As with other carcinogenesis studies, the duration of a photo cocarcinogenesis test is substantial and represents a considerable investment of resources. It should be noted that the FDA's National Center for Toxicological Research (NCTR) in Jefferson, Arkansas has developed a photobiology laboratory in an identical setup as existing in the Argus photobiology laboratories. The NCTR is presently conducting phototoxicity experiments on several appropriate test compounds that, in the future, could provide additional information toward validation of the photo cocarcinogenicity test.

At the present time, it is known that 8-MOP will reduce the time to papilloma formation in the hairless mouse following long-wave ultraviolet radiation (UVA) exposures. Furthermore, it has been recognized that 8-MOP and UVA (psoralen UVA) produced skin cancer several years after humans were treated for psoriasis (54). Furthermore, good correlations exist between human phototoxicity (i.e., sunburn) responses to light and drugs and tumor responses in the photo cocarcinogenicity test to the same drugs or to members of the same chemical class (55). While the photo cocarcinogenicity test is not recognized as validated at the present time, this test and emerging data contribute to a further understanding of the phototoxic response, its potential correlation with skin cancer production, and its regulatory utility.

Status and Level of Approval of Various Techniques, Tests, and Methods for Drug Safety Evaluation

Proteomics, metabonomics, combinatorial chemistry techniques, positron emission tomography (PET) labeling of anticancer drugs, use of in vitro skin substitutes, use of prodrugs, and applications of nanotechnology are all in the early stages of use and validation. In contrast, the Ames test, photo cocarcinogenesis testing, PET and magnetic resonance imaging (MRI), high throughput screening, use of monoclonals, micronucleus testing, and use of biomarkers such as troponins as well as utilization of QTC–IKR techniques for detecting potential for arrhythmias as well as other toxicity tests are far more mature in terms of use and acceptance. Of the latter, all but high throughput screening are the subject of FDA guidances and represent important tools currently used in the process of drug safety evaluation. The Ames test, the photo cocarcinogenicity test, the micronucleus test, and combinatorial chemistry techniques are to be found described in monographs as well as ICH guidances. The former two tests are often requested by FDA for safety evaluation. Table 3 describes these techniques, tests, and methods as well as their specific applications.

Table 3 Techniques, Tests, and Methods for Drug Approval: Status and Level of Approval

Test, method, or technique	Applications	Drug classes[a]	Early stages of use	FDA guidance	In a monograph	In ICH
Ames test	Genetox	Many		X	X	X
Micronucleus	Genetox	Many	X	X	X	X
PET imaging	Efficacy toxicity	Neuro drugs		X		
MRI	Imaging soft tissue injury			X		
High throughput screening	Receptor binding, drug discovery	All types	X			
PET labeling	Diagnosis and confirmation of drug delivery	Anti-cancer		X		
Monoclonals	Autoimmune diseases	Immunotherapy	X	X on immunotox		in progress
Genomics	Gene expression					
Proteomics	Protein expression	Drug toxicity	X	X		
Metabonomic	Metabolic products		X			
Photo cocarcinogenicity evaluation	Detect drug effect on UV-induced tumors	UV-absorbing chemicals[b]	As-needed basis	X		X
Biomarkers–troponins	Detect cardiac damage	Cardiac drugs		X		
QTC–IKR	Detect potential for arrythmia	All types of drugs		X		X
Combinatorial chemistry	SAR, new drug devel	All drug classes	X		X	X
In vitro substitutes	Cornea, skin, cardiac cells	Various classes	X			
Use of prodrugs	Active ingredient protection and delivery	Anticancer	X	X		
Nanotechnology	Drug formulations	Many drugs	X			

[a]Not inclusive.
[b]UV-blocking or skin modifying.
Note: A blank space in this table indicates a lack of activity in that area owing to the novelty or early stage of the technique or method listed or N/A.
Abbreviations: FDA, Food and Drug Administration; MRI, magnetic resonance imaging; PET, positron emission tomography; ICH, International Conference on Harmonisation; QT-IKR, QT segment of the electrocardiogram; I, current; K, potassium; R, rapidly activating delay rectifier; SAR, structure–activity relationships.

Accepted Older Methods No Longer Used

The evolution of methodology that is acceptable for use in the drug safety evaluation/approval arena is a reflection of not only the advance of basic science concepts and its accompanying cutting edge technologies, but also in the values and cultural atmosphere of the times. This is illustrated by the many examples of tests and methods (Table 4) that were once widely accepted but are no longer considered applicable or valid for evaluative purposes for a variety of reasons.

Table 4 Accepted Tests That Are No Longer Routinely Used

Name of test	Type of test	Problems or reasons for abandonment
Esophageal corrosion	Application of acids and bases to whole animal potential irritants and corrosives	Issues of animal welfare (pain, suffering, and tissue damage)
Eye and skin irritation	Application of acids and bases to whole animal potential irritants and corrosives	Issues of animal welfare (pain, suffering, and tissue damage)
Chick embryo assay	Injection of drugs into fertile chicken eggs	A variety of artifacts caused by physical factors; pressure, sand and water injection cause effects
Host mediated assay	In vivo mouse genotoxicity	Use of S-9 allowed in vitro assay
Dominant-lethal test	Chromosome damage in reproductive cells	Insensitive test
Classical LD-50s	3–5 dose levels; 10 animals per group; 14 day observation period	Use of too many animals; animal welfare issues
Rabbit head drop	Curare assay	Replaced by HPLC
Hormone bioassays	In vivo tests of potency and toxicity	Replaced by in vitro assays or chemical analysis
UDS	DNA repair assay	Detects limited set of chemical and damage; lack of universal validation
SCE	Chromatid rearrangements	Lots of false positives; no correlation with disease or biological endpoint
Comet assay	DNA strand break assay	Presently, unique utility not demonstrated for regulatory use
DNA adducts	Drug–DNA interactions	Cannot identify specific damage; no direct relationship with disease of biological endpoint
Rabbit pyrogen assay		Limulus Amoebocyte assay is the in vitro substitute

Abbreviations: HPLC, high-performance liquid chromatography; SCE, Sister chromatid exchange; UDS, unscheduled DNA synthesis.

Eye and skin irritation tests involve the application of potentially irritant or corrosive chemicals to the intact animal. They have been deemed generally unacceptable because of issues of animal welfare that strongly discourage or outright forbid pain and suffering through tissue damage in unanesthetized animals (8). Animal welfare groups and the enactment of animal welfare legislation have, in fact, encouraged the development of alternative methods that are often in vitro or in silico systems for evaluation of safety. The development of EPISKIN and EPIDERM are good examples of this; however, even these innovations do not incorporate the elements of the immune system which must still be examined in vivo for certain types of skin sensitivity or allergy testing. Furthermore, these tests do not allow for tissue repair and recovery following chemical insults.

Another important animal welfare issue involves the number of animals used for a particular test. The LD-50 and LC-50, as originally described by the Federal Insecticide Fungicide and Rodenticide Act of 1947, called for multiple dose levels (and controls), animals of both genders, and a 14 day observation period. The numbers of animals used for a single test were often in excess of 100 and gave little, if any, clues as to the cause of death. These tests have been abandoned in favor of a variety of dose-ranging tests that look at graded responses for effect (adverse, efficacious, etc.) with minimum morbidity. The Up and Down Method (described earlier) is one example. Some concerned people have argued that the absence of LD-50 values for modern drugs leaves safety gaps in areas of poison control or clinical toxicology, but by and large the modern principles of animal welfare prevail. Using these dose-ranging tests, approximate lethal dose estimations can be made, if needed.

In the area of teratology testing, a rather useful older method was the chick embryo assay, which involved the injection of drugs or food additives into the yolk sac or airspace of fertile chicken eggs. Once incubated and hatched, the chick was supposed to provide an early, inexpensive signal of teratology potential. Unfortunately, a variety of artifacts caused by physical factors such as pressure, granulated particles, and even control water or saline injections caused effects that produced false positives. This method is now largely abandoned in favor of mammalian systems that are part of a reproductive battery of tests (56). In silico databases are also being established that might predict reproductive and carcinogenic toxicants based on chemical structure (57).

In the area of hormone bioassay methodology, a large number of in vivo methods have been abandoned because suitable in vitro or solely chemical analysis have greatly reduced or abolished the use of animals. Biological tests such as the rabbit test for insulin potency are now being replaced by cell culture and high-performance liquid chromatography (HPLC) methods. The cat glucagon test is being replaced by cell culture; the growth hormone assay in rats (one month duration) is being replaced by cell culture or HPLC. The rabbit head drop assay for D-tubocurarine, rooster vasodepressor test for oxytocin, and rat pressor assay for vasopressin have been replaced by HPLC methods because of the marketing of the pure synthesized polypeptide rather than glandular extracts, etc. Occasionally, an in vivo test must be retained because the intact animal system is essential. An example of this is the use of the human chorionic gonadotropin assay in the rat in which the true potency is only determinable when the intact metabolic apparatus of the animal removes the glycosylation groups from the active component. Because the stoichiometry or extent of glycosylation varies from lot to lot, the intact animal is required; an HPLC method is not feasible at this point (58).

In the area of genetic toxicology, there are a number of tests or methods that have fallen into disuse. These include the dominant-lethal test which looks at chromosome damage in reproductive cells and has been deemed too insensitive; unscheduled DNA synthesis, a DNA repair assay, only detects a limited a set of chemicals and types of damage and lacks universal validation; sister chromatid exchange testing which looks at chromatid rearrangements is subject to many false positives and very little correlation with a disease state or a biological endpoint; and the comet assay which examines DNA strand breaks has not had its unique utility validated, and the DNA adducts method that looks at drug–DNA interactions but cannot identify specific damage or a direct relationship with a disease or a biological endpoint have not been accepted as part of a regulatory battery for drug development in the United States (59).

These are, with possible exceptions, a few examples of methods that may have largely outlived their usefulness; however, a very careful analysis of the shortcomings of these methods provides useful insight into the issues and criteria that have importance in the current evaluative efforts that constitute the validation process and the search for newer methods.

THE IMPACT OF FDA'S "CRITICAL PATH" CONCEPT ON FUTURE DEVELOPMENT OF DRUG SAFETY METHODOLOGY

In early 2004, the FDA issued a report titled "Innovation or Stagnation? Challenge and Opportunity on the Critical Path to New Medical Products" (60). It was written following analysis of drug development in the United States. The report identified a slowdown rather than an anticipated increase in the rate of new drug development. Furthermore, the report declared that new science was not guiding the drug development process in the same way that it was hastening the drug discovery process.

The critical path begins after basic research provides potential drug candidates for development. These candidates then face a more rigorous evaluation along the path including nonclinical development. Traditional methods used to assess drug safety using animal toxicity tests have changed little over the course of many decades. The recent gains in biological and toxicological knowledge have not shown much improvement in these tests. FDA's CDER has recognized that a new generation of predictive tests must aggressively be pursued, thus necessitating a collective action with academia, the pharmaceutical industry, and other government agencies to develop more useful and meaningful tests. As examples, in the past few years, the FDA

1. has worked with the scientific community to use human cell lines to characterize drug metabolic pathways and predict unfavorable human metabolism of candidate drugs,
2. has collaborated with the industry and scientific groups to develop data to allow international recognition of transgenic mouse models for use in testing drug candidates for potential carcinogenicity, and
3. has mined its databases to develop structure–activity relationship software to identify molecular substructures with potentially adverse toxicological properties early in drug development.

The agency has identified several opportunities for developing toxicological tools and methods that could reliably and more efficiently determine the safety of a potential drug. Some of these opportunities for development are

1. better predictors of human immune responses to foreign antigens,
2. new techniques for predicting drug–liver toxicities,
3. safety assessment techniques using advances in proteomics, metabolomics, and toxicogenomics, and
4. new tools to accurately assess the risk of drugs to cause heart arrhythmias.

In promoting the concept of a "critical path," the agency hopes that new methods will result in the development of more effective and safer drugs and expedite the availability of medical advances to improve the health and well being of individuals. This will require very robust methodology development for safety evaluation and an acceleration of the validation and adoption processes described in this chapter.

CRUCIAL ISSUES AND CHALLENGES

Currently, there are a number of crucial issues and challenges for emerging technologies and methodologies and these are summarized in Table 5. With respect to genomics, there are still issues relating to the comparability of the various platforms or arrays being used, but these questions are now being resolved in a recent (2004) collaborative study (61,62). Both the informatics and the statistical tools being developed are in early stages that will hopefully resolve the correlations between the genomic patterns observed and the specific organ toxicities. In the area of proteomics, correlation with toxicity is in its earliest stages and the work is extremely resource intensive. Metabonomics—analysis of the global pattern of metabolic status—is currently being used successfully to distinguish or separate various treatment groups in terms of dose, time course, gender, severity of lesions, etc., by the process of two-component analysis. Individual metabolites are not being identified, unless the process of metabolomic analysis is undertaken. Initial cost and resource outlay in this case is quite extensive. Similarly, the initial cost is quite expensive for MRI and PET imaging. MRI issues involve limitations on the approved field strength allowed in humans and its impact on resolution. PET imaging presents problems with resolution but offers unique opportunities for the observation of pharmacological and toxicological processes in real time. High throughput screening is very resource intensive and may only be affordable to very large institutions or commercial concerns, and cross-validation is difficult.

The use of biomarkers for toxicity prediction and detection raises issues of specificity, i.e., are we looking at a specific organ effect or nonspecific general inflammation? In the case of photo cocarcinogenesis techniques, these are quite expensive, have a long duration of testing, and use many animals. The in vitro substitutes such as artificial skin lack immunocompetency and repair capabilities and are, therefore, not a complete substitute for whole animal work, although observation of direct effects of agents on the in vitro model is a significant advancement. The use of nanotechnology is in its infancy, but it represents a much-needed tool to evaluate the potential for toxicity in future nanoformulated pharmaceuticals that will be applied topically or administered systemically. The shortcomings of the QTC–IKR methodology involve how well it relates to Torsade de Points versus the excellent correlation to K^+ channels test. Finally, issues with the Ames test are described elsewhere and in Table 5.

Table 5 Critical Issues and Challenges for Emerging Technologies and Methodologies

Genomics	Gene expression not necessarily always correlated with toxicity; comparison and validation of various platforms still in progress; informatics efforts to interpret data at an early stage; statistical models still being developed
Proteomics	Labor intensive; correlation with toxicity in early stages
Metabolomics	Initial cost and resource outlay is quite extensive; relies more on pattern recognition than recognition of individual metabolites
PET imaging	Issues involving sensitivity and resolution
MRI	Very expensive initial outlay; limitations on acceptable or approved field strength for human studies impacts resolution; still extremely expensive and not widely used by preclinical researchers
Biomarkers for toxicity	Issues of specificity; may be nonspecific, e.g., generalized inflammation; not many validated for safety toxicity evaluation
High throughput screening	Very resource intensive; used only by very large well-endowed organizations; cross-validation difficult
Photo cocarcinogenesis	Expensive; chronic duration; uses many hairless mice
Ames test	Does not detect the mutagenic effects of some substances, i.e., some antibiotics, metals, and proteins
QTC–IKR	Does not directly correlate with Torsade de Points, but relates to the K^+ channels which is the first step in production of arrhythmias
Combinatorial chemistry	Lack of sufficient information on novel chemical classes to make decisions; requires a large databases; often requires large amount of proprietary information that must be protected and/or is not available to all
In vitro substitutes	Lack of immunocompetency; repair mechanisms absent; still need to use animals for tissue and cell harvesting
Nanotechnology	Methodology for safety evaluation in its infancy; toxicity of nanoparticles and nanoparticlized drugs relatively unexplored

Abbreviations: MRI, magnetic resonance imaging; PET, positron emission tomography.

CONCLUSIONS

There are many principles that govern the evolutionary process of drug safety toxicology and not all of them are based solely on the progress and sophistication of science. Many drivers of this process are social and cultural as evidenced by the impact of the animal welfare movement in recent decades on a variety of in vivo methods and procedures and the recent ban on many kinds of stem cell research and associated technologies, and initiatives that have failed to be developed as a consequence. Many concepts and methodological approaches to drug safety toxicology have evolved or been abandoned for economic, moral, or practical reasons or the fact that they simply do not work or have no real validity in today's context.

The politics and economics involved in the globalization of drug safety toxicology has had a strong impact on negotiations that have led to important compromises in the interest of generating universally acceptable ICH guidances that set the current standard for drug safety submissions. These have shortened or minimized some of the original requirements of the most conservative governments and have added additional burdens for more extensive data generation to others. The net effect, for the most part, has been to shorten the numbers of chronic and subchronic studies and narrow the diversity of species and numbers of animals required for in vivo studies. However, many would argue that the variety of new and exciting in vitro and in silico options that are being provided by our scientific progress have compensated for the perceived loss in the extent of whole animal studies being conducted. It should be appreciated that regulators must exercise excellent judgment and obtain consultations from independent scientific bodies, before relying solely on these alternative assays in drug development until such time as their utilities are fully understood in terms of whole animal physiology. Finally, we have attempted to analyze some of the strengths and weaknesses of our newest and most exciting emerging technologies and predict what impact they have now and will have in the future on the utility of drug safety toxicology. Just as the buttons or controls for most of our chemical analytical detectors have moved from the hardwired chassis to the computer screen, so have our tools for conducting drug safety toxicology been computerized, automated, miniaturized, and integrated into high throughput systems that have become part of pattern recognition informatics modules that utilize artificial intelligence to manage high data traffic and optimize prediction and confirmation of the presence and absence of drug toxicity. One of the greatest contemporary challenges, in this area, is to harness and develop procedures for evaluating the validity of our high level technology so as to optimize and best serve our interests in the conduct of drug safety toxicology. The taming and channeling of innovation to optimize validation and adoption is the fundamental process that governs and supports the philosophy of drug safety toxicology development and its evolution.

REFERENCES

1. Sinclair U. The Jungle. New York: Grosset and Dunlap, 1906.
2. Lamb RD. American Chamber of Horrors. New York: Grosset and Dunlap, 1936.
3. Horton LR, Hoog TC, Stiff MJ, Levine AN. Compilation of Food and Drug Laws. Washington, D.C.: The Food and Drug law Institute, 1993.
4. The Editorial Committee. Appraisal of the Safety of Chemicals in Foods, Drugs and Cosmetics. Topeka: The Association of Food and Drug Officials of the United States, 1959.
5. Goldenthal EL. Current views on safety evaluation of drugs. FDA Papers 1968:2–8.
6. D'Aguanno W. Guidelines for reproduction studies II for safety evaluation of drugs for human use. U.S. Food and Drug Administration, Bureau of Medicine 1973:41–43.
7. U.S. Food and Drug Administration, Bureau of Foods. Toxicological Principles for the Safety Assessment of Direct Food Additives and Color Additives Used in Food (Redbook I). National Technical Information Service, Springfield, VA, 1982.
8. Silverman J, Suckow MA, Murthy S. The IACUC Handbook. New York: CRC Press, 2000.
9. http://www.wits.ac.za/bioethics/helsinki.htm (accessed January 2005).
10. http://www.fda.gov/OHRMS/DOCKETS/98fr/05–155.htm (accessed January 2005).
11. http://hesi.ilsi.org/.

12. Kenji Y, Sawaki W, Ohta R, et al. OECD validation of the Hershberger assay in Japan: phase 2 dose response of methyltestosterone, vinclozolin and p,p'-DDE. Environ Health Perspect 2003; 111:1912–1919.

13. Kanno J, Onyon L, Haseman J, Fenner-Crisp P, Ashby J, Owens W. The OECD program to validate the rat uterotrophic bioassay to screen compounds for in vivo estrogenic responses: phase 1. Environ Health Perspect 2001; 109:785–794.

14. Schechtman L. FDA's regulatory role in the ICCVAM process. Fourth World Congress, Alternatives Congress Trust, New Orleans, LA, August 11–15, 2002.

15. Interagency Coordinating Committee on the Validation of Alternative Methods (ICCVAM). The revised up and down procedure: a test method for determining the acute oral toxicity of chemicals. NIH Publication No. 20–450, 2002.

16. Interagency Coordinating Committee on the Validation of Alternative Methods (ICCVAM). Guidance document on using in vitro data to estimate in vivo doses for acute oral toxicity. NIH Publication No. 01–4500, 2001.

17. Interagency Coordinating Committee on the Validation of Alternative Methods (ICCVAM). ICCVAM evaluation of EPISKIN (EPI-200) and the rat skin transcutaneous electrical resistance (TER) assay: in vitro test methods for assessing the dermal corrosivity potential of chemicals. NIH Publication No. 02–4502, 2002.

18. Biennial progress report of the Interagency Coordinating Committee on the Validation of Alternative Methods (ICCVAM). NIH Publication No. 04–4509, 2003.

19. http://www.iivs.org/news/3t3.html (accessed January 2005).

20. European Centre for the Validation of Alternative Methods (ECVAM). Scientific Advisory Committee (ESAC). Statement of the scientific validity of the EPISKIN test as an in vitro test for skin corrosivity. 10th ECVAM Meeting, Ispra, Italy, March 31, 1998.

21. European Centre for the Validation of Alternative Methods (ECVAM). Scientific Advisory Committee (ESAC). Statement on the application of the EPIDERM™ human skin model for skin corrosivity testing. 14th ECVAM Meeting, Ispra, Italy, March 14–15, 2000.

22. http://www.emea.eu.int/htms/aboutus/emeaoverview.htm# (accessed January 2005).

23. http://www.jpma.or.jp/12english (accessed January 2005).

24. altweb.jhsph.edu (accessed January 2005).

25. http://vm.cfsan.fda.gov/~redbook/ (accessed January 2005).

26. Morgan KS. Current Protocols in Toxicology. Vol. 1. New York: John Wiley and Sons Inc, 1999.

27. Hayes WA. Principles and Methods of Toxicology. 4th ed. Philadelphia: Taylor and Francis, 2001.

28. http://www.epa.gov/OPPTS_Harmonized/ (accessed January 2005).

29. Ames BN, McCann J, Yamasaki E. Methods for detecting carcinogens and mutagens with the Salmonella/mammalian microsome mutagenicity test. Mutat Res 1975; 31(6):347–364.

30. Malling HV, Frantz CN. In vitro versus in vivo metabolic activation of mutagens. Environ Health Perspect 1973; 6:71–82.

31. Subcommittee on Environmental Mutagenesis. Approaches to determining the mutagenic properties of chemicals: risk to future generations. DHEW Committee to Coordinate Toxicology and Related Programs, Washington, 1977.

32. Perez MK. Human safety data collection and evaluation for the approval of new animal drugs. J Toxicol Environ Health 1977; 3:837–857.

33. U.S. Food and Drug Administration, Bureau of Veterinary Medicine. Chemical compounds in food-producing animals: availability of criteria for guideline. Federal Register 1982; 47(22).

34. International Conference on Harmonisation of Technical Requirements for Registration of Pharmaceuticals for Human Use (ICH). Guideline for industry: specific aspects of regulatory genotoxicity tests for pharmaceuticals (S2A), 1996 (http://www.fda.gov/cder/guidance/guidance.htm).

35. International Conference on Harmonisation of Technical Requirements for Registration of Pharmaceuticals for Human Use (ICH). Guideline for industry: S2B genotoxicity: a standard battery for genotoxicity testing of pharmaceuticals, 1997 (http://www.fda. gov/cder/guidance/guidance.htm).

36. Evans WE, Rolling MV. Pharmacogenomics: translating functional genomics into rational therapeutics. Science 1999; 286:487–491.

37. http://www.fda.gov/cder/guidance/index.htm (accessed January 2005).

38. Apple FS, Quist HE, Doyle PJ, Otto AP, Murakami MM. Plasma 99th percentile reference units for cardiac troponin and creatine kinase mb mass for use with European Society of Cardiology/American College of Cardiology consensus recommendations. Clin Chem 2003; 49:1331–1336.

39. Apple FS, Murakami MM. Serum 99th percentile reference cutoffs for serum cardiac troponin assays. Clin Chem 2004; 50:1477–1479.

40. Zhang J, Honchel R, Weaver JL, et al. An experimental PDIV inhibitor induced vascular injury (vi) associated with increased mast cell degranulation and elevated serum levels of acute-phase proteins in Sprague Dawley rats. 43rd Annual Society of Toxicology Meeting, Baltimore, MD, March 21–25, 2004.

41. Findlay GM. Ultraviolet light and skin cancer. Lancet 1928; 2:1070–1073.

42. Winkelman RK, Baldes EJ, Zollman PE. Squamous cell tumors induced in hairless mice with ultraviolet light. J Invest Dermatol 1960; 34:131–138.

43. Blum H. Carcinogenesis by Ultraviolet Light. Princeton: Princeton University Press, 1959.

44. Epstein JH. Comparison of the carcinogenic and cocarcinogenic effects of ultraviolet light in hairless mice. J Natl Cancer Inst 1965; 34:741–745.

45. Epstein JH. Animal models for studying photocarcinogenesis. In: Maibach H, Lowe N, eds. Models in Dermatology. Basel: Karger, 1985:303–312.

46. Forbes PD, Davies RE, Urbach F. Phototoxicity and photocarcinogenesis: comparative effects of anthracene and 8-methoxypsoralen in the skin of mice. Food Cosmet Toxicol 1975; 14:303–306.

47. Forbes PD, Davies RE, Urbach F, Dunnick J. Long-term toxicity of oral 8-methoxypsoralen plus ultraviolet radiation in mice. J Toxicol Cutaneous Ocul Toxicol 1990; 9:237–250.

48. Stern RS, Nichols KT, Vakeva LH. Malignant melanoma in patients treated for psoriasis with methoxsalen (psoralen) and ultraviolet radiation (PUVA). The PUVA follow up study. New Engl J Med 1997; 336:1041–1045.

49. Morison WI, Baughman RD, Day RM, et al. Consensus workshop on the toxic effects of long-term PUVA therapy. Arch Dermatol 1998; 134:595–598.

50. Forbes PD, Urbach F, Davies RE. Enhancement of experimental photocarcinogenesis by topical retinoic acid. Cancer Lett 1979; 7:85–90.

51. Klecak G, Urbach F, Urwyler H. Fluoroquinolone antibacterials enhance UVA-induced skin tumors. J Photochem Photobiol B 1997; 37:174–181.

52. US Food and Drug Administration. FDA Guidance for Photosafety Testing, 2003 (http://www.fda.gov/cder/guidance/guidance.htm).

53. Forbes PD, Beer JZ, Black HS, et al. Standard protocols for photocarcinogenesis safety testing. Front Biosci 2003; 8: d848–d854.

54. Stern RS, Lunder EJ. Risk of squamous cell carcinoma and methoxsalen (psoralen) and UV-A radiation (PUVA). A meta-analysis. Arch Dermatol 1998; 143:1582–1585.

55. Osterberg RE, Szarfman A. Assessment of risk for photocarcinogenesis: regulatory reviewer viewpoint. Photochem Photobiol 1996; 63(4):362–364.

56. Keller KK. Developmental and reproductive toxicology. In: Jacobson-Kram D, Keller KK, eds. Toxicology Testing Handbook. New York: Marcel Dekker Inc, 2001: 195–254.

57. Contrera JF, Matthews EJ, Benz R. Predicting the carcinogenic potential of pharmaceuticals in rodents using molecular structural similarity and E-state indices. Regul Toxicol Pharmacol 2003; 38:243–259.

58. U.S. Pharmacopeia-USP 23 NF 18. Rockville: U.S. Pharmacopeial Convention, Inc., 1995:1622.

59. Elespuru RK. Future approaches to genetic toxicology risk assessment. Mutat Res 1996; 365:191–204.

60. http://www.fda.gov/oc/initiatives/criticalpath/whitepaper.html (accessed January 2005).

61. Rosenzweig B, Mansfield E, Pine PS, Sistare FD, Fuscoe JC, Thompson KL. Formulation of RNA performance standards for regulatory toxicogenomic studies. 43rd Annual Society of Toxicology Meeting, Baltimore, March 21–25, 2004.

62. Thompson KL, Rosenzweig BA, Pine PS, et al. Use of a mixed tissue RNA design for performance assessments on multiple microarray formats. Nucleic Acids Res. In press. 2005.

15
Statins and Toxicity

John Farmer

Section of Cardiology, Baylor College of Medicine, Houston, Texas, U.S.A.

INTRODUCTION

The lipid hypothesis of atherosclerosis was proposed more than a century ago and is based on the premise that dyslipidemia is central to the initiation and progression of coronary and peripheral arterial disease. The corollary is that modification of the lipid profile [reduction of total, low-density lipoprotein (LDL), and very low-density lipoprotein] will reduce the risk for vascular morbidity and mortality by altering the inexorable progression of atherosclerosis. The lipid hypothesis has been verified in epidemiologic experimental, pathologic, and genetic studies. However, the statistical association between dyslipidemia and atherosclerosis does not necessarily prove that pharmacologic modification of various lipid subtypes will alter the course of vascular disease. The early clinical trial data investigated lipid lowering with bile acid sequestrants, fibric acid derivatives, or nicotinic acid compared to placebo and demonstrated statistically significant improvement in cardiovascular morbidity and mortality, although reductions in total mortality were not achieved. Despite the positive relative risk reduction, the absolute benefits were minimal and the pharmacologic agents employed were difficult to administer due to lack of patient compliance secondary to the inherent side-effect profile of the agents employed. The advent of 3-hydroxy-3-methylglutaryl coenzyme A (HMG-CoA) reductase inhibitors revolutionized the ability of the clinician to optimize the lipid profile via a predominant effect on LDL cholesterol levels. Statin therapy was subsequently demonstrated to reduce cardiovascular morbidity and mortality in both primary and secondary prevention trials. Additionally, in adequately powered studies, total mortality has also been decreased by the administration of inhibitors of cholesterol synthesis.

However, epidemiologic studies have demonstrated a U-shaped relationship between cholesterol levels and total mortality, which implied that the overzealous reduction of cholesterol below a theoretic threshold that may be achieved with statin therapy may have adverse effects on overall survival by increasing noncardiac mortality. The epidemiologic studies demonstrated an increase in subarachnoid hemorrhage, violent behavior, and certain forms of malignancy at low cholesterol levels. The question of a cause and effect relationship between low cholesterol and mortality cannot be answered in epidemiologic trials and must be considered as

273

hypothesis generating. However, to date, statin studies with aggressive lipid goals have lowered LDL cholesterol to the 60 to 80 mg/dL range without an apparent adverse effect on total mortality. Despite the clear clinical benefits, statin therapy is not without risk, and definite adverse hepatic and muscle toxicity have been demonstrated with statin therapy. Additionally, epidemiologic and early clinical trials coupled with proposed adverse effects of cholesterol reduction implicated central nervous system effects such as alteration of cognition, sleep, ocular acuity, and behavioral abnormalities, although the true prevalence and the causality are more controversial. This chapter will review the potential and documented toxic effects of statins as to their prevalence, mechanism, and clinical significance.

PATHOGENESIS OF ATHEROSCLEROSIS AND THE POTENTIAL OF STATIN TOXICITY

Age-adjusted cardiovascular mortality has been declining for the past several decades in the United States. The encouraging decline in mortality results from a complex interplay between improved medical therapies, recognition and modification of risk factors, enhanced diagnostic capabilities, and improved survival in acute coronary syndromes. Prior to the advent of coronary care units, hospital mortality for acute myocardial infarction approached 30%. The availability of antiarrhythmic therapy reduced the early (less than 24 hours) mortality associated with acute myocardial infarction, which was frequently due to the sudden onset of malignant ventricular rhythm disorders. However, mortality and atherosclerotic complications remained high due to irreversible loss of functioning myocardium. Significant attempts were made to limit infarct size, and a variety of medications (nitrates, beta blockers, etc.) were employed but with modest results. The advent of thrombolytic therapy revolutionized the management of acute coronary syndromes and reduced the death rates associated with serum transaminase elevation and myocardial infarction to less than 10%.

Reduction in early mortality associated with myocardial infarction has resulted in an increase in the prevalence of individuals with a significant atherosclerotic burden, and the absolute number of individuals with documented atherosclerosis is increasing in the United States. Thus, while the the acute management of a variety of ischemic syndromes has been dramatically improved with the advent of thrombolytic therapy, acute angioplasty with drug eluting stents, and advanced antiplatelet therapy, the key to successful management of ischemic syndromes lies in the prevention of atherosclerosis rather than acute management.

However, a unifying hypothesis which explains all of the known aspects of atherosclerosis and remains elusive a number of theories have been proposed which partially explain the pathogenesis of atherosclerosis. Statins had been postulated to play a potential beneficial role in several aspects of the major theories of atherosclerosis.

Monoclonal Hypothesis

Benditt suggested that the origin of the smooth muscle cells within the atherosclerotic plaque is secondary to an unregulated proliferation of a single clone of cells (1). Atherosclerosis may thus be similar to a neoplastic process with profuse cellular proliferation and subsequent vascular occlusion. Evidence in support of the monoclonal hypothesis has been substantiated by the finding of only one isozyme of glucose-6-phosphate dehydrogenase in atherosclerotic lesions isolated from individuals expressing the genetic deficiency of this enzyme. The single clone of cells is compatible with the premise that a monoclonal proliferation of cells is responsible for the hypercellular elements within the atherosclerotic plaque. The effect of statin therapy on cellular proliferation is variable. Pravastatin has been demonstrated to increase the migration of smooth muscle cells into the intima. The smooth muscle cells localized within the atherosclerotic plaque convert from a contractile to a synthetic phenotype, and thus may contribute to the proliferative cellular aspects of atherosclerosis. However, the role of cellular proliferation is controversial and increased smooth muscle elements may induce plaque stability. Simvastatin has the opposite effect on smooth muscle cell migration and has been demonstrated to reduce the movement and proliferation of smooth muscle cells within the atherosclerotic plaque (2). Additionally, smooth muscle cell apoptosis is enhanced by statins reducing the cellular aspects in the early phases of atherosclerosis (3,4), which may cause intraplaque necrosis and enhance the potential for intracoronary thrombus formation if fissuring or rupture of the fibrous cap allows contact with circulating platelets.

The Inflammatory Hypothesis

The hypothesis that chronic inflammation may play a role in the etiology of atherosclerosis has gained considerable impetus over the past decade, although the concept is not new. Early histologic studies of atherosclerotic lesions demonstrated increased concentration of inflammatory cells such as T-lymphocytes and monocytes within vascular beds associated with plaque rupture. Cytokines, which promote the migration of monocytes into the subendothelial space, are also modulated by the production of a variety of adhesion molecules that localize on the vascular endothelium and bind circulating cellular elements associated with the early stages of atherosclerosis (5).

Epidemiologic studies have correlated a variety of inflammatory markers (e.g., serum amyloid A, myeloperoxidase, and C-reactive protein) with the presence and severity of atherosclerosis. A considerable amount of clinical data concerning the role of C-reactive protein in atherosclerosis has been recently generated. The Physicians Health Study (PHS) evaluated the relationship between C-reactive protein and the risk for developing coronary ischemic disease and cerebrovascular accident (6). The PHS initially evaluated male subjects who were clinically free of atherosclerosis and also tabulated an extensive risk factor profile for the participants. The PHS cohort was generally considered to be at low risk by classical risk factor stratification and demographic data. C-reactive protein was determined and the cohort was subsequently stratified into quartiles. Individuals whose C-reactive protein fell in the highest quartile had a relative doubling of the risk for stroke and a tripling of the relative risk for myocardial infarction. The PHS was controlled for other risk factors and inflammation as measured by C-reactive protein was independent of lipid subfractions, fibrinogen, and the use of tobacco products. Statin therapy has been demonstrated to have a variety of pleiotropic effects that may alter the atherosclerotic process in nontraditional ways. Statins have anti-inflammatory effects and have been demonstrated to lower C-reactive protein, monocyte chemoattractant particle, serum amyloid A, and a number of other markers of inflammation (7). However, the role of anti-inflammatory therapy in acute or chronic coronary disease is controversial. Potent anti-inflammatory agents such as corticosteroids have the potential for reducing wound healing and the possibility of causing myocardial rupture. The Methylprednisolone in Unstable Angina (MUNA) trial addressed the role of anti-inflammatory therapy with a potent steroid (8). Methylprednisolone therapy significantly reduced C-reactive protein, which was employed as the marker of anti-inflammatory potency. However, no benefit to the administration of steroids could be demonstrated on recurrent angina, silent ischemia on Holter monitoring, myocardial infarction, or mortality. A meta-analysis of corticosteroid therapy in acute myocardial infarction evaluated the results from 16 studies involving 3793 subjects to determine whether a significant benefit could be obtained by potent anti-inflammatory interventions (9). The meta-analysis was performed due to the lack of large, adequately powered, randomized controlled trials with corticosteroid therapy in acute myocardial infarction. A possible benefit in mortality was achieved with corticosteroid therapy, although the results were not statistically significant when the analysis was limited to large studies or when only randomized controlled trials were performed. However, a definite adverse effect of corticosteroid therapy was not determined. The role of statins as primary anti-inflammatory agents is being prospectively evaluated in clinical trials.

The Lipid Hypothesis

The lipid hypothesis was first proposed more than 100 years ago and is based on the premise that dyslipidemia is central to the process of atherosclerosis. The role that dyslipidemia plays in coronary heart disease has been established by broadly based epidemiologic, genetic, pathologic, and controlled clinical trials. The epidemiologic data supporting the link between dyslipidemia and atherosclerosis are robust and best exemplified by the Framingham Heart Study. The Framingham data provide long-term epidemiologic data, which correlate dyslipidemia with the prevalence and severity of coronary heart disease, and have been extended into the Framingham Offspring Study (10). The Framingham database is now over 50 years old and has established the continuous and curvilinear relationship between dyslipidemia and coronary heart disease in addition to validating the central role of elevated LDL as a causal factor in atherosclerosis. The Framingham Heart Study is further substantiated by the large-scale, Multiple Risk Factor Intervention Trial which clearly demonstrated a curvilinear relationship between serum cholesterol and coronary heart disease in a cohort consisting of approximately 360,000 male subjects who were initially free of coronary disease (11). While lacking the decades-long evaluation period of the Framingham Heart Study, the six-year age-adjusted death rate for coronary disease demonstrated a continuous gradation that linked cholesterol levels and mortality following correction for the use of tobacco products and hypertension. The Multiple Risk Factor Intervention Trial did not demonstrate the threshold below which an excess risk for dyslipidemia could not be determined, although as total cholesterol fell below 200 mg/dL, the risk relationship became relatively flat. However, approximately 20% of acute myocardial infarctions occur in subjects whose cholesterol levels are below 200 mg/dL, implicating a significant residual risk even in individuals who previously were considered to have normal lipid levels. However, the statistical association between total cholesterol and coronary mortality did not establish that pharmacologic modification would lower vascular morbidity and mortality especially at lower LDL levels. The advent of statin therapy was a major advance in the role of the clinician to modify the risk associated with multiple factors in coronary heart disease, although the potential for a threshold beyond which cholesterol should not be lowered has not been definitely established. Post hoc analysis of the Cholesterol and Recurrent Events (CARE) trial implied no benefit to pravastatin therapy if the initial LDL level was less than 125 mg/dL, which is compatible with the premise that overzealous reduction of LDL may have detrimental effects (12).

STATIN TOXICITY

Toxicity of any pharmacologic agent may be inferred from statistical associations, but the definite documentation of a side effect that has clinical relevance requires large-scale clinical studies that support epidemiologic and experimental data. Statin toxicity may be divided into theoretic problems in which the data is conflicting and definite toxic effects that have been proven in clinical trials. The major potential toxic effects of statins relate to the central nervous system. Prior to the widespread use of statins, ocular toxicity had been a major concern owing to the large amount of experimental data that demonstrated adverse effects with early pharmacologic interventions that lowered serum cholesterol. Central nervous system effects such as alteration of sleep patterns, cognition, dementia, and behavioral changes have been ascribed to statins and remain controversial. Hepatotoxicity and abnormalities of muscle function have been definitely associated with the administration of statins and the potential and definite toxic effects of the drugs will be reviewed.

Central Nervous System Toxicity

Ocular

Exogenous administration of steroids has long been known to cause ocular defects including lens densities and frank cataracts. Pharmacologic agents, which interfere with the production of cholesterol, have the potential to cause ocular toxicity if the pathway is blocked beyond the formation of the basic steroid nucleus. The administration of inhibitors of cholesterol biosynthesis has been demonstrated to produce cataracts in experimental animals, which raised concern about the potential relationship between statin administration and cataract formation. An early pharmacologic agent, U1866A, was a potent inhibitor of cholesterol synthesis late in the production pathway and had been demonstrated to be associated with the risk of cataract formation (13). Experimental studies performed in rodents that were treated with this agent demonstrated a significantly increased risk of cataracts, which was characterized by a reduction in essential sterol formation and phospholipid content within the lens. Demonstration of the ocular toxicity with U1866A raised considerable concern about the relative safety of pharmacologic agents that would decrease the intraocular production of cholesterol. The potential ocular toxicity of the HMG-CoA reductase inhibitors related to lens opacities or frank cataract formation was closely monitored in preclinical and postrelease trials. The mechanism of action of statins is related to their partial inhibitory activity against HMG-CoA reductase, which is the rate-limiting enzyme in cholesterol synthesis. HMG-CoA reductase activity is an early step in the production of cholesterol and occurs well before the formation of the steroid nucleus. Concerns relative to statins and ocular toxicity also involved the potential for tissue penetration and a direct toxic drug effect. Lipophilic statins penetrate the lens tissue more readily and were considered to be associated with an increased potential risk for lens opacities or cataracts relative to the more hydrophilic agents. Early experimental studies were employed, utilizing a model that studied lenses explanted from experimental animals following the administration of relatively high-dose statin therapy (14). Statin administration resulted in marked reductions in circulating serum cholesterol levels, although a definite relationship between cataract formation and a decrease in circulating serum cholesterol levels was not clearly determined. The risk for the formation of cataracts, however, was correlated with the degree of achieved plasma drug levels of the different statins. The kinetics of drug appearance in the aqueous and visual cortex was established, and while the results were not statistically significant, a trend toward the relationship of tissue concentration of reductase inhibitor and a higher incidence of cataracts was demonstrated. The potential implication was that high-dose lipophilic statin administration may increase lenticular damage due to exposure to the drug via the aqueous humor following substantial systemic exposure to the inhibitors of cholesterol synthesis. Cholesterol production is critical in the outer cortical portion of the lens and its reduction may result in the development of opacities when the level of inhibitors of HMG-CoA reductase activity is higher. However, the dose levels of statins employed in these experimental studies were far in excess of the normal therapeutic levels of administration found in humans and suggested that the relative risk of lens opacities or cataracts may be lower with a more physiologic dose level. Pravastatin is a relatively hydrophilic inhibitor of HMG-CoA reductase and does not enter the tissues when compared to more lipophilic agents, such as simvastatin and lovastatin, which results in a 100-fold decrease in inhibition of cholesterol synthesis in the lens. The relative hydrophilicity of pravastatin raised the possibility that this class of statins may have a reduced potential for the generation of cataract or lens opacity when compared to the administration of reductase inhibitors with a high partition coefficient and increased concentration in the lens due to tissue penetration secondary to lipophilicity. However, in human studies, the potential for lens opacities as determined by slit lamp evaluation does not appear to be different when lipophilic or hydrophilic statins are employed, which may be due to the relatively lower doses used in human subjects. The original recommendations for the monitoring of statin safety included repeated slit lamp evaluations to determine the potential for the induction of ocular toxicity and intense scrutiny of patients who were administered statins. The Expanded Clinical Evaluation of Lovastatin (EXCEL) trial was a large-scale clinical study involving 8245 participants (15). The EXCEL study was a double-blind, placebo-controlled evaluation of dyslipidemic patients who were studied in a safety evaluation of lovastatin. Lovastatin was the first inhibitor of HMG-CoA reductase released into the clinical market and was characterized by a relatively high tissue penetration due to its lipophilic characteristics. Lovastatin was compared to placebo at a variety of dosing ranges and the potential for ocular toxicity was rigorously evaluated. Visual acuity assessments and slit lamp evaluations were serially performed over the initial 42-week trial period and during the subsequent extension of the study. Subjects were evaluated utilizing a biomicroscopic examination of the lens, and a standardized classification system was employed to determine the presence of lens opacities. Statistical analysis of the cortical, nuclear, and subcapsular opacities was performed and adjusted for age due to the increasing prevalence of lens opacities in elderly subjects. Additionally, a predrug examination was performed to determine the extent of baseline ocular abnormalities. Lovastatin was compared to placebo and was not demonstrated to be associated with an increased incidence of lens toxicity, alteration of visual acuity, or spontaneously recorded adverse ocular effects. The results of the EXCEL study implied that the lipophilic characteristics of lovastatin and the reduction of serum cholesterol did not result in a secondary effect on ocular toxicity. Simvastatin is more potent than lovastatin, has significant lipophilic characteristics, and also has been studied in dyslipidemic patients who were compared to an age- and gender-matched cohort of subjects with a normal baseline lipid profile (16). The presence of nuclear and cortical lens opacities was determined by baseline ophthalmologic evaluation in both the treatment and the control groups. The administration of simvastatin significantly improved the lipid profile over the two-year trial period and no difference in the incidence of ophthalmologic parameters could be demonstrated relative to placebo. The Oxford Cholesterol Study Group also evaluated simvastatin in 621 dyslipidemic subjects who were free of cataracts at the initiation of the trial. Simvastatin was employed at a variable dosing range of 20 or 40 mg/day. The induction of lens opacities, refractive condition of the eye, and mean ocular pressures were not altered by simvastatin therapy. The Oxford grading system was employed to evaluate measures of cataract formation and no differences were determined between the treatment and the control group. The grading system evaluated posterior subcapsular cataracts, white scatter, and morphologic features of the lens per se. Additionally, Scheimphflug slit lamp imaging and retro-illumination of the analysis of the percentage of cataracts within a defined region of the lens were performed at each visit. The effect of simvastatin therapy on the various ophthalmologic parameters was not different from placebo (17).

Large-scale epidemiologic studies utilizing case–control analysis have also implicated that the administration of statins is not associated with significant ocular toxicity. The United Kingdom General Practice Research Database was analyzed employing a case–control analysis based on initial theoretic concerns and findings based on experimental animals. The primary outcome of the study was the first-time diagnosis of cataract and/or cataract extraction in subjects whose age at baseline ranged from 40 to 79 years. Controls were matched as to age, gender, and duration of evaluation in the database. Subjects who had been administered lipid-lowering agents, which included statins, fibrates, and other compounds, were compared to subjects who had not been exposed to lipid-lowering drugs. The results were stratified by the dose level and exposure duration time. The large number of subjects in the United Kingdom General Practice Research Database provided significant statistical power. The control population consisted of 28,327 individuals who were compared to 7405 cases. The long-term administration of statin therapy was not associated with an increased risk of cataract extraction or incidence of diagnosis. Additionally, the administration of fibric acid derivatives or other lipid-lowering agents also did not appear to be associated with ocular toxicity. However, the possibility of drug interactions was raised in this trial and the coadministration of simvastatin and erythromycin, which share the P450-3A4 enzyme system for drug metabolism, was associated with an increase in cataract risk. The conclusion of this large-scale case–control epidemiologic study was that the isolated administration of statin therapy was not associated with an increased risk for the development of cataracts, although the potential drug interactions should be considered (18).

The United Kingdom General Practice Research Database was also evaluated purely for the administration of statins (19). The study included 15,479 subjects with documented cataracts who were compared to a similar cohort of matched control subjects. The database was controlled for body mass index, smoking status, diabetes, hypertension, and use of medications including aspirin, estrogen, and systemic corticosteroids. Analysis of the database suggested that at the dose range of statins administered in clinical practice, short-to-medium term (mean of 4.5 years) exposure is not associated with an increased risk of cataract formation. The larger database also evaluated the potential for drug interactions with inhibitors of the P450 enzyme system. The administration of calcium channel blockers, cyclosporine, azole antifungal and macrolide antibiotic usage was evaluated and no statistical difference could be demonstrated between the administration of statins and concomitant therapy with drugs with the potential for significant metabolic interaction via the P4503A4 enzyme system. The lack of demonstrable ophthalmologic toxicity with the original inhibitors of cholesterol production (pravastatin, lovastatin, and simvastatin) resulted in the U.S. Food and Drug Administration eliminating the requirement for intermittent ophthalmologic examinations to be performed on a routine basis in subjects receiving HMG-CoA reductase inhibitors.

Sleep

Insomnia is a common problem among the elderly who frequently have coexistent hypercholesterolemia. Anecdotal reports had suggested that the use of statin therapy may increase the prevalence of sleep disturbances. Statins cross the blood–brain barrier at variable rates, which is at least partially related to the relative degree of hydro- or lipophilicity. Due to the potential accumulation of the lipophilic statins within the central nervous system, it was postulated that hydrophilic statins such as pravastatin may have less potential for the induction of central nervous system side effects. Experimental studies performed in normal volunteers compared the central nervous system concentration of lovastatin and pravastatin utilizing pharmacologic levels within the cerebrospinal fluid as a marker for tissue concentration. Cerebrospinal fluid was removed and analyzed for levels of statins following a run-in period of drug administration. Lovastatin, which is relatively lipophilic, was detected at levels within the cerebrospinal fluid at which it was felt to potentially have a physiologic effect on central nervous system function. Conversely, pravastatin, which is hydrophilic, did not enter the cerebrospinal fluid and was felt potentially to be associated with a decreased risk for the induction of sleep disorders (20). The original reports of HMG-CoA reductase inhibitors on sleep disturbances were obtained from uncontrolled studies which were nonblinded and demonstrated conflicting results. The employment of uncontrolled, nonblinded studies on physiologic conditions as complex as sleep is not acceptable for definitive determinations of pharmacologic toxicity. Sleep disorders vary with socioeconomic class, associated medical conditions, age, gender, body mass index, and concomitant pharmacologic therapy. Large-scale pharmacologic studies are required for the determination of a causal effect of a pharmacologic intervention on sleep. The Oxford Clinical Trial Group performed a controlled study of the effect of simvastatin, which is highly lipophilic, on sleep parameters in 621 subjects (21). Volunteers received either 40 mg of simvastatin or 20 mg of simvastatin and their corresponding placebo for an average trial period of 88 weeks. The main outcome measurements were sleep-related problems and was analyzed via the quantification of the usage of sleep-enhancing medications and a questionnaire which evaluated the alteration of sleep duration and events. The use of simvastatin demonstrated the expected reduction of total cholesterol which averaged approximately 68 mg/dL. The study was unique in that it employed graded doses of simvastatin with a placebo control. No adverse effects of either dose of simvastatin were observed on perceived sleeping problems, duration of sleep, or the use of sleep-enhancing medications. Additionally, withdrawal parameters analyzed in the simvastatin and placebo groups were identical. Subjects allocated to simvastatin demonstrated a decreased frequency of waking during the night, although this may have occurred by chance. Furthermore, prolonged follow-up of up to five years reinforced the sleep questionnaire findings with no difference being demonstrable between simvastatin and placebo on any sleep-related problems.

Simvastatin and pravastatin have been compared to placebo in a smaller but more involved study which included electroencephalographic evoked potentials, power spectral analysis, Leeds sleep questionnaire, hospital anxiety depression scale, and digit symbol substitution testing (22). Pravastatin and simvastatin significantly reduced circulating cholesterol levels compared to placebo with a greater effect being demonstrated by the lipophilic agent simvastatin. The greater reduction of circulating cholesterol with simvastatin would imply that if plasma lipid levels determined the degree of induction of sleep disorders, a greater adverse effect would be demonstrable with simvastatin. However, no significant differences in the electroencephalographic evoked potentials could be demonstrated. Additionally, the questionnaire results revealed no difference between statin and placebo. While more subjects reported difficulty in initiating sleep while on simvastatin relative to pravastatin, neither score was significantly different from that of placebo. The study demonstrated that while both statins reduced cholesterol, no significant difference could be demonstrated compared to placebo on evoked potentials, mood, sleep, or cognitive performance in a relatively short-term trial.

While individual subjects may have altered sleep patterns, the body of data in clinical trials does not support a significant difference between cholesterol reduction, statin therapy (whether hydrophilic or lipophilic) and placebo, and other causes of sleep disorders should be sought and evaluated.

Cognition

Multi-infarct dementia and Alzheimer's disease are significant causes of impaired quality of life. Statins had been proposed to potentially alter cognitive function by reducing cerebral intracerebral cholesterol levels. In the United States, dementia is felt to be primarily secondary to Alzheimer's disease, which is a progressive neurodegenerative condition that appears to be increasing in incidence in the United States (23). The primary etiology of Alzheimer's disease is multifactorial and consists of both extraneuronal and intraneuronal abnormalities or a combination of both. Increasing evidence suggests that amyloid beta-peptides accumulate in the brain and initiate a cascade of anatomic and clinical events that account for the characteristics of Alzheimer's disease. The pathoschematic model for Alzheimer's disease was derived from an uncommon genetic form of the disorder, which is characterized by autosomal dominant transmission and is the basis for approximately 2% of all cases. The inherited forms are associated with missense mutations of genes that encode for the production of the amyloid precursor protein (APP) for degradative enzymes with the resultant accumulation of amyloid beta-peptide. However, Alzheimer's disease represents a clinical spectrum and the presenile form differs from late onset disease, which accounts for roughly 90% of all subjects. Alzheimer's disease is thus a heterogenetic disorder with significant differences in pathogenesis, anatomical findings, and risk factors. Alzheimer's disease may originate, at least partially, from cerebrovascular atherosclerosis, which results in progressive loss of brain matter due to progressive microinfarctions (24). The causal relationship between cholesterol and cerebrovascular disease had been controversial, possibly due to the multifactorial nature of stroke (embolic, hemorrhagic, ischemic, etc.). Statins would be unlikely to be associated with the microinfarct form of neurodegenerative disease due to the clear reduction in ischemic stroke demonstrated in clinical trials. The potential mechanisms involved in atherosclerosis and Alzheimer's disease suggest an interaction between these conditions which converge in later life and account for the clinical manifestations (25,26). Risk factors for atherosclerosis and Alzheimer's disease frequently coexist and include the use of tobacco products, inflammation, hypertension, elevated levels of homocysteine, diabetes, and apo E4 polymorphism. Cholesterol and systemic atherosclerosis are clearly linked in a causal pathogenetic relationship, which has been established by epidemiologic, pathologic, genetic, and clinical trials. Increasing evidence has linked the potential interrelationship between dyslipidemia and Alzheimer's disease, which would argue against a toxic effect of statins manifesting as dementia. Increased levels of cholesterol have been demonstrated to promote degeneration of amyloid beta-peptide, which is a dominant pathologic finding in Alzheimer's disease (27). The administration of lovastatin to human embryonic kidney cells that have been transfected with the APP decreases intracellular cholesterol levels by 50% and inhibits beta secretase cleavage of newly synthesized APP (dementia-4a). Additionally, the reduction of the cellular cholesterol level by lovastatin has been demonstrated to reduce the production of beta amyloid which is compatible with the premise that cholesterol is required for the production of this protein and implies a link between cholesterol, beta amyloid, and Alzheimer's disease and may provide a mechanism for protection as opposed to increased risk (25).

While the role of cholesterol lowering and drug therapy in the risk of dementia is controversial, epidemiologic data have linked the use of statins with a decrease in incidence of Alzheimer's disease. The United Kingdom General Practice Research Database has evaluated more than 3,000,000 subjects. Epidemiologic evaluation of a possible link between statins and dementia was performed in a case-controlled study that evaluated 284 cases of documented dementia compared to 1080 controls (28). Subjects above the age of 50 years who received statins appeared to have a 70% lower risk for the development of altered cognitive function, implying a possible protective effect for statin therapy on the subsequent alteration of mental functioning. However, prospective trials have not clearly demonstrated a benefit in cognitive function by statin therapy. The Pravastatin in Elderly Individuals at Risk of Vascular Disease (PROSPER) was a randomized controlled trial in 5804 subjects between the age of 70 and 82 years who were treated with 40 mg of pravastatin (29). A prospective substudy was done which involved testing of mental capabilities in the elderly population utilizing the Mini Mental State Examination. Additionally, the Stroop test for attention and a variety of other learning examinations were performed. Pravastatin did not alter the rate of cognitive decline that occurred at the same rate in both the treatment and the control groups. Significant differences between the two randomized patient subgroups could not be documented in any of the testing parameters, which implied that pravastatin was at least neutral relative to the alteration of mental function. Pravastatin also showed no difference in the reduction of strokes, although there was a significant decline in the number of transient ischemic attacks.

The role of dyslipidemia in cognition is controversial and while epidemiologic studies implicate a potential benefit in the reduction of dementia by statin administration, no clear evidence utilizing prospective clinical trials with statin therapy has demonstrated either a beneficial or adverse effect on the age-related cognitive decline or the risk of dementia.

Violent Behavior

The reduction of cholesterol has been clearly demonstrated to reduce cardiovascular morbidity and mortality. However, epidemiologic studies and the early intervention pharmacologic trials implicated that a low serum cholesterol level may be associated with violent behavior, including suicide and homicide. Epidemiologic evidence, which supports the correlation between low cholesterol level and violent behavior, has been published. Cholesterol measurements were obtained on 79,777 subjects who were enrolled in a health-screening project in Sweden (30). The study had been initiated due to experimental studies that linked low cholesterol level and aggressive behavior in primates. Additionally, meta-analyses had correlated low cholesterol levels and violent behavior in population surveys. Cholesterol levels were correlated with values from police records of those apprehended for violent crimes and adjusted for potential confounding influences. Nested case–control analysis was performed with violent criminals who had two or more arrests for aggressive behavior toward others being defined as subjects, and then compared to non-offenders with control for achieved educational level and alcohol utilization. A significant association was demonstrated between violent behavior and a cholesterol level that fell below the median. The results suggested that even after adjusting for a variety of potential confounding factors, a low cholesterol level was associated with increased subsequent violent behavior.

The Helsinki Heart Study was a trial of primary prevention which was conducted in 4081 male subjects with no clinical evidence of heart disease (31). The administration of gemfibrozil decreased cardiovascular mortality by 19%, which was statistically significant. However, no decrease in total mortality was demonstrable and subsequent analysis demonstrated an increase in suicide and homicide rates that negated the beneficial effect on cardiovascular mortality. Additionally, the Lipid Research Clinics Coronary Primary Prevention Trial (LRC-CPPT) was also a primary prevention trial which employed the utilization of the nonabsorbable bile acid resin, cholestyramine to placebo. The LRC-CPPT study also demonstrated a reduction in cardiovascular mortality but no change in total mortality. The failure to demonstrate decreased total mortality was associated with an increase in violent death, which was predominantly due to motor vehicle accidents (32). The potential relationship between violent behavior and cholesterol lowering in the Helsinki Heart Study could either have been related to lipid lowering or other drug effects as gemfibrozil is absorbed into the circulation. However, the results of the LRC-CPPT trial could not be attributed to a toxic effect of the drug employed, as resins do not enter the central circulation.

The deaths related to violent behavior in the Helsinki Heart Study and the LRC-CPPT trial were analyzed for the possibility of a causal relationship between violent behavior and cholesterol lowering (33). Suicides accounted for eight deaths and accidents were the cause of death in 10 subjects in these two trials. Additionally, two homicides were demonstrable in these studies. The 10 deaths due to accidents were analyzed for associated conditions and it was determined that two were dropouts from the trial and were not receiving medical therapy, while three others had high circulating blood ethanol levels which was detectable at the time of autopsy. Additionally, another three of the accidental deaths occurred in subjects who were known to have psychiatric disorders prior to entry within the trial. The deaths recorded as due to homicides were victims rather than offenders and one individual had discontinued therapy due to a myocardial infarction 12 months prior to his death. The trial was analyzed on an intent-to-treat basis, and thus these deaths were included in the therapy group. The individuals who committed suicide were characterized by five of the eight subjects having dropped out of the trial and not being on cholesterol-lowering medications for periods of months or years prior to committing suicide. The analysis of the deaths in these two trials revealed that when risk factors for these accidental deaths (e.g., pre-existent psychiatric disorder, alcohol intoxication, and actual use of lipid-lowering therapy) are considered, the clinical trial evidence to support that either cholesterol-lowering drugs or a reduction in serum cholesterol is a causal effect due to homicide, suicides, or accidents is difficult to support.

The results of the epidemiologic and early clinical trial data generated a hypothesis concerning low cholesterol and violent behavior, which was proposed to be related to deficient central nervous system serotonergic activity (34). Serotonin metabolism within the central nervous system is difficult to study in humans. However, experimental data from nonhuman primates demonstrated that animals which consumed a low cholesterol diet were more aggressive, less affiliative, and had lower cerebrospinal fluid concentrations of serotonin metabolites which was compatible with the concept that a reduced intake of dietary lipids may alter brain neural chemistry and behavior, and provided a potential explanation for suicide and violence-related deaths in cholesterol-lowering trials (35).

While the data from epidemiologic and experimental studies provide potential associations, a causal effect requires the prospective performance of placebo-controlled trials with predescribed end points. The two largest cholesterol trials were the Heart Protection Study (HPS) with simvastatin and the Pravastatin Pooling Project (PPP), which compiled the results of all of the major clinical trials utilizing pravastatin in a meta-analysis. The HPS enrolled 20,536 high-risk individuals whose cholesterol level at the time of randomization was above 140 mg/dL (36). Simvastatin reduced all-cause mortality by 12.9% and there was no increase in violent deaths in the group randomized to pharmacologic therapy. The PPP evaluated the role of pravastatin therapy in a number of trials which encompassed greater than 112,000 patient years of drug exposure in a double-blind, randomized trial (37). Data analysis did not reveal any category of noncardiovascular death in which the proportion of morbid events was different between the pravastatin and placebo-assigned groups. Thus, despite the implication by epidemiologic and experimental studies, altered central nervous system side effects characterized by violent behavior due to decreased intracerebral cholesterol levels, alteration of serotonin metabolism, or primary drug effect cannot be substantiated by double-blind placebo-controlled trials with statins.

Hepatoxicity

Statin therapy has the potential for altering liver function tests due to the hepatic metabolism of these drugs, which occurs primarily (with the exception of pravastatin) via the cytochrome P450 enzyme system. Drugs that undergo hepatic metabolism may alter circulating levels of hepatic enzymes or induce cholestatic changes within the liver. The effects of statin therapy on hepatic function have been evaluated in experimental studies using high nonphysiologic dosing levels and in clinical observation performed in controlled trials. Studies on guinea pigs that were administered simvastatin at a dose of 125 mg/kg/day demonstrated a significant hepatotoxic effect. However, the maximum dose for simvastatin in humans is roughly 1 mg/kg/day (38). Thus, this type of experimental study utilized a different species and employed nonphysiologic dosing of statin therapy. Hepatocellular necrosis was induced in some animals and biliary duct proliferation could also be demonstrated. The administration of the pharmacologic dose of simvastatin was associated with a 10-fold elevation in the hepatic enzymes alanine amino transferase (ALT) and aspartate amino transferase (AST) activity. The hepatotoxic effect of simvastatin is variable in different species and may be more pronounced in the guinea pig due to the intrinsic low basal level of HMG-CoA reductase activity and the prolonged inhibition of mevalonate synthesis secondary to marked inhibition of HMG-CoA reductase enzyme by simvastatin.

Lovastatin was the original inhibitor of HMG-CoA reductase activity and its hepatic effect has recently been reviewed (39). The hepatic toxicity of statins induced in experimental animals was documented early in the experimental trials and resulted in the Food and Drug Administration issuing labeling requirements for the monitoring of serum liver enzymes. However, following the release of lovastatin, it was realized that significant hepatotoxicity, defined as the presence of transaminitis, was relatively uncommon and, in the case of lovastatin, ranged from 2.6% to 5% when doses of 20 and 80 mg/day, respectively, were employed. The definition of hepatotoxicity was an increase in ALT levels of three times the upper limits of normal. The EXCEL study was a large-scale trial involving 8245 subjects with dyslipidemia (40). Lovastatin was compared to placebo in a randomized fashion while both groups received dietary therapy. Lovastatin was administered in 20, 40, and 80 mg doses for a trial duration of 48 weeks. Subjects with preexisting alteration of liver functional abnormalities were excluded from the trial and enzyme determinations were obtained every six weeks during the trial. The risk of developing an elevation of liver enzymes greater than three times the upper limits of normal was dose related and ranged from 0.1% at 20 mg/day to 1.9% at 80 mg/day. The placebo group had an overall risk of liver function abnormalities of 0.1%, indicating that at the lowest levels of lovastatin dosing the risk for the development of liver function problems was essentially as that of placebo.

The HPS also monitored subjects for liver function abnormalities (37). The HPS enrolled 20,536 subjects with a wide range of lipid values. The lowest cholesterol level which allowed randomization in the trial was 140 mg/dL. Thus, a large number of individuals who entered the trial with relatively low cholesterol levels were randomized to receive simvastatin. Hepatotoxicity was defined as an enzyme elevation four times the upper limit of normal. The concentration of ALT was measured at each follow-up visit, even if the participants were no longer continuing their study medication. Despite the large numbers of individuals enrolled in the trial, the number of elevations of ALT was minimal and there was no significant excess among those whose original allocation was to simvastatin. In such cases, the study medication was generally continued and another sample was collected within three weeks. Persistent elevation of ALT was found only rarely. In the HPS, a total of nine subjects with simvastatin compared to four patients who received placebo were found to have persistent ALT elevations greater than four times the upper limits of normal. On a percentage basis, the incidence of this degree of liver function abnormalities was 0.09% with simvastatin and 0.04% with placebo, which was not statistically significant. Additionally, there were no significant differences between the groups in the number of subjects whose study treatment was stopped due to elevation of liver enzymes (48 in the simvastatin group vs. 35 in the placebo group).

The PPP also monitored the incidence of elevations of liver enzymes in the major pravastatin trials which evaluated 9185 subjects (38). The incidence of liver function abnormalities, which exceeded 1.5 times the upper limits of normal, occurred in 8.8% of subjects who received pravastatin compared to 8.2% of individuals who were allocated to placebo, which was not statistically significant. Subjects in the PPP were also evaluated for the incidence of liver function abnormalities exceeding three, five, seven, and nine times the upper limits of normal. In no case was the incidence of liver function abnormalities significantly increased in the pravastatin group relative to placebo, indicating a significant safety profile for induced liver function problems with pravastatin. The quantitative role of the hydrophilic nature of pravastatin and the metabolism utilizing a non-P450 cytochrome pathway cannot be determined but does indicate a potential mechanism for the lack of liver function abnormalities induced by pravastatin.

The safety of atorvastatin has been analyzed in 44 trials which involved 9416 subjects (41). Persistent clinically relevant elevations in ALT and AST levels occurred in 0.5% of subjects who received atorvastatin at any dose when compared to a 0.3% incidence in the placebo groups. Approximately half of the individuals with induced liver function abnormalities continued their treatment and less than 1% experienced persistent elevations of transaminase levels. The risk of the development of liver abnormalities with atorvastatin is dose related and ranges from 0.2% at 10 mg/day to2.3% at 80 mg/day. Dose reduction or discontinuation of therapy resulted in a return of the enzyme levels to normal, indicating a lack of persistent liver toxicity.

The relative risk of hepatic toxicity in statin therapy has also been analyzed in a recent large-scale analysis involving 49,275 subjects who participated in 13 clinical trials (42). The meta-analysis demonstrated that the proportion of subjects who had definite liver function abnormalities was low in both treated and control populations. The overall risk of statin-associated hepatic toxicity was 1.14% as compared to the placebo group, which was 1.05% and did not reach statistical significance.

The role of screening for hepatotoxicity due to statin therapy is controversial. Statins may induce a degree of elevation of liver enzymes, which has been termed transaminitis to describe hepatic enzyme leakage without a clinically apparent hepatotoxic effect (43). The etiology of the enzyme leak is often unclear and may be related to fatty infiltration of the liver in addition to the potential alteration of hepatocellular membranes due to lipid lowering or direct drug toxicity. ALT levels are considered to be more specific for hepatotoxicity and are recognized as the standard. Extensive long-term trials that have occurred over 15 years have not demonstrated clinically significantly relevant hepatotoxicity and irreversible liver damage with statins appears to be extremely rare. The potential for relative safety of one statin over another is not clear, although it does appear that pravastatin may have pharmacologic features that render the risk of enzyme elevations to be reduced. However, the two most potent lipophilic statins (simvastatin and atorvastatin) have been utilized at high dose in clinical trials without evidence of permanent liver function abnormalities. A risk–benefit relationship must be established for the individual who would be potentially receiving statins, and in relatively high-risk subjects intermittent monitoring of ALT levels is recommended, although other conditions should be evaluated (toxins, fatty liver, viral hepatitis, etc.).

Muscle Toxicity

Myotoxicity was among the first major adverse drug reactions associated with statin therapy. The clinical spectrum of myotoxicity associated with statins is broad and ranges from a nonspecific syndrome primarily characterized by myalgias without definite abnormal physical signs or routine biochemical markers to life-threatening rhabdomyolysis.

Statins and Myotoxicity

Statins have proven to be safe and effective agents to improve dyslipidemia. Muscle toxicity was one of the first major adverse effects associated with the use of statin therapy and exists in a spectrum of disease characterized by mild myalgias without elevations of plasma creatine kinase (CK), myalgias plus enzymatic evidence of tissue damage, and frank rhabdomyolysis. The statin-associated myopathy with normal CK levels has recently been evaluated on a histopathologic basis (44). Statins had been proposed to be associated with a direct myotoxic effect which was below the threshold level required for the release of muscle-specific enzymes into the circulation, although clinical correlation with detailed histologic evidence was scanty. A pathologic study evaluated four patients enrolled in clinical trials who developed symptoms compatible with clinical myopathy during statin therapy despite normal CK levels. The patients underwent serial muscular strength evaluation with hip flexion and abduction measurements and percutaneous muscle biopsies. Patients who utilized macrolide antibiotics, cyclosporine, fluconazole, fibrates, alcohol, or grapefruit juice were excluded. The trial, albeit small, fulfilled accepted criteria for an adverse drug reaction. The muscle symptoms initially occurred following the institution of double-blind drug administration and normalized after conversion to placebo. Rechallenge with statins resulted in recurrent symptoms and characteristic pathologic and biochemical findings were noted on muscle biopsy. The histologic changes also reverted to normal following discontinuation of therapy. Muscle histopathology revealed accentuated lipid droplet accumulation and cytochrome oxidase–negative fibers that were consistent with a statin-related myopathy. The pathologists were not blinded to treatment status of the patient at the time of obtaining the percutaneous muscle sample. Plasma statin levels were obtained and were within the normal range, which suggested an undefined metabolic vulnerability to HMG-CoA reductase inhibition. CK levels remained normal before and during the trial. The study was not designed to address the precise mechanism involved in statin-related myopathy. However, normal levels of CK obtained during the initiation phase of statin therapy does not exclude underlying myopathic disease and the potential for long-term consequences, if not clinically recognized. The histologic features are compatible with a defect in mitochondrial respiratory chain function, which will require further evaluation. The accompanying editorial noted that despite the small number of patients and lack of appropriate controls, the trial represents a novel approach to the delineation of statin-associated myopathy with normal CK due to the utilization of quantitative measures of muscle strength and muscle biopsies which confirmed the clinical symptomatology (45). The prevalence and the clinical sequelae of statin-associated myopathy with normal CK levels are not known, but this important observation may provide insight into the future elaboration of the underlying mechanism and provide a means for patient selection.

Statin myotoxicity with symptoms and elevated CK levels also has an obscure mechanism, although genetic enzymatic defects, intrinsic pharmacologic properties, and potential interactions with coadministered drugs have been implicated. The relative lipophilic characteristics of statins determine the degree of tissue penetration and have the theoretic potential to increase myocardial toxicity if high intracellular levels are a significant determinant of the degree of induced myocyte damage (46). The partial inhibition of the rate-limiting enzyme in cholesterol synthesis by statins also is associated with a secondary intracellular depletion of a variety of metabolic intermediates that are normally generated within the cell. The depletion of these key metabolic intermediates, such as ubiquinone (coenzyme Q), farnesol, mevalonate, and geranylgeraniol, has been postulated to reduce the capacity for posttranslational modification of a variety of regulatory proteins. Pravastatin, which is relatively hydrophilic, has been demonstrated to penetrate striated smooth muscle cells poorly and cause minimal effects on intracellular cholesterol levels and metabolic intermediates. Interest has accumulated concerning the potential role of a reduction in ubiquinone or coenzyme Q as a cause of statin-mediated myopathies (47). Ubiquinone is a redox link between flavoproteins and the cytochrome system and is required for cellular energy production. Coenzyme Q is widely distributed in skeletal muscle and may also play a role in the stabilization of the myocyte cellular membrane. Decreased levels of coenzyme Q have been postulated to be associated with significant reductions in mitochondrial energy production and have the potential to contribute to myotoxicity. Gemfibrozil has also been demonstrated to reduce coenzyme Q in individuals with hyperlipidemia, which potentially may play a role in the statin–fibrate drug interaction. However, the definitive role that the intracellular depletion of coenzyme Q or other metabolic intermediates play in myotoxicity has not been clarified and remains controversial due to conflicting reports in the literature. Simvastatin, which is relatively lipophilic, has been administered to dyslipidemic subjects and has been demonstrated to reduce circulating plasma levels of coenzyme Q, but was paradoxically associated with an increase in muscle levels in short-term therapy (48). Pravastatin and atorvastatin, which differ considerably in tissue penetration, did not alter plasma levels of coenzyme Q in a direct comparative study, despite significant reductions in circulating cholesterol levels (49). Additionally, a recent analysis of the simvastatin megatrials revealed an overall incidence of definite muscle toxicity in 0.025% of the subjects, which indicates a low clinical risk despite a relatively high dosing regimen (50). All statins have been associated with rhabdomyolysis irrespective of lipophilic characteristics and the role of the intrinsic pharmacologic properties of the statins has not been elucidated (51).

Drug interactions between statins and other pharmacologic agents that also utilize the cytochrome P450 enzyme system have been recognized as a predisposing factor for statin-induced myopathy (52). The cytochrome P450 system is a ubiquitous group of related enzymes that oxidatively modifies drugs and results in a secondary conversion of drugs to water-soluble metabolites to facilitate renal excretion. The cytochrome 3A4 isoform accounts for the metabolism of approximately 50% of all commonly used drugs (53). Simvastatin, atorvastatin, and lovastatin are metabolized by the cytochrome P450 3A4 isoform. Fluvastatin is predominately metabolized by the 2C9 isoform, while cerivastatin has a dual mechanism of excretion, which utilizes both the 2C8 and the 3A4 pathway. The dual pathways of excretion for cerivastatin were postulated to increase the safety profile of this agent. Pravastatin is the only statin that does not utilize the cytochrome P450 system for drug metabolism and disposal. Pravastatin is thus unique among the statins in that it is secreted either in the bile or by renal mechanisms following the formation of a 3-alphahydroxy isomeric metabolite.

The potential for muscle toxicity with statins is significantly increased if coadministered with a drug that inhibits the activity of the specific P450 enzyme system that is involved in the metabolic pathway of the drug (54). Additionally, coadministration of drugs that share the same P450 enzyme system may result in high circulating levels due to altered metabolism of the compounds. However, despite the increase in the amount of data relative to pharmacokinetics, pharmacodynamics, and genetics, the mechanism by which statins cause myopathy has not been precisely delineated, and the overall incidence and risk quantification have relied on clinical observation in safety or efficacy trials and postrelease surveillance studies.

The interaction between statins and fibric acid derivatives has received intense scrutiny due to the increasing utilization of combination drug therapy to optimize lipid profiles in patients with complex phenotypes. Mixed dyslipidemia is common in diabetes, obesity, and hypertension and individual lipid subfractions may not be optimized with pharmacologic monotherapy or lifestyle modifications. The metabolism of the fibric acid derivatives is complex and the precise pathways involved in drug metabolism have not been clearly delineated. The fibrates have been postulated to utilize cytochrome P450 3A4 pathway (55). However, this is not universally accepted and the exact mechanisms involved in fibrate metabolism remains controversial. The risk of rhabdomyolysis utilizing combination statin and fibrate therapy varies with the population and ranges from 1% to 5%. The cause is multifactorial and probably related to a non–cytochrome P450 mechanism (56). The interaction may be due to pharmacodynamic considerations rather than a predominantly pharmacokinetic interaction. Fibric acid derivatives may alter hepatic function and result in reduced clearance of orally administered statins from the portal circulation with a resultant secondary increase in plasma concentrations. The combination of statin and fibrate therapy is feasible in select patients but is associated with increased potential for significant toxicity (57).

Rhabdomyolysis

Rhabdomyolysis is a life-threatening clinical syndrome with a number of potential precipitating causes including trauma, infection, toxins, genetically mediated enzyme deficiencies, and medications (58). Irrespective of etiology, the clinical manifestations of rhabdomyolysis share a final common pathway, which is the result of widespread myocyte sarcolemmal membrane destruction and resultant massive cell lysis. Myonecrosis is characterized by the release of a variety of intracellular constituents and enzymes into the circulation. The transmigration of myoglobin into the plasma compartment with subsequent renal clearance results in marked elevations of this protein in the glomerular filtrate and significantly increases the potential for tubular obstruction. Rhabdomyolysis is also characterized by diffuse intrarenal vasoconstriction, which is at least partially secondary to inhibition of nitric oxide synthase in the endothelial cells with resultant ischemic damage and the potential for acute renal failure. CK levels in excess of 5000 IU are associated with a greater than 50% risk for the development of acute renal failure. Rapid reduction of renal flow is associated with experimental rhabdomyolysis, which may reduce the effective arterial volume and activate the sympathetic nervous system and the renin angiotensin system, resulting in a vicious cycle of progressive vasoconstriction and organ ischemia.

The induced myocyte injury is accompanied by an influx of sodium and calcium from the extracellular fluid into the cytoplasm. Intracellular electrolytes and enzymes are also released into the plasma compartment and may cause hyperkalemia with an increased risk for complex cardiac arrhythmias.

Lipid peroxidation may also play a role in the tubular necrosis associated with rhabdomyolysis (59). The heme oxygenase enzyme system catalyzes the breakdown of heme into iron, carbon monoxide, and biliverdin. The renal failure that occurs in rhabdomyolysis and myoglobinuria is associated with an upregulation of the heme oxygenase enzyme system and facilitates the liberation of free iron. Ferric iron accumulation centers are localized within myoglobin and react with lipid hydroperoxides to form lipid peroxide radicals. Endogenous free radical scavenging compounds are consumed, and free radical mediated injury may play a role in the progression of renal tubular disease. Desferoxamine, which chelates and sequesters free iron, has been demonstrated in experimental models to protect the integrity of the nephron in induced rhabdomyolysis.

Special Case of Cerivastatin

The synthetic HMG-CoA reductase inhibitor, cerivastatin, was specifically designed to be a highly potent cholesterol-lowering agent that could be administered in microgram quantities. Cerivastatin is a pure enantiomeric HMG-CoA reductase inhibitor that had been evaluated in low dose (0.2–0.4 mg) clinical efficacy and safety studies since 1993 (60). Cerivastatin has a dual mechanism of excretion, which had been advocated as providing the basis for a low propensity for drug interactions. Cerivastatin is characterized by total absorption after oral administration and also exhibits moderate first-pass hepatic metabolism. The cytochrome P450 3A4 and 2C8 isozymes exclusively metabolize cerivastatin and the resultant breakdown products are cleared by both renal and biliary excretion (61). The early clinical safety trials did not demonstrate clinically significant drug interactions utilizing doses up to 0.4 mg/day. However, while the levels of cyclosporine, erythromycin, and itraconazole were demonstrated to be increased by coadministration of cerivastatin, the initial safety studies did not demonstrate an apparent increased clinical risk for rhabdomyolysis. Surveillance studies conducted following the release of cerivastatin documented an increased risk of severe muscle toxicity and rhabdomyolysis, especially when cerivastatin was administered either at a high initial (0.8 mg) dose as monotherapy or in combination with gemfibrozil. Despite a specific warning from the U.S. Food and Drug Administration concerning the danger of myotoxicity with the combination of gemfibrozil and cerivastatin, mortality continued to be reported and a total of 52 deaths were documented worldwide. The cerivastatin-associated mortality subsequently resulted in the removal of cerivastatin from the marketplace. The United States accounted for 31 fatalities due to rhabdomyolysis (62). In addition, 385 nonfatal cases were reported among the estimated 700,000 patients who received cerivastatin. Interestingly, a long-term efficacy and safety study that analyzed the effectiveness of the 0.8 mg dose of cerivastatin was published following withdrawal of the drug from the marketplace (63). Cerivastatin was administered to 1170 patients over a one-year period in a placebo-controlled trial which utilized dosing ranges from 0.4 to 0.8 mg/day. Pravastatin was substituted for placebo in the control group following an eight-week trial duration. Chemical and clinical evidence of potential myotoxicity were prospectively analyzed as a safety endpoint. CK elevations and symptomatology were divided into three categories: CK greater than 5–10 times the upper limits of normal without symptoms, CK greater than 10 times the upper limits of normal without symptoms, and CK greater than 10 times the upper limits of normal with symptoms. During the one-year trial period, subjects who were randomized to receive the initial eight weeks of placebo followed by pravastatin demonstrated no chemical evidence of myositis. Conversely, eight patients who received 0.8 mg/day of cerivastatin for 52 weeks had a CK elevation greater than 10 times the upper limit of normal associated with symptoms. Additionally, asymptomatic CK elevations 10 times the upper limit of normal occurred in one of the patients in the placebo/pravastatin group compared to 16 patients who received 0.8 mg of cerivastatin/day. The authors concluded that the long-term administration of cerivastatin at a dose level of 0.8 mg/day effectively and safely achieved National Cholesterol Education Goals. However, it would appear that a trend for myotoxicity as manifested by elevated CK and symptoms was associated with high dose cerivastatin administration despite the lack of clinical deterioration to frank rhabdomyolysis.

The case fatality rates associated with cerivastatin have prompted the European Medicine Evaluation Agency to undertake a comprehensive review of statins. Clinical risks and benefits will be correlated in an attempt to establish the risk–benefit role of these agents in people in primary or secondary prevention.

Myotoxicity in the Statin Megatrials

The large database accumulated in the primary and secondary prevention trials with first and second generation statins allow quantification of the risk of rhabdomyolysis associated with statin therapy and formulation of a clinical perspective. The first generation statins (lovastatin, pravastatin, and simvastatin) are structurally similar and are derived as metabolites from fungal cultures. The three original statins have now been evaluated in clinical trials that enrolled more than 50,000 patients in whom the occurrence of myotoxicity was monitored as a safety parameter. The largest experience is with simvastatin and pravastatin. The PPP accumulated a total of more than 112,000 patient years of monitored drug exposure by analyzing the CARE, West of Scotland Coronary Prevention Study, and the Long-Term Intervention with Pravastatin in Ischaemic Disease (LIPID) trials (38). In the combined analysis of these placebo-controlled trials, which analyzed more than 19,000 patients, a total of three subjects who received pravastatin were withdrawn for myositis associated with a CK level in excess of 10 times the upper limit of normal as compared to seven control patients. Rhabdomyolysis was not documented in either the pravastatin or control groups. The Scandinavian Simvastatin Survival Study (4S) randomized 4444 subjects and was the first large-scale statin trial to demonstrate a reduction in total mortality in concert with a significant reduction in cardiovascular end points (64). CK levels were determined in the 4S trial every six months on a routine basis and an increase of 10 times above the upper limits of normal occurred in six simvastatin patients, although the enzyme elevation was not associated with a symptom complex compatible with rhabdomyolysis. Rhabdomyolysis, which was associated with an increase in CK combined with a compatible clinical syndrome, occurred in one subject who received simvastatin and was reversible following discontinuation of the drug. Rhabdomyolysis did not occur in the placebo group, although one patient was withdrawn for asymptomatic but significant elevations of CK. The Air Force/Texas Coronary Atherosclerosis Prevention Study was a primary prevention trial which evaluated a relatively low risk cohort consisting of 6605 men and women (65). CK levels were monitored during the study and the incidence of CK elevation in excess of 10 times the upper limit of normal was identical in the lovastatin and placebo groups (0.6%). Rhabdomyolysis occurred in two patients who received placebo and one patient who was randomized to receive lovastatin. Interestingly, the single case of rhabdomyolysis attributed to statin therapy occurred when the subject was not taking active therapy but was included in the lovastatin group for an intention-to-treat analysis. The Medical Research Council/British Heart Foundation analyzed the effect of simvastatin with or without antioxidants in a two-by-two factorial analysis of 20,536 high-risk individuals (37). The potential for myotoxicity was analyzed on a prospective basis. The HPS documented that 6% of the participants reported unexplained muscle pain or weakness at each of the scheduled outpatient visits. However, no significant difference in muscular symptoms was determined between the treatment and control groups. During the entire trial, myalgias were reported by 32.9% of individuals who were randomized to receive simvastatin as opposed to 33.2% of individuals who received placebo. Simvastatin therapy was discontinued in 49 subjects due to muscle toxicity (0.5%) versus 50 individuals who were in the placebo group (0.5%). Simvastatin therapy was associated with a slight and not statistically significant increase in individuals who had elevated CK levels more than 10 times the upper limit associated with symptoms.

The Greek Atorvastatin and Coronary Heart Disease Evaluation (GREACE) Study was a secondary prevention trial that assessed the effect of atorvastatin in 1600 consecutive patients with established coronary heart disease over a two-year period (66). The previous major secondary prevention trials (4S, CARE, and LIPID) had clearly established the benefits of cholesterol lowering in secondary prevention. Thus, a placebo group was not included due to ethical reasons and atorvastatin therapy was compared to a control group who received usual care. A cholesterol goal of less than 100 mg/dL was established and atorvastatin therapy could be uptitrated from an initial dose of 10 to 80 mg/day. The mean dose of atorvastatin utilized in the GREACE trial was 24 mg/day. The usual care group received hypolipidemic therapy in 14% of the cases and only 3% reached the National Cholesterol Education Program LDL goal. CK levels were measured at 6, 12, and 24 weeks and thereafter every six months. The GREACE trial did not report myopathy defined as myalgia in the presence of CK levels 5 to 10 times the upper limits of normal or myalgia without CK elevation.

Muscle toxicity was one of the first major adverse side effects correlated with statin therapy. Rhabdomyolysis has occurred with all members of the statin class, although statin monotherapy is associated with a low (less than 0.1%) incidence rate. The risk for myotoxicity may be increased in combination with agents that either compete for metabolism with or inhibit the cytochrome P450 enzyme system. Pravastatin is unique among HMG-CoA reductase inhibitors in that it does not utilize the P450 enzyme system for metabolic processing and excretion. Further basic research which clarifies the mechanisms involved in statin-associated rhabdomyolysis is necessary and clarification of the implications of genetic, pharmacodynamic, pharmacokinetic, and structure–function relationships of these agents should be encouraged in light of the potential for statin usage to increase from 13 to 36 million recipients in the United States, if the recommendations of the Adult Treatment Panel III of the National Cholesterol Education Program are implemented.

REFERENCES

1. Benditt EP, Benditt JM. Evidence for a monoclonal origin of human atherosclerotic plaque. Proc Natl Acad Sci USA 1973; 70(6):1753–1756.
2. Tsiara S, Elisaf M, Mikhailidis DP. Early vascular benefits of statin therapy. Curr Med Res Opin 2003; 19(6):540–556.
3. Bellosta S, Bernini F, Ferri N, et al. Direct vascular effects of HMG-CoA reductase inhibitors. Atherosclerosis 1998; 137(suppl):S101–S109.
4. Erl W, Hristov M, Neureuter M, Yan ZQ, Hansson GK, Weber PC. HMG-CoA reductase inhibitors induce apoptosis in neointima-derived vascular smooth muscle cells. Atherosclerosis 2003; 169(2):251–258.
5. Ikeda U. Inflammation and coronary artery disease. Curr Vasc Pharmacol 2003; 1(1): 65–70.
6. Ridker PM, Cushman M, Stampfer MJ, Tracy RP, Hennekens CH. Inflammation, aspirin, and the risk of cardiovascular disease in apparently healthy men. N Engl J Med 1997; 336(14):973–979.
7. Jialal I, Stein D, Balis D, Grundy SM, Adams-Huet B, Devaraj S. Effect of hydroxymethyl glutaryl coenzyme a reductase inhibitor therapy on high sensitive C-reactive protein levels. Circulation 2001; 103(15):1933–1935.

8. Azar RR, Rinfret S, Theroux P, et al. A randomized placebo-controlled trial to assess the efficacy of anti-inflammatory therapy with methylprednisolone in unstable angina (MUNA trial). Eur Heart J 2000; 21(24):2026–2032.

9. Giugliano GR, Giugliano RP, Gibson CM, Kuntz RE. Meta-analysis of corticosteroid treatment in acute myocardial infarction. Am J Cardiol 2003; 91(9):1055–1059.

10. Kannel WB. Range of serum cholesterol values in the population developing coronary artery disease. Am J Cardiol 1995; 76(9):69C–77C.

11. Stamler J, Wentworth D, Neaton JD. Is relationship between serum cholesterol and risk of premature death from heart disease continuous and graded? Findings in 356, 222 primary screenees of the Multiple Risk Factor Intervention Trial (MRFIT). JAMA 1986; 256(20):2823–2828.

12. Sacks FM, Pfeffer MA, Moye LA, et al. The effect of pravastatin on coronary events after myocardial infarction in patients with average cholesterol levels. Cholesterol and recurrent events trial investigators. N Engl J Med 1996; 335(14):1001–1009.

13. Cenedella RJ. Source of cholesterol for the ocular lens, studied with U18666A: cataract-producing inhibitor of lipid metabolism. Exp Eye Res 1983; 37(1):33–43.

14. Gerson RJ, MacDonald JS, Alberts AW, et al. On the etiology of subcapsular lenticular opacities produced in dogs receiving HMG-CoA reductase inhibitors. Exp Eye Res 1990; 50(1): 65–78.

15. Laties AM, Shear CL, Lippa EA, et al. Expanded clinical evaluation of lovastatin (EXCEL) study results: assessment of the human lens after 48 weeks of treatment with lovastatin. Am J Cardiol 1991; 67(6):447–453.

16. Lundh BL, Nilsson SE. Lens changes in matched normals and hyperlipidemic patients treated with simvastatin for 2 years. Acta Ophthalmol (Copenh) 1990; 68(6):658–660.

17. Harris ML, Bron AJ, Brown NA, et al. Br J Ophthalmol 1995; 79(11):996–1002.

18. Schlienger RG, Haefeli WE, Jick H, Meier CR. Risk of cataract in patients treated with statins. Arch Intern Med 2001; 161(16):2021–2026.

19. Smeeth L, Hubbard R, Fletcher AE. Cataract and the use of statins: a case-control study. Q J Med 2003; 96:337–343.

20. Botti RE, Triscari J, Pan HY, Zayat J. Concentrations of pravastatin and Lovastatin in cerebrospinal fluid in healthy subjects. Clin Neuropharmacol 1991; 14(3):256–261.

21. Keech AC, Armitage JM, Wallendszus KR, et al. Oxford Cholesterol Study Group. Absence of effects of prolonged simvastatin therapy on nocturnal sleep in a large randomized placebo-controlled study. Br J Clin Pharmacol 1996; 42(4):483–490.

22. Harrison RW, Ashton CH. Do cholesterol-lowering agents affect brain activity? A comparison of simvastatin, pravastatin, and placebo in healthy volunteers. Br J Clin Pharmacol 1994; 37(3):231–236.

23. Minino AM, Arias E, Kochanek KD, Murphy SL, Smith BL. Deaths: final data for 2000. Natl Vital Stat Rep 2002; 50(15):1–119.

24. Vermeer SE, Prins ND, den Heijer T, Hofman A, Koudstaal PJ, Breteler MM. Silent brain infarcts and the risk of dementia and cognitive decline. N Engl J Med 2003; 48(13):1215–1222.

25. Frears ER, Stephens DJ, Walters CE, Davies H, Austen BM. The role of cholesterol in the biosynthesis of beta-amyloid. Neuroreport 1999; 10(8):1699–1705.

26. Casserly I, Topol E. Convergence of atherosclerosis and Alzheimer's disease: inflammation, cholesterol, and misfolded proteins. Lancet 2004; 363(9415):1139–1146.

27. Simons M, Keller P, De Strooper B, Beyreuther K, Dotti CG, Simons K. Cholesterol depletion inhibits the generation of beta-amyloid in hippocampal neurons. Proc Natl Acad Sci USA 1998; 95(11):6460–6464.

28. Jick H, Zornberg GL, Jick SS, Seshadri S, Drachman DA. Statins and the risk of dementia. Lancet 2000; 356:1627–1631.

29. Shepherd J, Blauw GJ, Murphy MB, et al. Pravastatin in elderly individuals at risk of vascular disease (PROSPER): a randomized controlled trial. Lancet 2002; 360:1623–1630.

30. Golomb BA, Stattin H, Mednick S. Low cholesterol and violent crime. J Psychiatr Res 2000; 34(4–5):301–309.

31. Frick MH, Elo O, Haapa K, et al. Helsinki Heart Study: primary-prevention trial with gemfibrozil in middle-aged men with dyslipidemia. Safety of treatment, changes in risk factors, and incidence of coronary heart disease. N Engl J Med 1987; 317(20):1237–1245.

32. The Lipid Research Clinics Coronary Primary Prevention Trial results. I. Reduction in incidence of coronary heart disease. JAMA 1984; 251(3):351–364.

33. Wysowski DK, Gross TP. Deaths due to accidents and violence in two recent trials of cholesterol-lowering drugs. Arch Intern Med 1990; 150(10):2169–2172.

34. Kaplan JR, Shively CA, Fontenot MB, et al. Demonstration of an association among dietary cholesterol, central serotonergic activity, and social behavior in monkeys. Psychosom Med 1994; 56(6):479–484.

35. Kaplan JR, Muldoon MF, Manuck SB, Mann JJ. Assessing the observed relationship between low cholesterol and violence-related mortality. Implications for suicide risk. Ann NY Acad Sci 1997; 836:57–80.

36. Heart Protection Study Collaborative Group. MRC/BHF Heart Protection Study of cholesterol lowering with simvastatin in 20,536 high-risk individuals: a randomized placebo-controlled trial. Lancet 2002; 360(9326):7–22.

37. Pfeffer MA, Keech A, Sacks FM, et al. Safety and tolerability of pravastatin in long-term clinical trials: prospective Pravastatin Pooling (PPP) Project. Circulation 2002; 105(20): 2341–2346.

38. Horsmans Y, Desager JP, Harvengt C. Biochemical changes and morphological alterations of the liver in guinea-pigs after administration of simvastatin (HMG CoA reductase-inhibitor). Pharmacol Toxicol 1990; 67(4):336–339.

39. Tolman KG. The liver and lovastatin. Am J Cardiol 2002; 89(12):1374–1380.

40. Bradford RH, Shear CL, Chremos AN, et al. Expanded Clinical Evaluation of Lovastatin (EXCEL) study results: two-year efficacy and safety follow-up. Am J Cardiol 1994; 74(7):667–673.

41. Newman CB, Palmer G, Silbershatz H, Szarek M. Safety of atorvastatin derived from analysis of 44 completed trials in 9,416 patients. Am J Cardiol 2003; 92(6):670–676.

42. de Denus S, Spinler SA, Miller K, Peterson AM. Statins and liver toxicity: a meta-analysis. Pharmacotherapy 2004; 24(5): 584–591.

43. Dujovne CA. Side effects of statins: hepatitis versus "transaminitis"-myositis versus "CPKitis". Am J Cardiol 2002; 89(12):1411–1413.

44. Phillips PS, Haas RH, Bannykh S, et al. Statin associated myopathy with normal creatine kinase levels. Ann Intern Med 2002; 137:581–585.

45. Grundy SM. Can statins cause chronic low grade myopathy? Ann Intern Med 2002; 137:617–618.

46. Farmer JA, Torre-Amione GT. Statins and myotoxicity: potential mechanisms and clinical implications. Klinik Forschung 2002; 8(3):87–91.

47. Hargreaves IP, Heales S. Statins and myopathy. Lancet 2002; 359(9307):711–712.

48. Laaksonen R, Jokelainen K, Sahi T, Tikkanen MJ, Himberg JJ. Decreases in serum ubiquinone concentrations do not result in reduced levels in muscle tissue during short-term simvastatin treatment in humans. Clin Pharmacol Ther 1995; 57(1):62–66.

49. Bleske BE, Willis RA, Anthony M, et al. The effect of pravastatin and atorvastatin on coenzyme Q-10. Am Heart J 2001; 142(2):E2.

50. Gruer PJ, Vega JM, Mercuri MS, Dobrinska MR, Tobert JA. Concomitant use of cytochrome P450 3A inhibitors and simvastatin. Am J Cardiol 1999; 87(7):811–815.

51. Omar MA, Wilson JP, Cox TS. Rhabdomyolysis and HMG Co A reductase Inhibitor. Ann Pharmacother 2001; 35:1096–1107.

52. Davidson MH. Does differing metabolism by cytochrome P450 have clinical importance?. Curr Atheroscler Rep 2000; 2(1):14–19.

53. Paoletti R, Corsini A, Bellosta S. Pharmacologic interactions with statins. Atheroscler Suppl 2002; 3:35–40.

54. Farmer JA, Torre-Amione G. Comparative tolerability of the HMG Co A reductase inhibitors. Drug Saf 2000; 23(3):197–213.
55. Miller DB, Spence JD. Clinical pharmacokinetics of fibric acid derivatives (fibrates). Clin Pharmacokinet 1998; 34(2):155–162.
56. Evans M, Rees A. Effects of HMG Co A reductase inhibitors on skeletal muscle: are all statins the same? Drug Saf 2002; 25(9):649–663.
57. Muscari A, Puddu GM, Puddu P. Lipid lowering drugs: are adverse effects predictable and reversible. Cardiology 2002; 97:115–121.
58. Holt SG, Moore KP. Pathogenesis and treatment of renal dysfunction in rhabdomyolysis. Intensive Care Med 2001; 27:803–811.
59. Holt S, Moore K. Pathogenesis of renal failure in rhabdomyolysis: the role of myoglobin. Exp Nephrol 2000; 8(2):72–76.
60. Ocose L, Luurila O, Eriksson J, Olsson A, Lithell H, Widgren B. Efficacy and safety of cerivastatin, 0.2 mg and 0.4 mg in patients with primary hypercholesterolemia. Curr Med Res Opin 2002; 15(3):228–240.
61. Evans N, Rees A. The myotoxicity of statins. Curr Opin Lipidol 2002; 13:415–420.
62. Furberg CD, Pitt B. Withdrawal of cerivastatin from the world market. Curr Control Trials Cardiovasc Med 2001; 2:205–207.
63. Isaacsohn J, Insull W, Stein E, et al. Long-term efficacy and safety of cerivastatin 0.8 mg in patients with primary hypercholesterolemia. Clin Cardiol 2001; 24(suppl 9):IV1–IV9.
64. Scandinavian Simvastatin Survival Study (4S). Randomized trial of 4,444 patients with coronary artery disease. Lancet 1994; 344:1383–1389.
65. Downs JR, Clearfield M, Weis S, et al. Primary prevention of acute coronary events with lovastatin in men and women with average cholesterol levels: results of the AFCAPS/TexCAPS. Air Force/Texas Coronary Atherosclerosis Prevention Study. JAMA 1998; 279:1615–1622.
66. Atyros BG, Papageogriou AA, Marcouris BR, et al. Treatment with atorvastatin to National Cholesterol Educational goal versus usual care in secondary coronary heart disease prevention. The Greek Atorvastatin and Coronary Heart Disease Evaluation (GREACE) study. Curr Med Res Opin 2002; 18:220–228.

16
Inhalation Toxicology

Roger O. McClellan
*Inhalation Toxicology and Human Health Risk Analysis, Albuquerque,
New Mexico, U.S.A.*

Michele A. Medinsky
Toxicon, Durham, North Carolina, U.S.A.

M. Burton Snipes
Inhalation Toxicology Research Institute, Tijeras, New Mexico, U.S.A.

INTRODUCTION

You, the reader, and we, the authors, have something in common. We are continuously breathing to provide oxygen to fuel metabolism and to eliminate CO_2 from our bodies. We must inhale whatever is in the air, including a wide range of gases and particles, in our home, workplace, or wherever we are, or take extraordinary measures to obtain purified air. Concern for the quality of air has prompted personal and collective actions to control air pollution in the workplace, in our homes, and in the ambient air. This has included both voluntary and legislated actions informed by scientific knowledge acquired from research. This chapter describes the research activities carried out in the interrelated fields of inhalation toxicology and respiratory toxicology. Inhalation toxicology, strictly defined, is the science of inhaled agents and how they interact with and affect the body. Respiratory toxicology, strictly defined, is the science of how agents may interact with and affect the respiratory tract. Obviously, the two fields are closely related and, frequently, the terms are used interchangeably.

This chapter is organized in a general fashion around the source to health response paradigm illustrated in Figure 1 (1). This basic paradigm was advocated by the Committee on Research Priorities for Particulate Matter (PM) (1). Because this is a toxicology text, emphasis is given to the portions of the paradigm linking (i) exposure to (ii) dose to critical tissues (iii) to health responses. However, it is important, when conducting and interpreting inhalation studies, to place this research within the context of the total paradigm shown. It is important to know the sources of the airborne materials, be it a pharmaceutical agent or an industrial facility or motor vehicles, and the likely exposure duration and concentrations to which people may be exposed. This also requires knowing the nature of the gases and particles in the air and the typical conventions for describing them. The structure

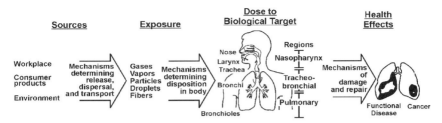

Figure 1 Schematic representation of the framework linking sources, exposure, dose, and health effects to provide an understanding of how airborne materials interact with the respiratory tract.

and functions of the respiratory tract must be understood with emphasis on both its role as a portal of entry and as a target organ for manifesting toxic effects. Building on this background, the processes involved in the uptake of inhaled gases and the deposition, clearance, translocation, and retention of inhaled particles are described. This is the linkage from exposure to dose to tissues—sometimes referred to as toxicokinetics, for toxic agents, or pharmacokinetics studies, for pharmaceutical agents. The chapter then proceeds to describe typical responses to inhaled agents, placing them in a dose–response context, sometimes referred to as toxicodynamic or pharmacodynamic studies.

There is substantial basic science that undergirds inhalation toxicology, and indeed, much research continues to be conducted to improve this science base. However, the vast majority of inhalation toxicology research is of an applied nature. This is what we refer to as "issue resolving science." Ultimately, the results of much of this research are used in assessing human health risks. It is not the intent of this chapter to describe the extensive and evolving field of risk analysis. Rather, a few basic concepts will be introduced to serve as background for understanding how the results of inhalation toxicology studies are used in risk assessment.

The basic risk assessment paradigm is shown in Figure 2 (2,3). The utility of inhalation toxicology studies can be substantially enhanced if the basic risk assessment paradigm is understood and research is planned so it can be used in assessing human health risks. All too often, inhalation toxicology studies are conducted in a manner such that the findings can only be used to characterize hazard (the first

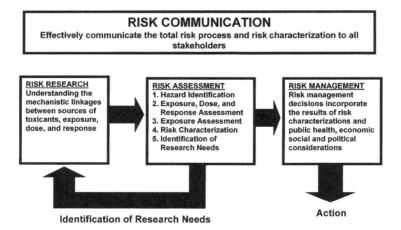

Figure 2 The risk paradigm. *Source*: From Refs. 2, 3.

element of risk assessment) and, even then, in a most general manner. Characterization of hazard may be adequate in some cases. For example, having identified hazard properties may rule out further development work on a prospective pharmaceutical or consumer product. More frequently, it is desirable to understand the exposure (dose)–response characteristics (the second element of risk assessment) for a range of exposure circumstances from brief exposures to exposure extending for a substantial portion of a lifetime. Such information, in conjunction with either measurements or estimates of exposure (the third element of risk assessment), allows for characterization of risk (the fourth element of risk assessment). Figure 2 also shows a fifth element of risk assessment, identification of research needs, that was advocated by the National Research Council (NRC) Committee that prepared the report "Science and Judgment in Risk Assessment" (3,4). Attention to this step in the risk assessment can have major impact in guiding the planning and conduct of research so that it will reduce uncertainties in subsequent risk assessments.

Throughout this text, emphasis is placed on relating the basic concepts of toxicology. We will recount some of those basic concepts here because of their importance in designing, conducting, and interpreting inhalation toxicology studies. As already noted, much of toxicology focuses on understanding human health risks. However, it is important to recognize related specialized areas of endeavor. One related field is veterinary toxicology where the focus is on domesticated animals, either as pets or commercial livestock. In both cases, there is an interest in the well-being of several species and, in the case of livestock, interest extends to their role in human exposure pathways, such as people eating meat from contaminated livestock. A second related field is environmental toxicology, where the focus may be on an array of species (i) that are pathways for transfer of toxic agents to humans, or (ii) whose well being is of concern as a part of an ecosystem that people value. In this chapter, the focus will be on the discussion of inhalation toxicology because it relates to assessing human risks.

A basic concept important to understanding toxicology of the respiratory tract, shared with the skin and gastrointestinal tract, is that it is a portal of entry and responds to inhaled materials. Moreover, it may also respond to agents that enter via other routes of entry and reach the respiratory tract tissues via the bloodstream. An example is the widely used herbicide paraquat (5). Other agents such as gases like NO_x, SO_2, bischloromethylether, and ozone and such particles as asbestos and silica may be inhaled and may directly affect the respiratory tract tissues. In other cases, as with the volatile organic chemicals such as benzene, trimethylpentane, and vinyl chloride, the agent may be inhaled and transferred to other organs via the blood, following absorption in the pulmonary (P) region. For benzene, the target is the hematopoietic tissue, for trimethylpentane, it is the kidney, and for vinyl chloride, it is the liver. For all three chemicals, it is not the chemical itself that causes the toxicity, but rather a metabolite. In recent years, increased evidence has developed for inhaled particles influencing not only the respiratory tract but also the heart, perhaps by particles or cytokines transported in the blood (6,7).

PLANNING INHALATION TOXICOLOGY RESEARCH

The starting point for planning the design and conduct of any investigation to address a particular issue is a review of the currently available scientific information on the topic. This, no doubt, may sound trite and be accepted as a clear given.

However, a review of recently published papers suggests that many investigators tend to ignore findings published more than a decade ago either out of ignorance of the literature, or with a view that findings developed with less sophisticated methodology are somehow not relevant. With an awareness of the issue of concern and knowledge of the literature in hand, it is appropriate to clearly articulate the hypothesis being tested. Although it has become fashionable to emphasize "hypothesis-testing" research, this does not negate the value of "issue resolving science" where the question may be stated directly, for example, as: "What is the lifetime risk of a noncancer or cancer effect from lifetime exposure to a specified substance?" The state of the available scientific information will help in forming the question or hypothesis.

The next issue to be addressed concerns the research system to be used. Figure 3 illustrates multiple sources of information used in evaluating the potential human risks of an airborne agent. Starting with the view that our interest is in humans, it is apparent that the most relevant information for evaluating human risks is that acquired from humans. If epidemiological data are available, they will be used. Unfortunately, positive epidemiological data stand as a testimonial to the past failure of society to control human exposures to the agent(s) in question. Obviously, our goal is to avoid such findings in the future. A difficulty in interpreting most epidemiological data is posed by the use of crude indices to describe exposure (i.e., years in a given job, a few industrial hygiene measurements, etc.), with actual dose data (i.e., blood levels of a chemical or some other measured biomarker of exposure) being rarely available. Nonetheless, whenever epidemiological data are available, they should be used to guide control strategies and scope future toxicological studies.

In some cases, it may be feasible and ethically appropriate to conduct studies with human subjects exposed under controlled conditions to the agent of interest (6,8). Many of the methods described later are appropriate for use in human studies;

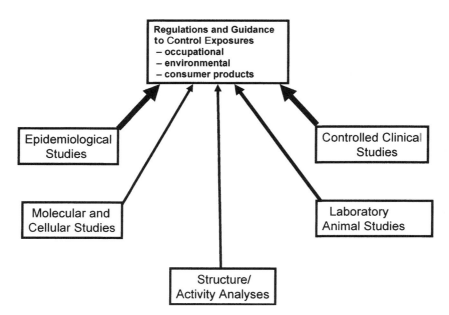

Figure 3 Sources of scientific information for developing regulations and guidance to control human exposures to airborne materials.

indeed, some may be more readily used with humans than with laboratory animals. Thus, the controlling consideration is whether it is ethically appropriate to conduct the study in humans and, indeed, use the data to evaluate human risk. This is a topic of current debate. Suffice it to say that in the past, two considerations served to guide decisions on the conduct of human studies. First, is there a high degree of confidence that any effects resulting from the exposures will be reversible? Second, are humans already exposed to the agents and, if so, are the exposure levels proposed for the experimental study similar to those people who have been exposed in the environment or workplace? One aspect of the current debate focuses on whether human subjects should be used to test new agents, such as pesticides, proposed for introduction in the marketplace. As a personal view, if assurance can be given that the biological effects will be reversible based on laboratory animal studies, we support the use of human studies to help ensure that appropriate exposure limits are set for the new product(s).

If the data from epidemiological studies or controlled exposure studies in humans are not adequate for assessing human risks, then it is necessary to conduct studies with intact laboratory animals, mammalian cells or tissues, cellular preparations, or nonmammalian cells. The issue of using various cell or tissue systems in toxicological studies has been covered in detail in other chapters. In this chapter, we focus on the unique issues faced in evaluating agents using inhalation exposure of experimental subjects. The use of in vitro systems for studying respiratory system toxicology has been reviewed (9). The entry of material via the respiratory system poses special challenges in selecting particular cell types for use in in vitro studies and in selecting exposure (dose) levels. The respiratory tract contains more than 40 different types of cells that have the potential to receive a dose. The cells lining the airways may be most directly exposed, and some cells may be exposed only to the metabolites of the inhaled agent produced in other cells. The interpretation of in vitro studies in which cells from the respiratory tract have been bathed with the test agent, typically at high concentrations, poses a special challenge. The target cells may be appropriate, but how does the dose concentration–time profile used with the cells compare to that seen by the cells from a relevant level of human exposure? Because of these challenges, the results of in vitro studies have generally served only as a crude index of the potential inhalation toxicity of materials. It follows that the results of in vitro studies have not been particularly useful in setting exposure levels for in vivo inhalation studies. The greatest value of in vitro studies relates to their use in interpreting metabolic patterns or biological responses observed in intact mammals exposed by inhalation, and only when the doses studied include levels likely to result from exposures potentially encountered by people. In short, how can one better understand and interpret the results observed in the intact mammal, and especially humans, exposed by the relevant route of exposure, inhalation?

A major issue in planning and designing inhalation studies is the selection of the species to be used (10–13). Fortunately, humans and mammalian laboratory animal species share many similarities that provide a sound scientific basis for using these species as surrogates for humans in investigations to develop toxicity information that is applicable to humans. However, there is no laboratory animal species that is identical to humans in all respects except for size. Moreover, the common laboratory animal species also differ from humans in various ways that must be considered in extrapolating the findings in laboratory animal species to humans. As will be related later in the discussion of the deposition and clearance of inhaled particles, there are important species differences that influence these parameters.

In summary, the design, conduct, and interpretation of toxicity studies inevitably involve careful consideration of a number of extrapolation issues: (i) from a specific population, such as workers who are subjects in an epidemiological investigation, to the general population, (ii) from a healthy population of humans studied in the laboratory under controlled conditions to populations with susceptible individuals due to genetic differences, preexisting disease, or the influence of age, (iii) from laboratory animal species to humans, (iv) from high levels of exposure (dose) to lower levels likely to be encountered by people, (v) from less-than-lifetime exposure to human exposures, and (vi) from cellular or tissue studies to the intact human. In this chapter, we will try to illustrate how these extrapolation issues may be addressed for airborne materials and show how they are considered in assessing human risks.

CHARACTERISTICS OF TOXICANTS AND TARGETS

Airborne Materials

A wide range of materials is of potential concern as inhaled toxicants, including gases, particles, and droplets. Gases and vapors are solutions of individual molecules in air where the movement of the individual molecules is practically unrestricted. The distinction between gases and vapors is based on temperature. A material in the gaseous state at a temperature below its boiling point is a vapor. A material in the gaseous state at a temperature above its boiling point is a gas. The distinction between a gas and vapor is rarely important in issues related to inhalation toxicology.

Aerosols is a term commonly encountered in the field; aerosols are defined as relatively stable suspensions of particles or droplets in gaseous media. The definition includes both the gaseous media and the contained particles or droplets. However, in common usage, aerosol frequently refers only to the suspended particles or droplets. Particles are solid materials and include fibers. Fibers are a special category of particles defined as elongated objects for which the aspect ratio, the ratio of the length of the object to the diameter of the object, is greater than 3.0. Droplets are liquids suspended in gaseous media. The term *ambient atmosphere* is used to describe the outdoor air as contrasted with indoor air. Indoor air includes the air in our homes, workplace, vehicles, and other locations occupied by people.

The atmosphere is a dynamic system influenced by changing inputs such as the environment (sea salt, volcanic eruptions, suspended soil, and gaseous emissions), industrial activities, agricultural activities, vehicles, and an array of other personal and societal activities. The composition and size of materials suspended in the atmosphere vary over a broad range, owing to the materials being produced by several processes, as presented in Figure 4. Materials emitted as gases may condense to a solid phase and undergo coagulation to form larger particles, and materials on particles may desorb and return to the gas phase. Moreover, chemical reactions are constantly occurring, including many influenced by sunlight. In addition, there are sinks with particles settling out and gases being absorbed or adsorbed by those particles.

Several terms used to describe different types of aerosols warrant definition. Within the trimodal distribution (Fig. 4), Wilson et al. (14) typically found in ambient air, the largest aerosol particles contained in the largest mode are typically considered dust, which is classically defined as fine particles of earth. However, in common usage, the term has been used to describe a wide range of airborne materials originating in the broadest sense from earth. Fumes are formed by combustion, sublimation, or condensation, usually with a change in chemical form. Metal oxide

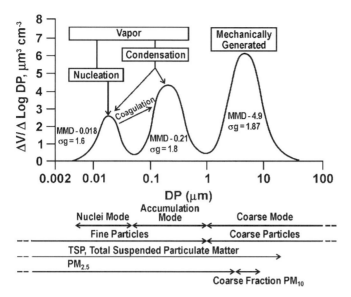

Figure 4 Schematic representation of the measured volume or mass size distribution of airborne particles illustrating the dynamic relationship between the suspending medium and particles of varied size. *Abbreviations*: PM, particulate matter; MMD, mass median diameter; DP, diameter particle. *Source*: From Ref. 14.

fumes are a good example; they typically start as very small particles of angstrom size and coagulate or flocculate to form large particles. Smokes are formed by combustion of organic materials, and their particles are usually smaller than 0.5 μm in diameter. Mists and fogs are liquid aerosols formed either by condensation of liquid on particulate nuclei in air, or by uptake of liquid by hygroscopic particles. The term *smog*, a contraction of smoke and fog, has usually been applied to a complex mixture of particles and gases in an atmosphere originating from combustion. The term *smog* dates back to episodes of pollution in London at the beginning of the industrial revolution. More recently, it has been used to describe polluted atmospheres arising from solar irradiation of hydrocarbons and NO_x from vehicle emissions, and other combustion processes and living plants.

The term *nuisance dust* was used for many years when referring to airborne particulate material of varied chemical toxicity, which produced toxic effect only when inhaled at high concentrations for long periods of time. The toxicity was attributed to the particulate form rather than any specific chemical toxicity. More recently, the term *particles not otherwise classified* has been used in referring to these materials (15).

In recent years, substantial attention has been given to nanoparticles, a term applied to a wide array of man-made particles less than 100 nm (or 0.1 μm) in diameter (16). Literature is just being developed on their potential hazard with much work yet to be done on potential workplace, consumer, or environmental exposures (Figs. 1 and 2) to fully characterize the potential exposure of humans to nanoparticles and the associated risks. In our opinion, the potential risk of these materials may substantially be influenced by rapid coagulation to form larger particles. Thus, they may have lower potential for reaching the respiratory tract as individual nanoparticles. Most of the research performed to date has focused on characterizing the potential hazard of nanoparticles. We will discuss the total suspended particulates (TSP), PM_{10}, and $PM_{2.5}$ nomenclature later.

Although particle mass is most commonly used in toxicity studies to describe exposures, it is important to recognize that other particle parameters may be of interest. This is illustrated in Figure 5 (17) showing particle mass, particle number, and particle surface area as a function of particle diameter. It is readily apparent that for particles larger than 1.0 µm in diameter, volume (or mass) is the dominant size parameter. For particles in the range of 0.1 to 1.0 µm, both the volume and surface area dominate. For the smallest particles, those termed *nanoparticles*, particle number dominates with a modest contribution from the surface parameter.

In considering particle size, it is important to distinguish between the real diameter of particles, such as might be measured using electron microscopy, and their aerodynamic diameters. The aerodynamic diameter of a particle is the diameter equivalent to that of a unit density particle with the same settling velocity. This is a measure of the particles inertial characteristics which will be shown later to be an important property influencing deposition. Particles less than about 0.5 µm have diffusion properties that strongly influence deposition.

The aerodynamic diameter of particles has been given increased prominence in the fields of occupational and environmental health and, thus, in inhalation toxicology since the 1980s. The shift to concern for characterizing the aerodynamic size of particles was heavily influenced by the improved understanding of the deposition and clearance of airborne particles in humans and laboratory animals that was facilitated by the use of radiolabeled particles. Industrial hygiene measurements and ambient aerosol measurements now are routinely made using methods calibrated to aerodynamic size. This is illustrated in Figure 4 with emphasis given to the characterization of TSP, PM_{10}, and $PM_{2.5}$. TSP refers to a sample collected with a high volume filter sampler, with which particles up to about 40 µm in size are collected,

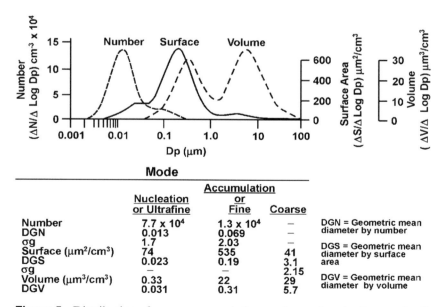

	Mode		
	Nucleation or Ultrafine	**Accumulation or Fine**	**Coarse**	
Number	7.7 x 10⁴	1.3 x 10⁴	–	DGN = Geometric mean diameter by number
DGN	0.013	0.069	–	
σg	1.7	2.03	–	
Surface (µm²/cm³)	74	535	41	DGS = Geometric mean diameter by surface area
DGS	0.023	0.19	3.1	
σg	–	–	2.15	
Volume (µm³/cm³)	0.33	22	29	DGV = Geometric mean diameter by volume
DGV	0.031	0.31	5.7	

Figure 5 Distribution of coarse, accumulation, or fine and nucleation or ultrafine mode particles by three characteristics: DGV, DGS, and DGN. *Abbreviations*: DGV, geometric mean diameter by volume; DGS, geometric mean diameter by surface area; DGN, geometric mean diameter by number. *Source*: From Ref. 17.

and essentially all particles larger than 30 μm are collected. "PM_{10}" is the term applied to PM collected with a sampler and calibrated to collect 50% of the particles that have an aerodynamic diameter of 10 μm. Above 10 μm, progressively fewer particles are collected and at sizes below 10 μm, progressively more particles are collected with 100% collected at 7 to 8 μm and smaller. "$PM_{2.5}$" is the term applied to a sample of PM collected with a device calibrated to collect 50% of the particles with an aerodynamic diameter of 2.5 μm, with progressively fewer large particles collected and progressively more smaller particles collected with essentially all particles collected at about 2 μm and smaller. Similar terms, "PM_{15}" and "$PM_{1.0}$," have been used for samples calibrated at 15 and 1.0 μm. Recently, attention has focused on samplers that will collect samples in the range between 2.5 and 10 μm, i.e., $PM_{10-2.5}$. A crude index for these "coarse" particles can be obtained by subtracting the $PM_{2.5}$ value from the PM_{10} value for samples collected concurrently. TSP served as the basis of ambient PM regulations from 1970 to 1987, PM_{10} from 1987 to the present, $PM_{2.5}$ from 1997 to the present and, in 2005, consideration is being given to a $PM_{10-2.5}$ standard (18). The latter would be the particulate mass with aerodynamic diameter less than 10 μm and greater than 2.5 μm.

Respiratory Tract Characteristics

The respiratory tract is an extremely complex organ system that has the principal function of delivering oxygen to the body and removing CO_2 and other gases. The P region is also involved in maintaining acid–base balance in circulating blood. A detailed description of the function and structure of the respiratory tract is beyond the scope of this chapter, thus, the focus here will be on key parameters that have the greatest influence in the design, conduct, and interpretation of inhalation toxicity studies.

The gross anatomy of the respiratory tract is illustrated in Figure 1. Typically, the respiratory tract is described as having three regions based on functional characteristics. The first is the nasopharyngeal (N-P) region (extending from the nares to the larynx), the second is the tracheobronchial (T-B) region (from the larynx to the respiratory bronchioles), and the third is the P region, which includes the alveoli. The N-P region serves as a channel for inspired and expired air and conditions the temperature and humidity of the inspired air. It also serves to protect the body by filtering out large particles, including bacteria and some viruses. The T-B region serves as a conduit with multiple branches distributing the air to the P region. The principal function of the P region is to provide for the exchange of gases between the inspired or expired air and the blood. In all three regions, phagocytes engulf deposited bacteria and viruses and destroy them.

The first portion of the airways inside the nostrils is lined by stratified epithelium which, a short distance from the nares, transitions to ciliated epithelium with interspersed mucus goblet cells. Portions of the pharynx and larynx are covered by stratified epithelium. The T-B region is lined by ciliated epithelium with interspersed mucus goblet cells. The P regions' principal anatomical characteristics are epithelial cells lining the alveoli, backed by an extensive network of capillaries lined by endothelial cells. Small amounts of connective tissue, muscle, and cartilage give structural support.

The highest velocities of the inspired and expired air are reached in the regions of the respiratory tract with the smallest cross-sectional diameter—the nose and the larynx. As the trachea divides into the bronchi and these in turn divide into bronchioles, the cross-sectional diameter of the total tract increases and the air velocity decreases until it becomes essentially stagnant in the alveoli.

It is important to recognize that many cells within the respiratory tract have the capacity to metabolize xenobiotic materials resulting in either less or more active metabolites (19–21). Macrophages, originating both from within the respiratory tract and from other sites and carried to the respiratory tract by the bloodstream, have the capacity to ingest and process foreign materials. Leukocytes in the blood, originating in the bone marrow, are constantly circulating through the respiratory tract tissues and are available to participate in inflammatory responses. In addition to the extensive vascular network containing blood, the respiratory tract contains a vascular network transporting lymph to regional lymph nodes.

Comparative Anatomical and Physiological Parameters

The respiratory systems of humans and experimental animals differ in anatomy and physiology in many quantitative and qualitative ways, beginning with the N-P region and extending into the P region of the lung. These differences affect airflow patterns in the respiratory tract, and in turn, the deposition and uptake of inhaled agents, and possibly the retention of those agents after they deposit in the respiratory tract.

In general, laboratory animals have much more convoluted nasal turbinate systems than do humans. Also, the length of the nasopharynx in relation to the entire length of the nasal passage differs among species, generally being longer in rodents. This greater complexity of the nasal passages, coupled with obligate nasal breathing of rodents, is generally thought to result in greater deposition in the upper respiratory tract of rodents than in that of humans breathing orally or even nasally, although only limited comparative data are available. Species differences in nasal airway epithelia (cell types and location) and the distribution and composition of mucous secretory products have also been noted (22,23). All of these factors contribute to differences among species in initial N-P deposition patterns for inhaled materials, as well as differences in dose–response patterns.

Conducting airways of the tracheobronchial (T-B) region also vary considerably among species with respect to physical dimensions and branching angles. These differences may result in significantly different patterns of transport and deposition for inhaled particles and gases. Airway length and diameter are important factors because they influence deposition due to sedimentation or diffusion; the time required for deposition to occur by both mechanisms is proportional to the distance the inhaled materials must travel. Airborne materials that have a short path length to travel between the trachea and terminal bronchioles have a higher probability of being deposited in the pulmonary (P) region. In addition to substantial differences in physical dimensions and branching patterns among species, conducting airways show a considerable degree of variability within species, which may be the primary factor responsible for variability in deposition seen within species (24). Larger airway diameter results in greater turbulence for the same relative flow velocity. Therefore, flow may be turbulent in the large airways of humans, whereas for an identical flow velocity, it would be laminar in the smaller airways of laboratory animals.

Differences in respiratory tract anatomy represent the structural basis for the species differences in gas uptake and particle deposition. In addition to the structure of the respiratory tract, regional thickness and composition of airway epithelium are also important factors in the absorption of deposited materials. Numbers and structural complexity of respiratory bronchioles and alveolar sizes also differ among species, which may affect deposition efficiencies because of variations in distances between the airborne particles or molecules' potential deposition sites.

There are also physiological differences that account for variations in deposition of inhaled materials among humans and laboratory animals. Breathing patterns and tidal volume determine the flow patterns and volume of airborne materials inhaled per respiratory cycle and per unit of time, and also determine the relative amounts of inhaled material available for deposition in the N-P, T-B, and P regions of the respiratory tract. Table 1 (25) presents a summary of significant respiratory parameters for selected laboratory animals and humans. There are clear differences among humans and these laboratory animal species in size, respiratory minute volumes, and respiratory minute volume per unit of body mass. The trend shown in Table 1 is that smaller animals inhale larger volumes per unit of body mass. Smaller animals have higher metabolic rates and must inhale relatively larger volumes to supply their systemic requirements for oxygen. This means that for the same exposure atmosphere, they also inhale larger amounts of airborne materials per unit body mass, as compared with larger animals and humans.

Changes in respiratory minute volumes can be expected to result in proportional changes in the amount of airborne materials entering the respiratory system. If increased respiratory minute volumes are induced, the effect is that more airborne material is drawn into the respiratory tract and deposition is increased. This was demonstrated in one study where CO_2 was used to cause experimental animals to breathe more deeply than normal, thereby increasing intake and deposition of an aerosol (26).

Respiratory parameters are a major source of variability. Lippmann (27) reviewed the literature available for human regional deposition and concluded that a considerable amount of intersubject variability exists for deposition fraction, as a function of inhaled particle size, in all regions of the respiratory tract. Most of the variability was ascribed to variations in tidal volumes, flow rates, and functional residual capacity. The respiratory pattern can markedly influence regional deposition. Shallow, rapid breathing or deep, slow breathing can produce the same respiratory minute volume. However, the shallow, rapid breathing (small tidal volume) results in increased deposition in the N-P and T-B regions and decreased deposition in

Table 1 Breathing Characteristics of Humans and Commonly Used Laboratory Animals

Species	Body mass (g)	Frequency (breaths/min)	Tidal volume (mL)	Minute volume (mL)	Minute volume per gram body mass (mL/g)
Human	70,000				
Resting		12	750	9000	0.129
Light exercise		17	1700	28,900	0.413
Dog	10,000	20	200	3600	0.360
Monkey	3000	40	21	840	0.280
Guinea pig	500	90	2.0	180	0.360
Rat	350	160	1.4	240	0.686
Mouse	30	180	0.25	45	1.500

Source: From Ref. 25.

the P region. Slow, deep breathing results in a larger fraction of the inhaled volume reaching the P region, and thereby enhances deposition in the P region. Increased flow rates through respiratory airways cause increased deposition due to impaction, especially at locations where sharp angles are encountered by the inhaled particles. Small functional residual capacity allows relatively more inhaled air to mix within the pulmonary air spaces, and thereby causes increased deposition in the P region; the opposite occurs with large functional residual capacity.

A factor that markedly influences respiratory tract deposition patterns is the mode of breathing. During nose breathing, deposition of inhaled materials in the head airways can be an effective way for the respiratory system to preferentially filter materials out of the inspired air. With mouth breathing, a potentially larger deposition of inhaled materials can occur in the P region. Rodents are obligatory nasal breathers unlike humans who breathe both through their nose and mouth.

An example of the effects of altered respiratory parameters on deposition can be seen with anesthetized animals. Hamsters under general anesthesia inhaled and deposited less of an aerosol composed of insoluble particles having an activity median aerodynamic diameter of 0.45 μm than did unanesthetized hamsters (28). The anesthesia reduced respiratory minute volume and thereby reduced the amount of aerosol drawn into the respiratory tract during the exposure.

A number of studies relating inhalation of irritant substances or stimulants to deposition have been reported. An important result of these studies is that they have provided a database demonstrating that deposition patterns are sensitive to factors that change respiratory patterns. Alarie (29) discussed this subject in detail and described ways to measure and compare the irritating properties of inhaled materials. Bruce et al. (30) reported effects of 10 ppm ozone exposure on respiratory parameters, metabolic rate, and rectal temperature in mice. Soon after initiation of the 90-minute exposures to ozone, these parameters all decreased, and the decreased respiratory parameters would have resulted in decreased deposition of aerosols inhaled along with the ozone. In a similar study, Silver et al. (31) exposed mice by inhalation to acrylate esters, which are respiratory tract irritants. Minute volume quickly dropped to one-third of normal resting values and body temperature decreased. Chang et al. (32) observed depressed minute volume in rats and mice exposed to formaldehyde and Medinsky et al. (33) noted the same result for methyl bromide inhalation by rats. These changes in respiratory parameters would have altered deposition of particles inhaled during the exposures to the irritant substances.

In addition to variability caused by respiratory parameters, some variability was ascribed to genetically related differences in airway and airspace morphometry. Deposition may therefore not be constant even for the same subject measured at different times, and any factor that influences anatomy or causes respiratory parameters to change with time will influence respiratory tract deposition patterns.

Species differences in anatomical and physiological characteristics dictate that the various animal species used in inhalation toxicology studies do not deposit the same relative amounts of inhaled materials in comparable respiratory tract regions when they are exposed to the same external particle or gas concentration. Differences among species in breathing parameters, as well as structural components and dimensions in the regions of the respiratory tract, result in species differences in airflow and deposition patterns. Biologic end points or health effects resulting from inhalation exposures, therefore, may be more directly related to the quantitative patterns of deposition within the respiratory tract than to the external exposure concentrations.

FATE OF INHALED MATERIALS

Particles

In this section, we briefly describe the general concepts that influence the fate of inhaled particles. For additional details, the reader is referred to the reviews of Snipes (34–36), Schlesinger (24,37) and Miller (38). In considering these concepts, it should be apparent that the fate of each particulate material in an agent requires characterization of that specific material as to its kinetics, or physical and chemical fate after deposition in the respiratory tract. Care should be taken in attempting to make quantitative extrapolations from one agent to another. While certain general concepts apply, as will be discussed below, the size distribution of the test agent and its chemical characteristics and, especially, solubility will determine its kinetics.

Deposition

The inertial properties, characterized by aerodynamic diameter, and diffusional properties, characterized by the particle diffusion diameter, are the major factors influencing the inhalability of particles and their ultimate deposition. Inhalability is defined as the probability of a particle being present in the breathing zone of the inhaling subject, i.e., taken into the nose or mouth of the individual upon inspiration. Rodents and lagomorphs are obligate nose breathers and, thus, oral breathing is not an issue for these species. However, many people regularly breathe in part through their mouth, especially when there is increased demand for oxygen, as when doing physical work or exercise.

The deposition fraction is defined as the portion of the inhaled particles deposited in each region of the respiratory tract or in the total respiratory tract. For most particle sizes, some portion of the inhaled particles will not be deposited but will be expired. For the smallest and largest inhalable particles, deposition may approach 100%. In considering deposition of inhaled particles in the respiratory tract, it is important to recognize that particles successively traverse the nares and nasal passage, then the pharynx and larynx, then the trachea, bronchia, and bronchioles, and finally may enter the alveoli. With expiration, a packet of air still containing airborne particles traverses these same structures in reverse order. A schematic rendering of deposition in the several regions of the human respiratory tract is shown in Figure 6 (36). Similar deposition curves for the most frequently used laboratory animal species have been compiled by Schlesinger (37). There are substantial differences among the species with rats and mice not being able to inhale particles with aerodynamic diameters greater than about 4 μm.

The major modes of deposition are shown in Figure 7. Sedimentation and impaction are largely influenced by the inertial properties of the particles which is reflected in the aerodynamic diameter of the particles. The nasal passages condition the temperature and humidity of the inspired air; past the nasal passages, the air has a high relative humidity. In this environment, hygroscopic particles such as NaCl and CsCl may grow in real and aerodynamic size. Interception is influenced by particle size and, in the case of fibers, by the fiber length. Diffusional deposition may be viewed as a special case of interception in which Brownian motion brings small particles into contact with respiratory tract surfaces.

Interception is an important mode of deposition for inhaled fibers. The aerodynamic diameter of a fiber may be relatively small because fibers align with the airflow stream. Envision a well-thrown javelin flying through the air in a manner not dissimilar to that of a golf ball. In the respiratory tract, as the air stream flows

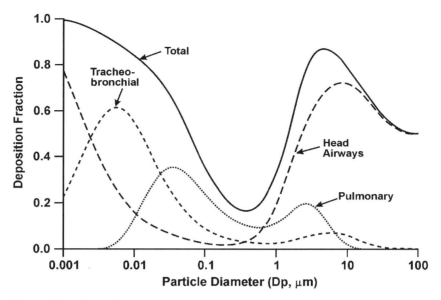

Figure 6 Schematic rendering of the fractional regional deposition of inhaled particles in the human. *Source*: From Ref. 36.

through bifurcations in the airways the length of the fiber provides increased opportunity for the ends of the fibers to be incepted by the wall of the airways. Electrostatic forces may also contribute to deposition of particles. However, in experimental studies, this can be minimized if the particles flow past a radioactive source such as [85]Kr to bring the charge on the particles to Boltzman equilibrium. A major issue with the charge on particles relates to losses to the walls of the exposure system.

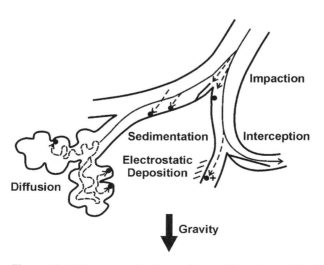

Figure 7 Primary mechanisms of deposition of particles in the lower respiratory tract.

Fate of Particles Postdeposition

The fate of the particles and constituents postdeposition is influenced by real size, surface area, and chemical characteristics of the particles. Once particles are deposited, the aerodynamic or diffusional properties of the particles are no longer a factor in the fate of the particles. Surface area and the chemical composition influence the dissolution rate of particles. A useful review of the role of particle solubility on retention of inhaled particulate material can be found in a National Council on Radiation Protection and Measurements (NCRP) report (39). Several modes by which particles are removed from the respiratory tract are illustrated in Figure 8. Clearance by mucociliary action is important in the nose and conducting airways. Mucociliary movement of particles from the nasal passages to the oral pharynx occurs in a matter of minutes and likely occurs with only a small fraction of the deposited particles being ingested by macrophages. Mucociliary movement from the distal bronchioles, through the bronchii, and up the trachea occurs in a matter of hours. The removal process from the airways is facilitated by the rapid phagocytosis of a large portion of the particles. Material reaching the oral pharynx is either expectorated (spit out) or swallowed. For certain-sized particles, there may be substantial deposition in the nasal and T-B regions and, thus, mucociliary clearance of the particles to the pharynx and ingestion. This material is available for absorption from the gastrointestinal tract.

Some rodents engage in coprophagy. Thus, relatively insoluble particles may be inhaled, cleared from the respiratory tract, ingested, excreted in the feces, and reingested. This phenomenon can lead to confusion in interpreting measurements of radiolabeled particles in the intact animal made by assaying radioactivity in

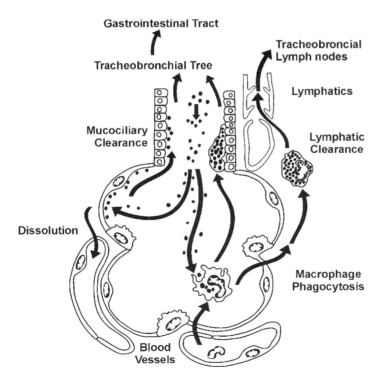

Figure 8 Schematic rendering of the several mechanisms involved in clearance of particles from the respiratory tract. *Source*: Courtesy of P.J. Haley.

whole-body counters. The whole-body retention pattern will not only reflect exclusively radiolabeled particles retained in the respiratory tract, but also material in the gastrointestinal tract being recycled by coprophagy.

Not all particles ingested by macrophages are cleared via the mucociliary escalator. Some of the macrophages carrying particles move to the interstitial areas of the lung. Radiolabeled particles in macrophages have been removed from the lungs by bronchoalveolar lavage more than eight months after a single brief inhalation exposure. This suggests that there is regular trafficking of macrophages between the alveoli and the interstitium. Other particles, most likely within macrophages, move to the lymphatics and are sequestered in regional lymph nodes. At long time after inhalation exposure, the quantity of very insoluble materials found in the regional lymph nodes may actually exceed the quantity remaining in the P region. There is also evidence for some small fraction of deposited particles directly reaching the bloodstream and being transported to other organs such as the heart, liver, or brain. It is not known if this is facilitated by macrophages. Particles that are highly soluble may dissolve in the lung fluids and move rapidly to the bloodstream. In other cases, the dissolution of particles may be facilitated by their ingestion by macrophages.

It is apparent that the quantity of test material present as a function of time in the three regions of the respiratory tract and other organs will be influenced by the duration of the exposure and by the aerodynamic and diffusional size characteristics governing deposition and by the real size, surface area, and chemical composition governing clearance from the respiratory tract and translocation to other organs. In some cases, the critical "dose" and effects will be in the respiratory tract, and in other cases, in organs remote from the respiratory tract.

The long-term fate of inhaled relatively insoluble particles following a single brief exposure is shown in Figure 9 (35) for the P region and lung-associated lymph nodes. A marked difference in the long-term clearance of particles in humans, monkeys, dogs, and guinea pigs versus that observed in the rat is readily apparent. In mice the clearance would be even more rapid than in rats. The biological basis for the more rapid clearance in the mice and rats versus the other species is not known. The difference in the pulmonary clearance in the several species is reflected in differences in accumulation of particles in the lung-associated lymph nodes. Quite obviously, if material is cleared from the P region and excreted, it is not available for clearance to the lymph nodes. An example of modeled accumulation of inhaled relatively insoluble particulate material in the P region of several species with chronic exposure is shown in Figure 10 (35).

A schematic rendering of the deposition and clearance of inhaled relatively insoluble particles is shown in Figure 11 (40). This figure emphasizes the distinction between exposure (the time profile of the air concentration of an agent), deposition, and clearance. The dose retained is equal to the dose deposited minus the amount cleared. When considering the figure, it is important to keep in mind the overlay of time as a critical dimension. As illustrated earlier, clearance processes that remove material from the respiratory tract via nose blowing, expectoration, ingestion, and fecal excretion have time dimensions extending from minutes to hours to even months. The rate of movement to blood and to other organs will vary depending upon the solubility of the particles but may extend over months or years.

The discussion of clearance mechanisms, as explained above, describes the mechanisms operative in a normal individual with low to moderate levels of exposure. It is important to recognize that both deposition and clearance pathways

(A)

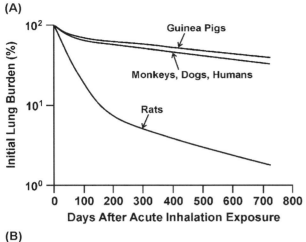

(B)

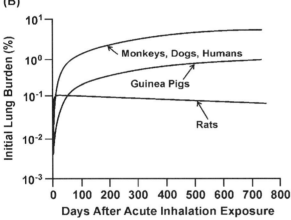

Figure 9 Schematic rendering of the clearance of relatively insoluble particles from the (**A**) P region and (**B**) accumulation in lung-associated lymph nodes in several species following a single brief inhalation exposure. *Abbreviation:* P region, pulmonary region. *Source*: From Ref. 35.

can be altered by disease and, certainly, in laboratory animals by long-term exposure to high concentrations of particulate material. This issue will be expanded on later.

It is well known that the kinetics of inhaled materials may be influenced by the exposure concentration and duration. Thus, it is important to conduct kinetic studies using multiple exposure concentrations and, in some cases, exposures of varied duration. If kinetic studies are being done as a prelude to studies to evaluate exposure–response relationships, it is important to use a range of exposure concentrations that will include the exposure concentrations likely to be used in the exposure–response study. Indeed, the results of the kinetic studies can be useful in selecting exposure concentrations to be used in the exposure–response study. In the case of studies with PM, it may be useful to include animals for serial termination points in the exposure–response study to provide insight into changes in the kinetics of the particulate material with protracted exposures.

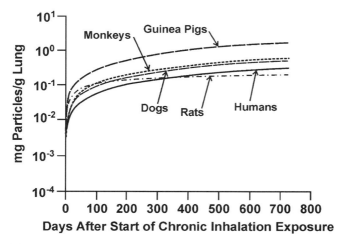

Figure 10 Simulation model results of the accumulation of particles in the P region after chronic exposure to an atmosphere containing 0.5 mg/m³ of PM. *Abbreviation*: P region, pulmonary region; PM, particulate matter. *Source*: From Ref. 35.

Gases and Vapors

Uptake of Gases and Vapors

One of the most important distinctions between the dosimetry of gases and vapors versus particles is that particles deposit at the interface between the airway lining and the respiratory air space, while gases and vapors can diffuse past the interface. Thus, we speak of uptake of a gas by the respiratory tract mucosa rather than deposition onto the surface of mucosa. The location for and extent of uptake of gases in the respiratory tract is influenced by the concentration in the air, the

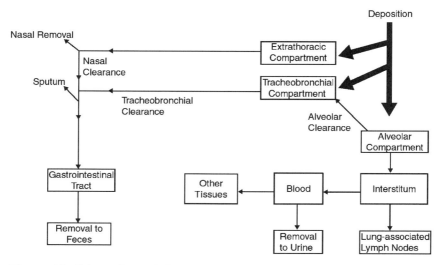

Figure 11 Schematic rendering of the deposition (*thick arrows*) and clearance of inhaled relatively insoluble particles. Transport within the body (*thin arrows*) (40). *Source*: From Ref. 41.

duration of exposure, the water and lipid solubility of the gas, and the reactivity of the molecule with biological components. For example, lipophilic gases and vapors diffuse rapidly from the alveolar spaces into the blood which distributes them throughout the body. In contrast, water-soluble gases and vapors are taken up extensively by the respiratory tract tissues. An understanding of the processes involved in uptake and distribution of volatile chemicals is important for predicting the amount of a chemical or its metabolite reaching various tissues in the body.

Determinants of Uptake into Mucosa

Six processes or properties determine the uptake of gases and vapors into the respiratory tract mucosa: convection, diffusion, dissolution, solubility, partitioning, and reaction. Convection and diffusion are transport processes. Dissolution and reactivity are chemical processes. Solubility and partition coefficients are physical properties. The rate of dissolution is very rapid for most gases and vapors. Convection moves molecules within the bulk airstream and in the blood. The rate at which gas moves from the bulk air to the mucosa is governed principally by solubility and rate of convection, diffusion, and reaction within each air or tissue phase.

Convection. Convection occurs when a fluid (air is a special case of a fluid) is in motion. Convection serves to bring inspired air and the molecules contained within it into contact with the lining of the respiratory tract. Further, convection removes air out of the respiratory tract during expiration. The rate at which a gas is taken up, i.e., the rate at which material is delivered to a region of the respiratory tract by airflow is limited by convection. The rate at which a gas is taken up by the lining tissue adjacent to the airway cannot be any faster than the product of the flow rate of the air entering the region and the concentration of the gas in the entering air stream. For gases that are largely absorbed in the upper respiratory tract, air near the nasal lining is quickly depleted of the gas. Thus, patterns of airflow set up a pattern of site-specific gas uptake by delivering more gas to some areas of the lining and less to other areas.

Diffusion. All individual molecules in a gas or liquid exhibit a random motion resulting from the fluctuation of forces exerted by surrounding molecules. This is called Brownian motion. Random Brownian motion results in a net migration of particles or molecules from regions of high concentration to regions of low concentration. The net migration is called diffusion. Diffusion occurs independent of convection. Convection carries material along the airways, while diffusion causes material to move from the center of the airways toward the walls and into the tissue. The layer of air immediately adjacent to the airway surface moves more slowly because of friction between the air and the surface. Diffusion is the primary process by which gases cross this slow moving layer of air. Once a material dissolves in the surface layer, the movement of the material occurs via diffusion. Each layer—the layer of air near the surface, the mucous layer, and the tissue per se—offers resistance. Resistance at each layer is a function of the ratio of the thickness of the layer to the ability of the gas to diffuse in that layer.

For gases or vapors with high aqueous solubility, the resistance resulting from diffusion across the layer of air plays an important role in determining the rate at which the gas is deposited because the subsequent uptake by the mucosa is relatively fast. The capacity of the mucus and the tissue lining to absorb the gas relative to the low total rate of mass delivery in inhaled air (a human breathing at a minute volume of 10 L/min inhales only 0.4 μmol/min of a gas present at 1 ppm) is such that the rate of delivery to the lining in the air is a primary determinant, or rate-limiting step, for gas uptake in most circumstances. For gases with low aqueous solubility, resistance caused by diffusion through the mucus and epithelium is more significant in determining the rate of uptake into the mucosa.

Dissolution, Solubility, and Partitioning. Dissolution is the process by which a gas becomes dispersed into a liquid or tissue. Dissolution can be thought of as a chemical process for which the associated energy barrier may be significant, and involves molecules moving across the surface between air and the underlying media. Solubility refers to the amount of one material that can be dissolved in a given amount of another, or the amount that exists in a saturated solution. Solubility has units of concentration.

A partition coefficient is the ratio of the concentration of a substance in one medium, or phase (C_1), to the concentration in a second phase (C_2) if the two concentrations are at equilibrium that is, partition coefficient $= (C_1/C_2)_{equil}$. A partition coefficient is essentially unitless in that it has units of concentration/concentration [e.g., (g/L)/(g/L)] and typically is defined at concentrations far from saturation. The constant for partitioning between a gas and a liquid at low concentrations is also called a Henry's law constant, $H = (C_1/C_2)_{equil}$. The Henry's law constant refers specifically to vapor–liquid equilibria. Commonly used partition coefficients in inhalation toxicokinetics are the blood–air partition coefficient for gases, and the tissue–blood partition coefficient for all substances. The tissue–air partition coefficient for a gas is an indicator of the extent to which the gas can be taken up by the respiratory tract tissues from the air.

Solubility and partitioning determine the site for uptake in the respiratory tract and the extent of uptake. For example, the pattern for uptake of gases in the respiratory tract is related to the aqueous solubility of the gas. Gases that are highly water soluble have large tissue–air partition coefficients because of the high water content of respiratory tract tissue and are thus readily taken up by respiratory tract mucosa. Additionally, a large fraction of the uptake of water-soluble gases occurs in the upper respiratory tract, because this is the first respiratory tract tissue that comes into contact with the inhaled gas and because it is a highly perfused tissue. On inspiration during nasal breathing, there is a pattern, or gradient, of gas concentration and uptake for soluble gases along the upper respiratory tract. Removal of a soluble gas from the air phase during inhalation results in a decrease in the air-phase concentration in the anterior to posterior direction. Because decreasing gas-phase concentration results in a decreased driving force for uptake, the gas-phase concentration gradient often can result in a similar gradient in the rate of uptake, although factors such as differences in tissue type and removal capacity in different regions may cause it to be otherwise. For water-soluble gases, airflow and, hence, nasal anatomy are important determinants for site of uptake.

In contrast, gases, such as ozone with low aqueous solubility are not extensively removed by the upper respiratory tract tissues, and thus gas concentration tends to be much more uniform throughout the airways, with airflow patterns and anatomy playing much less of a role. Because the concentration of the gas in the air space is uniform from region to region, the interaction of low-solubility gases with the mucosal surface in each region is roughly proportional to the area of airway surface in that region, and the rate of uptake is proportional to the blood flow in that region. As a result, the bulk of uptake occurs in the alveolar region, where both surface area and the rate of perfusion are high.

Reactivity. Reactivity refers to the rate at which a compound reacts, or tends to react, with substances in the lining of the respiratory tract. Chemical reactions include both enzyme-mediated and spontaneous reactions and influence uptake in the following manner. The presence of a gas in a tissue creates back pressure which reduces the rate of uptake. Chemical reactions that remove a gas from a tissue lower the gas concentration in the tissue and thus increase the net rate of uptake from the air. Because chemical reaction is a process of removal, it therefore also affects uptake. At the onset of exposure, when tissue concentrations are very low, the rate of uptake is highest and the rate of reaction is lowest. During exposure, the gas concentration in tissue rises, backpressure increases, the rate of uptake decreases, and the rate of reaction or removal increases. The rise in tissue concentration continues until the rate of uptake is exactly counterbalanced by the rate at which gas is removed from tissue by reaction, by the bloodstream, or by desorption into the exhaled air (Fig. 12) (41). If the partition coefficients, rates of dissolution, and diffusivity of two gases are equal, then the concentration of the more reactive gas in the tissue is lower than that of the less reactive gas. Reactivity usually is considered in relation to the rate at which the compound penetrates, or diffuses, through the epithelium (42).

Ozone is considered a reactive gas. It reacts with mucous components so rapidly that little of the ozone itself penetrates into the mucus-lined epithelium (43–45). Butadiene is not considered to be reactive because the rate of butadiene oxidation by cytochrome P450 is slow compared with the rate of diffusion and uptake into the blood. Most butadiene molecules pass through the airway lining without reacting. Viewed another way, a reactive gas is one for which a large portion of the molecules that enter the airway lining reacts before reaching the underlying capillary bed (blood exchange region). A comparison of the concentration gradients that might be expected across the respiratory tract tissue, from the airway to the capillary, for an inert and a reactive gas with identical tissue–air partition coefficients, is shown in Figure 13 (46). The rate at which a reactive gas is taken up by respiratory tract tissue is higher than that for a nonreactive gas, other things being equal.

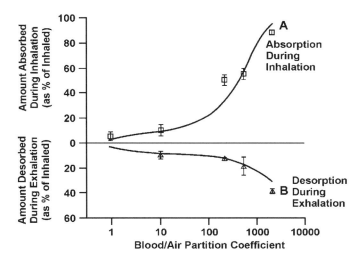

Figure 12 Uptake of vapors in the nasal airways of the dog during cyclic breathing as a function of the blood–air partition coefficient. Model simulation results (*solid lines*) are shown with experimental data points. Curve A represents nasal absorption in inhalation, and Curve B represents nasal desorption on exhalation. The data points left to right are for 2,4-dimethyl-pentane, propyl ether, butanone, dioxolane, and ethanol. *Source*: From Ref. 41.

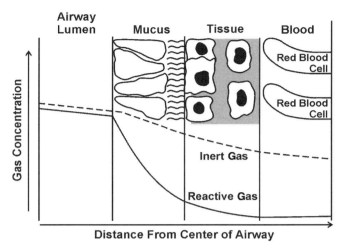

Figure 13 Theoretical concentrations of an inert and reactive gas in the various portions of an airway. *Source*: From Ref. 46.

This concept of higher uptake for reactive gases also applies to gases for which the reaction is enzyme mediated. The rate of uptake is higher for a gas if it is metabolized than if it is not.

Removal of Gases and Vapors

Material delivered as a gas or vapor to the tissues of the respiratory tract can be removed by three primary processes—uptake into the blood and transport to other parts of the body, elimination in the exhaled air (Fig. 12), and chemical reactions.

Uptake into Blood. Many gases and vapors are taken up into the blood, and blood circulation carries them to other tissues. The two primary factors that determine the extent to which a gas or vapor is taken up by the blood are the rate of blood perfusion to the tissue of interest in the respiratory tract and the tendency of the material to partition into blood. The extent to which molecules of a gas are taken up into blood depends upon the relative affinity of the gas for blood compared with air or respiratory tract tissues. The blood–air partition coefficient is one convenient predictor of how extensively gas molecules will be taken up into the blood. The alveolar region of the respiratory tract can be thought of as a simple permeable membrane across which gas molecules can diffuse. This simple approximation holds for gases that are not reactive, not extensively metabolized by the respiratory tract, and not significantly stored in that tissue. If diffusion of the molecules is fast relative to ventilation or perfusion, the concentration of the gas in blood exiting the respiratory tract tissues is assumed to be at equilibrium with the concentration of the gas in the alveolar spaces. For gases with large blood–air partition coefficients, the arterial blood concentration of the gas is large relative to the alveolar concentration and hence the rate at which the gas is taken up by the blood is also large. Increased uptake into the blood with increasing blood–air partition coefficients is shown in Figure 14 (47).

Exhalation. Not all gases and vapors that are inhaled are completely absorbed into the respiratory tract tissues or taken up into the blood. A fraction of most gases and vapors is exhaled unchanged (Fig. 12). Other gases, formed within the body, such as CO_2, are removed via exhalation. The same determinants that

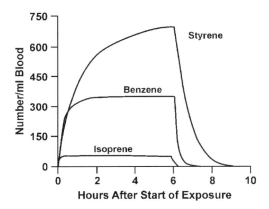

Figure 14 Effect of the blood–air partition coefficient on venous blood concentrations for three volatile organic chemicals during and after a six-hour exposure to 2000 mg/m³. *Source:* From Ref. 47.

affect absorption into respiratory tract tissue and uptake into the blood also regulate the extent to which the gas or vapor is exhaled.

Chemical Reaction. Chemical reaction is an important mechanism by which gases and vapors can be removed from respiratory tissues. Chemical reaction is considered a mechanism for removal of gases in the sense that the individual molecules are transformed into other chemical species. In some cases, this transformation may result in a product that is more toxic than the parent chemical. In other cases, chemical reaction is a detoxication mechanism, because the reaction product is less toxic than the parent chemical.

Many stable gases that are not sufficiently reactive to interact directly with mucosal components are substrates for the various enzymes in the cells of the respiratory tract. Metabolism can activate a stable gas by converting it to a much more reactive form. Differences in metabolic capacity among tissue types can determine site specificity for toxicity. For example, four types of epithelium are recognized in the nasal cavities—squamous, respiratory, transitional (between squamous and respiratory), and olfactory (48). Each type of epithelium is characterized by a different mix of cell types and different metabolic capacity. Clear differences in enzyme localization between tissue types have been observed using histochemical techniques (20). These differences in metabolism, in conjunction with the effects of airway geometry, play an important role in determining both dosimetry and toxicity.

Regional metabolism often explains why gases are toxic to the olfactory region but leave other areas relatively unaffected. In particular, the concentration of a gas in air adjacent to respiratory epithelium is at least as high as the concentration adjacent to olfactory epithelium, because air must pass over the respiratory epithelium before reaching the olfactory epithelium. Therefore, the rate at which a gas is delivered to olfactory tissue is no greater than the rate of delivery to respiratory tissue. Despite the higher delivery to respiratory tissue, the olfactory tissue may be targeted owing to the higher metabolic activity of this tissue. For example, the primary target for toxicity of certain dibasic esters, olfactory epithelial sustentacular cells (49), correlated strongly with regional carboxylesterase activity (19). A similar pattern was observed for vinyl acetate for which metabolic conversion to acetic acid appears to play a critical role (50).

Certain lung cells also contain significant quantities of xenobiotic-metabolizing enzymes (21). For example, the nonciliated Clara cells present in the bronchiolar epithelium have high metabolic activity and are targets for many toxicants. Interest in Clara cells was first generated by the observation that this cell type was damaged specifically by certain chemicals, even if these chemicals were not administered by inhalation (51). Immunochemical analysis of lung cells taken from rabbits showed that Clara cells contain measurable concentrations of certain cytochrome P450 isozymes. In the Clara cells, these concentrations exceed concentrations measured in the whole lung tissue by at least twofold.

Integrating Determinants for Uptake and Removal of Gases and Vapors

The discussion of mechanisms of uptake and removal for gases and vapors has focused on individual processes. The patterns of gas and vapor concentration in the airways of the respiratory tract and dosimetry in the lining of the respiratory tract are the result of dynamic balances between the various processes of transport, chemical reactions in respiratory mucosa, and removal via blood perfusion or desorption and exhalation. Mathematical models of gas and vapor uptake and dosimetry have been developed by linking together the equations for the individual processes to describe the overall mass balance. The type of model that is developed for a given gas or vapor depends on the physical and chemical properties of the material, the mathematical tools available, the end point, and the quality of the data or information available to define the model.

Dahl (52) has provided a useful schematic rendering for considering the fate of inhaled materials and the types of models that can be used to describe their kinetics (Fig. 15). He notes that the initial deposition of inhaled gases compared to inhaled particles are ascribable to differences in the size of the components. The deposition of gases, like very small particles, is dependent on a combination of diffusion and convection. For purposes of describing uptake, gases can be initially classified as being either stable or reactive. For stable gases having a water–air partition

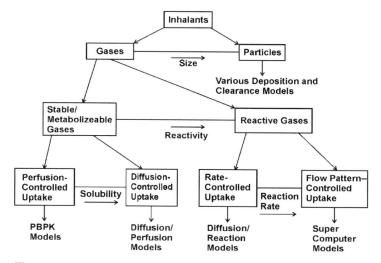

Figure 15   Schematic rendering of uptake of inhaled materials with emphasis on modeling the uptake of gases. *Abbreviation*: PBPK, physiologically based pharmacokinetic. *Source*: From Ref. 52.

coefficient of less than about 50, uptake takes place primarily in the P region. Physiologically based pharmacokinetic models can be used to describe total uptake of these gases. While metabolism in the respiratory tract may not contribute significantly to total uptake, local metabolism by P450 enzymes can cause local effects in the respiratory tract as is seen for certain volatile organic chemicals such as methylene chloride and styrene.

For stable gases having water–air (or blood–air) partition coefficients greater than 50, uptake is mainly in the nasal cavity and conducting airways. For stable gases, uptake in the respiratory tract via diffusion into the blood is dependent on the blood–air partition coefficient as shown in Figure 12. Dahl et al. (53) observed that uptake during the inspiratory phase approached 100%. Significant desorption of these gases from respiratory tract tissues can occur during exhalation (41). In contrast, highly reactive gases behave much like particles in that they react quickly and irreversibly on first contact with the surface of the airways. The uptake of these gases is flow dependent (42). This class of compounds which includes ozone, chlorine, and nitrogen dioxide is not dependent upon local or systemic metabolism to cause effects.

Conclusions

Different factors regulate the deposition of particles, gases, and vapors. Gases and vapors, like very small particles, deposit owing to diffusion. However, unlike particles, gases and vapors can desorb from the respiratory tract surfaces and reenter the airway. Solubility in and reactivity with mucus and the underlying respiratory tract tissues minimize desorption and increase uptake of gases and vapors. Mechanisms operating to remove gases and vapors are also different from that of particles. Dissolution, although important for particles, is not an important consideration for gases, because gases act as individual molecules. The reversible nature of gas uptake is another difference. Gas molecules can be desorbed from the airways and move further down the respiratory tract or are exhaled. Improved understanding of processes for uptake and removal of inhaled gases and vapors will result in better predictions of target tissue dosimetry and potential toxicity.

EXPOSURE SYSTEMS, GENERATION OF ATMOSPHERE, AND CHARACTERIZATION OF ATMOSPHERE

Exposure Systems

The emphasis in this chapter is on the study of airborne materials via inhalation exposure—the kind of exposures that most closely mimic human exposures. We appreciate that materials can also be administered via intracheal instillation, insufflation, or implantation directly into the lungs. However, we are concerned as to the relevance of information acquired from use of these nonphysiological modes of administration that bypass the normal protective mechanisms of the respiratory tract. These alternative exposure methods should be used with caution and careful attention given to how the results will be interpreted.

Integrated Exposure System

The conduct of studies with exposure of experimental subjects via inhalation to gases or particles requires the use of an integrated exposure system (54). Such a system must have provision for (i) generating the test material in the atmosphere,

(ii) delivering the test material to the experimental subjects in an atmosphere with appropriate humidity, temperature, and oxygen content, (iii) characterizing the test material in the atmosphere at the breathing zone of the subject(s), and (iv) disposing of the test atmosphere constituents. Such systems must usually be contained within specially designed facilities. In the ideal situation, the exposure system will be contained within an exposure room that is separate from a control room. The control room will have provision for readouts from the monitoring equipment and capability for adjusting the test material generator and airflows within the exposure system. In some cases, the potential hazardous properties of the test material being studied may require that the exposure system be contained within one or a series of sealed glove boxes. In short, it is crucial that appropriate precautions be taken to contain the test agent, minimize contamination of facilities, and, above all, avoid the exposure of research personnel.

Provision must exist for moving laboratory animals to and from the facilities where they are housed. In some cases, the animal housing and exposure facilities may be one and the same or may be contiguous to each other. For large operations, it is desirable to have necropsy facilities immediately adjacent to the exposure and housing facilities. The design and operation of inhalation exposure facilities and the related facilities for housing animals and conducting ancillary studies is not a trivial matter. It should always involve persons knowledgeable in health, environment and safety matters, and facilities design and operation and, especially, ventilation engineers, in addition to personnel knowledgeable about the experimental operations. Cheng and Moss (55) and Wong (54) have provided excellent reviews of the design and operation of exposure systems.

Nose-Only or Whole-Body Exposure

In planning every inhalation toxicity study, a critical decision must be made as to the approach to be used to nose-only or whole-body exposures. Each option has advantages and disadvantages (54–56). With nose-only exposure systems, animals are placed in restraining units that are inserted into the system delivering the test atmosphere. It is ideal, even for brief nose-only exposures, to acclimate the animals to the restraining tubes and the total system by conducting sham-exposures for a number of days prior to conducting exposures with the test agent. Multiple animal holders may be inserted into the aerosol delivery unit, which are usually designed to have the test atmosphere flow by each animal's nose. If the animal's body is sealed within the restraining unit, it is possible to have it function as a whole-body plethysmograph. Figure 16 (33) illustrates an example of a restraining unit modified for simultaneous use as a plethysmograph during exposures. In a plethysmograph, as the subject inspires the pressure increases in the body compartment and when the subject expires the pressure decreases. With appropriate calibration, the measurement of pressure changes provides a continuous measure of the respiratory rate and the volume of the test atmosphere inspired. Such data can be used to estimate the quantity of a test agent inhaled, this in conjunction with measurements of the quantity of test agents and metabolites in the respiratory tract and other tissues of animals sacrificed immediately after a short-term exposure can be used to calculate the fraction of the test agent deposited.

A major advantage of nose-only exposures is the elimination of contamination of the pelt by the test atmosphere and, thus, less opportunity for ingestion of the test agent by grooming. Another important advantage for some test agents is that much less of the test agent is required to conduct an inhalation exposure, as compared with

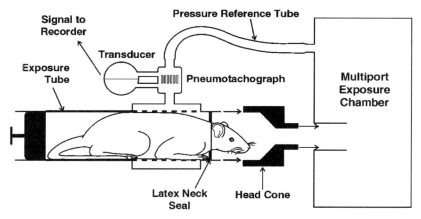

Figure 16 Schematic rendering of a nose-only exposure–restraining tube, modified for simultaneous use during exposure, as a plethysmograph. *Source*: From Ref. 33.

a whole-body exposure system. An obvious disadvantage of using nose-only exposure is the substantial effort required in handling each experimental subject. As a result, nose-only exposures are most frequently used for single exposures such as when conducting toxicokinetic or pharmacokinetic studies. However, there have been experiments conducted in which large numbers of rats were exposed nose-only for five days per week for extended periods of time to evaluate the carcinogenicity of fibers (57).

Whole-body exposures are the most typical mode exposure of subjects in inhalation toxicity studies. Exposure chambers used have ranged in size from as small as a bell jar placed in a laboratory hood to large rooms in which cage racks may be rolled in and out. Most frequently, the exposure chamber is on the order of 1 to $2\,m^3$ (Fig. 17) (55). The subjects may be housed individually or group housed in wire mesh cages. A major advantage of whole-body exposures is that a large number of subjects can be exposed with minimal handling. The H-2000 chamber illustrated can simultaneously house, in individual units, 144 rats (under 400 grams), or 360 mice, or 60 guinea pigs. A H-1000 chamber, half as large as the H-2000, is available and can house half as many animals. The chambers can be used to house the animals continuously irrespective of whether exposures are taking place. A major disadvantage

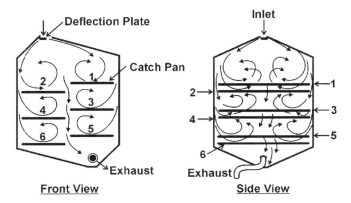

Figure 17 Schematic rendering of a whole-body exposure chamber. *Source*: From Ref. 55.

of the whole-body exposure mode is that the test atmosphere comes into contact with the animals' fur. This deposited material may be removed by grooming and be ingested. If the animals are group housed, this problem is further complicated by animals grooming each other. Grooming and ingestion of test material is a particular complication if the test agent is a particulate material and is readily absorbed from the gastrointestinal tract. Depending on the particle size of the test agent, it is possible for the amount of test agent to be ingested from grooming to exceed the quantity deposited in the respiratory tract. There is also potential for coprophagy—ingestion of feces—in which test material excreted in the feces may be ingested.

Test Agent Generation

Gases and Vapors

The approach used to generate each particular test material into the atmosphere delivered to the subjects must be customized; there is no universal approach available off the shelf. Wong (58) provides a review of methods applicable to the study of gases and vapors. For some gases, it may be as simple as generating the test atmosphere from a commercially available pressurized tank. However, even with a simple system, it is necessary to ensure that each test atmosphere is of appropriate concentration and purity. It is not appropriate to assume that the vendor has delivered a gas of the desired purity and concentration. For some chemicals, the gaseous test atmosphere may be generated, if the chemicals have the appropriate volatility, by passing the liquid chemical over glass beads, supported by a metal mesh, in a J-tube, to volatilize the chemical. The use of the glass beads increases the surface area available for volatilization. The rate of volatilization may be increased further by wrapping the J-tube with heating tape. In some cases, such as when generating an atmosphere containing ozone, a special generator must be used.

PM

The situation for generating particulate test atmospheres is usually much more complex than it is for agents in the gas phase. Moss and Cheng (59) provide an excellent review of methods applicable to studying particles and droplets. Numerous types of generators are available and the selection of a particular approach will be guided by the requirements of the specific study, especially the air concentrations needed, the nature of the exposure system (for example, nose-only versus whole-body), the duration of the exposures, and the animal species being studied.

Some test aerosols may be generated using nebulizers containing either solutions of soluble materials or liquid suspensions of insoluble particles. In the case of the soluble material, the liquid droplets from the nebulizer dry and the resulting particles contain the dried solute; examples are cesium chloride or sodium chloride. A radioactive tracer such as ^{137}Cs can be added to help determine exposure concentrations and deposition patterns. The liquid droplets from the nebulizer are polydisperse in distribution and, hence, the dried particles are polydisperse in distribution. The terms "monodisperse" and "polydisperse" refer to the size distribution of particles or droplets. *Monodisperse* particles refers to essentially uniform size while *polydisperse* particles refers to varied size. A liquid suspension of insoluble particles can be nebulized as liquid droplets containing particles. The number of particles in each liquid droplet will be a function of the concentration of the particles in the liquid suspension. If the suspended particles are in low concentration, there is a high probability that each liquid droplet will contain either no particles or a single particle. However, as the concentration of particles in the liquid suspension increases, more of the droplets will contain multiple particles, i.e., doublets, triplets, quadruplets, etc. Once the liquid droplets dry, the particles present in the droplet tend to adhere to one another. Thus, the resulting aerosol will be polydisperse in size distribution even if the particles were monodisperse. A very important point to make is that if the suspended particles are monodisperse in size, then the resulting aerosol will contain monodisperse or polydisperse distributions of particles, depending on the concentration of particles in the liquid suspension.

In most situations, the experimentalist is provided a sample of the particulate material of interest. In some cases, this may be a commercial product. It is even possible that it is a commercial product originally manufactured as large particles to minimize the potential for an inhalation hazard. In such a case, the experimentalist should avoid the temptation, even if pressured by regulatory authorities, to grind or otherwise treat the product to yield an increased fraction of inhalable particles. If this is done, the resulting assessment of hazard is no longer applicable to the original product. The goal of the researcher should be to characterize the hazard or risk of the product, as contrasted to demonstrating that health effects can be produced by manipulating the particle size of the material being studied. The hazardous properties of the product are a function of both its chemical composition and particle size.

The key performance standard of the PM generator should be to mimic the aerosol characteristics likely to be observed during: (i) the manufacture of the product if occupational exposures are of concern, (ii) the use of the product, as intended, if consumer or applicator exposures are of concern, and (iii) the presence of the particles in the environment if environmental exposures are of concern. To meet this performance standard, it is obvious that there must be close collaboration between multiple parties—those familiar with the manufacture and use of the product, industrial hygienists, and inhalation toxicologists. In some situations, it may be appropriate to expose experimental subjects to complex atmospheres such as those from motor vehicles. Hesterberg et al. (60) have recently reviewed the extensive research conducted on exposure of laboratory animal species to vehicle emissions, emphasizing the results of studies of diesel exhaust. During the late 1990s, studies began to be conducted using concentrated fine ambient aerosols (61).

Test Atmosphere Delivery

Once the test atmosphere is generated, it must be delivered to the breathing zone of the experimental subjects. In some cases, the test atmosphere may need to be diluted and conditioned to achieve the desired temperature and humidity. Special attention must be given to the internal surface characteristics of the delivery system and the flow rates used. Reactive gases such as ozone may interact with some surfaces and not with others. Particles less than $0.5\,\mu m$ may be lost to surfaces by diffusion. Very small particles less than $0.5\,\mu m$, and especially those less than $0.1\,\mu m$, have a high potential for coagulation. Large particles may deposit through settling in long delivery or sampling tubes and by inertial impact, i.e., at points in the system with abrupt changes in the direction of flow.

Test Atmosphere Characterization

Sampling at the Breathing Zone

The test atmosphere should be sampled for its chemical characteristics, size (if a particulate material), and concentration as close to the breathing zone of the subjects as possible. This requires sampling directly from where the cages housing the animals are located in the exposure chamber if whole-body exposures are being conducted or at the location of the subject's nose in a nose-only exposure unit. Samples collected at the test material generator are a poor surrogate for what the subjects are likely to breathe. Likewise, the use of a long sampling tube between the chamber or nose port and the monitoring device may yield erroneous measurements. Instrumentation that allows both instantaneous measurements and integrated measurements of concentration are particularly useful. A few very brief measurements of concentration may not be representative of exposures that are hours in duration and are certainly not appropriate for exposures that extend over months. Excellent reviews of methods used to characterize test atmospheres are found in the cited references (58,62–65).

Characterizing PM

With PM exposures, indirect measurements of particle mass may be made using light-scattering monitors. However, those measurements should always be validated with gravimetric measurements made on samples collected on filters. This requires accurate weighing of filters before and after collection of the sample under controlled conditions, especially of humidity. In addition, it is necessary to have accurate measurements of airflow through the filter to calculate the air volume sampled.

For exposures involving particulate material, it is crucial that the particle size distribution be measured in addition to PM mass. A variety of instruments and approaches are available to make such measurements. The validity of any measurements made will depend on careful calibration of the unit. Such calibrations must take into account the chemical or physical properties of the material being studied, for example, its specific density and how this influences aerodynamic size. Many monitors are calibrated by the manufacturer using polystyrene particles that have a specific density close to $1 \, \text{g/cm}^3$. Frequently, the investigator may be working with a material having a specific density of $2 \, \text{g/cm}^3$ or more. The aerodynamic diameter is heavily influenced by the specific density of the material. Hinds (63) provides an excellent review of the concepts of particle sampling and sizing.

A frequent approach to characterizing particle size distribution in inhalation toxicity studies is to use a cascade impactor. With a cascade impactor, particles are collected by aerodynamic size on a series of collection plates and a backup filter. The airflow rate through the device is constant; each stage has one or more openings that are progressively smaller, hence, the air velocity increases at each stage. The particles with the largest aerodynamic size are deposited in the first stage with successively smaller aerodynamic sized particles collected on each successive stage. The smallest particles are collected on the backup filter. An example of size distribution data collected with a cascade impactor is shown in Figure 18 (59). In Figure 18A, note that the polydisperse size distribution frequency has a bell shape. When plotted as a logarithmic probability, the data points fall on a straight line, i.e., it is log normal in distribution (Fig. 18B). In this example, the test material represents a single mode, the coarse mode, of the trimodal distribution illustrated earlier in Figure 4. This is typical of the particle sizes for test materials used in toxicity studies. If data had been shown for particles in diesel exhaust, the distribution would have corresponded to the smaller accumulation mode in Figure 4.

When gases are studied, care must be taken to make certain that the monitor is accurately calibrated for the specific gas being studied at the concentrations expected in the exposure system. Only a small fraction of the test material generated is actually inhaled and deposited by the experimental subjects; most of the test agent simply

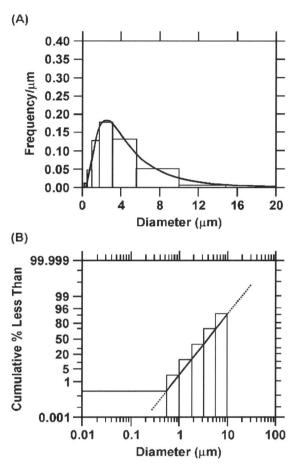

Figure 18 Schematic rendering of size distribution data collected with a cascade impactor. (A) Mass fraction: impactor data. (B) Cumulative mass fraction: impactor data. *Source*: From Ref. 59.

flows through the system. In most cases, provision must be made for disposal of residual test material and test material collected in sampling devices. The disposal system must be tailored to the test agent and physical states of residual and collected samples of the test agent.

STUDY DESIGN

Toxicokinetics or Pharmacokinetics Studies

Toxicokinetic studies provide a quantitative assessment of the temporal dimensions of deposition, retention, excretion, and tissue distribution of a potentially toxic test material. When the test material is a pharmaceutical agent, the term *pharmacokinetic* is more appropriate. The approach to conducting and interpreting kinetic studies is described in detail in other chapters in this text. In this chapter, we will briefly describe unique aspects of conducting kinetic studies with inhaled materials. A schematic representation of the experimental design for a toxicokinetic study is shown in Figure 19.

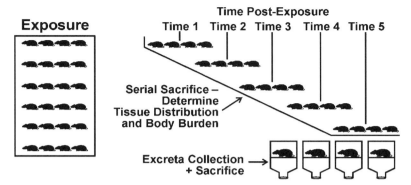

Figure 19 Schematic rendering of the design of an inhalation toxicokinetic or pharmacokinetic study.

The most challenging aspect of conducting kinetic studies with inhaled materials, compared with other routes of administration of test agents, relates to estimating the quantity of test agent received by the subjects. For particles, this involves estimating the amount of the test agent that is inhaled and deposited in the respiratory tract and is available for clearance from the body or distribution to other tissues or for excretion. For gases, this involves estimating the amount of the test agent that is inhaled, and either retained in the respiratory tract or distributed throughout the body. In both situations, it is necessary to accurately estimate the quantity of agent inhaled. This requires knowledge of both the air concentration of the test agent and the volume of air inhaled by the subject(s). With particulate material, there is the additional challenge of determining what portion of the inhaled material was deposited and where it was deposited in the respiratory tract. There is the further complication as to what portion of the deposited material is cleared to the pharynx and ingested. For gases, it is necessary to determine the portion of the inhaled material that is retained and not eliminated upon exhalation.

As illustrated in Figure 19, provision may be made for termination of some exposed subjects immediately after cessation of exposure. The total quantity of material recovered in the tissues may be used as an estimate of the initial body burden. With particulate material and reactive gases, it may be appropriate to sample the various regions or subregions of the respiratory tract. If the concentration of the test material in the test atmosphere and the volume of air inspired by the subjects is known, it may be feasible to calculate the fractional deposition. Obviously, such measurements are greatly facilitated when a radioactive or stable isotope label is incorporated into the test agent.

Depending on the test agent being studied, it may be appropriate to have some exposed subjects housed in sealed metabolism units with provision for trapping the agent or metabolites in expired air and for collecting urine and feces. The termination of animals and the evaluation of the tissue burden of the test agent or metabolites, in combination with the measurements described above, allow for "mass balance" determinations to be made. In our experience, recovery of material equivalent to 80–120% of the estimated intake is quite good. It is appropriate to be wary of "mass balance" data or tissue recovery that equals 100%; the data are likely to be too good to be true and probably reflect the use of a normalization procedure.

Termination of animals at multiple times post a single exposure allows the development of a dataset that can be evaluated to provide insight into the temporal and spatial kinetics of the test agent. The specific times selected for terminating animals need to reflect the expected kinetics of the particular test agents. For some rapidly metabolized compounds such as vinyl chloride, the period of interest may be minutes. For relatively insoluble particulate material, the period of interest may extend over hundreds of days.

Exposure (Dose)–Response Studies

A key issue in evaluating the potential toxicity of any agent, including inhaled agents, is to identify both the critical health responses associated with exposure to the agent and the relationship of these responses to exposure. Ultimately, there is a desire to characterize the exposure–response relationship for the agent. A schematic representation of an experimental design for evaluating the exposure–response relationship for any inhaled material is shown in Figure 20.

Earlier reference was made to toxicokinetic studies, providing a linkage between exposure and tissue dose. Counterpart studies linking dose and health responses are called toxicodynamic studies, with toxic agents, or pharmacodynamic studies, with pharmaceuticals. In the case of inhaled materials, there are special challenges to defining tissue doses, as discussed earlier, that complicate any consideration of the mechanisms by which the agent may cause adverse health responses. This is the case for the various parts of the respiratory tract as well as other organs.

The schematic representation in Figure 20 is applicable to understanding potential adverse health responses from long-term exposures including lifetime or near-lifetime exposures. Evaluation of response with increasing duration of exposure has traditionally been used to evaluate the potential of an agent to cause cancer. Increasingly, a similar approach has been used to evaluate the potential of agents to cause noncancer health effects. A key aspect of the design of such studies is to include three or more exposure levels and sham-exposed controls to be observed for two years or longer. There are multiple issues involved in selecting exposure levels. It is obviously desirable to have the lowest level of exposure anchored to likely levels of human exposure. Traditionally, in toxicology, some level greater than likely, human exposure is studied to maximize the potential for observing

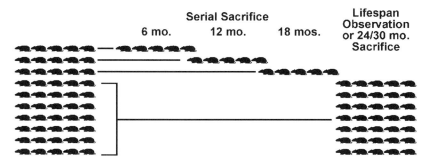

Single level illustrated—preferable to use 3 exposure levels, chamber controls and colony controls

Figure 20 Schematic rendering of the design of an inhalation exposure–response study such as might be used to evaluate end points of chronic diseases including cancer.

effects attributed to exposure to the agent. It has generally been advocated that the highest level of exposure should be at a "maximum tolerated dose" (MTD) to again maximize the potential for observing effects. The MTD is generally viewed as a level less than a level that would produce generalized, nonspecific toxicity. The NRC has addressed this general issue (66), and a committee formed by the National Toxicology Program (NTP) has specifically addressed the issue for inhaled materials (67); Hesterberg et al. (68) have addressed the MTD issue for studies with inhaled fibers.

An additional key element of the experimental design advocated is the provision for serial termination of subjects to provide material for detailed study. Such observations are important to understanding the temporal dynamics of the pathobiological processes leading to the critical health responses. It is important to recognize that the same approach can be used to understand adverse health responses, other than cancer, that may be observed with short periods of exposure or following lifetime exposure. One school of thought is that it is not necessary to include provision for serial termination in the initial long-term bioassay, but rather merely observe the results of the bioassay and then determine if it is appropriate to do follow-on studies involving serial termination. We hold to a view that it is more appropriate to include provision for serial termination in the original core study. The results of the serial terminations inevitably help in interpreting the observations in the animals dying late in the study or in those euthanized at study termination. Serial measurements of the tissue burden of material may be especially useful if the test material, or its metabolites, have a long retention time and accumulate with protracted exposure.

RESPIRATORY TRACT RESPONSES

Multiple Responses

The various regions of the respiratory tract respond to inhaled materials with the same generic responses observed in other organs, namely; irritation, inflammation, cell death, hypertrophy, hyperplasia, fibrosis, metaplasia, and neoplasia. The complex structure of the respiratory tract, the many cell types present, the diversity of toxic insults, and the influence of exposure duration and intensity of exposure to toxicants alter the occurrence of a myriad of diseases of the respiratory tract arising from multiple etiologies. It is not surprising that in human medicine, two major medical specialties focus on diagnosis and treatment of diseases of the respiratory tract—ear, nose, and throat specialists, and chest specialists.

Disease responses in the respiratory tract, as in all tissues and organs, are manifest at all levels of biological organizations from molecular and biochemical changes, to cellular responses, at the tissue and organ level and finally with impact on the individual. It follows then that a range of approaches can be used to evaluate normal versus alterations in structure and function and that the study of responses of the respiratory tract to toxic agents is best approached as a collaborative, multidisciplinary effort. The purpose of this section of the chapter is to outline some of the important concepts for evaluating respiratory tract responses.

A starting point for evaluating responses to inhaled materials is to consider the general clinical response of the exposed subjects, directing special attention to those responses related to the respiratory tract. Is there evidence of altered respiratory rate or the nature of respiration? Is there evidence of any nasal secretions? Are there any enlargements of the nose that may serve as an indication of developing nasal tumors?

Pulmonary Function

If evidence exists from short-term studies that the test agent does have potential for injuring the lungs, there is certainly merit in assessing pulmonary function. This is a specialized area of endeavor requiring specialized equipment and experienced personnel. A useful review of this subject has been prepared by Mauderly (69). A major value of pulmonary function tests is that they provide a nondestructive and largely noninvasive means of assessing the functional status of the respiratory tract with regard to whether or not function is impaired, the nature of the impairment, and the magnitude of any loss of function. These nondestructive assessments have been carried out serially in two-year studies of rodents and in long-term studies of larger species. The range of pulmonary function tests that can be performed in laboratory animals is remarkably similar to those performed in a human clinical laboratory, concerning evaluating respiratory disease. This is of value in understanding the potential human relevance of findings in laboratory animals. The value of pulmonary function findings is enhanced where they can be related to serial histopathological observations.

Pulmonary function tests are used to obtain information on ventilation, gas distribution, alveolar capillary gas exchange, and perfusion. The goal of respiration at the pulmonary level is to bring air and blood continually into opposition across the thin alveolar–capillary boundary so that gas exchange can take place by diffusion. The ventilation parameters of primary interest are the respiratory rate, lung volumes, and lung mechanics. Measurements of minute volume, total lung capacity, vital capacity, fractional residual capacity, and residual volume can be made. These measurements can provide insight into the size of the lungs and incidence of restrictive or obstructive lung disease. The elastic properties of lung tissue may be evaluated by measuring pressure–volume characteristics and expressing the results as compliance, the slope of the pressure–volume curve.

Forced or maximal effort inhalation has routinely been used in human medicine to assess dynamic lung mechanics with the aim of detecting airflow obstruction with greater sensitivity and descriptive value than that achieved during tidal breathing. Perhaps you have had the test procedure conducted in this manner—"take a deep breath and exhale as rapidly and completely as possible." The same basic procedure can be carried out in apneic laboratory animals and induced with negative pressures. In humans, the results are usually expressed as the forced expiratory volume in one second. For animals, the usual approach is to plot the maximal expiratory flow volume curve and measure peak expiratory flow, flow at different lung volumes or percentages of the forced vital capacity, and the mean flow over the midvolume range (usually 75–25% of forced vital capacity).

The distribution of gas in the lung is important. Optimally, the inhaled gas is distributed proportionately to the area of the alveolar capillary membrane perfused by pulmonary arterial blood. However, not all areas of the lungs, even when normal, are ventilated or perfused equally at all times. One approach to evaluating this parameter is to study the washout of gases such as helium or nitrogen by evaluating the fractional end-tidal concentration of the gas. It is also possible in larger species to study the distribution of inhaled radioactively labeled gas using a gamma camera to assess the distribution of inhaled gas in the lungs. Diffusion of oxygen and CO_2 across the alveolar capillary membrane can be assessed indirectly by evaluating expired gas. It is also feasible to measure the partial pressures of oxygen and CO_2 in arterial blood, and the blood pH.

The carbon monoxide diffusion capacity is a sensitive method for detecting impairment of alveolar capillary gas transfer at rest in both laboratory animals and people. Several methods have been adapted for use in laboratory animals. Perfusion of the alveolar capillary bed is as critical to gas exchange as is the ventilation of the alveoli. Perfusion may be evaluated in humans by administering radiolabeled albumin, and imaging the lungs with a gamma camera. The use of the procedure in small laboratory animals is obviously limited by resolution of the gamma cameras available.

In summary, pulmonary function tests have the potential for more frequent use in evaluating inhaled toxicants in laboratory animals. The ease of extrapolation of the findings from laboratory animals to humans contrasts with recent attention given to the use of more "reductionist" measurements at the molecular or biochemical level for which the significance of the findings in humans may be open to question.

Pulmonary function evaluations have been especially useful in evaluating the irritant properties of airborne materials. Early work in this area by Alarie (29) has been extended by him and others so that a large database on sensory irritants exists (70,71).

Immunological Studies

The evaluation of immune phenomena in the respiratory tract is an area that has received substantial attention with regard to the increased incidence of asthma. However, the amount of research conducted in laboratory animals has been limited and generally related to the development of hypersensitivity (72–74).

Bronchoalveolar Lavage Markers

Valuable insight into the status of the terminal airways and alveolar region can be gained from studying cellular and biochemical markers in recovered bronchoalveolar lavage fluid (75). In humans, the procedure involves the clinician using a fiber optic bronchoscope to introduce a small volume of saline into a peripheral airway and then recovering the fluid along with cells and biochemical markers contained therein. The procedure is especially useful in studying responses to inhaled agents in people because serial measurements can be made on the fully conscious individual. Similar approaches have been used in anesthetized large laboratory animals. With small laboratory animals such as rats, the procedure is usually done on animals that have been euthanized at various times after initiation of exposure to the test agent. A single lobe may be used for lavage, with the remainder of the lung being available for evaluation using histopathology or other methods.

The recovered lavage fluid may be examined to characterize the cells present, especially the numbers of inflammatory cells versus macrophages. A wide array of biochemical parameters can be measured, limited only by the interests of the investigator and the assays available. Frequently, protein is measured to assess damage to the capillary-alveolar barrier, and lactate dehydrogenase is measured as an indicator of cytotoxicity. An increasing number of cytokines involved in development of pulmonary inflammation and fibrosis have been measured in humans and are also being studied in laboratory animals.

Pathological Evaluations

Both gross and histopathological evaluations remain important approaches to evaluating the response of the respiratory tract to toxic agents (11,48,76,77). Careful gross examination of the entire respiratory tract from the nares to the alveoli can give insight into the presence of both noncancer effects and tumors. Because of the structure of the respiratory tract, special care must be taken in collecting specimens for processing for histopathological evaluation.

Because the nasal cavity is complex and changes in its characteristics along its major dimension, it has become standard practice to section the noses at multiple levels. Likewise, it is important to collect and process the larynx as a separate specimen. The P region is best examined in an "inflated" state as contrasted to a "collapsed" state in which the normal structure has been changed. To provide suitable fixed specimens, the trachea, bronchia, and lungs are usually fixed at constant pressure and a ligature is placed around the trachea or, alternatively, around a bronchus if only a single lobe is to be used for histopathological examination. A pathologist should be consulted in advance with regard to the most appropriate fixative and fixation time, the choice being dependent on the evaluation procedures. It is customary to section the lung in a manner that allows the pathologist to observe on a single section all the structures from the terminal airways to the pleura of the lung.

Recent research on inhaled particulate material has emphasized the importance of evaluating the complete cardiorespiratory system rather than focusing on the respiratory tract and heart as separate entities. Thus, it is important even in "mechanistic" studies that may focus on the respiratory tract to also evaluate the heart.

Biochemical Studies

Much can be learned from evaluating the response of various biochemical parameters in respiratory tract tissues to toxic agents either inhaled or that reaching the respiratory tract via the bloodstream (52,78,79). The planning and conduct of such studies needs to recognize the diverse characteristics of various portions of the respiratory tract including areas that are anatomically close together. For example, if one is interested in biochemical changes in the alveoli, care should be taken to not include large portions of conducting airway tissue in the specimen being homogenized for evaluation. Likewise, it would be inappropriate to include the alveolar tissue in a sample focusing on changes in the airways. This admonishment is equally appropriate for studies examining molecular or "genomic" markers. Potential responses in one microanatomical area may well be masked by a lack of responses in closely adjacent tissues.

Effects in Other Tissues from Inhaled Materials

It has already been emphasized that the effects of inhaled agents are not restricted to the respiratory tract. Hence, it is important in planning and interpreting any inhalation study to consider the likely nature of nonrespiratory tract end points. Other chapters in this book address many of those end points. In addition, it is useful to consider the relevant Environmental Protection Agency (EPA) guidelines for developmental toxicity (80), neurotoxicity (81), carcinogenicity (82), gene cell mutations (83), and reproductive toxicity (84). The application of these guidelines to the inhalation mode of exposure may require considerable ingenuity, certainly more than that required in the case of other modes of exposure.

ASSESSING HUMAN RISKS OF INHALED PARTICLES

Use of Inhalation Toxicology Data to Assess Human Risks

Data derived from inhalation toxicology studies play a vital role in developing work-place practices, voluntary guidance, or regulations protecting workers and the public from exposures to airborne materials that may cause adverse health impacts (85). The approaches used by both private and government organizations around the world are, in principle, quite similar. As discussed earlier, maximum use is made of data obtained from human studies, either epidemiological or from controlled human exposures. In the absence of adequate human data, findings from laboratory animal studies are required. Data obtained from studies in cell and tissue systems can be useful in interpreting findings in intact mammals or can be used in a precaution-ary manner to identify potential human hazards. However, the current lack of under-standing of the relationship between dose–response relationships in tissue and cellular systems and exposure (dose)–response relationships in intact mammals for specific agents precludes the use of cell and tissue data in setting quantitative expo-sure limits for people.

A critical issue in evaluating the potential human health risks of an inhaled tox-icant is the nature of the exposure–response relationship. Typically, the data from any individual study are inadequate to provide a basis for selection from several alternative models (Fig. 21), necessitating the need to extrapolate from the exposure region where an excess of effects over background incidence can be observed. To a considerable extent, the challenge is to maximize the data that can be used to define exposure–response relationships in the observable range and provide a scientific basis for extrapolating to lower exposure levels with acceptable levels of risk. Approaches to assessing the health risks from airborne materials can be divided into (i) those for which the primary concern is cancer, and (ii) those for which the con-cerns are adverse functional changes and diseases other than cancer. Traditionally,

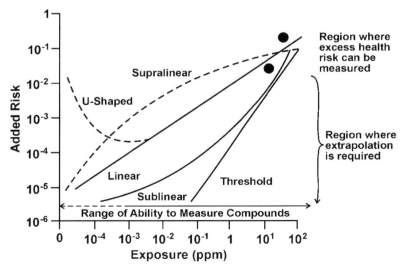

Figure 21 Schematic rendering of potential exposure–response relationships for inhaled tox-icants. The solid circles represent the results of a cancer bioassay with formaldehyde. A control value is not shown.

the exposure–response relationship for functional or clinically discernible changes has been viewed as having a threshold level for exposure that must be exceeded before responses are observed. Below the threshold, the adverse effect was presumed to not occur or be at a minimum level relative to the background incidence of disease. Above the threshold, the severity of the adverse effect was considered to increase with increasing exposure. The threshold level may vary for subpopulations differing in sensitivity. With functional changes, the response variable is continuous. In contrast, exposure–response relationships between exposure to a pollutant and cancer, traditionally, have been assumed to not have a threshold. The disease outcome, cancer, is dichotomous for individuals in a population. The relationship between exposure and incidence of cancer in the population is often described in probabilistic terms and extrapolated, as a linear function of exposure, to the lowest levels of exposure. The use of linear exposure–response relationships also makes it easy to calculate population risk if the population exposure is known or can be estimated.

Noncancer Responses

Threshold Exposure–Response Relationships

The earliest risk assessments for airborne materials, which were much less formal than those conducted today, focused on alterations in function and structure. The American Conference of Governmental Industrial Hygienists (ACGIH), since 1938, has provided leadership for the development of uniform guidance for limiting exposure to industrial chemicals. The ACGIH approach is shown in Table 2 (86) along with other approaches to evaluating noncancer effects. A cornerstone of the ACGIH approach has been that for adverse effects, a threshold was assumed to exist for exposure–response relationships such that a level of exposure could be defined below which no effect would be observed. This is shown schematically in Figure 22. Hence, the term threshold limit value (TLV) is used, which is usually expressed as a time-weighted average for a normal eight-hour workday and a 40-hour work week (87).

The introduction to the ACGIH TLV document (2005) defines TLVs as airborne concentrations of substances [that] represent conditions under which it is believed that nearly all workers may be repeatedly exposed day after day without adverse health effects. Because of wide variation in individual susceptibility, however, a small percentage of workers may experience discomfort from some substances at concentrations at or below the threshold limit; a smaller percentage may be affected more seriously by aggravation of a preexisting condition or by development of an occupational illness. Smoking of tobacco is harmful not only because of the cancer and cardiorespiratory diseases it causes, but for several additional reasons. Smoking may act to alter the biological effects of chemicals encountered in the workplace and may reduce the body's defense mechanisms against other substances.

Individuals may also be hypersusceptible or otherwise unusually responsive to some industrial chemicals because of genetic factors, age, personal habits (e.g., smoking, consumption of alcohol, or use of other drugs), medication, or previous exposures. Such workers may not be adequately protected from adverse health effects from certain chemicals at concentrations at which such workers require additional protection.

TLVs are based on available information from industrial experience, from experimental human and animal studies; and, when possible, from a combination

Table 2 Comparison of Risk-Assessment and Risk-Management Estimates

Agency/guidance	Use	Exposure scenario	Effect level	Population	Database	Dosimetry	SF or UF
ACGIH TLV-TWA	Management	8 hr/day, 40 hr/wk, 40 yr	Impairment of health or freedom from irritation, narcosis, nuisance	Healthy worker experience, experimental human and animal	Industrial	No	SF
NIOSH REL	Characterization	10 hr/day, 40 hr/wk, "Working lifetime"	Impairment of health or functional capacity and technical feasibility	Healthy worker	Medical, biological, chemical trade	No	SF
OSHA PEL	Management	8 hr/day, 40 hr/wk, 45 yr	Impairment of health or functional capacity and technical feasibility	Healthy worker	Medical, biological, chemical trade	No	SF
EPA RfC	Dose-response	24 hr/day, 70 yr	NOAEL	General population, including susceptible	Occupational, experimental human and animal	Yes	UF
ATSDR MRL	Dose-response	24 hr/day, 70 yr	NOAEL	General population, including susceptible	Occupational, experimental human and animal	No	UF

Abbreviations: ACGIH, American Conference of Governmental Industrial Hygienists; ATSDR, Agency for Toxic Substances and Disease Registry; EPA, Environmental Protection Agency; MRL, minimal risk level; NIOSH, National Institute of Safety and Health; NOAEL, no observed adverse effect level; OSHA, Occupational Safety and Health Administration; PEL, permissible exposure limit; REL, recommended exposure limit; RfC, reference concentration; SF, safety factor; TLV-TWA, threshold limit value-time-weighted average; UF, uncertainty factor for explicit extrapolations applied to data.
Source: From Ref. 86.

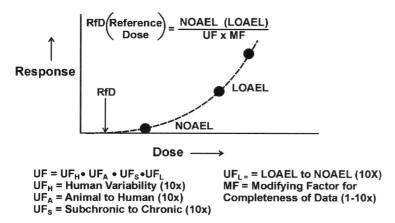

$$\text{RfD} \binom{\text{Reference}}{\text{Dose}} = \frac{\text{NOAEL (LOAEL)}}{\text{UF x MF}}$$

Response

RfD

LOAEL

NOAEL

Dose ⟶

UF = UF$_H$• UF$_A$ • UF$_S$•UF$_L$ UF$_{L=}$ = LOAEL to NOAEL (10X)
UF$_H$ = Human Variability (10x)   MF = Modifying Factor for
UF$_A$ = Animal to Human (10x)   Completeness of Data (1-10x)
UF$_S$ = Subchronic to Chronic (10x)

Figure 22 Schematic representation of a threshold exposure (or dose)–response relationship typically used for evaluating noncancer end points. *Abbreviations*: LOAEL, lowest observed adverse effect level; NOAEL, no observed adverse effect level; UF, uncertainty factor.

of three. The basis on which the values are established may differ from substance to substance; protection against impairment of health may be a guiding factor for some, whereas reasonable freedom from irritation, narcosis, nuisance, or other forms of stress may form the basis for others. Health impairments considered include those that shorten life expectancy, compromise physiological function, impair the capability for resisting other toxic substances or disease processes, or adversely affect reproductive function or developmental processes. The ACGIH emphasizes that the TLVs are intended for use in the practice of industrial hygiene as guidelines in the control of workplace health hazards and are not intended for other use. The ACGIH documents also emphasize that the limits are not fine lines between safe and dangerous concentrations, nor are they relative indices of toxicity.

Both the National Institute of Safety and Health (NIOSH) and the Occupational Safety and Health Administration (OSHA) develop guidance or regulations that are similar to the ACGIH TLVs (Table 2). NIOSH, as a research organization within the Department of Health and Human Services, develops recommended exposure limits (RELs) (88). A *Pocket Guide to Chemical Hazards* is also published regularly (89). The RELs are guidance values that are subsequently used by OSHA and Mine Safety and Health Administration in promulgating limits expressed as permissible exposure limits (PELs).

In addition to the RELs and PELs developed by government agencies and the TLVs developed by the independent ACGIH, another set of guidance values have been developed by the NRC. The NRC Committee on Toxicology has, for many years, developed PELs and emergency exposure guidance levels for selected airborne contaminants (90). These values have been set for selected contaminants, with specific consideration given to the needs of those in the military services. This includes the development of short-term public emergency guidance levels, continuous exposure guidance levels, occupational exposure guidance levels for repeated exposure of military personnel during training exercises, and long-term public exposure guidance levels for repeated exposure of the public residing or working near military facilities. In addition, the committee has provided guidance to the National Aeronautics and Space Administration for the setting of spacecraft maximum allowable concentrations (91).

In addition to the activities described here, many industrial corporations that are producers and users of chemicals have active internal programs for the appropriate control of airborne contaminants. Such programs, which provide guidance for use within the corporation or by its customers, complement the regulatory mandates of the government and the guidance values of the ACGIH. In addition, there is a need to provide guidance for new and specialized materials for which formal guidance is not available.

Because TLVs involve threshold exposure–response relationship, the establishment of a TLV quite naturally is based on the use of safety factors. A threshold level for some identified hazard and for the lowest observed adverse effect may be established in humans or laboratory animal populations and then safety factors added to extrapolate to lower levels with an associated lower probability for hazard (Table 3).

If human data of sufficient quality are available on a particular chemical, they are used to set the TLV, REL, or PEL. In many cases, however, human data are not available. This may be the result of adequate control measures such that adverse effects have not been observed in workers despite the chemical having been in commerce for many years. By definition, a newly synthesized material has not been available, so human exposure could not have occurred. In these cases, data from laboratory animals are essential for establishing a TLV for humans. This is the current situation with nanomaterials and hence, the need for research across the full paradigm shown in Figure 1.

Traditionally, in the absence of adequate human data, the information needed for establishing a TLV has been obtained by conducting studies in one or more laboratory animal species (usually rats and mice) with groups of animals exposed at various concentrations and then a range of observations made in life and at necropsy as illustrated in Figure 20. Customarily, studies progress from an assessment of the concentrations required to cause acute injury and death to studies involving exposures of longer duration, such as 2 weeks, 90 days, and 2 years. The experience gained with the short-term exposures provides information on target organs and the possibility of chemical accumulation. Most importantly, the information helps guide the establishment of exposure concentrations for the long-term studies. In the ideal situation, having animals killed at intermediate time points to provide specimens for detailed evaluation is advantageous. The number of animals used per group can vary from as few as 5 to 10 for the shortest studies to 50 or more in the two-year studies. Multiple exposure levels, usually at least three levels of the test agent, and a control are used.

Beyond evaluating various responses to the inhaled material, it is desirable to establish the dose of the toxicant that reaches critical tissues and cells as described earlier under toxicokinetic studies. For compounds that are metabolized, it is desirable to establish the dose of metabolites. This broadens the orientation from a focus on exposure–response relationships to that of exposure-dose–response relationships. This broader orientation is especially critical because various mammalian species differ significantly in how they handle specific chemicals. The relationship between exposure and tissue dose (i.e., the disposition of the toxicant) in humans should not be assumed to be the same as that observed in rats or mice. If significant differences in the exposure-to-dose linkage are observed, then there will also be an influence on the overall linkage of exposure-tissue dose–response. Indeed, studying and understanding the basis for interspecies differences in exposure–dose relationship can give valuable insight into understanding the exposure–response linkage.

Table 3 Guidelines for the Use of UFs in Deriving Inhalation RfC

Standard UFs	Processes considered in UF purview
H = Human to sensitive human. Use a 10-fold factor when extrapolating from valid experimental results from studies using prolonged exposure to average healthy humans. This factor is intended to account for the variation in sensitivity among the members of the human population	Pharmacokinetics–pharmacodynamics Sensitivity Differences in mass (children, obese) Concomitant exposures Activity pattern Does not account for idiosyncrasies Pharmacokinetics–pharmacodynamics
A = Animal to human. Use a threefold factor when extrapolating from valid results of long-term studies on experimental animals when results of studies of human exposure are not available or are inadequate. This factor is intended to account for the uncertainty in extrapolating animal data to the case of average healthy humans. Use of a UF of 3 is recommended with default dosimetric adjustments. More rigorous adjustments may allow additional reduction. Conversely, judgment that the default may not be appropriate could result in an application of a factor	Relevance of laboratory animals model Species sensitivity
S = Subchronic to chronic Use a factor when extrapolating from less-than-chronic results on experimental animals or humans when there are no useful long-term human data. This factor is intended to account for the uncertainty in extrapolating from less-than-chronic NOAEL to chronic NOAEL	Accumulation-cumulative damage Pharmacokinetics-pharmacodynamics Severity of effect Recovery Duration of study Consistency of effect with duration Severity
L = LOAEL$_{HEC}$ to NOAEL$_{HEC}$ Use a factor when deriving an RfC from a LOAEL$_{HEC}$, instead of a NOAEL$_{HEC}$. This factor is intended to account for the uncertainty in extrapolating from LOAEL$_{HEC}$ to NOAEL$_{HEC}$	Pharmacokinetics–pharmacodynamics Slope of dose–response curve Trend, consistency of effect Relationship of end points Functional vs. histopathological evidence Exposure uncertainties
D = Incomplete to complete database Use up to a factor when extrapolating from valid results in experimental animals when the data are "incomplete." This factor is intended to account for the inability of any single animal study to adequately address all possible adverse outcomes in humans	Quality of critical study Data gaps Power of critical study and supporting studies Exposure uncertainties
MF = Modifying factor Use professional judgment to determine whether another uncertainty factor MF that is ≤10 is needed. The magnitude of the MF depends upon the professional assessment	

(Continued)

Table 3 Guidelines for the Use of UFs in Deriving Inhalation RfC (*Continued*)

Standard UFs	Processes considered in UF purview
of scientific uncertainties of the study and database not explicitly treated above (e.g., the number of animals tested or quality of exposure characterization). The default value of the MF is 1	

Note: Assuming that the range of the UF is distributed log-normally, reduction of a standard UF by half (i.e., 10^{-5}) results in a UF of 3. Composite UF for derivation involving four areas of uncertainty is 3000 in recognition of the lack of independence of these factors. Inhalation reference concentrations are not derived if all five areas of uncertainty are invoked.
Abbreviations: HEC, human equivalent concentration; LOAEL, lowest observed adverse effect level; MF, modifying factor; NOAEL, no observed adverse effect level; RfC, reference concentration; UF, uncertainty factor.
Source: From Ref. 86.

The information acquired from the animal studies, the results of which are shown schematically in Figure 22, is used to provide guidance for controlling human exposures. The key determinations are the establishment of a no observed adverse effect level (NOAEL), or, in the absence of such a determination, a lowest observed adverse effect level (LOAEL). The selection of exposure levels for study dictates the specific NOAEL and LOAEL values that can be observed. To state the obvious, observations can be made only at the exposure levels studied. Selection of these levels can have a dramatic impact on the calculated TLV. Extrapolations from the NOAEL or LOAEL to the TLV have been made for occupational exposures using safety factors. As a default assumption, laboratory animal species and humans are assumed to have similar exposure–response relationships. Thus, the addition of safety factors in extrapolating from laboratory animals to humans is viewed as increasing the likelihood of human safety. As is described subsequently, the U.S. EPA and the Agency for Toxic Substances and Disease Registry (ATSDR) use basically the same numerical values and only identify them as uncertainty factors (UF) for developing environmental exposure limits (Table 2). The EPA's and ATSDR's use of the term *UF* places emphasis on the uncertainty of the extrapolation.

As noted previously, the TLVs, RELs, and PELs are intended to provide guidance for occupational exposure and are not intended for use in establishing exposure limits for the general population. To provide guidance applicable to the general population for evaluating noncancer health effects of inhaled materials, the EPA has developed an inhalation reference concentration (RfC) methodology (86,92–95). Jarabek (86) reviewed the methodology in detail and defined an RfC as "an estimate (with uncertainty spanning perhaps an order of magnitude) of a continuous inhalation exposure to the human population (including sensitive subgroups) that is likely to be without appreciable risk of deleterious noncancer health effects during a lifetime." The RfCs values are comparable to the referenced dose values used for oral intake.

Jarabek (86) compared the RfC approach used by the EPA with approaches used by other agencies for providing guidance for exposures to limit health effects other than cancer (Table 2). All the approaches have in common the use of safety factors or UFs for making extrapolations from laboratory animals to humans if sufficient human data of adequate quality are not available (Table 3). In sharp contrast to the no-threshold approach that the EPA has traditionally used in assessing cancer risks, all the approaches assume a threshold in the exposure–response relationship. Of the five approaches listed in Table 2, only the EPA RfC methodology has provision for using dosimetry data to make extrapolations. Recognizing the extent to which there are marked species differences in exposure (dose)–response relationships, the dosimetry adjustment provision in the RfC methodology is a significant advance over other approaches that do not have a provision for such adjustments. The RfC methodology is intended to provide guidance for noncancer toxicity, that is, for adverse health effects or toxic end points such as changes in the structure or function of various organ systems. This includes effects observed in the respiratory tract as well as extrarespiratory effects related to the respiratory tract as a portal of entry.

In developing and implementing the RfC methodology, special attention has been directed to the nature of the database available, with provision made for various levels of confidence related to the quantity of available data (Table 4). The system also has a provision for using range-of-effects levels (Table 5). Jarabek (86) provides an excellent discussion of the difficulties involved in distinguishing between adverse and nonadverse effects and assigning levels of adversity. The RfC is intended to estimate a benchmark level for continuous exposure. Thus, normalization procedures are used to adjust less-than-continuous exposure data to 24 hours per day for a lifetime of 70 years.

A wide range of dosimetric adjustments are accommodated within the RfC methodology (95) to take into account differences in exposure–dose relationships among species. Regional differences (extrathoracic, T-B, and P) are taken into

Table 4 Minimum Animal Bioassay Database for Various Levels of Confidence in the Inhalation RfC

Mammalian database[a]	Confidence	Comments

1. A. Two inhalation bioassays[b] in different species
   B. One two-generation reproductive study
   C. Two developmental toxicity studies in different species
	High	Minimum database for high confidence

2. 1A and 1B, as above	Medium to high	
3. Two of three studies, as above in 1A and 1B; one or two developmental toxicity studies	Medium to high	
4. Two of three studies, as above in 1A and 1B	Medium	
5. One of three studies, as above in 1A and 1B; one or two developmental toxicity studies	Medium to low	
6. One inhalation bioassay[c]	Low	Minimum database for estimation of a RfC

[a]Composed of studies published in refereed journals, final quality assured and quality checked and approved contract laboratory studies, or core minimum Office of Pesticide Programs rated studies. It is understood that adequate toxicity data in humans can form the basis of a RfC and yield high confidence in the RfC without this database. Pharmacokinetic data indicating insignificant distribution occurring remote to the respiratory tract may decrease requirements for reproductive and developmental data.
[b]Chronic data.
[c]Chronic data preferred but subchronic acceptable.
Abbreviation: RfC, reference concentration.
Source: From Ref. 86.

Table 5 Effect Levels Considered in Deriving RfC in Relationship to Empirical Severity
Rating Values

Effect or no-effect level	Rank	General effect
NOAEL	0	No observed effects
NOAEL	1	Enzyme induction or other biochemical change, consistent with possible mechanism of action, with no pathologic changes and no change in order weight
NOAEL	2	Enzyme induction and subcellular proliferation or other changes in organelles, consistent with possible mechanism of action but no other apparent effects
NOAEL	3	Hyperplasia, hypertrophy, or atrophy but no change in organ weights
NOAEL–LOAEL	4	Hyperplasia, hypertrophy, or atrophy with changes in organ weights
LOAEL	5	Reversible cellular changes, including cloudy swelling, hydropic change, or fatty changes
(LO)AEL[a]	6	Degenerative or necrotic tissue changes with no apparent decrement in organ function
(LO)AEL–FEL	7	Reversible slight changes in organ function
FEL	8	Pathological changes with definite organ dysfunction that are unlikely to be fully reversible
FEL	9	Pronounced pathologic changes with severe organ dysfunction with long-term sequelae
FEL	10	Death or pronounced life shortening

Note: Ranks are from lowest to highest severity.
[a]The parentheses around the LO refer to the fact that any study may have a series of doses that evoke toxic effects of rank 5 through 7. All such doses are referred to as (AEL). The lowest AEL is the (LO)AEL.
Abbreviations: AEL, adverse effect levels; FEL, functional effect level; LOAEL, lowest observed adverse effect level; NOAEL, no observed adverse effect level.
Source: From Ref.(86). De Rosa et al. (1985), and Hartung (1986).

account, as are adjustments for particles versus gases and adjustments within gases for three categories based on degree of reactivity (including both dissociation and local metabolism) and degree of water solubility. Provision is also made for using more detailed, experimentally derived models if they are available.

A key step in arriving at the RfC is the development of the entire toxicity profile or data array which is examined to select the prominent toxic effect. The toxicity profile is defined as the critical effect pertinent to the mechanism of action of the chemical that is at or just below the threshold for more serious effects. The study that best characterizes the critical effect is identified as the principal study. The critical effect chosen is generally characterized by the lowest NOAEL, adjusted to the human equivalent concentration (HEC) that is representative of the threshold region for the entire data array. The use of the HEC provides a means for explicitly considering laboratory animal to human differences in exposure–tissue dose relationships.

The RfC is then derived from the NOAEL (HEC) for the critical effect by application of UFs, as shown in Table 3. A specific UF may be included if observed effects are considered to be related to exposure duration rather than concentration. In addition, a modifying factor (MF) also may be applied if scientific uncertainties in the principal study are not explicitly addressed by the standard UFs shown in Table 3.

Inhalation RfC Example

The following is an example of how the EPA's RfC methodology has been recently used. The chemical substance was hydrogen sulfide (H_2S; CASRN 7783–06-4) and the RfC was last revised 07/28/2003 as cited in the EPA Integrated Risk Information System (IRIS) (96). The revision made use of the results of a subchronic inhalation study by Brenneman et al. (97). Brenneman et al. (97) exposed Sprague–Dawley rats to 0, 10, 30, or 80 ppm H_2S for 6 hr/day, 7 days/wk for 10 weeks. The "critical effect" determined for this study was "nasal lesions of the olfactory mucosa" with a NOAEL of 10 ppm. The following steps were taken by the EPA emphasizing the results of the Brenneman et al. (97) study to calculate RfC for H_2S:

- Calculation of NOAEL, conversion from ppm to mg/m^3. Conversion factors and assumptions: $MW = 34.08$, assuming $25°C$ and $760\,mm\,Hg$. NOAEL $(mg/m^3) = 10\,ppm \times (34.08/24.45) = 13.9\,mg/m^3$
- Conversion of NOAEL to NOAEL (ADJ), to normalize exposures to 24 hr/day, 7 days/wk. $13.9\,mg/m^3 \times (6\,hr/24\,hr) \times (7\,days/7\,days) = 3.48\,mg/m^3$
- Calculation of the NOAEL (HEC) was for a gas–respiratory effect in the thoracic region. A regional gas deposition for the extrathoracic region ($RGDR_{ET}$) was calculated taking into account the differences in minute volumes (V) and surface area (SA) for humans and rats to calculate the (HEC) for NOAEL (ADJ). $V_{E(rat)} = 0.19\,L/min$, $V_{E(human)} = 13.8\,L/min; SA_{(rat)} = 15\,cm^2$, $SA_{human} = 200\,cm^2$; $RGDR_{ET} = (V_E/SA_{ET})_{rat}/(V_E/SA_{ET})_{human} = (0.19/15)/(13.8/200) = 0.184$; NOAEL(HEC) $=$ NOAEL(ADJ) $\times RGDR_{ET} = 3.48\,mg/m^3 \times 0.184 = 0.64\,mg/m^3$.
- A UF of 300 and MF of 1 were applied to the NOAEL (HEC) to determine the RfC. The UF of 300 consisted of 10 for sensitive populations, 10 for subchronic exposure, and 3 ($10^{1/2}$) for interspecies extrapolation rather than 10 because of the dosimetric adjustment already made for rat to humans. $(0.64\,mg/m^3)/(300 \times 1) = 0.002\,mg/m^3$.

The RfC value of 0.002 mg/m^3 represents the EPA's "estimate (with uncertainty spanning perhaps an order of magnitude) of a daily inhalation exposure of the human population (including sensitive subgroups) [to H_2S] that is likely to be without an appreciable risk of deleterious effects during a lifetime exposure."

The RfC values developed by the EPA can be found in the Agency's Integrated Risk Information System (IRIS) that can be readily accessed on the Internet. Each IRIS entry provides a review of the available scientific information on the chemical, the RfC value and its derivation, and the chemical's carcinogenicity classification.

The RfC methodology focuses on the establishment of either an LOAEL or an NOAEL as the starting point for deriving exposure limits. One criticism of the approach is that it does not make use of all the available data on a chemical. An alternative approach is to identify specific features of the existing data such as an effective dose associated with a given level of response, that is, a benchmark dose or benchmark concentration (BMC). The BMC approach fits a dose–response curve to the data in the observed range. A lower bound on the dose causing some specified level of risk above background is calculated. In one of the early papers on this approach, Crump (98) proposed using the below 95% confidence limit on dose at a given level of response to establish the benchmark dose. The BMC is then used as a point of departure for the application of UFs in place of the LOAEL or NOAEL. This approach allows for use of a standardized measure of the dose level near the point at which responses would no longer be expected to be observed using standard experimental designs. Thus, the BMC does not depend on a single data point such as the LOAEL or NOAEL and uses the entire dataset and also accounts for sample size. By using all the data in calculating a BMC, account can be taken of the steepness of the dose–response relationship. The steepness of the dose–response curve in the dose region from which extrapolations are made can markedly influence the calculated RfC.

Criteria Pollutants

In concluding this section on evaluating noncancer risks, it is appropriate to briefly describe the approach used in setting National Ambient Air Quality Standards (NAAQS) for criteria pollutants. To date, the NAAQS have all been set based on the intent to limit noncancer risks. For more detail, the reader is referred to two reviews of the criteria pollutant methods used to develop the NAAQS (99,100). The criteria pollutants (Table 6) are air pollutants that arise from multiple sources and are widely distributed, hence the need for national standards.

For criteria pollutants, the primary standards, which are intended to protect against health effects, "shall be ambient air quality standards, the attainment and maintenance of which in the judgment of the Administrator, based on such criteria and allowing an adequate margin of safety, are required to protect the public health" (101,102). The primary standards are intended to protect against "adverse effects," not necessarily against all identifiable changes produced by a pollutant. As noted earlier, the issue of what constitutes an "adverse effect" has been a matter of debate. Although the Clean Air Act (CAA) did not specifically characterize an "adverse effect," it provided some general guidance. It noted concern for effects ranging from cancer, metabolic and respiratory diseases, and impairment of mental abilities, to headaches, dizziness, and nausea.

In developing the CAA, Congress also noted concern for sensitive populations in setting the NAAQS. Specifically, it was noted that the standard should protect "particularly sensitive citizens as bronchial asthmatics and emphysematics who, in the normal course of daily activity, are exposed to the ambient environment." This has been interpreted to exclude individuals who are not performing normal activities, such as hospitalized individuals. Guidance was given noting that the standard is sta-tutorily sufficient whenever there is "an absence of adverse effect on the health of a statistically related sample of persons in sensitive groups from exposure to ambient air." A statistically related sample has been interpreted as "the number of persons necessary to test in order to detect a deviation in the health of any persons within such sensitive groups which is attributable to the condition of the ambient air." In setting NAAQS, the EPA, while recognizing the need to consider sensitive or suscep-tible groups or subpopulations, has also recognized that it is impractical to set NAAQS at a level to protect the most sensitive individual. In setting the NAAQS, the administrator must also incorporate an "adequate margin of safety." This is intended to protect against effects that have not yet been uncovered by research, and effects whose medical significance is a matter of disagreement. In setting the health-based standards, the administrator cannot consider the cost of achieving the standards. However, costs can be considered in planning for when the NAAQS must be achieved.

The CAA, in addition to calling for primary or health-based standards, requires the promulgation of secondary or welfare-based NAAQS. The welfare

Table 6 Summary Information on Criteria Air Pollutants

Pollutant	Sources	Key effects of concern	Subpopulations of concern	NAAQS[a]
Ozone	Photochemical oxidation of nitrogen oxides (primarily from combustion and volatile organic compounds) from stationary and mobile sources	Decreased pulmonary function, lung inflammation, increased respiratory hospital admissions	Children, people with preexisting lung disease, outdoor-exercising health people	1-hour average 0.12 ppm ($235\,\mu g/m^3$), 8-hour average 0.08 ppm ($157\,\mu g/m^3$)
Nitrogen dioxide	Photochemical oxidation of nitric oxide (primarily from combustion of fossil fuels) and direct emissions (primarily from combustion of natural gas)	Respiratory illness, decreased pulmonary function	Children, people with preexisting lung disease	Annual arithmetic mean 0.053 ppm ($100\,\mu g/m^3$)
Sulfur dioxide	Primarily combustion of sulfur-containing fossil fuels; also smelters, refineries, and others	Respiratory injury or death, decreased pulmonary function	Children, people with preexisting lung disease (especially asthma)	Annual arithmetic mean 0.03 ppm ($80\,\mu g/m^3$), 24-hour average 0.14 ppm ($365\,\mu g/m^3$)
Particulate matter	Direct emission of particles during combustion, industrial processes, gas reactions and condensation and coagulation, natural sources	Injury and death	People with preexisting heart and lung disease, children	PM_{10} annual arithmetic mean $50\,\mu g/m^3$, 24-hour average $150\,\mu g/m^3$, $PM_{2.5}$ annual arithmetic mean $15\,\mu g/m^3$, 24-hour average $85\,\mu g/m^3$

(*Continued*)

Table 6 Summary Information on Criteria Air Pollutants (*Continued*)

Pollutant	Sources	Key effects of concern	Subpopulations of concern	NAAQS[a]
Carbon monoxide	Combustion of fuels, especially by mobile sources	Shortening of time to onset angina and other heart effects	People with coronary artery disease	8-hour average 9 ppm (10 mg/m^3), 1-hour average 35 ppm (40 mg/m^3)
Lead	Leaded gasoline (prior to phase-out in gasoline); point sources, such as Pb mines, smelters, and recycling operations	Developmental neurotoxicity	Children	Quarterly average 1.5 µg/m^3

[a]NAAQSs are specified by indicator concentration, averaging time, and statistical form. The latter used to determine compliance and is too complex for coverags here.
Abbreviation: NAAQS, National Ambiant Air Quality Standard.

standards are intended to protect against effects such as damage to crops and eco-systems, soiling of buildings, and impairment of visibility.

The process by which the NAAQS are set is well established and involves multiple phases. The first phase involves the EPA's Office of Research and Development preparing a criteria document on the pollutant. These documents are essentially encyclopedias of all that is known about the particular pollutant. This is followed by the development of a staff position paper on the pollutant by the EPA's Office of Air Quality Planning and Standards. The staff paper reviews the studies contained with the criteria document, which are most germane to setting the four elements of a NAAQS. The four elements are (a) the indicator, (b) the averaging time over which ambient measurements must be made, i.e., 8 hours, 24 hours or annual, (c) the numerical level, in ppm or mass per m^3, and (d) the statistical form, a 98th percentile. The indicator for specific chemicals is the chemical (SO_2, CO, NO_2, O_3, and Pb) while for PM it is the mass size fraction such as $PM_{2.5}$, not specified as to chemical composition. At the final stage, the EPA publishes proposed rules to reaffirm or change the NAAQS; this includes the rationale for decisions on each of the four elements of the NAAQS. The criteria document and staff papers and, on some occasions, the proposed rule, are reviewed and commented on by an independent committee, the Clean Air Scientific Advisory Committee (CASAC). The letters from the CASAC to the administrator are a matter of public record.

Extensive scientific databases including substantial data from epidemiological and controlled human exposure studies exist for all the criteria air pollutants. Readers interested in learning more about the individual criteria air pollutants are encouraged to review the latest criteria document or staff papers prepared by the EPA. Recent examples are the criteria document on PM (103) and the related staff paper (18). The documents also provide substantial background material that may be relevant to evaluating the health risks of other air pollutants, i.e., the carbon monoxide documents for information on inhaled nonreactive gases, the O_3 documents for information on reactive gases, and the PM documents for information relevant to all kinds of inhaled particles. For an up to date assessment of research needs related to PM, the reader is also referred to the NRC report (104) and to the published work of McClellan and Jessiman (7).

Cancer Risks

Characterization of Carcinogen Hazards

The earliest risk assessments for cancer focused on whether a compound or occupation posed a carcinogenic hazard based on epidemiological data. Later, the carcinogen assessment process was broadened to include consideration of data from laboratory animal studies including inhalation studies. This gave rise to formalized criteria for evaluating the carcinogenic risks to humans such as that pioneered in 1969 by the International Agency for Research on Cancer (IARC) (105).

The IARC approach is described in the preamble to each monograph with individual monographs typically covering several chemicals, biological agents, or occupations. The preamble in a recent monograph on formaldehyde is an example (106). The monographs and the carcinogen categorization results are developed by international working groups of experts. These working groups carry out five tasks—(i) to ascertain that all appropriate references have been collected, (ii) to select the data relevant for the evaluation on the basis of scientific merit, (iii) to prepare accurate summaries of the data to enable the reader to follow the reasoning of the working group, (iv) to evaluate the results of experimental and epidemiological studies, and (v) to make an overall evaluation of the carcinogenicity of the agent to humans.

In the monographs, the term *carcinogen* is used to denote an agent that is capable of increasing the incidence of malignant neoplasms. Traditionally, IARC has evaluated the evidence for carcinogenicity without regard to specific underlying mechanisms involved. In 1991, IARC convened a group of experts to consider how mechanistic data could be used in the classification process. This group suggested a greater use of mechanistic data including information relevant to extrapolation between laboratory animals and humans (107,108). The use of mechanistic data can be used to either "downgrade" or "elevate" the carcinogen classification of a chemical. Examples are found in the monograph on "some chemicals that cause tumors of the kidney or urinary bladder in rodents and some other substances" (109) and in "man-made vitreous fibers" (110).

The IARC evaluation process considers three types of data: (i) human carcinogenicity data, (ii) experimental carcinogenicity data, and (iii) supporting evidence of carcinogenicity. Definitive evidence of human carcinogenicity can only be obtained from epidemiological studies. The epidemiological evidence is classified into four categories: (i) sufficient evidence of carcinogenicity is used if a causal relationship has been established between exposure to the agent and human cancer, (ii) limited evidence of carcinogenicity is used if a positive association between exposure to an agent and human cancer is considered to be credible, but chance, bias, or confounding cannot be ruled out with reasonable confidence, (iii) inadequate evidence of carcinogenicity is used if the available studies are of insufficient quality, consistency, or statistical power to permit a conclusion regarding the presence or absence of a causal association, and (iv) evidence suggesting lack of carcinogenicity is used if there are several adequate studies covering the full range of doses to which human beings are known to be exposed, which are mutually consistent in not showing a positive association between exposure and any studied cancer at any observed level of exposure.

The IARC evaluation process gives substantial weight to carcinogenicity data from laboratory animals. IARC has concluded "in the absence of adequate data in humans, it is biologically plausible and prudent to regard agents for which there is sufficient evidence of carcinogenicity in experimental animals as if they presented a carcinogenic risk to humans." IARC classifies the strength of the evidence of carcinogenicity in experimental animals in a fashion analogous to that used for the human data. The evidence of carcinogenicity from laboratory animals is classified into four categories: (i) sufficient evidence of carcinogenicity is used if a working group considers that a causal relationship has been established between the agent and an increased incidence of malignant neoplasms or an appropriate combination of benign and malignant neoplasms in (a) two or more species of animals, or (b) two or more independent studies in one species carried out at different times, in different laboratories, or under different protocols. A single study in one species might be considered under exceptional circumstances to provide sufficient evidence when malignant neoplasms occur to an unusual degree with regard to incidence, site, type of tumor, or age at onset; (ii) limited evidence of carcinogenicity is used if the data suggest a carcinogenic effect but are limited for making a definitive evaluation; (iii) inadequate evidence of carcinogenicity is used if studies cannot be interpreted as showing either the presence or the absence of a carcinogenic effect because of major qualitative or quantitative limitations; and (iv) evidence suggesting lack of carcinogenicity is used if adequate studies involving at least two species are

available that show that, within the limits of the tests used, the agent is not carcinogenic. Such a conclusion is inevitably limited to the species, tumors, and doses of
exposure studied.

Supporting evidence includes a range of information such as structure–activity
correlations, toxicological information, and data on kinetics, metabolism, and genotoxicity. This includes data from laboratory animals, humans, and lower levels of
biological organization such as tissues and cells. In short, any information that
may provide a clue as to the potential for an agent causing cancer in humans is
reviewed and presented.

Finally, all of the relevant data are integrated and the agent categorized on the
basis of the strength of the evidence derived from studies in humans and experimental animals and from other studies as shown in Table 7. As noted, the IARC categorization scheme does not address the potency of carcinogens. This poses serious
constraints on the utility of IARC carcinogen classifications beyond hazard identification. In short, a carcinogen is a carcinogen irrespective of potency. This "lumping"
of carcinogens irrespective of potency can be misleading to nonspecialists including
the lay public. The IARC is not a regulatory agency; its cancer classification findings
are strictly advisory. However, the IARC classifications are used around the world
by many national, state, and local agencies.

The EPA in 1986 adopted a carcinogen classification scheme as part of a set of
guidelines for cancer risk assessment that codified the Agency's practices (111). The
EPA scheme is used as a tool for regulation of chemicals under all of the various laws
it administers. The 1986 EPA guidelines used an approach very similar to that of the
IARC in categorizing agents based on the weight of the evidence of carcinogenicity,
except for designating an A, B, or C categorization rather than a numeric designation
of the different categories. The origins of the EPA's carcinogen risk assessment practices have been described by Albert (112). Almost as soon as the 1986 cancer guidelines
were adopted, the EPA began revising them. Revised guidelines were finally promul-
gated in 2005 (113). They include provision for using mechanistic data, emphasis on
use of a narrative description, and attempt to harmonize the cancer guidelines with
the approach used in evaluating noncancer responses as will be discussed later.

A third carcinogen classification scheme is in use by the U.S. National Toxicology Program (NTP) (114). The NTP is mandated by the U.S. Congress to publish
reports that designate the carcinogenicity of chemicals, biennially. The designation
scheme is even simpler than that used by the IARC or EPA. If a chemical is listed,
it is placed in one of the two categories: (i) known to be a human carcinogen, or (ii)
reasonably anticipated to be a human carcinogen. The requirements for evidence for
designation of a chemical in the second category are quite broad, especially based on
recent changes that allow the use of mechanistic evidence of carcinogenicity in the
absence of direct evidence of cancer in animals or humans. The designation of a large
number of chemicals as reasonably anticipated to be human carcinogens based on
variable evidence may be confusing to the public and regulators because the chemicals vary widely in their cancer-causing potency and, indeed, ultimately may prove to
not even be carcinogens when subjected to more rigorous evaluation. The issue of
classifying chemicals on the basis of carcinogenicity or other disease-producing
potential such as neurotoxicity or reproductive effects deserves additional research
and discussion. This is especially important in view of the attention being given to
developing alternative toxicological methods that refine, reduce, or replace the use
of laboratory animals. The "dose" of a chemical evaluated in the alternative system
may be substantially greater than would be found in the tissues of animals given

Table 7 IARC Classification Scheme for Human Carcinogens

Category	Human evidence	Experimental evidence
Group 1. The agent (mixture) is carcinogenic to humans. The exposure circumstance entails exposures that are carcinogenic to humans	(a) Sufficient, (b) Less than sufficient	(a) No animal evidence required, (b) Sufficient evidence and strong evidence in exposed humans that the agent (mixture) acts through a relevant mechanism of carcinogenicity
Group 2A. The agent (mixture) is probably carcinogenic to humans. The exposure circumstance entails exposures that are probably carcinogenic to humans	(a) Limited, (b) Limited, (c) Inadequate	(a) None, (b) Sufficient, (c) Sufficient and strong evidence that the carcinogenesis is mediated by a mechanism that also operates in humans
Group 2B. The agent (mixture) is possibly carcinogenic to humans. The exposure circumstances entails exposures that are possibly carcinogenic to humans	(a) Limited, (b) Inadequate, (c) Inadequate	(a) Less than sufficient, (b) Sufficient, (c) Limited together with supporting evidence from other relevant data
Group 3. The agent (mixture or exposure circumstance) is not classifiable as to its carcinogenicity	(a) Inadequate, (b) Inadequate	(a) Inadequate, (b) Limited
Group 4. The agent (mixture) is probably not carcinogenic to humans	(a) Lack of carcinogenicity, (b) Inadequate	(a) Lack of carcinogenicity, (b) Lack of carcinogenicity consistently and strongly supported by a broad range of other relevant data

Note: Abstract from the Preamble of IARC Monographs, Vol. 84 (IARC, 2004).
Abbreviation: IARC, International Agency for Research on Cancer.

realistic exposures, and certainly much higher than likely to be found in humans under plausible conditions of exposure.

Yet, a fourth scheme to categorize chemicals as to their carcinogenicity has been used by ACGIH (87). The ACGIH scheme has five categories:

A1. Confirmed human carcinogen
A2. Suspected human carcinogen
A3. Animal carcinogen
A4. Not classifiable as a human carcinogen
A5. Not suspected as a human carcinogen

Substances for which no human or experimental animal carcinogenicity data have been reported are assigned a "no carcinogenicity" designation.

It has been noted that the IARC, EPA, and NTP have taken steps to incorporate mechanistic information into the carcinogen classification or categorization process. For inhaled agents, the development of a strong mechanistic database requires understanding the exposure dose to critical biological targets—health effects paradigm illustrated in Figure 1. This is the case because of the unique nature of the exposure–dose linkage for inhaled agents and how it may be influenced by interspecies considerations in extrapolating from laboratory animal species to humans and the perturbing influence of high exposure concentrations on exposure (dose)–response relationships.

Importance of Understanding Species Differences

The importance of considering species differences is illustrated by the studies of the Chemical Industry Institute of Toxicology team with inhaled formaldehyde. This large body of research has been reviewed (115–117). A key finding related to the exposure–dose linkage for formaldehyde is shown in Figure 23 (118,119). From the data in this figure, it is readily apparent that there are significant differences in

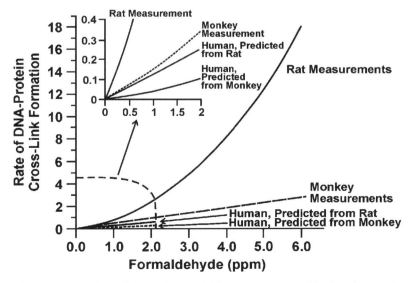

Figure 23 Delivered dose of formaldehyde, as measured by DNA–protein cross-links, measured in rats and monkeys, and estimated for humans. *Source*: From Refs. 118, 119.

measure of dose, DNA–protein cross-links per ppm of formaldehyde exposure, for rats versus monkeys and that the extrapolated values for humans are lower than for either rats or monkeys. Moreover, the exposure–dose relationship is clearly not linear over the range studied. Thus, it would be inappropriate to directly extrapolate the rat findings of an excess of nasal cancer at 6 ppm exposure to humans.

The second example relates to the observation of an excess of lung tumors in rats exposed for two years to high concentrations of either diesel exhaust or carbon black. With both agents, the exposure–response relationship was similar, suggesting that the excess lung tumor incidence was related to the carbonaceous PM with no added effect attributable to the polycyclic aromatic hydrocarbons present in the diesel exhaust and not present with the carbon black. Most importantly, a large body of data provide strong evidence that the excess of lung tumors is related to the high exposure concentration, impaired clearance of the carbonaceous material, chronic inflammation, increased cell proliferation, and development of epithelial cell mutations by an epigenetic mechanism. The mechanistic linkages are illustrated in Figure 24 and have been discussed by several authors (60,120–122). The body of evidence now available indicates that while protracted high concentration exposure to diesel exhaust and carbon black may produce lung tumors in rats, the findings should not be extrapolated to people exposed at much lower concentrations. The extensive research conducted with formaldehyde and diesel exhaust emphasizes the importance of using comprehensive research strategies when investigating the potential human health risks of airborne materials. The research can serve as a prototype for research with other agents.

Quantitative Estimates of Carcinogenic Potency

The 1986 EPA guidelines for carcinogen risk assessment went beyond the IARC approach in that guidance was provided for developing quantitative estimates of carcinogen potency, that is, cancer risk per unit of exposure. Because

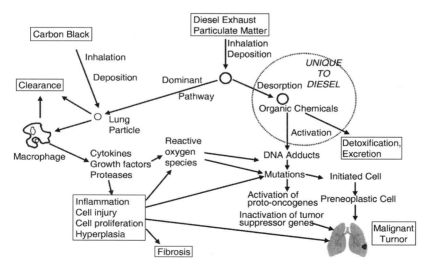

Figure 24 Schematic representation of the pathogenesis of lung tumors in rats with prolonged exposure to high concentrations of diesel exhaust or carbon black particles. *Source:* From Ref. 120.

the information bases for individual chemicals vary markedly and are never fully complete, the guidelines included a number of "default" options for use in the assessment process unless there are compelling scientific data to be used in place of the default position. Some of the key "default" assumptions are (i) humans are as sensitive as the most sensitive laboratory animal species, strain, or sex evaluated, (ii) chemicals act like radiation at low doses in inducing cancer with a linearized multistage model appropriate for estimating dose–response relationships below the range of experimental observation, (iii) the biology of humans and laboratory animals, including the rate of metabolism of chemicals, is a function of body surface area, (iv) a given unit of intake of chemical has the same effect irrespective of the intake time or duration, and (v) as discussed for the IARC classification scheme, laboratory animals are a surrogate for humans in assessing cancer risks, with positive cancer bioassay results in laboratory animals taken as evidence of the chemical's potential to cause cancer in people. The validity of all of the default assumptions has been vigorously debated. The key to resolving the issue of their individual validity is to acquire a larger body of scientific data that obviates the need for using the defaults (3).

From the beginning, the EPA's carcinogen risk assessment procedures were controversial. The U.S. Congress, in amending the CAA in 1990, recognized the continuing controversy and requested a review of the carcinogen risk assessment process by a committee of the National Academy of Sciences/NRC. The result was the report *Science and Judgment in Risk Assessment*. The report supported the basic risk assessment structure advocated in 1983 (2). However, it emphasized the utility of using the risk assessment to more fully identify uncertainties in risk characterization and then use these uncertainties to help guide research that would over time serve to reduce the uncertainties. In addition, the report emphasized the value of developing specific science to reduce uncertainties in risk assessments. The 1994 NRC recommendations stimulated the EPA to continue with revisions to the cancer risk assessment guidelines published in 1986. The old guidelines called, in all cases, for use of a linearized multistage model that included a linear component extrapolated to zero dose–zero exposure. Thus, even exposure to one molecule of chemical had an associated calculable risk.

This regimented approach imposed by the use of the linearized multistage model has been very controversial with recognition that a range of potential exposure–cancer response relationships might exist. A major difficulty is the blunt nature of epidemiological and experimental animal approaches, which makes it very difficult to detect small increases in cancer incidence above background. In both kinds of studies, a doubling of cancer incidence or an observed ratio of 2.0 usually is required to have a high degree of confidence that the effect is real (123). Increases in the range of 20% to 100% or an observed ratio of 1.2 to 2.0 are frequently challenged. These statistical limitations are of profound importance, recognizing the need to estimate risk impact in the range of 1 excess cancer in 10 thousand to 1 million individuals. Under the old EPA guideline, a linearized multistage model was used to calculate upper-bound unit risk estimates (URE) or potency estimates (111). A URE represents an estimate of the increased cancer risk from a lifetime (70-year) exposure to a concentration of one unit of exposure. In practice, the URE was typically expressed as risk per microgram per cubic meter for air contaminants. The EPA has defined the URE as a plausible upper-bound estimate of the risk, with the actual risk not likely to be higher but possibly lower and maybe zero.

The new EPA cancer risk assessment guidelines (113) provide for alternative approaches dependent on the available data on the mode of action by which the chemical causes cancer. The new guidelines are accompanied by supplemental guidance for considering susceptibility to early life exposures to carcinogens (124). Butterworth et al. (125) have defined the mode of action of a chemical as the key obligatory process governing the action of the chemical in causing cancer. For example, it has been known for some time that the mode of action for some chemicals involves direct interaction between the chemical or its metabolites, and DNA. Genetic damage is the result. More recently, alternative modes of action have been identified, including cytotoxicity at high doses of chemical resulting in compensatory cell regeneration, mitogenic stimulation, or receptor-mediated interactions with genetic control processes. Each of these modes of action may yield exposure–cancer response relationships that are nonlinear at low doses and not accurately described by the linear multistage model.

The new EPA approach is illustrated schematically in Figure 25. The new EPA approach envisions a two-step process. The first step is to model the actual observed data using an approach similar to the BMC for noncancer responses to arrive at an LEC_{10}, the 95% lower confidence limit in the concentration associated with the estimated 10% increase in tumor or tumor-related response. This value then serves as a point of departure for extrapolation to lower exposure concentrations. The specific method of extrapolation from the point of departure depends on the agent's mode of action and whether it indicates a linear or nonlinear mode of action. For example, the EPA has suggested that a chemical interacting with DNA and causing mutations be modeled with a linear low-dose component. In such a case, a straight line is drawn to zero and the risk at any concentration or the concentration at any risk level can be interpolated along the line. A linear mode of action is used as a default option. Thus, for an increment in exposure above zero, there is a calculable cancer risk above the background cancer incidence. This approach is based on the linear no-threshold dose–response model that has dominated the field of radiation carcinogenesis and was recently reaffirmed by the NRC (126).

A nonlinear model can be used if there is sufficient evidence to support a nonlinear mode of action. Several models might be appropriate, depending upon the

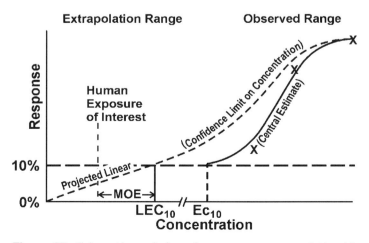

Figure 25 Schematic rendering of exposure–response relationships for cancer with alternative approaches for estimating risks at low levels of exposure. *Abbreviations*: EC_{10}, effective concentration, 10%; LEC_{10}, lowest effective concentration; MOE, margin of exposure.

mode of action. If the cancer response is secondary to toxicity or an induced physiological change that has a threshold, a threshold model could be used.

Another option has been proposed for agents acting via a nonlinear mode of action (Fig. 25). This is a margin-of-exposure approach in which the point of departure (i.e., the LEC) is compared with the actual human exposure level. An acceptable margin of exposure can be arrived at on a case-by-case basis using expert judgment and taking into account all of the available data on the agent's mode of action and exposure–response end points.

QUALITY CONTROL AND QUALITY ASSURANCE

It would be inappropriate to conclude this chapter without briefly noting the importance of quality control and quality assurance in the planning, conduct, interpretation, and reporting of the results of inhalation toxicity studies. Wong (54) has briefly addressed these important topics. The level of these activities may be varied depending upon the anticipated use of the data resulting from the experimentation. For hypothesis-generating and exploratory studies, it may be sufficient to conduct the research using experimental designs, calibrations, methodology, results, and interpretations documented in standard laboratory notebooks and computer files. Even at this level, the investigator should take care to be certain that all critical details are carefully documented, dated, and the responsible individual identified in the records. If one feels compelled to limit documentation because it is viewed as a nuisance and, after all, not required by journal editors, they may find it useful to have a short conversation with a patent attorney to clarify what is required to support a patent application for a new discovery. Inadequate documentation has stymied the filing, and ultimately, the defense of many patent applications.

It is now standard practice, when it is anticipated that information will be used in regulatory proceedings, for research to be conducted with full adherence to formal quality control and quality assurance procedures. The use of such procedures does require additional effort. However, modest additional efforts such as doing the research right and having appropriate documentation are a small price for having data accepted and for avoiding the need to attempt to replicate a poorly documented study.

ADDITIONAL REFERENCES

In this chapter we have only been able to briefly review the basic concepts and techniques of the field of inhalation toxicology. The reader interested in additional detail is referred to several books published during the last several decades. A book entitled *Concepts in Inhalation Toxicology* (second edition) authored by McClellan and Henderson (127) provides broad coverage of the field and was originally developed from a series of workshops given by the Lovelace organization in Albuquerque, New Mexico. *Toxicology of the Lung* (third edition) edited by Gardner et al. (128) includes a series of chapters highly relevant to the field of inhalation toxicology that complement or extend material presented in the first two editions of *Toxicology of the Lung* that is a part of the Target Organ Toxicology Series. Phalen (129,130) has authored two excellent handbooks on inhalation toxicology which have a strong methods orientation. Another useful reference is the volume on *Toxicology of the Respiratory System* (131) in the series—Comprehensive Toxicology. More than 150 volumes have been published by Marcel Dekker, now Taylor & Francis, in a series entitled *Lung Biology in Health and Disease* under the executive editorship of Claude Lenfant. These volumes contain a treasure trove of information including frequent reference to inhalation toxicology studies. Examples are the volumes titled *Air Pollutants and the Respiratory Tract* (132), *Particle-Lung Interactions* (133), *Cytokines in Pulmonary Disease* (134), and *Disease Markers in Exhaled Breath* (135). Beyond these books, the interested reader will find numerous papers published each year in the peer-reviewed literature with several journals focusing on each of the interrelated fields of aerosol science, inhalation toxicology, respiratory biology and pathobiology, and the diagnosis, treatment and pathogenesis of respiratory diseases.

SUMMARY

This chapter has provided an overview of the basic concepts and techniques used in inhalation toxicology. Emphasis has been given to two guiding paradigms—the first linking sources of airborne materials to exposure, dose and health responses, and the second stressing the role of scientific information in assessing the human health risks of airborne materials. The reader should have gained an appreciation for the distinction between exposure (the time profile of air concentrations of agents) and dose (the time profile of concentrations of the agent and its metabolites in the respiratory tract and other tissues).

The complexities of conducting inhalation toxicology studies has been emphasized not to intimidate or discourage newcomers to the field, but to serve as a reminder that it is necessary to use carefully developed experimental designs, research approaches, and analyses if the research is going to meet the high standards of the field and be relevant and used in human risk assessment. It is relatively easy to conduct simple experiments using large quantities of material, perhaps administered by nonphysiological modes to cells, tissues, or intact animals and evaluate a plethora of molecular and cellular end points with a goal of observing effects. In short, the creation of information that indicates a particular airborne material is hazardous. It is much more difficult to develop information that will help characterize the potential risks to people from specific, and frequently, very low levels of exposure in the home, workplace, during use of a consumer product, or from a complex mixture of materials in the environment. It is important to recall that characterization of risk requires knowledge of both the measured or estimated concentration of the agent over time and the potency of the agent. Exposure assessment studies can provide the former and inhalation toxicology studies can provide the latter information. The conduct of high-quality inhalation toxicity studies that will yield information relevant to evaluating human risks typically requires expensive facilities and equipment and a multidisciplinary team of investigators and supporting personnel. The cost of conducting such research is a sound investment recognizing the importance of the inhalation route of entry of materials and the need to use a science-based approach to controlling exposures and limiting human disease.

ACKNOWLEDGMENT

The authors gratefully acknowledge the word processing assistance of Mildred Morgan.

REFERENCES

1. NRC, National Research Council. Research Priorities for Airborne Particulate Matter. I. Immediate Priorities and a Long-Range Research Portfolio. Washington, D.C.: National Academies Press, 1998.
2. NRC, National Research Council. Risk Assessment in the Federal Government: Managing the Process. Washington, D.C.: National Academy Press, 1983.
3. NRC, National Research Council. Science and Judgment in Risk Assessment. Washington, D.C.: National Academy Press, 1994.
4. McClellan RO. A commentary on the NRC report "Science and judgment in risk assessment." Regul Toxicol Pharmacol 1994; 20:S142–S168.
5. Smith LL. Paraquat. In: Roth RA, ed. Toxicology of the Respiratory System. Vol. 8. In: Sipes IG, McQueen CA, Gandolfi AJ, eds. Comprehensive Toxicology. Oxford, U.K.: Elsevier, 1997:581–589.
6. Utell MJ, Frampton MW, Zareka W, Devline RB, Cascio WE. Cardiovascular effects associated with air pollution: potential mechanisms and methods of testing. Toxicology 2002:101–117.
7. McClellan RO, Jessiman B. Health context for management of particulate matter. In: McMurry PFH, Sheppard MF, Vickery JS, eds. Particulate Matter Science for Policymakers: A NARSTO Assessment. Cambridge, U.K.: Cambridge Press, 2004:69–101 (chap. 2).
8. Frampton MW, Utell MJ. Clinical studies of airborne pollutants. In: Gardner DE, Crapo JD, McClellan RO, eds. Toxicology of the Lung. 3rd ed. Philadelphia: Taylor & Francis, 1999:455–481.
9. Postlethwait EM, Bidani A. In vitro systems for studying respiratory system toxicology. In: Roth RA, ed. Toxicology of the Respiratory System. Vol. 8. In: Sipes IG, McQueen CA, Gandolfi AJ, eds. Comprehensive Toxicology. Oxford, U.K.: Elsevier, 1997: 249–263.
10. Cantor JO. CRC Handbook of Animal Models of Pulmonary Disease. Boca Raton, FL: CRC Press, 1989.
11. Hahn FF. Chronic inhalation bioassays for respiratory tract carcinogenesis. In: Gardner DE, Crapo JD, McClellan RO, eds. Toxicology of the Lung. 3rd ed. New York: Raven Press, 1999:241–268.
12. Mauderly JL. Animal models for the effect of age on susceptibility to inhaled particulate matter. Inhal Toxicol 2000; 12:863–900.
13. Muggenburg BA, Tilley L, Green FHY. Animal models of cardiac disease: potential usefulness for studying effects of inhaled particles. Inhal Toxicol 2000; 12:901–925.
14. Wilson WE, Spiller LL, Ellestad TG, et al. General Motors sulfate dispersion experiments. Summary of EPA measurements. J Air Pollut Control Assoc 1977:46–51.
15. McClellan RO. Nuisance dusts (particles not otherwise classified). In: Roth RA, ed. Toxicology of the Respiratory System. Vol. 8. In: Sipes IG, McQueen CA, Gandolfi AJ, eds. Comprehensive Toxicology. Oxford, U.K.: Elsevier, 1997:495–520.
16. Karn B, Masciangioli T, Zheng W, Calvin V, Alivisatos P, eds. Nanotechnology and the Environment. Oxford, U.K.: Oxford University Press, 2005.
17. Whitby KT. The physical characteristics of sulfur aerosols. Atmos Environ 1978; 12:135–159.
18. U.S. Environmental Protection Agency. Review of the National Ambient Air Quality Standards for Particulate Matter: policy assessment of scientific and technical information. Office of Air Quality Planning and Standards, U.S. Environmental Protection Agency, 2005.
19. Bogdanffy MS, Randall HW, Morgan KT. Biochemical quantification and histochemical localization of carboxylesterase in the nasal passage of the Fischer 344 rat and $B_6C_3F_1$ mouse. Toxicol Appl Pharmacol 1987; 88:183–194.

20. Bogdanffy MS. Biotransformation enzymes in the rodent nasal mucosa: the value of a histochemical approach. Environ Health Perspect 1990; 85:117–186.

21. Bond JA. Metabolism of xenobiotics by the respiratory tract. In: Gardner DE, Crapo JD, McClellan RO, eds. Toxicology of the Lung. 2d ed. Philadelphia: Taylor & Francis, 1993:187–215.

22. Harkema JR. Comparative aspects of nasal airway anatomy: relevance to inhalation toxicology. Toxicol Pathol 1991; 19:321–336.

23. Guilmette RA, Wicks JD, Wolff RK. Morphometry of human nasal airways in vivo using magnetic resonance imaging. J Aerosol Med 1989; 2:365–377.

24. Schlesinger RB. Comparative deposition of inhaled aerosols in experimental animals and humans: a review. J Toxicol Environ Health 1985; 15:197–214.

25. Boggs DF. Comparative biology of the normal lung. In: Parent R, ed. Treatise in Pulmonary Toxicology. Vol. 1. Boca Raton, FL: CRC Press, 1992.

26. Newton PE, Pfledderer C. Measurement of the deposition and clearance of inhaled radiolabeled particles from rat lungs. J Appl Toxicol 1986; 6:113–119.

27. Lippmann M. Recent advances in respiratory tract particle deposition. In: Esmen NA, Mehlman MA, eds. Advances in Modern Environmental Toxicology. Occupational and Industrial Hygiene, Concepts and Methods. Vol. 8. Princeton: Princeton Scientific Publishers Inc., 1984:75–103.

28. Sweeney TD, Brain JD, LeMott S. Anesthesia alters the pattern of aerosol retention in hamsters. J Appl Physiol 1983; 54:37–44.

29. Alarie Y. Irritating properties of airborne materials to the upper respiratory tract. Arch Environ Health 1966; 13:433–439.

30. Bruce MC, Bruce EN, Leith DE, Murphy SD. Diethyl maleate and/or ozone (10 ppm) reduce ventilation by 60–80% in awake mice [abstr]. Physiologist 1979; 22:16.

31. Silver EH, Leith DE, Murphy SD. Potentiation by triorthotolyl phosphate of acrylate ester-induced alterations in respiration. Toxicology 1981; 22:193–203.

32. Chang JCF, Steinhagen WH, Barrow CS. Effect of single or repeated formaldehyde exposure on minute volume of B6C3F mice and F344 rats. Toxicol Appl Pharmacol 1981; 61:451–459.

33. Medinsky MA, Dutcher JS, Bond JA, et al. Uptake and excretion of [^{14}C]methyl bromide as influenced by exposure concentration. Toxicol Appl Pharmacol 1985; 78:215–225.

34. Snipes MB. Long-term retention and clearance of particles inhaled by mammalian species. Crit Rev Toxicol 1989:175–211.

35. Snipes MB. Species comparisons for pulmonary retention of inhaled particles. In: McClellan RO, Henderson RF, eds. Concepts in Inhalation Toxicology. New York, NY: Hemisphere Publishing, 1989:193–227.

36. Snipes MB. Biokinetics of inhaled radionuclides. In: Raabe OG, ed. Internal Radiation Dosimetry, Health Physics Society 1994 Summer School. Madison, WI: Medical Physics Publishing, 1994:181–196.

37. Schlesinger RB. Deposition and clearance of inhaled particles. In: McClellan RO, Henderson RF, eds. Concepts in Inhalation Toxicology. 2d ed. Washington, D.C.: Taylor & Francis, 1995:191–224.

38. Miller FJ. Dosimetry of particles in laboratory animals and humans. In: Gardner DE, Crapo JD, McClellan RO, eds. Toxicology of the Lung. 3rd ed. Philadelphia: Taylor & Francis, 1999:513–555.

39. NCRP, National Council on Radiation Protection and Measurements. Deposition, retention, and dosimetry of inhaled radioactive substances. NCRP Report SC-72, Bethesda, MD, 1997.

40. Muhle H, McClellan RO. Respiratory tract. In: Marquardt H, Schafer SG, McClellan RO, Welsch F, eds. Toxicology. San Diego, CA: Academic Press, 1999:331–347.

41. Gerde P, Dahl AR. A model for the uptake of inhaled vapors in the nose of the dog during cyclic breathing. Toxicol Appl Pharmacol 1991; 109:276–288.

42. Dahl AR. Contemporary issues in toxicology: dose concepts for inhaled vapors and gases. Toxicol Appl Pharmacol 1990; 103:185–197.

43. Pryor WA. How far does ozone penetrate into the pulmonary air/tissue boundary before it reacts? Free Radic Biol Med 1992; 12:83–88.

44. Miller FJ, Kimbell JS. Regional deposition of inhaled reactive gases. In: McClellan RO, Henderson RF, eds. Concepts in Inhalation Toxicology. 2d ed. Washington D.C.: Taylor & Francis, 1995:257–288.

45. Kimbell JS, Miller FJ. Regional respiratory tract absorption of inhaled reactive gases. In: Gardner DE, Crapo JD, McClellan RO, eds. Toxicology of the Lung. 3rd ed. New York: Raven Press, 1999:557–597.

46. Ultman JS. Watson AY, Bates RR, Kennedy D, eds. Air Pollution, the Automobile and Public Health. Washington, D.C.: National Academy Press, 1988:323–366.

47. Medinsky MA. Critical determinants in the systemic availability and dosimetry of volatile organic chemicals. In: Gerrity TR, Henry CJ, eds. Principles of Route-to-Route Extrapolation for Risk Assessment. Englewood Cliffs, NJ: Prentice Hall, 1990:155–171.

48. Harkema JR. Comparative structure, function and toxicity of the nasal airways. In: Gardner DE, Crapo JD, McClellan RO, eds. Toxicology of the Lung. 3rd ed. New York: Raven Press, 1999:25–83.

49. Trela BA, Frame SR, Bogdanffy MS. A microscopic and ultrastructural evaluation of dibasic esters (DBE) toxicity in rat nasal explants. Exp Mol Pathol 1992; 56:208–218.

50. Kuykendall JR, Taylor ML, Bogdanffy MS. Cytotoxicity and DNA-protecin crosslink formation in rat nasal tissue exposed to vinyl acetate are carboxylesterase-mediated. Toxicol Appl Pharmacol 1993; 123:283–292.

51. Boyd MR. Evidence for the Clara cell as a site for cytochrome P450-dependent mixed function oxidase activity in lung. Nature 1977; 269:713–715.

52. Dahl AR. Metabolic characteristics of the respiratory tract. In: McClellan RO, Henderson RF, eds. Concepts in Inhalation Toxicology. 2d ed. Washington, D.C.: Taylor & Francis, 1995:175–190.

53. Dahl AR, Snipes MB, Gerde P. Sites for uptake of inhaled vapors in beagle dogs. Toxicol Appl Pharmacol 1991; 109:263–275.

54. Wong BA. Inhalation exposure systems design, methods, and operation. In: Gardner DE, Crapo JD, McClellan RO, eds. Toxicology of the Lung. 3rd ed. Philadelphia: Taylor & Francis, 1999:1–53.

55. Cheng YS, Moss OR. Inhalation exposure systems. In: McClellan RO, Henderson RF, eds. Concepts in Inhalation Toxicology. 2nd ed. Washington, D.C.: Taylor & Francis, 1995:25–66.

56. Gardner DE, Kennedy GL Jr. Methodologies and technology for animal inhalation toxicology studies. In: Gardner DE, Crapo JD, McClellan RO, eds. Toxicology of the Lung. 2nd ed. New York: Raven Press, 1993:1–30.

57. Hesterberg TW, Hart GA. Synthetic vitreous fibers: a review of toxicology research and its impact on hazard classification. Crit Rev Toxicol 2001; 31(1):1–53.

58. Wong BA. Generation and characterization of gases and vapors. In: McClellan RO, Henderson RF, eds. Concepts in Inhalation Toxicology. 2d ed. Washington, D.C.: Taylor & Francis, 1995:67–90.

59. Moss OR, Cheng YS. Generation and characterization of test atmospheres: particles and droplets. In: McClellan RO, Henderson RF, eds. Concepts in Inhalation Toxicology. 2nd ed. Washington, D.C.: Taylor & Francis, 1995:91–128.

60. Hesterberg TW, Bunn WB, McClellan RO, Hart GA, Lapin CA. Carcinogenicity studies of diesel engine exhausts in laboratory animals: a summary of past studies and a discussion of future research needs. Crit Rev Toxicol 2005; 35:379–411.

61. Sioutas C, Koutrakis P, Burton RM. A technique to expose animals to concentrated fine ambient aerosols. Environ Health Perspect 1995; 103:172–177.

62. Moss OR, Cheng YS. Generation and characterization of test atmospheres: particles. In: McClellan RO, Henderson RF, eds. New York, NY: Hemisphere Publishing, 1989:85–119.

63. Hinds WC. Aerosol technology: properties, behavior and measurements of airborne particles. New York, NY: Wiley-Interscience, 1999.

64. Baron PA, Willeke K. Aerosol Measurements: Principles Techniques and Applications. 2nd. New York, NY: Wiley, 2001.

65. NIOSH, National Institute for Occupational Safety and Health. Manual of Analytical Methods. 4th ed. NIOSH, 1994.

66. NRC, National Research Council. Issues in Risk Assessment. I. Use of the Maximum Tolerated Dose in Animal Bioassays for Carcinogenicity. Washington, D.C.: National Academy Press, 1993.

67. Lewis TR, Morrow PE, McClellan RO, et al. Contemporary issues in toxicology: establishing aerosol exposure concentrations for inhalation toxicity studies. Toxicol Appl Pharmacol 1989; 99:377–383.

68. Hesterberg TW, Miller WC, McConnell EE, et al. Use of lung toxicity and lung particle clearance to estimate the maximum tolerated dose (MTD) for a fiber glass chronic inhalation oxicity study in the rat. Fundam Appl Toxicol 1996; 32:31–44.

69. Mauderly JL. Assessment of pulmonary function and the effects of inhaled toxicants. In: McClellan RO, Henderson RF, eds. Concepts in Inhalation Toxicology. 2d ed. Washington, D.C.: Taylor & Francis, 1995:355–412.

70. Kane LE, Barrow CS, Alarie Y. A short-term test to predict acceptable levels of exposure to airborne sensory irritants. Am Ind Hyg Assoc J 1979; 40:207–229.

71. Schaper M. Development and database for sensory irritants and its use in establishing occupational exposure limits. Am Ind Hyg Assoc J 1993; 54:488–544.

72. Karol MH, Stadler J, Magreni C. Immunologic evaluation of the respiratory system: animal models for immediate and delayed hypersensitivity. Fundam Appl Toxicol 1985; 5:459–472.

73. Karol MH. Assays to evaluate pulmonary sensitivity. Method in Immunotoxicity 1995; 2:401–409.

74. Bice DE. Immunologic responses of the respiratory tract to inhaled materials. In: McClellan RO, Henderson RF, eds. Concepts in Inhalation Toxicology. 2d ed. Washington, D.C.: Taylor & Francis, 1995:413–440.

75. Henderson RF. Biological markers in the respiratory tract. In: McClellan RO, Henderson RF, eds. Concepts in Inhalation Toxicology. 2d ed. Washington, D.C.: Taylor & Francis, 1995:441–470.

76. Dungworth DL, Hahn FF, Nikula KJ. Non-carcinogenic responses of the respiratory tract to inhaled toxicants. In: McClellan RO, Henderson RF, eds. Concepts in Inhalation Toxicology. 2d ed. Washington, D.C.: Taylor & Francis, 1995:533–578.

77. Feron VJ, Arts JHE, Kupen CF, Slootweg PJ, Wontersen RA. Health risks associated with inhaled nasal toxicants. Crit Rev Toxicol 2001; 31:313–347.

78. Fisher AB. Lung biochemistry and intermediary metabolism. In: McClellan RO, Henderson RF, eds. Concepts in Inhalation Toxicology. 2d ed. Washington, D.C.: Taylor & Francis, 1995:151–174.

79. Driscoll KE. Cytokines and regulation of pulmonary inflammation. In: Gardner DE, Crapo JD, McClellan RO, eds. Toxicology of the Lung. 3rd ed. Philadelphia: Taylor & Francis, 1999:149–172.

80. U.S. Environmental Protection Agency. Guidelines for developmental toxicity risk assessment. Fed Reg 1991; 56:63798–63826.

81. U.S. Environmental Protection Agency. Proposed guidelines for neurotoxicity risk assessment. Fed Reg 1995; 60:52032–52056.

82. U.S. Environmental Protection Agency. Health effects test guidelines carcinogenicity. OPPTS 870,4200, EPA 712-C-96–211, Public Draft, 1996.

83. U.S. Environmental Protection Agency. Health effects test guidelines. OPPTS 870.5300, Detection of gene mutations in somatic cells in culture, EPA712-C-96–221, 1996.

84. U.S. Environmental Protection Agency. Proposed guidelines for reproductive toxicity risk assessment. EPA-630/R-96–009, 1996.

85. McClellan RO. Developing risk assessments for airborne materials. In: Gardner DE, Crapo JD, McClellan RO, eds. Toxicology of the Lung. 3rd ed. Philadelphia: Taylor & Francis, 1999:599–650.

86. Jarabek AM. Inhalation RfC methodology: dosimetry adjustments and dose-response estimation of noncancer toxicity in the upper respiratory tract. In: Miller F, ed. Nasal Toxicity and Dosimetry of Inhaled Xenobiotics: Implications for Human Health. Washington, D.C.: Taylor & Francis, 1995:301–325.

87. American Conference of Governmental Industrial Hygienists. Threshold limit values and biological exposure indices for chemical substances and physical agents. Cincinnati, OH, 2005.

88. NIOSH, National Institute for Occupational Safety and Health. Recommendations for Occupational Safety and Health: Compendium of Policy Documents and Statements. DHHS (NIOSH) Publication No. 92–100. Cincinnati, OH: NIOSH Publications, 1992.

89. NIOSH, National Institute for Occupational Safety and Health. Recommendations for Occupational Safety and Health. Pocket Guide to Chemical Hazards. DHHS (NIOSH) Publication No. 94–11B. Cincinnati, OH: NIOSH Publications, 1997.

90. NRC, National Research Council. Acute Exposure Guideline Levels for Selected Air Chemicals. Vol. 4. Washington, D.C.: National Academies Press, 2004.

91. Gardner DE. Space toxicology: assessing human health hazards during space flight. In: Gardner DE, Crapo JD, McClellan RO, eds. Toxicology of the Lung. 3rd ed. Philadelphia: Taylor & Francis, 1999:343–364.

92. Barnes DG, Dourson M. Reference dose (RfD): description and use in health risk assessments. Regul Toxicol Pharmacol 1988; 8:471–486.

93. Jarabek AM, Menache MG, Overton JH Jr, Dourson ML, Miller FJ. The U.S. Environmental Protection Agency's inhalation RFD methodology: risk assessment for air toxics. Toxicol Ind Health 1990; 6:279–301.

94. Shoaf CR. Current assessment practices for noncancer end points. Environ Health Perspect 1991; 95:111–119.

95. U.S. Environmental Protection Agency. Methods for Derivation of Inhalation Reference Concentrations and Application of Inhalation Dosimetry. Washington, D.C., 1994.

96. U.S. EPA. Integrated Risk Information System (IRIS). Accessed on 7/14/05 at http:// www.epa.gov/IRIS/, 2005.

97. Brenneman KA, James RA, Gross EA, Dourman DC. Olfactory loss in adult CD rats following inhalation exposure to hydrogen sulfide. Toxicol Pathol 2000; 28:326–333.

98. Crump KS. An improved procedure for low-dose carcinogenic risk assessment from animal data. J Environ Pathol Toxicol 1984; 5:339–348.

99. Padgett J, Richmond H. The process of establishing and revising National Ambient Air Quality Standards. J Air Pollut Control Assoc 1983; 33:13–16.

100. Grant LD, Jordan BC. Basis for Primary Air Quality Criteria and Standards. Springfield: National Technical Information Service, NTIS PB88–180070. Research Triangle Park, NC: U.S. Environmental Protection Agency, Environmental Criteria and Assessment Office, 1988.

101. Clean Air Act, Public Law No. 91–604; 84 STAT. 1676, 1970.

102. Clean Air Act, Public Law No. 101–549; 104 STAT. 2399, 1990.

103. U.S. Environmental Protection Agency. Air Quality Criteria for Particulate Matter. Research Triangle Park, NC: Office of Research and Development, National Center for Environmental Assessment, U.S. Environmental Protection Agency, 2004.

104. NRC, National Research Council. Research Priorities for Airborne Particulate Matter. IV. Continuing Research Progress. Washington, D.C.: National Academies Press, 2004.

105. IARC, International Agency for Research on Cancer. IARC Monographs on the Evaluation of Carcinogenic Risks of Chemicals to Man. Vol. 1. Lyon, France, 1972.

106. IARC, International Agency for Research on Cancer. Formaldehyde, 2-Butoxyethanol and 1-tert-Butoxy-2-propanol. Vol. 88. Monograph Series, 2005.

107. IARC, International Agency for Research on Cancer. A Consensus Report of an IARC Monographs Working Group on the Use of Mechanisms of Carcinogenesis in Risk Identification. IARC Internal Technical Report No. 91/002, Lyon, France, 1991.

108. Vainio H, Heseltine E, McGregor D, Tomatis L, Wilbourn J. Working group on mechanisms of carcinogenesis and evaluation of carcinogenic risks. Cancer Res 1992; 52:2357–2361.

109. IARC, International Agency for Research on Cancer. Some Chemicals that Cause Tumours of the Kidney or Bladder in Rodents and Some Other Substances. IARC, Monograph Vol. 73, 1999.

110. IARC, International Agency for Research on Cancer. Man-Made Vitreous Fibers. Monograph Vol. 81. IARC, 2002.

111. U.S. Environmental Protection Agency. Guidelines for carcinogen risk assessment. Fed Reg 1986; 51:33992–34003.

112. Albert RE. Carcinogen risk assessment in the U.S. Environmental Protection Agency. Crit Rev Toxicol 1994:75–85.

113. U.S. Environmental Protection Agency. Guidelines for carcinogen risk assessment. Risk assessment forum. U.S. Environmental Protection Agency, 2005.

114. NTP, National Toxicology Program. Report on Carcinogens, 11th ed. Washington, D.C.: U.S. Department of Health and Human Services, Public Health Service, 2004.

115. Heck HD, Casanova M, Starr TB. Formaldehyde toxicity-new understanding. Crit Rev Toxicol 1990; 20:397–426.

116. Heck HD, Casanova M. The implausibility of leukemia induction by formaldehyde: a critical review of the biological evidence on distant-site toxicity. Regul Toxicol Pharmacol 2004; 40:92–106.

117. Conolly RB, Kimbell JS, Janszen D, et al. Human respiratory tract cancer risks of inhaled formaldehyde: dose-response predictions derived from biologically-motivated computational modeling of a combined rodent and human dataset. Toxicol Sci 2004; 82:279–296.

118. Casanova M, Deyo DF, Heck HD. Covalent binding of inhaled formaldehyde to DNA in the nasal mucosa of Fischer-344 rats: analysis of formaldehyde and DNA by high performance liquid chromatography and provisional pharmacokinetic interpretation. Fundam Appl Toxicol 1989:397–417.

119. Casanova M, Morgan KT, Stinhagen WH, Everitt JI, Popp JA, Heck HD. Covalent binding of inhaled formaldehyde to DNA in the respiratory tract of rhesus monkeys: pharmacokinetics rat to monkey interspecies scaling and extrapolation to man. Fundam Appl Toxicol 1991; 17:409–428.

120. Nauss KM, HEI Working Group. Critical issues in assessing the carcinogenicity of diesel exhaust: a synthesis of current knowledge. In: Diesel Exhaust: A Critical Analysis of Emission Exposure and Health Effects. A Special Report of the Diesel Working Group of the Health Effects Institute, Cambridge, MA. Health Effects Institute, 1995:11–61.

121. Mauderly JL. Relevance of particle-induced rat lung tumors for assessing lung carcinogenic hazard and human lung cancer risk. Environ Health Perspect 1997; 105: 1337–1346.

122. McClellan RO. Lung cancer in rats from prolonged exposure to high concentrations of particles: implications for human risk assessment. Inhal Toxicol 1996; 8(suppl):193–226.

123. Stayner L, Smith RJ, Gilbert S, Bailar AJ. Epidemiologic approaches to risk assessment. Inhal Toxicol 1998; 11:593–601.

124. U.S. Environmental Protection Agency. Supplemental guidance for assessing susceptibility from early-life exposure to carcinogens. Risk assessment forum. U.S. Environmental Protection Agency, 2005.

125. Butterworth BE, Connolly RB, Morgan KT. A strategy for establishing mode of action of chemical carcinogens as a guide for approaches to risk assessments. Cancer Lett 1995; 93:129–146.
126. NRC, National Research Council. Health risks from exposure to low levels of ionizing radiation: In: Biological Effects of Ionizing Radiation VII – Phase 2. Washington, D.C.: National Academies, 2005.
127. McClellan RO, Henderson RF, eds. Concepts in Inhalation Toxicology. 2d ed. Washington, D.C.: Taylor & Francis, 1995.
128. Gardner DE, Crapo JD, McClellan RO, eds. Toxicology of the Lung. 3rd ed. Philadelphia: Taylor & Francis, 1999.
129. Phalen RF. Inhalation Studies: Foundation and Techniques. Boca Raton, FL: CRC Press, 1984.
130. Phalen RF, ed. Methods in Inhalation Toxicology. Boca Raton, FL: CRC Press, 1997.
131. Roth RA, ed. Toxicology of the Respiratory Tract. Vol. 8. In: Sipes IG, McQueen CA, Gandolfi AJ, eds. Comprehensive Toxicology Series. Oxford, U.K.: Elsevier, 1997.
132. Swift DL, Foster WM. Air Pollutants and the Respiratory Tract. New York, NY: Marcel Dekker, 1999.
133. Gehr P, Heyder J. Particle-Lung Interactions. New York, NY: Marcel Dekker, 2000.
134. Nelson S, Martin TR. Cytokines in Pulmonary Disease: Infection and Inflammation. New York, NY: Marcel Dekker, 2000.
135. Marczin N, Kharitonov S, Yacoub MH, Barnes PJ. Disease Markers in Exhaled Breath. New York, NY: Marcel Dekker, 2002.

17

Modern Gas Chromatography–Mass Spectrometry in Human Forensic Toxicology

Beat Aebi and Werner Bernhard
Institute of Legal Medicine, University of Berne, Berne, Switzerland

INTRODUCTION

Human forensic toxicology focuses on the determination of pharmacological effects of alcohol, abused drugs, medications, or other possibly toxic compounds. In most cases, police or prosecutors give the order for a scientific investigation in a forensic case. The analytical results from the laboratory and its scientific expertise will in most cases be used in court. Consequently, all technical, analytical, and expert work needs to be fulfilled with the utmost professional care. A very important part is the chain of custody of evidence, where at any moment in time the origin of every sample and any analytical result must be unequivocally clear. Fields of action comprise the analytical investigation and evaluation of biological samples or suspect items from living or dead persons. Drug abuse and influence of substances that act on the human central nervous system during or just before a certain event (such as traffic accidents, homicide, and submission of sedating drugs) are the most common issues, belonging to the field of behavioral forensic toxicology. To fulfill these tasks, the two biological fluids, urine and blood, are the most frequently used specimens. Recent scientific research also involves new specimens such as sweat and saliva (oral fluid). When investigating on a presumably unnatural death, additional samples from the autopsy (e.g., gastric content and organ tissues) are used. In this field of postmortem toxicology, the scientific goal usually is to confirm or exclude intoxication as the most plausible cause of death. This task is often complicated by various biological, chemical, and physical effects that can occur postmortem (1–7).

The main task of the expert in forensic toxicology is to answer questions from the principal investigator (8). Such questions can be rather easy, such as: Has the accused driver consumed alcohol before being stopped by the police? The questions can also be more difficult to answer: Was the accused person acting under the influence of commonly encountered drugs? Or even more complex: Can death be attributed to the action of a toxic substance? This last question definitely needs more precision, because no analytical laboratory can check for all possibly toxic compounds. Besides that, *sola dosis facit venenum*, the dose makes the poison. In any case, the objective must be clear before starting to analyze samples. All biological samples are always present in limited amounts, and care must be taken regarding which sample to choose, because the extraction and its subsequent analysis by gas chromatography–mass spectrometry (GC–MS) are destructive actions. Before the analyses, the forensic toxicologist must try to gather all case relevant information to make his or her decisions. He or she will have to set priorities regarding analyses and samples. Immunological tests on samples of urine and blood are common preliminary tests. They are commercially available for many frequently abused drugs and medications. GC–MS is used for confirmation or exclusion of preliminary test results, for all kinds of qualitative and quantitative tests as well as for multidrug screenings. Pretests such as immunological tests cannot give definite proof of the presence or the absence of a certain compound and thus offer not conclusive but circumstantial evidence. GC–MS is a combination of two different analytical methods, chromatography and spectrometry, and has a high degree of selectivity and sensitivity. GC–MS results give conclusive evidence and currently have the status of a gold standard within forensic toxicology (9–24).

The goal of this publication is to give an overview of modern GC–MS technology in human forensic toxicology, with examples from daily routine work and current information on the quality of analytical laboratory results in forensic toxicology. It is thought to be of interest for professionals and students of toxicology, forensic sciences and other related fields.

TYPES OF GC–MS AND EXPERIMENTAL SETUPS

The first research GC–MS instruments were developed around 1960. Within a few following years, problems with rugged interfacing and data acquisition could be solved, and GC–MS quickly became the analytical method of choice for environmental, nutritional, chemical, pharmaceutical, quality control, doping, toxicological analyses, etc. Together with miniaturization, computer development, and increasing numbers of sold units, GC–MS became more affordable. Several thousand scientific reports involving GC–MS have been published so far. For every year since 1996, the number of published scientific articles about utilizing GC–MS in toxicology has exceeded 200.

GAS CHROMATOGRAPHY

The instrumental analysis of a sample begins with its introduction into the GC–MS system. The sample itself can be gaseous, liquid, or even solid. Gaseous substances or evaporated liquids can be analyzed by head-space technology (25–29). It is frequently used in forensic toxicology, for instance, for the detection of ethyl alcohol in blood. The newer technology of solid phase microextraction can be used on gaseous, volatile, and nonvolatile liquid biological samples and extracts (30–33). A coated microcapillary is hereby introduced into the sample and then into the injection port of the GC–MS. By the use of a convenient capillary coating, the deposition of the target compounds can be enhanced, and the introduction of unwanted compounds can be reduced (34). In forensic toxicology, the most common use of GC–MS is the injection of diluted extracts from biological samples. In the beginning of GC–MS, these systems were equipped with packed GC columns. Today, these columns have mostly been replaced by fused silica capillary columns. In the splitless-injection mode, all of the evaporated sample reaches the separation column. With the split-injection, only a defined portion of the sample's vapor is actually analyzed. Newer injection modes such as programmed temperature vaporization and large volume injection (LVI) are described in only a few publications in forensic toxicology (35,36). LVIs bear the risk of rapid system contamination. The separation and the detection of chiral drugs of abuse (37) using GC–MS and medications (38) using GC–flame ionization detector have been published. Pesticides and other environmental contaminants have been successfully determined by isomer separation GC–MS (39–41). In the analysis of narcotics, the identification of illicit drug samples can be done without a chromatographic separation by the use of a direct insertion probe. In forensic polymer analysis of solid samples, pyrolysis GC is one of the methods of choice (42–44). Other published research on GC–MS is based on new GC injection (45), separation (46), and detection techniques. The recent developments in GC have been reviewed by several authors (47–49).

MASS SPECTROMETRY WITH QUADRUPOLE ANALYZERS

Electron impact ionization with 70 eV electrons is the standard for routine GC–MS analyses. Some of the modern desktop quadrupole instruments are capable of chemical ionization. This is a softer ionization technique, resulting in fewer molecule fragmentations, and thus, detection limits for various compounds can be drastically lowered (50–55). It proved to be a well-suited technique for the detection of abused drugs and steroids in hair (56–59). But it can also give assistance for the molecular weight determination and substance confirmation (60). Ions inside quadrupole instruments need a certain time to obtain a stable flight path. As a result, such analyzers cannot undergo strong miniaturization, and speeding up the scan rates is difficult. It must further be kept in mind that by scanning the quadrupole mass filter only the selected ions have a stable flight path and can thus reach the detector. So with a full scan of more than typically 500 m/z values, only a minute amount of all the ions entering the quadrupole mass filters are finally collected by the detector. All this results in a small duty cycle for quadrupole instruments.

For advanced compound identification and structure elucidation, two-dimensional mass spectrometry (MS–MS) proved to be very helpful. Triple stage quadrupole MS can be used to detect chemical compounds at very low levels (61,62). Product ion scans and multiple reaction monitoring largely help in increasing the selectivity, increasing the signal of interest, and reducing the background signals simultaneously.

Quadrupole ion traps are capable of storing and gradually detecting ions. Some of these instruments are also capable of performing MS–MS and MS^n experiments (63–65). The produced ions are first stored and only then detected, giving higher ion yields than linear quadrupole systems. The duty cycle for ion detection is usually around 50%. But still, ion traps can have some major drawbacks: The total number of ions inside the ion trap is limited, and space charging may occur. Because the ion trap cannot distinguish between "wanted" and "unwanted" ions, a very high chromatographic separation and a careful selection of sample concentration are necessary for trace analyses. In recent years, the development of MS has brought the merging of triple stage quadrupoles and ion traps, yielding the new and very promising technology of linear ion traps. The physical space for ion trapping and the total number of stored ions are increased, making these systems more sensitive and more efficient (66–68).

SECTOR FIELD INSTRUMENTS, TANDEM AND HYBRID MASS SPECTROMETERS

The most common technical modifications to enhance the specificity of GC–MS or liquid chromatography–mass spectrometry (LC–MS) analyses (i.e., to differentiate the analyte from background compounds) are high-resolution mass spectrometry and tandem MS. Tandem MS was traditionally achieved by large and expensive double focusing magnetic sector field instruments. Nowadays, improved resolution can also be obtained from time-of-flight instruments (TOF-MS). Tandem mass spectrometry is characterized by the combinations of almost any type of mass spectrometers. When sector analyzers (magnetic, electrostatic) are coupled to nonsector analyzers (quadrupole, ion traps, and TOF), these combinations are then called hybrid instruments. The benefit of high-resolution is the potential to differentiate between ions of identical nominal mass. The efficiency of this technique depends on the mass defect of the chemical elements present in the molecule of interest. Substances containing heteroatoms (halogens in particular) can be detected quite specifically. They can be differentiated from highly abundant endogenous compounds that mainly consist of carbon, hydrogen, and oxygen.

There are two typical objectives of high-resolution MS experiments in forensic science. The use of high-resolution MS for improving the peak to noise ratio (by suppression of matrix compounds with identical nominal mass but different elemental composition) is shown in Figure 1. Another application is the chemical structure elucidation of unknown compounds by evaluating its elemental composition, as depicted in Figure 2. This possibility is primarily based on a high mass accuracy of the MS, but also requires enhanced resolution to diminish the otherwise

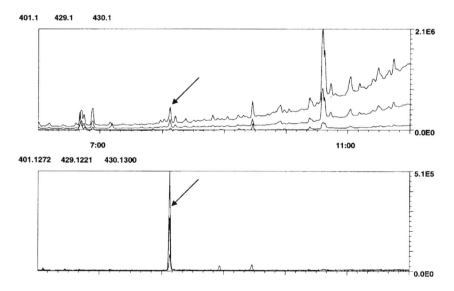

Figure 1 Improvement of peak-to-noise ratio of oxandrolone (bis-TMS) detected by high-resolution (*bottom*) compared to conventional MS (*top*). The significant effect is mainly due to the mass defect of the chlorine atom in the molecule. *Source*: From Detlef Thieme, Ph.D., Institute of Legal Medicine, University of Munich, Germany.

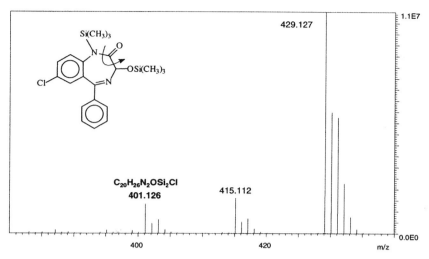

Figure 2 Interpretation of fragment ions by means of accurate mass measuring. Additional restrictions (potential elements, double bonds, etc.) are required for the estimation of the fine mass (301.126 Da ± 10 ppm) to predict the elemental composition. *Source*: From Detlef Thieme, Ph.D., Institute of Legal Medicine, University of Munich, Germany. *Abbreviation*: MS, mass spectrometry.

important matrix interference. However, even in case of highly accurate data, hundreds of structural proposals might still result. In this case, additional information (e.g., number and kind of heteroatoms and double bound equivalents) is needed to reduce the number of analytical matches.

Tandem and multidimensional LC–MS instruments have outnumbered the analogous GC–MS instruments. Tandem MS experiments usually involve the selection and isolation of an abundant precursor ion in the first analyzer (MS1), the fragmentation in a collision cell, and the analyses of the product ions in the second analyzer (MS2). This technique may result in various applications such as product ion scans for the identification of unknowns and multiple reaction monitoring for quantitative trace analyses (Fig. 3). When MS1 is scanned and MS2 is set to a defined m/z value, a precursor ion scan results and related compounds (e.g., metabolites) can be detected. When both MS1 and MS2 are scanned with a constant offset, losses of neutral (uncharged) compounds can be observed. With this setup, all molecules that loose the same neutral fragment upon collision (e.g., loss of an uncharged methoxy-group with a mass of 31 Da) can be selectively monitored.

TOF-MS

TOF-MS make use of the velocity of ions in a (field free) drift tube under high vacuum. Externally generated ions are strongly accelerated and then detected after having passed the flight region. TOF systems can be operated at very high scan rates, up to 10 kHz. With a fast GC–MS data system, up to 200 mass scans/sec can be acquired. Another possibility is the high mass resolution, making a molecular formula estimation possible (69–72).

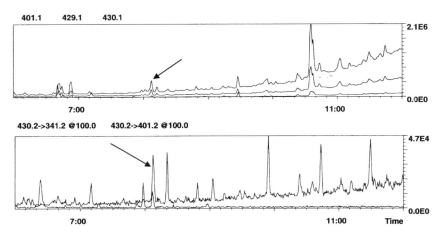

Figure 3 Improvement of the peak-to-noise ratio of oxandrolone (bis-TMS) detected by tandem MS. The effectiveness of tandem MS is limited by the abundance of the precursor ion and intensity and specificity of the fragmentation reaction. *Abbreviation*: MS, mass spectrometry. *Source*: From Detlef Thieme, Ph.D., Institute of Legal Medicine, University of Munich, Germany.

SCREENING ANALYSES WITH GC–MS

The sample preparation of the biological material for analysis is a very important task. Most extracts from biological samples exhibit strong chemical background because of the high concentration of endogenous compounds. The extraction (liquid–liquid extraction or solid phase extraction) should yield as little unwanted compounds and as much wanted compounds as possible. If the analyst knows what compounds to look for, then the confirmation or quantitation is a targeted type of analysis. But quite often it is not known what medications, drugs of abuse, or potentially toxic substances are present or to be expected. In this case, a general unknown analysis or systematic toxicological analysis must be performed (19,73–78).

GC–MS combines a high-resolution chromatographic separation with high sensitivity detection. Together with a powerful library search program and up-to-date MS libraries, GC–MS is the most frequently used method in forensic toxicology (79–81). The most comprehensive work in forensic toxicology is without doubt from Pfleger et al. (79) where the latest version not only contains over 6000 mass spectra, but also many extraction and derivatization procedures as well as different cross-reference lists. The spectral library is commercially available for different MS data systems (82). Other commercial GC–MS libraries are the designer drugs library from DigiLab Software GmbH, Scharnhagen, Germany (83) as well as the large and multipurpose libraries from the U.S. National Institute of Standards and Technology (NIST) (84) and Wiley (85). Non-commercial libraries are available for members of the American Academy of Forensic Sciences (86) and The International Association of Forensic Toxicologist (87). The Pfleger, Maurer, and Weber library in combination with at least one of the two large libraries from either NIST or Wiley is estimated to be the minimum requirement for routine GC–MS data evaluation in forensic toxicology. Whereas most GC–MS library search systems are bound to the GC–MS manufacturer data formats, the earlier described software program

MassLib is system independent (80). MassLib offers the possibility of searches within multiple libraries simultaneously, an advanced structure editor and search, a fragmentation editor, the search for identical or similar spectra (with or without automatic background spectra), and most lately, the possibility of overlapping two GC–MS runs and comparing the spectra simultaneously. MassLib uses its own data format, but GC–MS chromatograms and mass spectra from all current GC–MS data formats can be imported (88). Other software products for the up-to-date evaluation of GC–MS data are also available (89–91).

The addition of a nitrogen–phosphorus selective detector (NPD) to the GC–MS has proven very helpful (92,93). Apart from ethyl alcohol, acetyl salicylic acid, cannabinoids, and others, the large majority of the substances of interest in forensic toxicology contain at least one nitrogen atom. The NPD can detect minute traces of nitrogen or phosphorus containing compounds and at the same time discriminate all other organic compounds.

The GC–MS–NPD run shown in Figure 4 is a basic extract of a native urine sample. A female driver was suspected of driving her car under the influence of drugs, and a general unknown screening for basic, acidic, and neutral drugs in urine was performed. The MS library search suggested the presence of caffeine (A), tramadol and its metabolites (B, C, D, G), metoprolol plus metabolites, and artifacts (E, F, H, I) as well as codeine (J). The above mentioned peaks have much higher intensities than the internal standard used (proadifen, not annotated). Even with purely qualitative analyses, we use internal chemical standards to estimate the extraction efficiency. Human urine is a relatively constant matrix and its extracts usually do not exhibit strong chemical background signals, in contrast to extracts from other specimens such as blood, serum, gastric content, or even organ tissues. The injected sample is parted at a ratio of nearly 1:1 inside the GC injector, where they reach two capillary separation columns. The same type of GC column is connected to both detectors. The column that is connected to the MS (e.g., 30 m) is about 20% longer than the one connected to the NPD (e.g., 25 m). With this setup, retention time differences of 20 seconds or less can easily be achieved (92). Some analytical systems

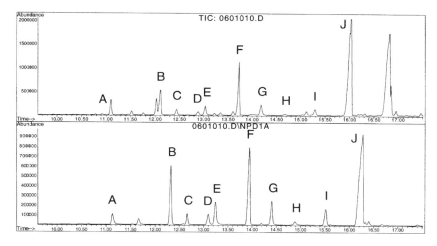

Figure 4 Reconstructed TIC (*above*) and NPD trace (*below*) from the same gas chromatographic run. *Abbreviations*: TIC, total ion chromatogram; NPD, nitrogen–phosphorus selective detector.

do support the so-called "retention time locking," making a complete retention time overlapping possible. One must be aware that the correct retention time is the most important criteria for the identification or exclusion of compounds with GC–MS. So far, we have not made use of the any retention time locking system.

ALTERNATIVE MATRICES AND SAMPLES

Besides urine, blood, and tissue, human scalp hair has become the matrix of choice for the determination of a previous consumption of frequently abused drugs and medications (94–102). Wennig describes in a very practical way the potential problems with the interpretation of hair analysis results (103). Kintz et al. (104) very recently proposed the detection of a single dose of gamma hydroxybutyrate in hair. Hair and other specimen have successfully been analyzed for markers of excessive alcohol consumption (105–107). For other alternative matrices, such as sweat and saliva (oral fluid), on-site tests are commercially available. The main purpose of these tests is to check on-site for a prior consumption of abused drugs, where saliva and sweat might have some advantages over urine samples. In the analytical laboratory of forensic toxicology, however, saliva and sweat do not currently have major advantages over urine. For the estimation of the actual pharmacological effects, blood, plasma, or serum is still needed. Numerous publications exist on alternative matrices, both review articles (108–110) and experimental publications (111–118).

QUALITY OF ANALYTICAL DATA

In earlier times, the quality of laboratories in forensic toxicology was estimated on the basis of a successful participation at external quality control programs. Nowadays, these controls are still in use and of prime importance, but as a part of a much more comprehensive quality management. The professional organizations involved have published guidelines and requirements for forensic toxicology (119–122).

The normative guidelines of ISO 17025 comprise the management of a laboratory and the quality of all of its data and final products (e.g., report and expertise). The norm ISO 17025 is named as general requirements for the competence of testing and calibration laboratories (ISO/IEC 17025:1999). The five chapters (scope, normative references, terms and definitions, management requirements and technical requirements) give a practical frame on all the work that is done in an analytical laboratory. Based on this frame, the laboratory must prepare and maintain documents that describe all actions and responsibilities from the receipt of the samples and the order to investigate to the final product, usually being the forensic toxicology report or expertise. Besides the competence of all personnel and the qualification of the instrumentation, a special focus is put on the validation of analytical methods (123–125). The laboratory must define its scientific scope and types of the analytical procedures in use. Every analytical method and each test (qualitative, quantitative, presumptive, confirming, targeted, and untargeted) must be validated. The process of method validation consists of the thorough determination of all the analytical parameters of the method and may comprise selectivity, linearity, range, recovery, repeatability, precision, limit of detection, lower limit of quantitation (quantitative analyses only), robustness, routine method performance tests, statistical process (method) control, and estimation of the measurement uncertainty. The measurement uncertainty is the newly adopted term used for the sum of all imprecisions connected to the specific analytical result. One can wonder if the police, the lawyers, and the judges can make more sense out of "uncertain" analytical results than of "biased" or "imprecise" results. Following the norm ISO 17025, all quantitative results must be written as "value ± measurement uncertainty." To keep the effort of method validation limited, one can validate retrospectively by using earlier analytical results. If the analytical method is based on a previously validated and published method, several parts of the in-house validation might be omitted. The preparation for accreditation following ISO 17025 gives the laboratory the unique opportunity to review and reassess its documentation and responsibilities. But the efforts to obtain accreditation with ISO 17025 are important, the literature describes the additional efforts to be between 10% and 20%, both in personnel and all other costs.

For GC–MS, written up-to-date procedures for the testing, use and qualification/clearing of all analytical instruments and methods must be present and strictly followed by the personnel. Regular performance tests must be undertaken, documented, and evaluated on a documented basis, as an important part of the statistical process (method) control. Only if the instrument and method used are viable and the personnel is skilled and trained, the result can also be viable.

LIMITATIONS OF GC–MS, TRAPS AND PITFALLS

Problems concerning sample identity, storage conditions, sample workup and extraction, matrix effects, contamination, interpretation, and documentation of results can occur in all analytical domains and will not be specifically discussed in this chapter. GC–MS can be used to analyze a wide range of different chemical compounds. But there are a few limitations that are GC–MS specific. The most important subject is the need for volatility. In many cases, the GC–MS sample consists of an organic extract of a biological sample. The introduction of the sample occurs with the injection of microliter volumes into the heated injector block. Common injector temperatures range from about 250°C to 300°C. The substances to be chromatographically separated must evaporate in the injector, otherwise it will condense or decompose. The injector plays a crucial role within the whole GC–MS system. Nowadays, chemically inert quartz injector liners are used to protect both the GC–MS and the sample itself from excessive contamination. Free silanol groups with a high chemical reactivity need to be chemically disactivated. A precolumn can help to protect the main analytical column. To reduce the risk of cross-contamination, it is a good practice to inject at least one sample of pure solvent after each extract or test solution.

Besides volatility, thermal stability is another important issue in GC–MS. This is particularly true when performing a general unknown screening. If a decomposition or other chemical reaction takes place inside the injector, the GC–MS will not detect the substance itself but, at best, the decomposition products and artifacts. By the use of MS libraries, these artifacts can sometimes be detected. But most of the time, the analyst does not know that decomposition had occurred and that artifacts are present. When no good match is found in the MS library, then the spectrum is probably not further checked. GC–MS analyses of amphetamines can easily yield such artifacts. As an example, the new designer drug 2C-T-7 (2,5-dimethoxy-4-thiopropyl-phenylethylamine) is shown in Figure 5 .

The methanolic extract of an illicit tablet was suspected to contain designer amphetamines, presumably 2C-T-7. Once injected into the GC–MS, the lower

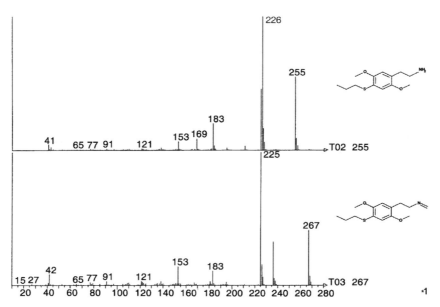

Figure 5 Spectrum of 2C-T-7 (*upper spectrum*) and artifact thereof (*lower spectrum*).

spectrum of Figure 5 resulted. The chromatogram showed high concentrations of this compound and only smallest amounts of the original spectrum of 2C-T-7 (upper spectrum in Fig. 5). While searching the GC–MS libraries for this unknown compound was unsuccessful, the formation of an artifact was soon thought to be a possible cause. It is known that amines tend to form imino-type of artifacts when dissolved in methanol and introduced into the GC–MS. The effect is even enhanced, when the compound is introduced into the GC–MS in its salt form, e.g., as hydrochloride salt. The introduction of salts into the GC–MS should also be avoided because of the deposition of unvolatile components and the formation of gases, for instance, hydrochloric acid. It is always preferable not to directly inject the dissolved sample but rather to first convert the unknown compounds to its form of a free base (or free acid). This can be achieved by adding a base such as ammonia to a weighted portion of the sample and then extract the free base with an organic solvent such as ethyl acetate. To prevent the injection of water into the GC–MS, one should dry the extract with a small amount of drying agent, for instance, anhydrous sodium sulfate, before injection. Many different designer drugs exist today, and these analyses can sometimes be a real challenge to the forensic toxicologist (126–129).

Care must be taken with the chemicals that are injected into the GC–MS. Water, strong acids or bases, metals, and salts may instantly damage the stationary phase of the capillary column immediately. Poor GC separation power and shifting retention times will be the result. Extracts from biological samples such as blood, plasma, or serum and especially organ tissues may contribute important chemical background and in the worst case, even change the chromatographic properties of the column. Strong underground signals are able to mask less intense signals from compounds of interest. Several endogenous compounds (e.g., fatty acids and esters) can produce signals at almost every m/z value and by this exclude other compounds from detection.

Even without strong matrix interference, the elution of two or more compounds at more or less the same retention time can always happen. This well-known phenomenon is the coelution of two or more compounds and can be a serious problem. When trying to confirm or quantitate a compound by GC–MS, coelution is usually not a major problem. The retention time and the expected ions and their ratios are known, and the compound should be found even if coelution occurs. This is not the case when screening for unknowns. With coelution, the spectra of the compounds involved will be added, and the library search in most cases will be unsuccessful. With increasing experience, the analyst will know the most prominent background spectra and be able to find most cases of coelution. One good point to start is looking at the chromatographic shape of the suspected peak. Usually the coelution is not complete, meaning that the peak maxima are at least slightly different, giving the composite peak a distorted shape. A normal GC peak should have a Gaussian type of shape with a slight tailing toward higher retention times. All other peak shapes are suspect and could be due to coelution or sample decomposition. Only when the coelution is complete, the peak maximum of the total ion current and also every ion trace are the same, and overlapping is complete. Because Gaussian shape plus Gaussian shape equals Gaussian shape again, peaks with complete coelution are sometimes hard to detect. When detected, coeluting peaks need to be background corrected and will only by then yield the pure spectra. Plotting the separate ion chromatograms can be a big help here. But one needs to be careful with background subtraction and any other spectrum arithmetics. The pitfall of generating an "artificial" spectrum must be kept in mind.

The analyst should have the possibility to backup new and unknown compounds. When new data or new libraries are available, the unknowns should be checked. Spectra that have been hard to interpret, e.g., because of coelution, can also be saved into that library of unknowns. With this action, the chance of finding the coeluting compounds on the next occasion will be increased. It is good practice to save these data in separate libraries and not in existing libraries, to avoid a mix-up of external, internal, confirmed, and unconfirmed spectra. The chances for a substance identification are largely increased by the use of multiple GC–MS libraries (80). Even with the standard electron impact ionization using 70 eV electrons, mass spectra can be somewhat different from one instrument to the other. Quite a few compounds in forensic toxicology have a large molecule body and a small amine side chain. The ionization will give a small, positively charged amine side chain, and the large molecule body will pass the detector as an undetectable radical. To illustrate this behavior, a typical example is shown in Figure 6.

All of the three spectra in Figure 6 are rather similar, because the ion with m/z 58 is always the base peak, a signal at m/z 73 and clusters around m/z 165 are present. The top spectrum is that of diphenhydramine, a commonly encountered antihistaminic. The middle spectrum belongs to bromdiphenhydramine and the bottom spectrum originates from the skeletal muscle relaxant and antihistaminic orphenadrine (*ortho*-methylated diphenhydramine). The identification or exclusion of such compounds can in fact be difficult. One needs to be aware that positive hits from MS library searches do not finally prove the presence of a certain compound, but rather indicate its presence. To obtain definite proof and establish full evidence, the retention time as well as the correct ions and ion ratios must be checked. This is usually not a difficult task, when the pure substance is commercially available. But many compounds like minor metabolites do not meet this requirement, and the presence of such a compound cannot be finally proven. The analyst can still try to find

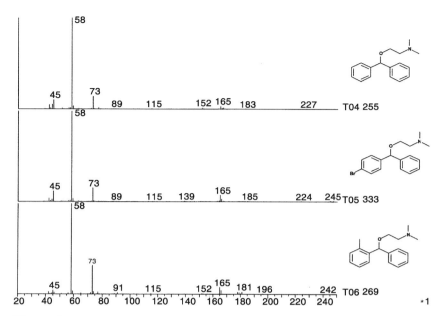

Figure 6 Mass spectra of diphenhydramine and related compounds.

the main compound or other related compounds, maybe also in another biological specimen.

Whereas most GC–MS runs and also MS library entries have a mass range starting at m/z 50, this might be sufficient for most applications. On the other hand, it might also be quite a pitfall. As an example, Figure 7 shows two mass spectra from the very same compound amphetamine, a frequently abused drug.

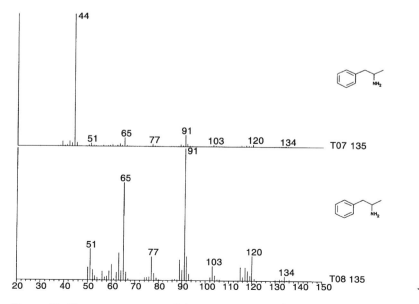

Figure 7 Two mass spectra of the same compound (amphetamine base), above starting at m/z 40 (correct) and below starting at m/z 50 (incorrect, artificially generated).

Although both spectra in Figure 7 look quite different at first sight, they are from the same compound, amphetamine. The base peak of a mass spectrum is (by definition) always the highest peak within the selected mass range and will automatically be normalized to 100%. But when the selected mass range is too narrow, the "true base peak" might be missed, and another peak becomes "base peak." The search within the MS libraries will likely be unsuccessful, if the library does not contain a reference with the incomplete spectrum. At the worst, wrong hits with high matching factors will be attributed to this incomplete spectrum. As a consequence, one must be aware of the fact that important ions below or above the selected mass range could exist and also that the library entries might on some rare occasions be incomplete. A broad range of different libraries can be of help here. Other publications on possible traps and pitfalls are found in the recent literature (130–143).

SUMMARY AND OUTLOOK, EXPANDING LC–MS, NEW TECHNOLOGIES

As any other analytical technique, GC–MS has many advantages but also some disadvantages when compared to other techniques. By derivatizing polar and thermally instable compounds, the field of applications of GC–MS could be much expanded. But the derivatization step also means more work and costs. LC combined with MS (LC–MS and LC–MS–MS) is complementary to GC–MS and steadily conquering analytical forensic toxicology, especially in the confirmation or quantitation of selected compounds (144–148). The important step of sample cleanup and extraction will undergo important technological development, especially toward further automation and miniaturization (149–151). For systematic toxicological screenings, GC–MS is for the moment still the best choice. This can be attributed to the fact that GC–MS spectra are mostly independent of the GC–MS system and its analytical parameters chosen, whereas LC–MS is strongly dependent on these factors. Additionally, GC–MS libraries are commonly used and contain many thousands of different spectra and compounds. Several researchers have lately begun to define LC–MS setup parameters for screening analyses and started to collect appropriate LC–MS library spectra (67,152). It can be expected that these approaches will become routine practice in the next years to follow. Some approaches also involve LC–MS–MS with spectral libraries of product ion spectra of collision-induced dissociations, where product ion spectra are expected to be less instrument parameter dependent. Recently developed instrumentation, for instance, TOF-MS, is currently available both as GC–MS and LC–MS versions. But LC–MS is rapidly gaining importance and conquering one analytical field after the other. The current efforts in the development of new MS technologies are now in favor of new LC–MS systems and no longer in favor of GC–MS systems. One stronghold of GC–MS is the wide-range of toxicological screenings. But again, new technologies involving highly exact mass determinations and powerful molecular composition estimation might soon bypass the need for large spectra libraries and GC–MS. For the moment, most substances of interest in forensic toxicology can be detected with common desktop quadrupole GC–MS systems. But when it comes to higher throughput, reduced sample workup, higher degrees of automation, and research on new compounds and their metabolites, one needs more sophisticated instrumentation. The costs for acquisition of an MS–MS or high-resolution capable instrument have dropped in the last years. LC–MS is in direct competition with GC–MS and has replaced GC–MS in many instances. The transition from GC–MS to LC–MS seems to be slower in forensic toxicology than in other analytical fields. But even without large spectral libraries, LC–MS will rapidly gain further attention and will replace many GC–MS instruments for routine confirmation and quantitation in biological samples. High-end instrumentation such as ion cyclotron resonance mass spectrometry (153) will probably find its way into forensic toxicology research, but is unlikely to be used in routine forensic analyses in the near future.

ACKNOWLEDGMENTS

We thank Detlef Thieme, PhD, from the Institute of Legal Medicine, University of Munich, Germany, for the kind support with know-how, data, and examples for publication. We also thank Winfried Wagner Redeker, PhD (Spectronex, Basle, Switzerland), for his valuable input on the future of GC–MS in forensic toxicology. This article is dedicated to our deceased friend Dr.phil.nat. Felix Friedli.

REFERENCES

1. Alunni-Perret V, Kintz P, Ludes B, Ohayon P, Quatrehomme G. Determination of heroin after embalmment. For Sci Int 2003; 134:36–39.
2. García-Repetto R, Moreno E, Soriano T, Jurado C, Giménez MP, Menéndez M. Tissue concentrations of MDMA and its metabolite MDA in three fatal cases of overdose. For Sci Int 2003; 134:110–114.
3. Hino Y, Ojanperä I, Rasanen I, Vuori E. Performance of immunoassays in screening for opiates, cannabinoids and amphetamines in post-mortem blood. For Sci Int 2003; 131: 148–155.
4. Moriya F, Hashimoto Y. Tissue distribution of nitrazepam and 7-aminonitrazepam in a case of nitrazepam intoxication. For Sci Int 2003; 131:108–112.
5. Moriya F, Hashimoto Y. Postmortem diffusion of drugs from the bladder into femoral venous blood. For Sci Int 2001; 123:248–253.
6. Skopp G, Klinder K, Pötsch L, et al. Postmortem distribution of dihydrocodeine and metabolites in a fatal case of dihydrocodeine intoxication. For Sci Int 1998; 95:99–107.
7. Takayasu T, Kondo T, Sato Y, Oshima T. Determination of triazolam by GC-MS in two autopsy cases: distribution in body fluids and organs. Legal Med 2000; 4:206–211.
8. Simpson D. The analytical chemist as expert witness. Trend Anal Chem 1996; 15(10): 494–500.
9. Amendola L, Colamonici C, Mazzarino M, Botrè F. Rapid determination of diuretics in human urine by gas chromatography–mass spectrometry following microwave assisted derivatization. Anal Chim Acta 2003; 475:125–136.
10. Amendola L, Garribba F, Botrè F. Determination of endogenous and synthetic glucocorticoids in human urine by gas chromatography–mass spectrometry following microwave-assisted derivatization. Anal Chim Acta 2003; 489:233–243.
11. Bosman IJ, Lusthof KJ. Forensic cases involving the use of GHB in The Netherlands. For Sci Int 2003; 133:17–21.
12. Chou S-L, Yang M-H, Ling Y-C, Giang Y-S. Gas chromatography-isotope dilution mass spectrometry preceded by liquid–liquid extraction and chemical derivatization for the determination of ketamine and norketamine in urine. J Chromatogr B 2004; 799:37–50.
13. Elsirafy AA, Ghanem AA, Eid AE, Eldakroory SA. Chronological study of diazinon in putrefied viscera of rats using GC/MS, GC/EC and TLC. For Sci Int 2000; 109: 147–157.

14. Hindmarch I, ElSohly M, Gambles J, Salamone S. Forensic urinalysis of drug use in cases of alleged sexual assault. J Clin For Med 2001; 8:197–205.

15. Ishii A, Tanaka M, Kurihara R, et al. Sensitive determination of pethidine in body fluids by gas chromatography–tandem mass spectrometry. J Chromatogr B 2003; 792: 117–121.

16. Kudo K, Kiyoshima A, Ohtsuka Y, Ikeda N. Simultaneous determination of bromvaler-ylurea and allylisopropylacetylurea in human blood and urine by gas chromatography–mass spectrometry. J Chromatogr B 2003; 791:171–177.

17. Lora-Tamayo C, Tena T, Rodríguez A, Sancho JR, Molina E. Intoxication due to 1,4-butanediol. For Sci Int 2003; 133:256–259.

18. Meadway C, George S, Braithwaite R. A rapid GC–MS method for the determination of dihydrocodeine, codeine, norcodeine, morphine, normorphine and 6-MAM in urine. For Sci Int 2002; 127:136–141.

19. Moeller MR, Steinmeyer S, Kraemer Th. Determination of drugs of abuse in blood. J Chromatogr B 1998; 713:91–109.

20. Müller IB, Willads Petersen H, Johansen SS, Theilade P. Fatal overdose of the herbicide bentazone. For Sci Int 2003; 135:235–236.

21. Namera A, Yashiki M, Hirose Y, Yamaji S, Tani T, Kojima T. Quantitative analysis of tropane alkaloids in biological materials by gas chromatography–mass spectrometry. For Sci Int 2002; 130:34–43.

22. Nishida M, Namera A, Yashiki M, Kojima T. On-column derivatization for determination of amphetamine and methamphetamine in human blood by gas chromatography–mass spectrometry. For Sci Int 2002; 125:156–162.

23. Romain N, Giroud C, Michaud K, Mangin P. Suicide by injection of a veterinarian bar-biturate euthanasia agent: report of a case and toxicological analysis. For Sci Int 2003; 131:103–107.

24. Stillwell ME. A reported case involving impaired driving following self-administration of xylazine. For Sci Int 2003; 134:25–28.

25. Coopman VAE, Cordonnier JACM, De Letter EA, Piette MHA. Tissue distribution of trichloroethylene in a case of accidental acute intoxication by inhalation. For Sci Int 2003; 134:115–119.

26. Kintz P, Doray S, Cirimele V, Ludes B. Testing for alpha-chloralose by headspace-GC/MS: a case report. For Sci Int 1999; 104:59–63.

27. Liu J, Hara K, Kashimura S, et al. Headspace solid-phase microextraction and gas chromatographic–mass spectrometric screening for volatile hydrocarbons in blood. J Chromatogr B 2000; 748:401–406.

28. Snow NH. Head-space analysis in modern gas chromatography. Trend Anal Chem 2002; 21(9–10):608–617.

29. Wasfi IA, Al-Awadhi AH, Al-Hatali ZN, Al-Rayami FJ, Al Katheeri NA. Rapid and sensitive static headspace gas chromatography–mass spectrometry method for the ana-lysis of ethanol and abused inhalants in blood. J Chromatogr B 2004; 799:331–336.

30. De Martinis BS, Santos Martin CC. Automated headspace solid-phase microextraction and capillary gas chromatography analysis of ethanol in postmortem specimens. For Sci Int 2002; 128:115–119.

31. Lucas ACS, Bermejo AM, Tabernero MJ, Fernandez P, Strano-Rossi S. Use of solid-phase microextraction (SPME) for the determination of methadone and EDDP in human hair by GC–MS. For Sci Int 2000; 107:225–232.

32. Sporkert F, Pragst F, Hübner S, Mills G. Headspace solid-phase microextraction with 1-pyrenyldiazomethane on-fibre derivatisation for analysis of fluoroacetic acid in biolo-gical samples. J Chromatogr B 2002; 772:45–51.

33. Van Hout MWJ, van Egmond WMA, Franke JP, de Zeeuw RA, de Jong GJ. Feasibility of the direct coupling of solid-phase extraction–pipette tips with a programmed-temperature vaporiser for gas chromatographic analysis of drugs in plasma. J Chroma-togr B 2001; 766:37–45.

34. Sporkert F, Pragst F. Determination of methadone and its metabolites EDDP and EMDP in human hair by headspace solid-phase microextraction and gas chromatography–mass spectrometry. J Chromatogr B 2000; 746:255–264.

35. Teske J, Putzbach K, Engewald, Müller RK. Determination of cannabinoids by gas chromatography–mass spectrometry and large-volume programmed-temperature vaporizer injection using 25 ml of biological fluid. J Chromatogr B 2002; 772:299–306.

36. Van Hout MWJ, de Zeeuw RA, de Jong GJ. Coupling device for desorption of drugs from solid-phase extraction-pipette tips and on-line gas chromatographic analysis. J Chromatogr A 1999; 858:117–122.

37. Liu J-T, Liu RH. Enantiomeric composition of abused amine drugs: chromatographic methods of analysis and data interpretation. J Biochem Biophys Meth 2002; 54: 115–146.

38. Tao QF, Zeng S. Analysis of enantiomers of chiral phenethylamine drugs by capillary gas chromatography/mass spectrometry/flame–ionization detection and pre-column chiral derivatization. J Biochem Biophys Meth 2002; 54:103–113.

39. Leone AD, Ulrich EM, Bodnar CE, Falconer RL, Hites RA. Organochlorine pesticide concentrations and enantiomer fractions for chlordane in indoor air from the US cornbelt. Atmos Environ 2000; 34:4131–4138.

40. Vetter W, Alder L, Kallenborn R, Schlabach M. Determination of Q1, an unknown organochlorine contaminant, in human milk, Antarctic air, and further environmental samples. Environ Pollut 2000; 110:401–409.

41. Watson WP, Cottrell L, Zhang D, Golding BT. Metabolism and molecular toxicology of isoprene. Chem-Biol Interact 2001; 135–136:223–238.

42. Haken JK. Pyrolysis gas chromatography of synthetic polymers—a bibliography. J Chromatogr A 1998; 825:171–187.

43. Wampler TP. Introduction to pyrolysis–capillary gas chromatography. J Chromatogr A 1999; 842:207–220.

44. Wampler TP, Bishea GA, Simonsick WJ. Recent changes in automotive paint formulation using pyrolysis-gas chromatography/mass spectrometry for identification. J Anal Appl Pyrol 1997; 40–41:79–89.

45. Takayasu T, Ohshima T, Kondo T. Rapid analysis of pesticide components, xylene, o-dichlorobenzene, cresol and dichlorvos, in blood and urine by pulsed heating-gas chromatography. Legal Med 2001; 3:157–161.

46. Fialkov AB, Gordin A, Amirav A. Extending the range of compounds amenable for gas chromatography–mass spectrometric analysis. J Chromatogr A 2003; 991:217–240.

47. Marriott PJ, Haglund P, Ong RCY. A review of environmental toxicant analysis by using multidimensional gas chromatography and comprehensive GC. Clin Chim Acta 2003; 328:1–19.

48. Ragunathan N, Krock KA, Klawun C, Sasaki TA, Wilkins CL. Gas chromatography with spectroscopic detectors. J Chromatogr A 1999; 856:349–397.

49. Santos FJ, Galceran MT. Modern developments in gas chromatography–mass spectrometry based environmental analysis. J Chromatogr A 2003; 1000:125–151.

50. Akrill P, Cocker J. Determination of nitroglycerin and its dinitrate metabolites in urine by gas chromatography–mass spectrometry as potential biomarkers for occupational exposure. J Chromatogr B 2002; 778:193–198.

51. Ugland HG, Krogh M, Rasmussen KE. Automated determination of 'Ecstasy' and amphetamines in urine by SPME and capillary gas chromatography after propylchloroformate derivatisation. J Pharmaceut Biomed 1999; 19:463–475.

52. Libong D, Bouchonnet S. Collision-induced dissociations of trimethylsilylated lysergic acid diethylamide (LSD) in ion trap multiple stage mass spectrometry. Int J Mass Spectrum 2002; 219:615–624.

53. Moore C, Guzaldo F, Hussain MJ, Lewis D. Determination of methadone in urine using ion trap GC/MS in positive ion chemical ionization mode. For Sci Int 2001; 119:155–160.

54. Pellegrini M, Rosati F, Pacifici R, Zuccaro P, Romolo FS, Lopez A. Rapid screening method for determination of Ecstasy and amphetamines in urine samples using gas chromatography–chemical ionisation mass spectrometry. J Chromatogr B 2002; 769: 243–251.

55. Pirnay S, Ricordel I, Libong D, Bouchonnet S. Sensitive method for the detection of 22 benzodiazepines by gas chromatography–ion trap tandem mass spectrometry. J Chromatogr A 2002; 954:235–245.

56. Baptista MJ, Monsanto PV, Marques EGP, et al. Hair analysis for delta9-THC, delta9-THC-COOH, CBN and CBD, by GC/MS-EI, comparison with GC/MS-NCI for delta9-THC-COOH. For Sci Int 2002; 128:66–78.

57. Cirimele V, Kintz P, Ludes B. Testing of the anabolic stanozolol in human hair by gas chromatography–negative ion chemical ionization mass spectrometry. J Chromatogr B 2000; 740:265–271.

58. Höld KM, Crough DJ, Wilkins DG, Rollins DE, Maes RA. Detection of alprazolam in hair by negative ion chemical ionization mass spectrometry. For Sci Int 1997; 84: 201–209.

59. Shen M, Xiang P, Wu H, Shen B, Huang Z. Detection of antidepressant and antipsychotic drugs in human hair. For Sci Int 2002; 126:153–161.

60. Gimeno P, Besacier F, Chaudron-Thozet H, Girard J, Lamotte A. A contribution to the chemical profiling of 3,4-methylenedioxymethamphetamine (MDMA) tablets. For Sci Int 2002; 127:1–44.

61. Hernández F, Pitarch E, Beltran J, López FJ. Headspace solid-phase microextraction in combination with gas chromatography and tandem mass spectrometry for the determination of organochlorine and organophosphorus pesticides in whole human blood. J Chromatogr B 2002; 769:65–77.

62. Rashed MS. Clinical applications of tandem mass spectrometry: ten years of diagnosis and screening for inherited metabolic diseases. J Chromatogr B 2001; 758:27–48.

63. Uhl M. Tandem mass spectrometry: a helpful tool in hair analysis for the forensic expert. For Sci Int 2000; 107:169–179.

64. Frison G, Favretto D, Tedeschi L, Ferrara SD. Detection of thiopental and pentobarbital in head and pubic hair in a case of drug-facilitated sexual assault. For Sci Int 2003; 133:171–174.

65. Terada M, Masui S, Hayashi T, et al. Simultaneous determination of flunitrazepam and 7-aminoflunitrazepam in human serum by ion trap gas chromatography–tandem mass spectrometry. Legal Med 2003; 5:96–100.

66. Hager JW. A new linear ion trap mass spectrometer. Rapid Commun Mass Spectr 2002; 16:512–526.

67. Marquet P, Saint-Marcoux F, Gamble TN, Leblanc JCY. Comparison of a preliminary procedure for the general unknown screening of drugs and toxic compounds using a quadrupole-linear ion-trap mass spectrometer with a liquid chromatography–mass spectrometry reference technique. J Chromatogr B 2003; 789:9–18.

68. Schwartz JC, Senko MW, Syka JEP. A two-dimensional quadrupole ion trap mass spectrometer. J Am Soc Mass Spectr 2002; 13:659–669.

69. Aebi B, Sturny-Jungo R, Bernhard W, Blanke R, Hirsch R. Quantitation using GC-TOF-MS: example of bromazepam. For Sci Int 2002; 128:84–89.

70. Adahchour M, van Stee LLP, Beens J, Vreuls RJJ, Batenburg MA, Brinkman UAT. Comprehensive two-dimensional gas chromatography with time-of-flight mass spectrometric detection for the trace analysis of flavour compounds in food. J Chromatogr A 2003; 1019:157–172.

71. Audinot J-N, Yegles M, Labarthe A, Ruch D, Wennig R, Migeon H-N. Detection and quantitation of benzodiazepines in hair by ToF-SIMS: preliminary results. Appl Surf Sci 2003; 203–204:718–721.

72. Decaestecker TN, Clauwaert KM, Van Bocxlaer JF, et al. Evaluation of automated single mass spectrometry to tandem mass spectrometry function switching for comprehensive

drug profiling analysis using a quadrupole time-of-flight mass spectrometer. Rapid Commun Mass Spectr 2000; 14:1787–1792.

73. Drummer OH. Chromatographic screening techniques in systematic toxicological analysis. J Chromatogr B 1999; 733:27–45.

74. Maurer HH. Systematic toxicological analysis procedures for acidic drugs and/or metabolites relevant to clinical and forensic toxicology and/or doping control. J Chromatogr B 1999; 733:3–25.

75. Polettini A. Systematic toxicological analysis of drugs and poisons in biosamples by hyphenated chromatographic and spectroscopic techniques. J Chromatogr B 1999; 733:47–63.

76. Polettini A, Groppi A, Vignali C, Montagna M. Fully-automated systematic toxicological analysis of drugs, poisons, and metabolites in whole blood, urine, and plasma by gas chromatography–full scan mass spectrometry. J Chromatogr B 1998; 713:265–279.

77. Tarbah FA, Mahler H, Temme O, Daldrup T. An analytical method for the rapid screening of organophosphate pesticides in human biological samples and foodstuffs. For Sci Int 2001; 121:126–133.

78. Thieme D, Sachs H. Improved screening capabilities in forensic toxicology by application of liquid chromatography–tandem mass spectrometry. Anal Chim Acta 2003; 492:171–186.

79. Pfleger K, Maurer HH, Weber A. Mass Spectral and GC Data of Drugs, Poisons, Pesticides, Pollutants and their Metabolites. Part 4. 2d ed. Weinheim, Germany: Wiley-VCH, 2000.

80. Aebi B, Bernhard W. Advances in the use of mass spectral libraries for forensic toxicology. J Anal Toxicol 2002; 26:149–156.

81. Stimpfl T, Demuth W, Varmuza K, Vycudilik W. Systematic toxicological analysis: computer-assisted identification of poisons in biological materials. J Chromatogr B 2003; 789:3–7.

82. Pfleger K, Maurer HH, Weber A. Mass Spectral Library of Drugs, Poisons, Pesticides, Pollutants and their Metabolites. 3rd rev. Palo Alto, CA, USA: Agilent Technologies, 2000.

83. http://www.chemograph.de/de/start.htm (accessed in June 2004).

84. http://www.nist.gov/public_affairs/releases/tn6191.htm (accessed in June 2004).

85. http://www.wiley.com/WileyCDA/WileyTitle/productCd-0471440973.html (accessed in June 2004).

86. http://www.ualberta.ca/~gjones/mslib.htm (accessed in October 2004, look for the August 2004 update).

87. http://www.tiaft.org/main/mslib.html (accessed in June 2004).

88. http://www.masslib.com (accessed in June 2004).

89. Wylie PL, Szelewski MJ, Meng C-K, Sandy CP. Comprehensive Pesticide Screening by GC/MSD using Deconvolution Reporting Software. Application Report, Agilent Technologies, 2004.

90. http://chemdata.nist.gov/mass-spc/amdis/ (accessed in June 2004).

91. Kagan MR. Analysis of Verapamil Microsomal Incubation using Metabolite ID and Mass Frontier. Application Report 320, Thermo Finnigan, 2004.

92. Aebi B, Bernhard W. Gas chromatography with dual mass spectrometric and nitrogen–phosphorus specific detection: a new and powerful tool for forensic analyses. For Sci Int 1999; 102:91–101.

93. Aebi B, Bernhard W. Modern GC–MS technology in the forensic laboratory. Chimia 2002; 56:48–52.

94. Hadidi KA, Almasad JK, Al-Nsour T, Abu-Ragheib S. Determination of tramadol in hair using solid phase extraction and GC–MS. For Sci Int 2003; 135:129–136.

95. Jurado C, Sachs H. Proficiency test for the analysis of hair for drugs of abuse, organized by the Society of Hair Testing. For Sci Int 2003; 133:175–178.

96. Montagna M, Polettini A, Stramesi C, Groppi A, Vignali C. Hair analysis for opiates, cocaine and metabolites. Evaluation of a method by interlaboratory comparison. For Sci Int 2002; 128:79–83.

97. Musshoff F, Lachenmeier DW, Kroener L, Madea B. Automated headspace solid-phase dynamic extraction for the determination of cannabinoids in hair samples. For Sci Int 2003; 133:32–38.

98. Pujadas M, Pichini S, Poudevida S, et al. Development and validation of a gas chromatography-mass spectrometry assay for hair analysis of amphetamine, methamphetamine and methylenedioxy derivatives. J Chromatogr B 2003; 798:249–255.

99. Romano G, Barbera N, Spadaro G, Valenti V. Determination of drugs of abuse in hair: evaluation of external heroin contamination and risk of false positives. For Sci Int 2003; 131:98–102.

100. Romolo FS, Rotolo MC, Palmi I, Pacifici R, Lopez A. Optimized conditions for simultaneous determination of opiates, cocaine and benzoylecgonine in hair samples by GC–MS. For Sci Int 2003; 138:17–26.

101. Skender L, Karačić V, Brčić I, Bagarić A. Quantitative determination of amphetamines, cocaine, and opiates in human hair by gas chromatography/mass spectrometry. For Sci Int 2002; 125:120–126.

102. Skopp G, Pötsch L, Moeller MR. On cosmetically treated hair—aspects and pitfalls of interpretation. For Sci Int 1997; 84:43–52.

103. Wennig R. Potential problems with the interpretation of hair analysis results. For Sci Int 2000; 107:5–12.

104. Kintz P, Cirimele V, Jamey C, Ludes B. Testing for GHB in hair by GC/MS/MS after a single exposure application to document sexual assault. J For Sci 2003; 48(1):195–200.

105. Hartwig S, Auwärter V, Pragst F. Effect of hair care and hair cosmetics on the concentrations of fatty acid ethyl esters in hair as a markers of chronically elevated alcohol consumption. For Sci Int 2003; 131:90–97.

106. Musshoff F. Chromatographic methods for the determination of markers of chronic and acute alcohol consumption. J Chromatogr B 2002; 781:457–480.

107. Wurst FM, Kempter C, Metzger J, Seidl S, Alt A. Ethyl glucuronide: a marker of recent alcohol consumption with clinical and forensic implications. Alcohol 2000; 20:111–116.

108. Rivier L. Techniques for analytical testing of unconventional samples. Bailliere Clin Endoc 2000; 14(1):147–165.

109. Staub C. Chromatographic procedures for determination of cannabinoids in biological samples, with special attention to blood and alternative matrices like hair, saliva, sweat and meconium. J Chromatogr B 1999; 733:119–126.

110. Daniel CR, Piraccini BM, Tosti A. The nail and hair in forensic science. J Am Acad Dermatol 2004; 50(2):258–261.

111. Fucci N, De Giovanni N, Chiarotti M. Simultaneous detection of some drugs of abuse in saliva samples by SPME technique. For Sci Int 2003; 134:40–45.

112. De Giovanni N, Fucci N, Chiarotti M, Scarlata S. Cozart Rapiscan System: our experience with saliva tests. J Chromatogr B 2002; 773:1–6.

113. Kidwell DA, Kidwell JD, Shinohara F, et al. Comparison of daily urine, sweat, and skin swabs among cocaine users. For Sci Int 2003; 133:63–78.

114. Kintz P, Cirimele V, Ludes B. Detection of cannabis in oral fluid (saliva) and forehead wipes (sweat) from impaired drivers. J Anal Toxicol 2000; 24:557–560.

115. Presley L, Lehrer M, Seiter W, et al. High prevalence of 6-acetylmorphine in morphine-positive oral fluid specimens. For Sci Int 2003; 133:22–25.

116. Samyn N, De Boeck G, Wood M, et al. Plasma, oral fluid and sweat wipe ecstasy concentrations in controlled and real life conditions. For Sci Int 2002; 128:90–97.

117. Schütz H, Gotta JC, Erdmann F, Risse M, Weiler G. Simultaneous screening and detection of drugs in small blood samples and bloodstains. For Sci Int 2002; 126:191–196.

118. Thieme D, Anielski P, Grosse J, Sachs H, Mueller RK. Identification of anabolic steroids in serum, urine, sweat and hair comparison of metabolic patterns. Anal Chim Acta 2003; 483:299–306.
119. http://www.ilac.org/downloads/Ilac-g19.pdf (accessed in June 2004).
120. http://www.soft-tox.org/docs/Guidelines.2002.final.pdf (accessed in June 2004).
121. Laboratory Guidelines for Toxicological Analysis. TIAFT Bull 1994; 24(1):7–16.
122. http://www.gtfch.org (accessed in June 2004, in German).
123. Peters FT, Maurer HH. Bioanalytical method validation and its implications for forensic and clinical toxicology—a review. Accredit Qual Assur 2002; 7:441–449.
124. Phillips JE, Bogema S, Fu P, et al. Signify® ER Drug Screen Test evaluation: comparison to Triage® Drug of Abuse Panel plus tricyclic antidepressants. Clin Chim Acta 2003; 328:31–38.
125. Steinmeyer S, Bregel D, Warth S, Kraemer T, Moeller MR. 9 Improved and validated method for the determination of Δ9-tetrahydrocannabinol (THC), 11-hydroxy-THC and 11-nor-9-carboxy-THC in serum, and in human liver microsomal preparations using gas chromatography–mass spectrometry. J Chromatogr B 2002; 772:239–248.
126. Liau A-S, Liu J-T, Lin L-C, et al. Optimization of a simple method for the chiral separation of methamphetamine and related compounds in clandestine tablets and urine samples by β-cyclodextrine modified capillary electrophoresis: a complementary method to GC–MS. For Sci Int 2003; 134:17–24.
127. Lua AC, Lin HR, Tseng YT, Hu AR, Yeh PC. Profiles of urine samples from participants at rave party in Taiwan: prevalence of ketamine and MDMA abuse. For Sci Int 2003; 136:47–51.
128. Springer D, Fritschi G, Maurer HH. Metabolism of the new designer drug α-pyrrolidinopropiophenone (PPP) and the toxicological detection of PPP and 4′-methyl-α-pyrrolidinopropiophenone (MPPP) studied in rat urine using gas chromatography-mass spectrometry. J Chromatogr B 2003; 796:253–266.
129. Staack RF, Maurer HH. Toxicological detection of the new designer drug 1-(4-methoxyphenyl)piperazine and its metabolites in urine and differentiation from an intake of structurally related medicaments using gas chromatography–mass spectrometry. J Chromatogr B 2003; 798:333–342.
130. Cirimele V, Villain M, Pépin G, Ludes B, Kintz P. Screening procedure for eight quaternary nitrogen muscle relaxants in blood by high-performance liquid chromatography–electrospray ionization mass spectrometry. J Chromatogr B 2003; 789:107–113.
131. El Haj BM, Al Ainri AM, Hassan MH, Bin Khadem RK, Marzouq MS. The GC/MS analysis of some commonly used non-steroidal anti-inflammatory drugs (NSAIDs) in pharmaceutical dosage forms and in urine. For Sci Int 1999; 105:141–153.
132. Essien H, Lai SJ, Binder SR, King DL. Use of direct-probe mass spectrometry as a toxicology confirmation method for demoxepam in urine following high-performance liquid chromatography. J Chromatogr B 1996; 683:199–208.
133. Greenhill B, Valtier S, Cody JT. Metabolic profile of amphetamine and methamphetamine following administration of the drug famprofazone. J Anal Toxicol 2003; 27: 479–484.
134. Impens S, De Wasch K, Cornelis M, De Brabander HF. Analysis on residues of estrogens, gestagens and androgens in kidney fat and meat with gas chromatography–tandem mass spectrometry. J Chromatogr A 2002; 970:235–247.
135. Kraemer T, Maurer HH. Determination of amphetamine, methamphetamine and amphetamine-derived designer drugs or medicaments in blood and urine. J Chromatogr B 1998; 713:163–187.
136. Libong D, Pirnay S, Bruneau C, Rogalewicz F, Ricordel I, Bouchonnet S. Adsorption-desorption effects in ion trap mass spectrometry using in situ ionization. J Chromatogr A 2003; 1010:123–128.
137. Meadway C, George S, Braithwaite R. Interpretation of GC–MS opiate results in the presence of pholcodine. For Sci Int 2002; 127:131–135.

138. Mozayani A, Schrode P, Carter J, Danielson TJ. A multiple drug fatality involving MK-801 (dizocilpine), a mimic of phencyclidine. For Sci Int 2003; 133:113–117.
139. Polettini A, Huestis MA. Simultaneous determination of buprenorphine, norbuprenorphine, and buprenorphine–glucuronide in plasma by liquid chromatography–tandem mass spectrometry. J Chromatogr B 2001; 754:447–459.
140. Toennes SW, Fandiño AS, Hesse F-J, Kauert GF. Artifact production in the assay of anhydroecgonine methyl ester in serum using gas chromatography–mass spectrometry. J Chromatogr B 2003; 792:345–351.
141. Toennes SW, Kauert GF. Importance of vacutainer selection in forensic toxicological analysis of drugs of abuse. J Anal Toxicol 2001; 25:339–343.
142. Van Thuyne W, Van Eenoo P, Delbeke FT. Urinary concentrations of morphine after the administration of herbal teas containing Papaveris fructus in relation to doping analysis. J Chromatogr B 2003; 785:245–251.
143. Vékey K. Mass spectrometry and mass-selective detection in chromatography. J Chromatogr A 2001; 921:227–236.
144. Dienes-Nagy A, Rivier L, Giroud C, Augsburger M, Mangin P. Method for quantification of morphine and its 3- and 6-glucuronides, codeine, codeine glucuronide and 6-monoacetylmorphine in human blood by liquid chromatography–electrospray mass spectrometry for routine analysis in forensic toxicology. J Chromatogr A 1999; 854:109–118.
145. Raith K, Neubert R, Poeaknapo C, Boettcher C, Zenk MH, Schmidt J. Electrospray tandem mass spectrometric investigations of morphinans. J Am Soc Mass Spectr 2003; 14:1262–1269.
146. Rivier L. Criteria for the identification of compounds by liquid chromatography–mass spectrometry and liquid chromatography–multiple mass spectrometry in forensic toxicology and doping analysis. Anal Chim Acta 2003; 492:69–82.
147. Smyth WF. Electrospray ionisation mass spectrometric behaviour of selected drugs and their metabolites. Anal Chim Acta 2003; 492:1–16.
148. Williams JD, Burinsky DJ. Mass spectrometric analysis of complex mixtures then and now: the impact of linking liquid chromatography and mass spectrometry. Int J Mass Spectrom 2001; 212:111–133.
149. Kataoka H. New trends in sample preparation for clinical and pharmaceutical analysis. Trend Anal Chem 2003; 22(4):232–244.
150. Psillakis E, Kalogerakis N. Developments in liquid-phase microextraction. Trend Anal Chem 2003; 22(10):565–574.
151. Smith RM. Before the injection—modern methods of sample preparation for separation techniques. J Chromatogr A 2003; 1000:3–27.
152. Weinmann W, Wiedemann A, Eppinger B, Renz M. Screening for drugs in serum by electrospray ionization/collision-induced dissociation and library searching. J Am Soc Mass Spectr 1999; 10:1028–1037.
153. http://www.thermo.com (last accessed in October 2004).

Index